Selected Titles in This Series

13 **Paul C. Shields,** The ergodic theory of discrete sample paths, 1996
12 **N. V. Krylov,** Lectures on elliptic and parabolic equations in Hölder spaces, 1996
11 **Jacques Dixmier,** Enveloping algebras, 1996 Printing
10 **Barry Simon,** Representations of finite and compact groups, 1996
9 **Dino Lorenzini,** An invitation to arithmetic geometry, 1996
8 **Winfried Just and Martin Weese,** Discovering modern set theory. I: The basics, 1996
7 **Gerald J. Janusz,** Algebraic number fields, second edition, 1996
6 **Jens Carsten Jantzen,** Lectures on quantum groups, 1996
5 **Rick Miranda,** Algebraic curves and Riemann surfaces, 1995
4 **Russell A. Gordon,** The integrals of Lebesgue, Denjoy, Perron, and Henstock, 1994
3 **William W. Adams and Philippe Loustaunau,** An introduction to Gröbner bases, 1994
2 **Jack Graver, Brigitte Servatius, and Herman Servatius,** Combinatorial rigidity, 1993
1 **Ethan Akin,** The general topology of dynamical systems, 1993

An Introduction to Gröbner Bases

William W. Adams
Philippe Loustaunau

Graduate Studies in Mathematics

Volume 3

American Mathematical Society

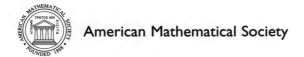

1991 *Mathematics Subject Classification*. Primary 13P10.

ABSTRACT. Gröbner bases are the primary tool for doing explicit computations in polynomial rings in many variables. In this book we give a leisurely introduction to the subject and its applications suitable for students with a little knowledge of abstract and linear algebra. The book contains not only the theory over fields, but also, the theory in modules and over rings.

Library of Congress Cataloging-in-Publication Data

Adams, William W., 1937-
An introduction to Gröbner bases/William W. Adams, Philippe Loustaunau.
p. cm. —(Graduate studies in mathematics, ISSN 1065-7339; 3)
Includes bibliographical references and index.
ISBN 0-8218-3804-0
1. Gröbner bases. I. Loustaunau, Philippe, 1958- . II. Title. III. Series.
QA251.3.A32 1994
512′.4–dc20 94-19081
CIP

Printed in the United States of America.

♾ The paper used in this book is acid-free and falls within the guidelines
established to ensure permanence and durability.
This publication was typeset by the authors, with assistance
from the American Mathematical Society, using AMS-TEX,
the American Mathematical Society's TEX macro system.

10 9 8 7 6 5 4 3 2 01 00 99 98 97 96

To my wife Elizabeth
and our daughters Ruth and Sarah
WWA

To my wife Yvonne
and our children Eileen, Gareth, and Manon
PL

Contents

Preface

We wrote this book with two goals in mind:

(i) To give a leisurely and fairly comprehensive introduction to the definition and construction of Gröbner bases;

(ii) To discuss applications of Gröbner bases by presenting computational methods to solve problems which involve rings of polynomials.

This book is designed to be a first course in the theory of Gröbner bases suitable for an advanced undergraduate or a beginning graduate student. This book is also suitable for students of computer science, applied mathematics, and engineering who have some acquaintance with modern algebra. The book does not assume an extensive knowledge of algebra. Indeed, one of the attributes of this subject is that it is very accessible. In fact, all that is required is the notion of the ring of polynomials in several variables (and rings in general in a few places, in particular in Chapter 4) together with the ideals in this ring and the concepts of a quotient ring and of a vector space introduced at the level of an undergraduate abstract and linear algebra course. Except for linear algebra, even these ideas are reviewed in the text. Some topics in the later sections of Chapters 2, 3, and 4 require more advanced material. This is always clearly stated at the beginning of the section and references are given. Moreover, most of this material is reviewed and basic theorems are stated without proofs.

The book can be read without ever "computing" anything. The theory stands by itself and has important theoretical applications in its own right. However, the reader will not fully appreciate the power of, or get insight into, the methods introduced in the book without actually doing some of the computations in the examples and the exercises by hand or, more often, using a Computer Algebra System (there are over 120 worked-out examples and over 200 exercises). Computing is useful in producing and analyzing examples which illustrate a concept already understood, or which one hopes will give insight into a less well understood idea or technique. *But the real point here is that computing is the very essence of the subject.* This is why Gröbner basis theory has become a major research area in computational algebra and computer science. Indeed, Gröbner basis theory is generating increasing interest because of its usefulness in providing computational tools which are applicable to a wide range of problems in mathematics, science, engineering, and computer science.

Gröbner bases were introduced in 1965 by Bruno Buchberger[1] [**Bu65**]. The basic idea behind the theory can be described as a generalization of the theory of polynomials in one variable. In the polynomial ring $k[x]$, where k is a field, any ideal I can be generated by a single element, namely the greatest common divisor of the elements of I. Given any set of generators $\{f_1, \ldots, f_s\} \subseteq k[x]$ for I, one can compute (using the Euclidean Algorithm) a single polynomial $d = \gcd(f_1, \ldots, f_s)$ such that $I = \langle f_1, \ldots, f_s \rangle = \langle d \rangle$. Then a polynomial $f \in k[x]$ is in I if and only if the remainder of the division of f by d is zero. Gröbner bases are the analog of greatest common divisors in the multivariate case in the following sense. A Gröbner basis for an ideal $I \subseteq k[x_1, \ldots, x_n]$ generates I and a polynomial $f \in k[x_1, \ldots, x_n]$ is in I if and only if the remainder of the division of f by the polynomials in the Gröbner basis is zero (the appropriate concept of division is a central aspect of the theory).

This abstract characterization of Gröbner bases is only one side of the theory. In fact it falls far short of the true significance of Gröbner bases and of the real contribution of Bruno Buchberger. Indeed, the ideas behind the abstract characterization of Gröbner bases had been around before Buchberger's work. For example, Macaulay [**Mac**] used some of these ideas at the beginning of the century to determine certain invariants of ideals in polynomial rings and Hironaka [**Hi**], in 1964, used similar ideas to study power series rings. But the true significance of Gröbner bases is the fact that they can be computed. Bruno Buchberger's great contribution, and what gave Gröbner basis theory the status as a subject in its own right, is his algorithm for computing these bases.

Our choice of topics is designed to give a broad introduction to the elementary aspects and applications of the subject. As is the case for most topics in commutative algebra, Gröbner basis theory can be presented from a geometric point of view. We have kept our presentation algebraic except in Sections 1.1 and 2.5. For those interested in a geometric treatment of some of the theory we recommend the excellent book by D. Cox, J. Little and D. O'Shea [**CLOS**]. The reader who is interested in going beyond the contents of this book should use our list of references as a way to access other sources. We mention in particular the books by T. Becker and V. Weispfenning [**BeWe**] and by B. Mishra [**Mi**] which contain a lot of material not in this book and have extensive lists of references on the subject.

Although this book is about computations in algebra, some of the issues which might be of interest to computer scientists are outside the scope of this book. For example, implementation of algorithms and their complexity are discussed only briefly in the book, primarily in Section 3.3. The interested reader should consult the references.

[1]Wolfgang Gröbner was Bruno Buchberger's thesis advisor.

In Chapter 1 we give the basic introduction to the concept of a Gröbner basis and show how to compute it using Buchberger's Algorithm. We are careful to give motivations for the definition and algorithm by giving the familiar examples of Gaussian elimination for linear polynomials and the Euclidean Algorithm for polynomials in one variable. In Chapter 2 we present the basic applications to algebra and elementary algebraic geometry. We close the chapter with three specialized applications to algebra, graph theory, and integer programming. In Chapter 3 we begin by using the concept of syzygy modules to give an improvement of Buchberger's Algorithm. We go on to show how to use Gröbner bases to compute the syzygy module of a set of polynomials (this is solving diophantine equations over polynomial rings). We then develop the theory of Gröbner bases for finitely generated modules over polynomial rings. With these, we extend the applications from the previous chapter, give more efficient methods for computing some of the objects from the previous chapter, and conclude by showing how to compute the Hom functor and free resolutions. In Chapter 4 we develop the theory of Gröbner bases for polynomial rings when the coefficients are now allowed to be in a general Noetherian ring and we show how to compute these bases (given certain computability conditions on the coefficient ring). We show how the theory simplifies when the coefficient ring is a principal ideal domain. We also give applications to determining whether an ideal is prime and to computing the primary decomposition of ideals in polynomial rings in one variable over principal ideal domains.

We give an outline of the section dependencies at the end of the Preface. After Chapter 1 the reader has many options in continuing with the rest of the book. There are exercises at the end of each section. Many of these exercises are computational in nature, some doable by hand while others require the use of a Computer Algebra System. Other exercises extend the theory presented in the book. A few harder exercises are marked with (*).

This book grew out of a series of lectures presented by the first author at the National Security Agency during the summer of 1991 and by the second author at the University of Calabria, Italy, during the summer of 1993.

We would like to thank many of our colleagues and students for their helpful comments and suggestions. In particular we would like to thank Beth Arnold, Ann Boyle, Garry Helzer, Karen Horn, Perpetua Kessy, Lyn Miller, Alyson Reeves, Elizabeth Rutman, Brian Williams, and Eric York. We also want to thank Sam Rankin, Julie Hawks and the AMS staff for their help in the preparation of the manuscript.

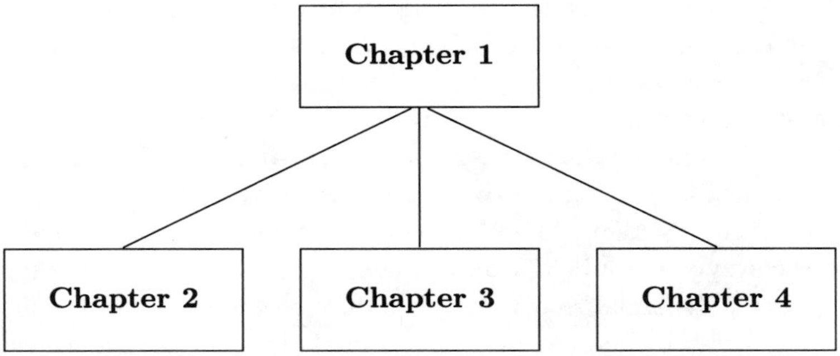
Chapter 1
Chapter 2
Chapter 3
Chapter 4

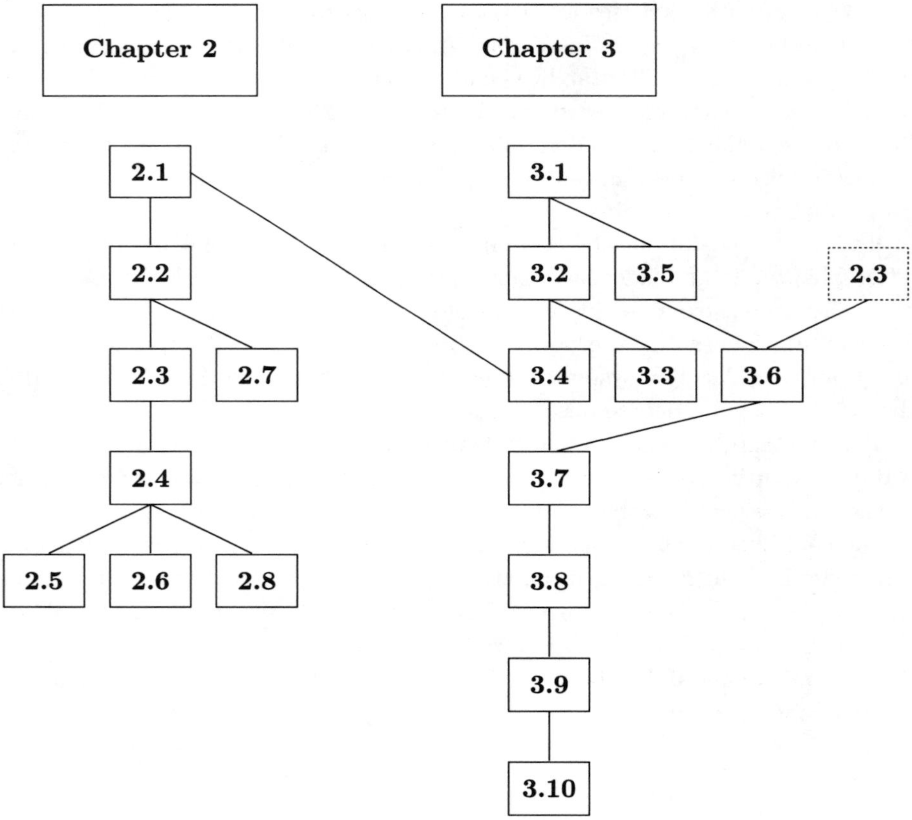
Chapter 2
Chapter 3
2.1
2.2
2.3
2.7
2.4
2.5
2.6
2.8
3.1
3.2
3.5
2.3
3.4
3.3
3.6
3.7
3.8
3.9
3.10

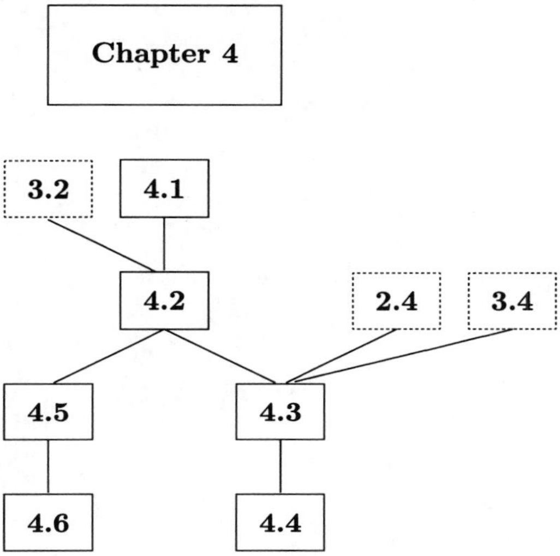
Chapter 4
3.2
4.1
4.2
2.4
3.4
4.5
4.3
4.6
4.4

Chapter 1. Basic Theory of Gröbner Bases

In this chapter we give a leisurely introduction to the theory of Gröbner bases. In the first section we introduce the reader to the kinds of problems we will be interested in throughout this book. In the next two sections we motivate the method of solution of these problems by presenting the method of solution in two familiar special cases, namely the row reduction of matrices of systems of linear polynomials, and the division algorithm for polynomials in one variable. The basic method in both cases is to use the leading term of one polynomial to subtract off a term in another polynomial. In Section 1.4 we introduce what we mean by the leading term of a polynomial in n variables. In Section 1.5 we go on to generalize the ideas in Sections 1.2 and 1.3. This leads us in Section 1.6 to defining the central notion in this book, namely the notion of a Gröbner basis. In Section 1.7 we present the algorithm due to Bruno Buchberger which transformed the abstract notion of a Gröbner basis into a fundamental tool in computational algebra. We refine the definition of a Gröbner basis in an important way in Section 1.8 and summarize what we have done in Section 1.9.

1.1. Introduction. Let k be any field (e.g., the rational numbers, $\mathbb{Q}$, the real numbers, $\mathbb{R}$, or the complex numbers, $\mathbb{C}$). We consider polynomials $f(x_1, \ldots, x_n)$ in n variables with coefficients in k. Such polynomials are finite sums of *terms* of the form $ax_1^{\beta_1} \cdots x_n^{\beta_n}$, where $a \in k$, and[1] $\beta_i \in \mathbb{N}, i = 1, \ldots, n$. We call $x_1^{\beta_1} \cdots x_n^{\beta_n}$ a *power product.* For example, $f = x_1^2 + x_2^2 - 1$ and $g = x_1 - 3x_2^2 + \frac{1}{2}x_1x_3$ are polynomials in three variables. We let $k[x_1, \ldots, x_n]$ denote the set of all polynomials in n variables[2] with coefficients in the field k. Note that in $k[x_1, \ldots, x_n]$ we have the usual operations of addition and multiplication of polynomials, and with respect to these operations $k[x_1, \ldots, x_n]$ is a commutative ring. Also, $k[x_1, \ldots, x_n]$ is a k-vector space with basis the set, $\mathbb{T}^n$, of all power products,

$$\mathbb{T}^n = \{x_1^{\beta_1} \cdots x_n^{\beta_n} \mid \beta_i \in \mathbb{N}, i = 1, \ldots, n\}.$$

[1]We denote by $\mathbb{N}$ the set of non-negative integers, that is, $\mathbb{N} = \{0, 1, 2, 3, \ldots\}$.

[2]Most of the time, from now on, whenever we work with just one, two, or three variables, we will not use variables with subscripts, but instead will use the variables x, y, or z as needed. For example, $f = x^2 + y^2 - 1$ is a polynomial in $\mathbb{Q}[x, y]$ and $g = x - 3y^2 + \frac{1}{2}xz$ is a polynomial in $\mathbb{Q}[x, y, z]$.

For a positive integer n we define the *affine n-space*

$$k^n = \{(a_1, \dots, a_n) \mid a_i \in k, i = 1, \dots, n\}.$$

(For example, if $k = \mathbb{R}$, then $k^n = \mathbb{R}^n$ is the usual Euclidean n-space.) A polynomial $f \in k[x_1, \dots, x_n]$ determines a function $k^n \longrightarrow k$ defined by

$$(a_1, \dots, a_n) \longmapsto f(a_1, \dots, a_n), \text{ for all } (a_1, \dots, a_n) \in k^n.$$

This function is called *evaluation.* We thus have two ways of viewing a polynomial $f \in k[x_1, \dots, x_n]$. One is as a formal polynomial in $k[x_1, \dots, x_n]$ and the other is as a function $k^n \longrightarrow k$ (it should be noted that if k happens to be a finite field then two different polynomials can give rise to the same function; however this need not concern us here).

This "double identity" of polynomials is the bridge between algebra and geometry. For $f \in k[x_1, \dots, x_n]$ we define $V(f)$ to be the set of solutions of the equation $f = 0$. More formally,

$$V(f) = \{(a_1, \dots, a_n) \in k^n \mid f(a_1, \dots, a_n) = 0\} \subseteq k^n.$$

$V(f)$ is called the *variety* defined by f. For example, $V(x^2 + y^2 - 1) \subseteq \mathbb{R}^2$ is the circle in the xy-plane with center $(0, 0)$ and radius 1.

More generally, given $f_1, \dots, f_s \in k[x_1, \dots, x_n]$, the *variety* $V(f_1, \dots, f_s)$ is defined to be the set of all solutions of the system

$$f_1 = 0, f_2 = 0, \dots, f_s = 0. \tag{1.1.1}$$

That is,

$$V(f_1, \dots, f_s) = \{(a_1, \dots, a_n) \in k^n \mid f_i(a_1, \dots, a_n) = 0, i = 1, 2, \dots, s\}.$$

Note that $V(f_1, \dots, f_s) = \bigcap_{i=1}^{s} V(f_i)$. For example, the variety $V(x^2 + y^2 - 1, x - 3y^2) \subseteq \mathbb{R}^2$ is the intersection of the circle $x^2 + y^2 = 1$ and the parabola $x = 3y^2$ in the xy-plane. More generally still, if $S \subseteq k[x_1, \dots, x_n]$, we define

$$V(S) = \{(a_1, \dots, a_n) \in k^n \mid f(a_1, \dots, a_n) = 0 \text{ for all } f \in S\}.$$

There are many numeric algorithms for solving non-linear systems such as (1.1.1). These algorithms solve for one solution at a time, and find an "approximation" to the solution. They ignore the geometric properties of the solution space (the variety), and do not take into consideration possible alternate descriptions of the variety (using a different system). Indeed, as we will see below, a variety can be the solution set of a number of systems such as (1.1.1), and the computation of the solutions can drastically improve if the given system of equations is transformed into a different system that has the same solutions but is "easier" to solve. To illustrate this, recall that the Gauss-Jordan elimination method transforms a system of linear equations into the so-called row echelon form (see Section 1.2). The system thus obtained has exactly the same solutions as the original system, but is easier to solve; this example will be discussed more thoroughly in the next section. We will develop an analogous procedure for System (1.1.1) which will give us algebraic and geometric information about the entire solution space of System (1.1.1).

The method for obtaining this information is to find a better representation for the corresponding variety. This will be done by considering the ideal generated by polynomials $f_1, \ldots, f_s$, denoted $\langle f_1, \ldots, f_s \rangle$:

$$\langle f_1, \ldots, f_s \rangle = \left\{ \sum_{i=1}^{s} u_i f_i \mid u_i \in k[x_1, \ldots, x_n], i = 1, \ldots, s \right\}.$$

It is easy to check that $I = \langle f_1, \ldots, f_s \rangle$ is an *ideal* in $k[x_1, \ldots, x_n]$; that is, if $f, g \in I$, then so is $f + g$ and if $f \in I$ and h is any polynomial in $k[x_1, \ldots, x_n]$, then $hf \in I$. The set $\{f_1, \ldots, f_s\}$ is called a *generating set* of the ideal I. The desired "better" representation for the variety $V(f_1, \ldots, f_s)$ will be a better generating set for the ideal $I = \langle f_1, \ldots, f_s \rangle$.

To see how this might help, we consider the variety $V(I)$, that is, the solutions of the infinite system of polynomial equations

$$f = 0, f \in I; \tag{1.1.2}$$

and contrast it with the solutions of the finite system

$$f_1 = 0, f_2 = 0, \ldots, f_s = 0. \tag{1.1.3}$$

A solution of System (1.1.2) will clearly be a solution of System (1.1.3), since $f_i \in I$ for $i = 1, \ldots, s$. Conversely, if $(a_1, \ldots, a_n) \in k^n$ is a solution of System (1.1.3), and if f is any element of I, then $f(a_1, \ldots, a_n) = 0$, since $f = \sum_{i=1}^{s} u_i f_i$, for some $u_i \in k[x_1, \ldots, x_n]$. Hence $(a_1, \ldots, a_n)$ is a solution of System (1.1.2). Thus we have that $V(I) = V(f_1, \ldots, f_s)$. We note that an ideal may have many different generating sets with different numbers of elements. For example, in $k[x, y]$, $\langle x + y, x \rangle = \langle x, y \rangle = \langle x + xy, x^2, y^2, y + xy \rangle$. Now, if we have $I = \langle f_1, \ldots, f_s \rangle = \langle f_1', \ldots, f_t' \rangle$, then $V(f_1, \ldots, f_s) = V(I) = V(f_1', \ldots, f_t')$. This means that the system $f_1 = 0, \ldots, f_s = 0$ has the same solutions as the system $f_1' = 0, \ldots, f_t' = 0$, and hence a variety is determined by an ideal, not by a particular set of equations. So, if we obtain a "better" generating set for the ideal $I = \langle f_1, \ldots, f_s \rangle$, we will have a "better" representation for the variety $V(f_1, \ldots, f_s)$. And by "better" we mean a set of generators that allows us to understand the algebraic structure of $I = \langle f_1, \ldots, f_s \rangle$ and the geometric structure of $V(f_1, \ldots, f_s)$ better. The remainder of this chapter is devoted to finding this "better" generating set for I (which will be called a Gröbner basis for I). In the case of linear polynomials this "better" generating set is the one obtained from the row echelon form of the matrix of the system.

We will now look at the problem from a different perspective. Consider a collection, V, of points of the affine space k^n. We define the set $I(V)$ of polynomials in $k[x_1, \ldots, x_n]$ by

$$I(V) = \{f \in k[x_1, \ldots, x_n] \mid f(a_1, \ldots, a_n) = 0 \text{ for all } (a_1, \ldots, a_n) \in V\}.$$

It is easy to verify that the set $I(V)$ is an ideal in $k[x_1, \ldots, x_n]$. It would seem that this ideal is very different from the ideal $\langle f_1, \ldots, f_s \rangle$. The latter ideal is defined algebraically as the set of all linear combinations of $f_1, \ldots, f_s$ with polynomial coefficients, while the former ideal is defined by the geometric condition that f is in $I(V)$ if and only if $f(a_1, \ldots, a_n) = 0$ for all $(a_1, \ldots, a_n) \in V$. We will examine the exact relationship between these two descriptions later. For now, we note that the ideal $I(V)$ can be put in the form $\langle f_1, \ldots, f_s \rangle$ for some $f_1, \ldots, f_s \in k[x_1, \ldots, x_n]$. Indeed, the Hilbert Basis Theorem (Theorem 1.1.1) states that any ideal I in $k[x_1, \ldots, x_n]$ (in particular the ideal $I(V)$) has a finite generating set. We will prove the Hilbert Basis Theorem at the end of the section. Another consequence of this result is that if Λ is an infinite set and for all $\lambda \in \Lambda$ we have a polynomial $f_\lambda \in k[x_1, \ldots, x_n]$, then the solution set of the infinite system

$$f_\lambda = 0, \lambda \in \Lambda$$

is, in fact, the solution set of a finite system, namely, of a finite generating set for the ideal $\langle f_\lambda \mid \lambda \in \Lambda \rangle$ (this ideal is defined to be the set of all *finite* linear combinations of the $f_\lambda, \lambda \in \Lambda$, with polynomial coefficients).

The construction of the ideal $I(V)$ above is a very important one. It is the bridge from geometry back to algebra since, in addition to the map

$$\begin{array}{ccc} \{\text{ Subsets of } k[x_1, \ldots, x_n]\} & \longrightarrow & \{\text{ Varieties of } k^n\} \\ S & \longmapsto & V(S), \end{array}$$

we now have a map

$$\begin{array}{ccc} \{\text{ Subsets of } k^n\} & \longrightarrow & \{\text{ Ideals of } k[x_1, \ldots, x_n]\} \\ V & \longmapsto & I(V). \end{array}$$

Understanding the relationship between these two maps allows us to go back and forth between algebraic and geometric questions. In particular, we will be interested in the exact relationship between the ideal I and the ideal $I(V(I))$. It is easy to see that $I \subseteq I(V(I))$, but equality does not always hold. For example, if $I = \langle x^2, y^2 \rangle \subseteq k[x, y]$, then $V(I) = \{(0,0)\}$, and so x and y are in the ideal $I(V(I))$, but they are not in I. For more on the relationship between I and $I(V(I))$, see Section 2.2.

In order to find the "better" generating set discussed above, we will need to determine whether two finite sets of polynomials in $k[x_1, \ldots, x_n]$ give rise to the same ideal. More specifically, given $f_1, \ldots, f_s \in k[x_1, \ldots, x_n]$, and $f'_1, \ldots, f'_t \in k[x_1, \ldots, x_n]$, we will need to determine whether $\langle f_1, \ldots, f_s \rangle = \langle f'_1, \ldots, f'_t \rangle$. For this reason and many others, it is desirable to solve the following problems: given $I = \langle f_1, \ldots, f_s \rangle$ and $f \in k[x_1, \ldots, x_n]$,

PROBLEM 1. Determine whether f is in I. This is the so-called "ideal membership problem."

PROBLEM 2. If $f \in I$, determine $u_1, \ldots, u_s \in k[x_1, \ldots, x_n]$ such that $f = u_1 f_1 + u_2 f_2 + \cdots + u_s f_s$.

REMARK: In this book, the word "determine" is informally understood to mean that one can give an algorithm that can be programmed on a computer.

The discussion above is related to another problem that deals with a certain algebraic construction. Let I be an ideal of $k[x_1, \ldots, x_n]$, and let $f \in k[x_1, \ldots, x_n]$. We saw earlier that f determines an evaluation function $k^n \longrightarrow k$ defined by $(a_1, \ldots, a_n) \longmapsto f(a_1, \ldots, a_n)$. We now consider the restriction of this function to $V(I)$; that is, we consider the evaluation function $V(I) \longrightarrow k$ defined by $(a_1, \ldots, a_n) \longmapsto f(a_1, \ldots, a_n)$ for all $(a_1, \ldots, a_n) \in V(I)$.

We would like to answer the following question: for f, g in $k[x_1, \ldots, x_n]$, when are the corresponding evaluation functions $V(I) \longrightarrow k$ equal? We note that this is related to the ideal $I(V(I))$ introduced earlier. Indeed, if $f - g$ is in the ideal $I(V(I))$, then the evaluation function $V(I) \longrightarrow k$ defined by $f - g$ is identically zero, and hence the evaluation functions $V(I) \longrightarrow k$ determined by f and g are equal. Recall that given f and g in $k[x_1, \ldots, x_n]$, and an ideal J of $k[x_1, \ldots, x_n]$, we say that f is *congruent* to g modulo J, denoted $f \equiv g \pmod{J}$, if $f - g \in J$. Observe that "$\equiv$" is an equivalence relation on $k[x_1, \ldots, x_n]$. We denote the set of equivalence classes by $k[x_1, \ldots, x_n]/J$. Elements of $k[x_1, \ldots, x_n]/J$ are of the form $f + J$ and are called *cosets* of J. Also, $k[x_1, \ldots, x_n]/J$ is a commutative ring with the usual operations of addition and multiplication inherited from $k[x_1, \ldots, x_n]$ and is called the *quotient ring* of $k[x_1, \ldots, x_n]$ by J. It is also a vector space over k.

In connection with this construction, we would like to solve the following problems:

PROBLEM 3. Determine a set of coset representatives of $k[x_1, \ldots, x_n]/J$.

PROBLEM 4. Determine a basis for $k[x_1, \ldots, x_n]/J$ as a vector space over k (which may or may not be finite).

We now turn our attention to the Hilbert Basis Theorem. This result is crucial in everything we will be doing throughout this book. It guarantees the termination of our algorithms and also, as pointed out above, it guarantees that every variety is the solution set of a finite set of polynomials.

THEOREM 1.1.1 (HILBERT BASIS THEOREM). *In the ring $k[x_1, \ldots, x_n]$ we have the following:*

(i) *If I is any ideal of $k[x_1, \ldots, x_n]$, then there exist polynomials $f_1, \ldots, f_s \in k[x_1, \ldots, x_n]$ such that $I = \langle f_1, \ldots, f_s \rangle$.*

(ii) *If $I_1 \subseteq I_2 \subseteq I_3 \subseteq \cdots \subseteq I_n \subseteq \cdots$ is an ascending chain of ideals of $k[x_1, \ldots, x_n]$, then there exists N such that $I_N = I_{N+1} = I_{N+2} = \cdots$.*

Before we go on to the proof we would like to make a couple of definitions. An ideal I in a general ring R which satisfies Condition (i) is said to be *finitely generated*, or to have a *finite generating set.* Condition (ii) is sometimes referred to as the *Ascending Chain Condition*, and any commutative ring R satisfying that condition is called a *Noetherian ring.*

In the next two sections we will illustrate the discussion of this section using two examples: linear systems and polynomials in one variable. These will be fundamental motivations for the general constructions we will develop in the remainder of this chapter.

The remainder of this section is devoted to the proof of Theorem 1.1.1. The reader may skip the proof and proceed directly to the next section.

It turns out that if either of the two conditions in Theorem 1.1.1 holds, then the other also holds; this is the content of the next theorem.

THEOREM 1.1.2. *The following conditions are equivalent for a commutative ring R:*

(i) *If I is any ideal of R, then there exist elements $f_1, \ldots, f_s \in R$ such that $I = \langle f_1, \ldots, f_s \rangle$.*

(ii) *If $I_1 \subseteq I_2 \subseteq I_3 \subseteq \cdots \subseteq I_n \subseteq \cdots$ is an ascending chain of ideals of R, then there exists N such that $I_N = I_{N+1} = I_{N+2} = \cdots$.*

That is, the ring R is Noetherian if and only if every ideal in R has a finite generating set.

PROOF. Let us first assume Condition (i), and let

$$I_1 \subseteq I_2 \subseteq I_3 \subseteq \cdots \subseteq I_n \subseteq \cdots$$

be an ascending chain of ideals of R. Consider the set $I = \bigcup_{n=1}^{\infty} I_n$. Since the ideals I_n are increasing, it is easy to see that I is an ideal of R. By Condition (i), $I = \langle f_1, \ldots, f_s \rangle$, for some $f_1, \ldots, f_s \in R$. Since for $i = 1, \ldots, s$, f_i is in I, there exists N_i such that $f_i \in I_{N_i}$. Let $N = \max_{1 \leq i \leq s} N_i$; then $f_i \in I_N$ for all $i = 1, \ldots, s$, and so $I \subseteq I_N$. Thus $I = I_N$, and Condition (ii) follows.

For the reverse implication, assume to the contrary that there exists an ideal I of R that is not generated by a finite set of elements of R. Let $f_1 \in I$. Then there exists $f_2 \in I$ with $f_2 \notin \langle f_1 \rangle$. Thus $\langle f_1 \rangle \subsetneq \langle f_1, f_2 \rangle$. We continue in this fashion, and we get a strictly ascending chain of ideals of R which contradicts Condition (ii). □

We now state and prove a more general version of the Hilbert Basis Theorem.

THEOREM 1.1.3. *If R is a Noetherian ring, then so is $R[x]$.*

PROOF. Let R be a Noetherian ring, and let J be an ideal of $R[x]$. By Theorem 1.1.2, it is enough to show that J is finitely generated. For each $n \geq 0$, define $I_n = \{r \in R \mid r \text{ is the leading coefficient of a polynomial in } J \text{ of degree } n\} \cup \{0\}$ (that is, r is the coefficient of x^n). It is easy to see that I_n is an ideal of R and that $I_n \subseteq I_{n+1}$, for all $n \geq 0$. Since R is Noetherian, there exists N such that $I_n = I_N$ for all $n \geq N$. Also, by Theorem 1.1.2, each I_i is finitely generated, say $I_i = \langle r_{i1}, \ldots, r_{it_i} \rangle$. Now for $i = 1, \ldots, N$ and $j = 1, \ldots, t_i$, let f_{ij} be a polynomial in J of degree i with leading coefficient r_{ij}. To complete the proof of the theorem it suffices to show that $J = \langle f_{ij} \mid 1 \leq i \leq N, 1 \leq j \leq t_i \rangle$.

So let $J^* = \langle f_{ij} \mid 1 \leq i \leq N, 1 \leq j \leq t_i \rangle$. Clearly $J^* \subseteq J$. Conversely, let $f \in J$, and let the degree of f be n. We prove by induction on n that $f \in J^*$. If $f = 0$ or $n = 0$, then $f \in I_0$, and hence $f \in J^*$. Now let $n > 0$, and assume that all the elements of J of degree at most $n-1$ are in J^*. Let r be the leading coefficient of f. If $n \leq N$, then, since $r \in I_n$, we have $r = \sum_{j=1}^{t_n} s_j r_{nj}$, for some $s_j \in R$. Then the polynomial $g = \sum_{j=1}^{t_n} s_j f_{nj}$ is of degree n, has leading coefficient r, and is in J^*. Thus $f - g$ has degree at most $n-1$ and is in J. By induction, $f - g$ is in J^*, and hence f is also in J^*. If $n > N$, then $r \in I_n = I_N$, and $r = \sum_{j=1}^{t_N} s_j r_{Nj}$, for some $s_j \in R$. The polynomial $g = \sum_{j=1}^{t_N} s_j x^{n-N} f_{Nj}$ has degree n, leading coefficient r, and is in J^*. Thus $f - g$ has degree at most $n-1$ and, by induction, $f - g \in J^*$. Therefore f is in J^*. □

Using a simple induction on n and the above result, we can easily show that $k[x_1, \ldots, x_n]$ is Noetherian (first noting that the field k is trivially Noetherian). That is, Theorem 1.1.1 is true.

1.2. The Linear Case. In this section we consider the system

$$f_1 = 0, \ldots, f_s = 0, \text{ where each } f_i \text{ is linear.} \tag{1.2.1}$$

In this case, the algorithmic method to answer all the questions raised in Section 1.1 is the well-known row reduction which changes System (1.2.1) to row echelon form. Consider the following examples.

EXAMPLE 1.2.1. Let $f_1 = x+y-z$ and $f_2 = 2x+3y+2z$ be linear polynomials in $\mathbb{R}[x, y, z]$. We consider the ideal $I = \langle f_1, f_2 \rangle$ and the variety $V(f_1, f_2)$, that is, the solutions to the system

$$\left\{ \begin{array}{rcrcrcl} x & + & y & - & z & = & 0 \\ 2x & + & 3y & + & 2z & = & 0. \end{array} \right. \tag{1.2.2}$$

We now perform row reduction on the matrix associated with this system:

$$\begin{bmatrix} 1 & 1 & -1 \\ 2 & 3 & 2 \end{bmatrix} \longrightarrow \begin{bmatrix} 1 & 1 & -1 \\ 0 & 1 & 4 \end{bmatrix}.$$

The last matrix is in row echelon form. The solutions of System (1.2.2) are the same as those of the following system

$$\left\{ \begin{array}{rcrcrcl} x & + & y & - & z & = & 0 \\ & & y & + & 4z & = & 0. \end{array} \right. \tag{1.2.3}$$

and are easily obtained parametrically as: $x = 5z$ and $y = -4z$.

The row reduction process is, in fact, a method to change a generating set for the ideal $I = \langle f_1, f_2 \rangle$ into another generating set. We subtracted twice the first row from the second row and replaced the second row by this new row. This amounts to creating a new polynomial, $f_3 = f_2 - 2f_1 = y + 4z$, and replacing f_2

by f_3. The original ideal I is equal to the ideal $\langle f_1, f_3\rangle$. Indeed, since $f_3 = f_2 - 2f_1$ we see that $f_3 \in I = \langle f_1, f_2\rangle$, and since $f_2 = 2f_1 + f_3$ we see that $f_2 \in \langle f_1, f_3\rangle$ and so $I = \langle f_1, f_2\rangle = \langle f_1, f_3\rangle$. This process simplifies the generating set of the ideal I and allows for an easy resolution of System (1.2.2), that is, it makes it easy to determine $V(I)$.

The process by which the polynomial f_2 was replaced by f_3 using f_1 is called *reduction* of f_2 by f_1, and we write

$$f_2 \xrightarrow{f_1} f_3.$$

The new polynomial f_3 that was created can be viewed as a remainder of a certain division: we used the first term of f_1, namely x, to eliminate a term from f_2, namely $2x$. Since this first term of f_1 cannot eliminate any other terms, the division stops and the remainder is exactly f_3. This can be written in long division form

$$\begin{array}{r|l} & \quad 2 \\ \cline{2-2} x+y-z & 2x+3y+2z \\ & 2x+2y-2z \\ \cline{2-2} & \qquad y+4z \end{array}$$

which gives us $f_2 = 2f_1 + f_3$.

When the system has more than two equations, the division (or reduction) of a polynomial may require more than one polynomial.

EXAMPLE 1.2.2. Let $f_1 = y - z$, $f_2 = x + 2y + 3z$, and $f_3 = 3x - 4y + 2z$ be linear polynomials in $\mathbb{Q}[x, y, z]$. We consider the ideal $I = \langle f_1, f_2, f_3\rangle$ and the variety $V(f_1, f_2, f_3)$, that is, the solutions to the system

$$\left\{ \begin{array}{rcrcrcl} & & y & - & z & = & 0 \\ x & + & 2y & + & 3z & = & 0 \\ 3x & - & 4y & + & 2z & = & 0. \end{array} \right. \tag{1.2.4}$$

The row reduction is as follows:

$$\begin{bmatrix} 0 & 1 & -1 \\ 1 & 2 & 3 \\ 3 & -4 & 2 \end{bmatrix} \longrightarrow \begin{bmatrix} 0 & 1 & -1 \\ 1 & 2 & 3 \\ 0 & -10 & -7 \end{bmatrix} \longrightarrow \begin{bmatrix} 0 & 1 & -1 \\ 1 & 2 & 3 \\ 0 & 0 & -17 \end{bmatrix}.$$

This says that a new generating set for $I = \langle f_1, f_2, f_3\rangle$ is $\{f_1, f_2, -17z\}$. Note that the polynomial $-17z$ is obtained by the following reductions:

$$f_3 \xrightarrow{f_2} -10y - 7z \xrightarrow{f_1} -17z. \tag{1.2.5}$$

This amounts to a division, similar to that in Example 1.2.1, of f_3 by f_2 and f_1 in succession.

Repeated use of the reduction steps, as in the above, will be denoted by $f_3 \xrightarrow{f_1, f_2}_+ -17z$.

Note that we have

$$f_3 = -10f_1 + 3f_2 - 17z. \tag{1.2.6}$$

The coefficient "3" of f_2 is the multiple of f_2 used in the first reduction in (1.2.5) and the coefficient "-10" of f_1 is the multiple of f_1 used in the second reduction in (1.2.5).

We would like to "extract" from these examples some general ingredients that will be used in the general situation of non-linear polynomials. We will concentrate on Example 1.2.2.

First, we imposed an order on the variables: we chose to eliminate x first from the third equation of (1.2.4) and then we chose to eliminate y from the new third equation. That is, when we row-reduce a matrix there is an order on how to proceed to introduce zeros: first we introduce zeros into the first column (that is, we eliminate x), and then we introduce zeros in the second column (we eliminate y) etc. We could have written the variables in the polynomials in a different order, say $f_1 = -z + y$, $f_2 = 3z + 2y + x$ and $f_3 = 2z - 4y + 3x$. We would have used the same row reduction method, but would have eliminated z first, then y. We would have wound up with a different set of equations in row echelon form, but they would have been just as good for our purpose of solving System (1.2.4). So the order does not matter, but there must be an order. This issue becomes essential in our generalization of these ideas. We note that in our example the order is such that x is first followed by y and then z and so the leading term of f_1 is y, the leading term of f_2 is x, and the leading term of f_3 is $3x$.

Second, the reductions in (1.2.5) were obtained by subtracting multiples of f_1 and f_2. This had the effect of using the leading terms of f_1 and f_2 to eliminate terms in f_3 and in $-10y - 7z$ leaving the remainder of $-17z$ and giving us Equation (1.2.6). Note that $-17z$ cannot be reduced further using the leading terms of f_1 and f_2.

The process of row reduction viewed in this light gives us a way to solve the problems posed in Section 1.1. Let us concentrate on Example 1.2.1. First, we have a very clear description of the solution space:

$$V(I) = V(f_1, f_2) = V(f_1, f_3) = \{\lambda(5, -4, 1) \mid \lambda \in \mathbb{R}\};$$

it is a line in $\mathbb{R}^3$. We next turn to the question of determining whether a polynomial $f \in k[x, y, z]$ is in I and, if so, express it as a linear combination of the elements in the generating set. In our case, because the leading term of f_1 is x and the leading term of f_3 is y, any polynomial f can be reduced to a polynomial in z alone by the division process using both f_1 and f_3 in a way similar to that used in (1.2.5). Also, any polynomial in z alone cannot be reduced using division by f_1 and f_3. The division process allows us to write f as a linear combination of f_1 and f_3 plus a remainder in a similar fashion to Equation (1.2.6) (the remainder is in z alone). It is not too hard to see that $f \in I = \langle f_1, f_2 \rangle = \langle f_1, f_3 \rangle$ if and only if this remainder is zero. Finally one could also check that the basis of the vector space $k[x, y, z]/I$ is the set of all cosets of powers of z. The statements made in this paragraph may be a little difficult to verify or appreciate at this point but will become clear later.

Exercises

1.2.1. Prove the last statement made about Example 1.2.1 in the last paragraph of the section. Namely, prove that a basis of the vector space $\mathbb{Q}[x, y, z]/I$ is the set of all cosets of powers of z. Assume that we now eliminate z first, then x, then y. What is the row echelon form of the matrix? Use this to give another basis for the vector space $\mathbb{Q}[x, y, z]/I$.

1.2.2. Following what was done for Example 1.2.1, solve the problems posed in Section 1.1 for Example 1.2.2. Repeat this eliminating y first, x second and z last.

1.2.3. Consider the following polynomials in $\mathbb{Q}[x, y, z, t]$, $f_1 = x - 2y + z + t$, $f_2 = x + y + 3z + t$, $f_3 = 2x - y - z - t$, and $f_4 = 2x + 2y + z + t$. Solve the problems posed in Section 1.1 for this set of polynomials.

1.2.4. Let A be an $s \times n$ matrix with entries in a field k. Let $f_1, \ldots, f_s$ be the linear polynomials in $k[x_1, \ldots, x_n]$ corresponding to the rows of A, as in Example 1.2.2. Let B be a row echelon form for the matrix A and assume that B has t non-zero rows. Let $g_1, \ldots, g_t$ be the polynomials corresponding to the non-zero rows of B. Prove that $\langle f_1, \ldots, f_s\rangle = \langle g_1, \ldots, g_t\rangle$. Use the polynomials $g_1, \ldots, g_t$ to obtain a basis of the k-vector space $k[x_1, \ldots, x_n]/\langle f_1, \ldots, f_s\rangle$.

1.3. The One Variable Case. In this section we consider polynomials in $k[x]$, that is, polynomials in one variable. In this context we will use the well-known Euclidean Algorithm to solve the problems mentioned in Section 1.1. In doing this we will present some of the standard material concerning $k[x]$ but will present this material using notation that will be more immediately generalizable to the study of polynomials in many variables. The theory of polynomials in one variable is a good illustration of the more general theory that will be presented in the remainder of this chapter.

For $0 \neq f \in k[x]$, we recall that the *degree* of f, denoted $\deg(f)$, is the largest exponent of x that appears in f. The *leading term* of f, denoted $\mathrm{lt}(f)$, is the term of f with highest degree. The *leading coefficient* of f, denoted $\mathrm{lc}(f)$, is the coefficient in the leading term of f. So, if $f = a_n x^n + a_{n-1}x^{n-1} + \cdots + a_1 x + a_0$, with $a_0, \ldots, a_n \in k$ and $a_n \neq 0$, then $\deg(f) = n$, $\mathrm{lt}(f) = a_n x^n$ and $\mathrm{lc}(f) = a_n$.

The main tool in the Euclidean Algorithm is the Division Algorithm (also known as long division of polynomials) which we illustrate in the next example.

EXAMPLE 1.3.1. Let $f = x^3 - 2x^2 + 2x + 8$, and $g = 2x^2 + 3x + 1$ be in $\mathbb{Q}[x]$. We divide f by g to get the quotient $\frac{1}{2}x - \frac{7}{4}$ and the remainder $\frac{27}{4}x + \frac{39}{4}$ as follows:

$$
\begin{array}{r|l}
 & \frac{1}{2}x - \frac{7}{4} \\ \hline
2x^2 + 3x + 1 & x^3 - 2x^2 + 2x + 8 \\
 & x^3 + \frac{3}{2}x^2 + \frac{1}{2}x \\ \cline{2-2}
 & \quad -\frac{7}{2}x^2 + \frac{3}{2}x + 8 \\
 & \quad -\frac{7}{2}x^2 - \frac{21}{4}x - \frac{7}{4} \\ \cline{2-2}
 & \qquad \frac{27}{4}x + \frac{39}{4}
\end{array}
$$

and so we have $f = (\frac{1}{2}x - \frac{7}{4})g + (\frac{27}{4}x + \frac{39}{4})$.

Let us analyze the steps in the above division. We first multiplied g by $\frac{1}{2}x$ and subtracted the resulting product from f. The idea was to multiply g by an appropriate term, namely $\frac{1}{2}x$, so that the leading term of g times this term canceled the leading term of f. After this first cancellation we obtained the first remainder $h = f - \frac{1}{2}xg = -\frac{7}{2}x^2 + \frac{3}{2}x + 8$. In general if we have two polynomials $f = a_nx^n + a_{n-1}x^{n-1} + \cdots + a_1x + a_0$ and $g = b_mx^m + b_{m-1}x^{m-1} + \cdots + b_1x + b_0$, with $n = \deg(f) \geq m = \deg(g)$, then the first step in the division of f by g is to subtract from f the product $\frac{a_n}{b_m}x^{n-m}g$. Using the notation introduced above, we note that the factor of g in this product is $\frac{\operatorname{lt}(f)}{\operatorname{lt}(g)}$ and so we get $h = f - \frac{\operatorname{lt}(f)}{\operatorname{lt}(g)}g$ as the first remainder. We call h a *reduction* of f by g and the process of computing h is denoted

$$f \xrightarrow{g} h.$$

Going back to Example 1.3.1, after this first cancellation we repeated the process on $h = -\frac{7}{2}x^2 + \frac{3}{2}x + 8$ by subtracting $\frac{\operatorname{lt}(h)}{\operatorname{lt}(g)}g = -\frac{7}{2}x^2 - \frac{21}{4}x - \frac{7}{4}$ from h to obtain the second (and in this example the final) remainder $r = \frac{27}{4}x + \frac{39}{4}$. This can be written using our reduction notation

$$f \xrightarrow{g} h \xrightarrow{g} r.$$

Repeated use of reduction steps, as in the above, will be denoted

$$f \xrightarrow{g}_+ r.$$

We note that, in the reduction $f \xrightarrow{g} h$, the polynomial h has degree strictly less than the degree of f. When we continue this process the degree keeps going down until the degree is less than the degree of g. Thus we have the first half of the following standard theorem.

THEOREM 1.3.2. *Let g be a non-zero polynomial in $k[x]$. Then for any $f \in k[x]$, there exist q and r in $k[x]$ such that*

$$f = qg + r, \text{ with } r = 0 \text{ or } \deg(r) < \deg(g).$$

Moreover r and q are unique (q is called the quotient and r the remainder).

PROOF. The proof of the existence of q and r was outlined above. The proof of the uniqueness of q and r is an easy exercise (Exercise 1.3.3). $\square$

Observe that the outline of the proof of Theorem 1.3.2 gives an algorithm for computing q and r. This algorithm is the well-known Division Algorithm, which we present as Algorithm 1.3.1.

INPUT: $f, g \in k[x]$ with $g \neq 0$

OUTPUT: q, r such that $f = qg + r$ and
$r = 0$ or $\deg(r) < \deg(g)$

INITIALIZATION: $q := 0; r := f$

WHILE $r \neq 0$ **AND** $\deg(g) \leq \deg(r)$ **DO**

$$q := q + \frac{\mathrm{lt}(r)}{\mathrm{lt}(g)}$$
$$r := r - \frac{\mathrm{lt}(r)}{\mathrm{lt}(g)}g$$

ALGORITHM 1.3.1. *One Variable Division Algorithm*

The steps in the WHILE loop in the algorithm correspond to the reduction process mentioned above. It is repeated until the polynomial r in the algorithm satisfies $r = 0$ or has degree strictly less than the degree of g. As mentioned above this is denoted

$$f \xrightarrow{g}_{+} r.$$

EXAMPLE 1.3.3. We will repeat Example 1.3.1 following Algorithm 1.3.1.

INITIALIZATION: $q := 0, r := f = x^3 - 2x^2 + 2x + 8$

First pass through the WHILE loop:

$q := 0 + \frac{x^3}{2x^2} = \frac{1}{2}x$

$r := (x^3 - 2x^2 + 2x + 8) - \frac{x^3}{2x^2}(2x^2 + 3x + 1) = -\frac{7}{2}x^2 + \frac{3}{2}x + 8$

Second pass through the WHILE loop:

$q := \frac{1}{2}x + \frac{-\frac{7}{2}x^2}{2x^2} = \frac{1}{2}x - \frac{7}{4}$

$r := (-\frac{7}{2}x^2 + \frac{3}{2}x + 8) - \frac{-\frac{7}{2}x^2}{2x^2}(2x^2 + 3x + 1) = \frac{27}{4}x + \frac{39}{4}$

The WHILE loop stops since $\deg(r) = 1 < 2 = \deg(g)$.

We obtain the quotient q and the remainder r as in Example 1.3.1.

Now let $I = \langle f, g \rangle$ and suppose that $f \xrightarrow{g} h$. Then, since $h = f - \frac{\mathrm{lt}(f)}{\mathrm{lt}(g)}g$, it is easy to see that $I = \langle h, g \rangle$, so we can replace f by h in the generating set of I. This idea is similar to the one presented for linear polynomials studied in Section 1.2. Using this idea repeatedly (that is, using Theorem 1.3.2 repeatedly) we can prove the following result.

THEOREM 1.3.4. *Every ideal of* $k[x]$ *is generated by one element*[3].

PROOF. Let I be a non-zero ideal of $k[x]$. Let $g \in I$ be such that $g \neq 0$ and $n = \deg(g)$ is least. For any $f \in I$ we have, by Theorem 1.3.2, that $f = qg + r$ for some $q, r \in k[x]$, with $r = 0$ or $\deg(r) < \deg(g) = n$. If $r \neq 0$, then $r = f - qg \in I$, and this contradicts the choice of g. Therefore $r = 0$, $f = qg$, and $I \subseteq \langle g \rangle$. Equality follows from the fact that g is in I. □

Observe that the polynomial g in the proof of Theorem 1.3.4 is unique up to a constant multiple. This follows from the fact that if $I = \langle g_1 \rangle = \langle g_2 \rangle$, then g_1 divides g_2 and g_2 divides g_1.

We see that the polynomial g in the proof of Theorem 1.3.4 is the "best" generating set for the ideal $I = \langle f_1, \dots, f_s \rangle$. For example, the system of equations

$$f_1 = 0, \dots, f_s = 0 \text{ with } f_i \in k[x], i = 1, \dots, s, \tag{1.3.1}$$

has precisely the same set of solutions as the single equation $g = 0$, where $\langle f_1, \dots, f_s \rangle = \langle g \rangle$.

We now investigate how to compute the polynomial g of Theorem 1.3.4. We will first focus on ideals $I \subseteq k[x]$ generated by two polynomials, say $I = \langle f_1, f_2 \rangle$, with one of f_1, f_2 not zero. We recall that the *greatest common divisor* of f_1 and f_2, denoted $\gcd(f_1, f_2)$, is the polynomial g such that:

- g divides both f_1 and f_2;
- if $h \in k[x]$ divides f_1 and f_2, then h divides g;
- $\mathrm{lc}(g) = 1$ (that is, g is monic).

We further recall

PROPOSITION 1.3.5. *Let* $f_1, f_2 \in k[x]$, *with one of* f_1, f_2 *not zero. Then* $\gcd(f_1, f_2)$ *exists and* $\langle f_1, f_2 \rangle = \langle \gcd(f_1, f_2) \rangle$.

PROOF. By Theorem 1.3.4, there exists $g \in k[x]$ such that $\langle f_1, f_2 \rangle = \langle g \rangle$. Since g is unique up to a constant multiple, we may assume that $\mathrm{lc}(g) = 1$. We will show that $g = \gcd(f_1, f_2)$. Since $f_1, f_2 \in \langle g \rangle$, g divides both f_1 and f_2. Now, let h be such that h divides both f_1 and f_2. Since g is in the ideal $\langle f_1, f_2 \rangle$, there exist $u_1, u_2 \in k[x]$ such that $g = u_1 f_1 + u_2 f_2$. Thus h divides g, and we are done. □

As a consequence, if we have an algorithm for finding gcd's, then we can actually find a single generator of the ideal $\langle f_1, f_2 \rangle$. The algorithm for computing gcd's is called the *Euclidean Algorithm.* It depends on the Division Algorithm discussed above and the following fact.

LEMMA 1.3.6. *Let* $f_1, f_2 \in k[x]$, *with one of* f_1, f_2 *not zero. Then* $\gcd(f_1, f_2) = \gcd(f_1 - qf_2, f_2)$ *for all* $q \in k[x]$.

---

[3]Recall that an ideal generated by one element is called a *principal ideal,* and an integral domain for which every ideal is principal is called a *principal ideal domain,* or *PID*. Therefore Theorem 1.3.4 says that $k[x]$ is a *PID*.

PROOF. It is easy to see that $\langle f_1, f_2\rangle = \langle f_1 - qf_2, f_2\rangle$. Therefore, by Proposition 1.3.5,

$$\langle \gcd(f_1, f_2)\rangle = \langle f_1, f_2\rangle = \langle f_1 - qf_2, f_2\rangle = \langle \gcd(f_1 - qf_2, f_2)\rangle.$$

Thus since the generator of a principal ideal is unique up to constant multiples, and since the gcd of two polynomials is defined to have leading coefficient 1, we have $\gcd(f_1, f_2) = \gcd(f_1 - qf_2, f_2)$. □

We give the Euclidean Algorithm as Algorithm 1.3.2. The reader should note that the algorithm terminates because the degree of r in the WHILE loop is strictly less than the degree of g, which is the previous r, and hence the degree of r is strictly decreasing as the algorithm progresses. Also, the algorithm does give $\gcd(f_1, f_2)$ as an output, since at each pass through the WHILE loop, we have $\gcd(f_1, f_2) = \gcd(f, g) = \gcd(r, g)$, by Lemma 1.3.6, as long as $g \neq 0$. When $g = 0$, then $\gcd(f_1, f_2) = \gcd(f, 0) = \frac{1}{\operatorname{lc}(f)} f$. The last step in the algorithm ensures that the final result has leading coefficient 1 (that is, is monic).

INPUT: $f_1, f_2 \in k[x]$, with one of f_1, f_2 not zero

OUTPUT: $f = \gcd(f_1, f_2)$

INITIALIZATION: $f := f_1, g := f_2$

WHILE $g \neq 0$ **DO**

  $f \xrightarrow{g}_+ r$, where r is the remainder of the division of f by g

  $f := g$

  $g := r$

$f := \frac{1}{\operatorname{lc}(f)} f$

ALGORITHM 1.3.2. *Euclidean Algorithm*

To illustrate this algorithm, consider the following

EXAMPLE 1.3.7. Let $f_1 = x^3 - 3x + 2$ and $f_2 = x^2 - 1$ be polynomials in $\mathbb{Q}[x]$.

INITIALIZATION: $f := x^3 - 3x + 2, g := x^2 - 1$

First pass through the WHILE loop:

$x^3 - 3x + 2 \xrightarrow{x^2-1} -2x + 2$

$f := x^2 - 1$

$g := -2x + 2$

Second pass through the WHILE loop:

$x^2 - 1 \xrightarrow{-2x+2} x - 1 \xrightarrow{-2x+2} 0$

$f := -2x + 2$

$g := 0$
The WHILE loop stops
$f := \frac{1}{\mathrm{lc}(f)} f = \frac{1}{-2} f = x - 1$

Therefore $\gcd(f_1, f_2) = x - 1$.

We now turn our attention to the case of ideals generated by more than two polynomials, $I = \langle f_1, \dots, f_s \rangle$, with not all of the f_i's zero. Recall that the *greatest common divisor* of s polynomials $f_1, \dots, f_s$, denoted $\gcd(f_1, \dots, f_s)$, is the polynomial g such that:

- g divides $f_i, i = 1, \dots, s$;
- if $h \in k[x]$ divides $f_i, i = 1, \dots, s$, then h divides g;
- $\mathrm{lc}(g) = 1$ (that is, g is monic).

PROPOSITION 1.3.8. *Let $f_1, \dots, f_s$ be polynomials in $k[x]$. Then*
(i) $\langle f_1, \dots, f_s \rangle = \langle \gcd(f_1, \dots, f_s) \rangle$;
(ii) *if* $s \geq 3$, *then* $\gcd(f_1, \dots, f_s) = \gcd(f_1, \gcd(f_2, \dots, f_s))$.

PROOF. The proof of statement (i) is similar to the proof of Proposition 1.3.5. To prove statement (ii), let $h = \gcd(f_2, \dots, f_s)$. Then, by (i), $\langle f_2, \dots, f_s \rangle = \langle h \rangle$, and hence $\langle f_1, \dots, f_s \rangle = \langle f_1, h \rangle$. Again, by (i),

$$\gcd(f_1, \dots, f_s) = \gcd(f_1, h) = \gcd(f_1, \gcd(f_2, \dots, f_s)),$$

as desired. □

With the ideas developed in this section we can now solve all the problems raised in Section 1.1 for the special case of polynomials in one variable. As noted before, to solve System (1.3.1) we first compute $g = \gcd(f_1, \dots, f_s)$. It then suffices to solve the single equation $g = 0$. The computation of $\gcd(f_1, \dots, f_s)$ is done by induction, a polynomial at a time, as is easily seen from part (ii) of Proposition 1.3.8. To decide whether a polynomial f is in the ideal $I = \langle f_1, \dots, f_s \rangle$, we first compute $g = \gcd(f_1, \dots, f_s)$. We then use the Division Algorithm to divide f by g. The remainder of that division is zero if and only if f is in the ideal $I = \langle f_1, \dots, f_s \rangle = \langle g \rangle$. Using the notation introduced earlier:

$$f \in I = \langle g \rangle \text{ if and only if } f \xrightarrow{g}_+ 0.$$

Also, the coset representative of the element $f + I$ in the quotient ring $k[x]/I$ is $r + I$, where r is the remainder of the division of f by g (that is, $f \xrightarrow{g}_+ r$, with $r = 0$ or $\deg(r) < \deg(g)$). Finally, the cosets of $1, x, x^2, \dots, x^{d-1}$, where $d = \deg(g)$, form a basis for the k-vector space $k[x]/I$ (Exercise 1.3.6).

In the last section (the linear case) we saw that there were two ingredients for our solution method: a reduction algorithm (in that case it was row reduction) and an order among the terms. In the current section we saw that the concept of reduction leading to the Division Algorithm (Algorithm 1.3.1) was the key to solving the problems mentioned in Section 1.1. We have not yet stressed the importance of the ordering of the terms in the one variable case, even though we have already used the notion of ordering in the concepts of degree and leading term. In effect, the ordering is forced upon us. Indeed, in the Division Algorithm, when we compute $r - \frac{\text{lt}(r)}{\text{lt}(g)}g$, the terms that we introduce (coming from $\frac{\text{lt}(r)}{\text{lt}(g)}g$) must be smaller than the leading term of r which has been canceled, in order for the algorithm to terminate. This can only occur if the powers of x are ordered so that $x^n < x^m$ if and only if $n < m$ (Exercise 1.4.2). We note that the condition $n < m$ is equivalent to the statement that x^n divides x^m.

Exercises

1.3.1. Follow Algorithm 1.3.1 (as in Example 1.3.3) to divide $f = 2x^5 - 4x^3 + x^2 - x + 2$ by $g = x^2 + x + 1$.

1.3.2. Find a single generator for the ideal $I = \langle x^6 - 1, x^4 + 2x^3 + 2x^2 - 2x - 3\rangle$. Is $x^5 + x^3 + x^2 - 7 \in I$? Show that $x^4 + 2x^2 - 3 \in I$ and write $x^4 + 2x^2 - 3$ as a linear combination of $x^6 - 1$ and $x^4 + 2x^3 + 2x^2 - 2x - 3$.

1.3.3. Prove that q and r obtained in Theorem 1.3.2 are unique.

1.3.4. Compute $\gcd(f_1, f_2, f_3)$ using Proposition 1.3.8, where $f_1 = x^5 - 2x^4 - x^2 + 2x$, $f_2 = x^7 + x^6 - 2x^4 - 2x^3 + x + 1$, and $f_3 = x^6 - 2x^5 + x^4 - 2x^3 + x^2 - 2x$.

1.3.5. Modify Algorithm 1.3.2 to output $f, u_1, u_2 \in k[x]$ such that $f = \gcd(f_1, f_2)$ and $f = u_1 f_1 + u_2 f_2$. Apply your algorithm to the polynomials $f_1 = x^6 - 1$, $f_2 = x^4 + 2x^3 + 2x^2 - 2x - 3 \in \mathbb{Q}[x]$ of Exercise 1.3.2.

1.3.6. Let $g \in k[x]$ be of degree d. Prove that $\{1 + \langle g\rangle, x + \langle g\rangle, \ldots, x^{d-1} + \langle g\rangle\}$ is a k-vector space basis for $k[x]/\langle g\rangle$.

1.3.7. Show that in $k[x, y]$, Theorem 1.3.4 is false. In particular, show that the ideal $\langle x, y\rangle \subseteq k[x, y]$ cannot be generated by a single element. Show that, in general, $k[x_1, \ldots, x_n]$ is not a PID.

1.3.8. Prove that a system of equations $f = 0$, $g = 0$ with two relatively prime polynomials $f, g \in k[x, y]$ has at most finitely many solutions. [Hint: View f and g in $k(x)[y]$ and use the Gauss Lemma: f and g are relatively prime when viewed in $k[x, y]$ if and only if they are relatively prime in $k(x)[y]$, where we recall that $k(x)$ denotes the field of fractions of $k[x]$, i.e. $k(x) = \{\frac{a}{b} \mid a, b \in k[x], b \neq 0\}$.]

1.3.9. Let $g \in k[y]$ be irreducible, and let $f \in k[x_1, \ldots, x_n, y]$ be such that $f \notin \langle g\rangle$. Prove that $\langle f, g\rangle \cap k[x_1, \ldots, x_n] \neq \{0\}$. [Hint: Use the hint of Exercise 1.3.8 with $k(x_1, \ldots, x_n)[y]$, where we recall that $k(x_1, \ldots, x_n)$ denotes the field of fractions of $k[x_1, \ldots, x_n]$, i.e. $k(x_1, \ldots, x_n) = \{\frac{a}{b} \mid a, b \in k[x_1, \ldots, x_n], b \neq 0\}$.]

1.3.10. Let $f, g \in \mathbb{C}[x, y]$. Prove that if f and g have a non-constant common factor in $\mathbb{C}[x, y]$, then $V(f, g)$ is infinite. That is, show that if $h \in \mathbb{C}[x, y]$ and h is not in $\mathbb{C}$, then the equation $h = 0$ has infinitely many solutions. Generalize this exercise to the case where $h \in \mathbb{C}[x_1, \ldots, x_n]$, for $n \geq 2$.

1.3.11. Let $f_1, f_2, h \in k[x]$. We consider the equation $u_1 f_1 + u_2 f_2 = h$, with unknowns u_1, u_2, to be polynomials in $k[x]$.

a. Show that the above equation is solvable if and only if $g = \gcd(f_1, f_2)$

divides h.

b. Prove that if $g = \gcd(f_1, f_2)$ divides h, then there exist unique $u_1, u_2 \in k[x]$ that satisfy the equation above and such that $\deg(u_1) < \deg(f_2) - \deg(g)$. Moreover, if $\deg(h) < \deg(f_1)+\deg(f_2)-\deg(g)$, then $\deg(u_2) < \deg(f_1) - \deg(g)$. Give an algorithm for computing such u_1 and u_2.

c. Let $f_1 = x^3 - 1, f_2 = x^2 + x - 2, h = x^2 - 4x + 3 \in \mathbb{Q}[x]$. Find u_1, u_2 which satisfy **b**.

d. Use **b** to show that if f_1 and f_2 are relatively prime, then for every $h \in k[x]$ such that $\deg(h) < \deg(f_1)+\deg(f_2)$, there exist $u_1, u_2 \in k[x]$ such that

$$\frac{h}{f_1 f_2} = \frac{u_1}{f_1} + \frac{u_2}{f_2},$$

with $\deg(u_1) < \deg(f_1)$ and $\deg(u_2) < \deg(f_2)$. (This is the partial fraction decomposition of rational functions.)

e. Use **d** to compute the partial fraction decomposition of

$$\frac{x-3}{x^3 + 3x^2 + 3x + 2}.$$

f. Generalize **a** and **b** to the case of s polynomials $f_1, \dots, f_s \in k[x]$.

1.3.12. When the coefficients of polynomials in one variable are not in a field k, the Division Algorithm (Algorithm 1.3.1) has to be modified. In this exercise we present a "pseudo" division algorithm for polynomials in $R[x]$, where R is a unique factorization domain (UFD).

a. Let $f, g \in R[x]$ be such that $g \neq 0$ and $\deg(f) \geq \deg(g)$. Prove that there exist polynomials $q, r \in R[x]$ such that $\mathrm{lc}(g)^\ell f = gq + r$, where $r = 0$ or $\deg(r) < \deg(g)$, and $\ell = \deg(f) - \deg(g) + 1$.

b. Give an algorithm for computing q and r. The polynomials q and r are called the *pseudo-quotient* and the *pseudo-remainder* respectively.

c. Use this algorithm to find q and r in the following cases:

(i) $f = 6x^4 - 11x^3 - 3x^2 + 2x, g = 10x^3 - 23x^2 - 20x - 3 \in \mathbb{Z}[x]$;

(ii) $f = (-2+4i)x^3 + (5+3i)x^2 - 2ix + (-1+i), g = 2x^2 + (1+i)x + (1+i) \in (\mathbb{Z}[i])[x]$, where $i^2 = -1$.

d. A polynomial $f \in R[x]$ is called *primitive* if its coefficients are relatively prime. Let f, g be primitive polynomials in $R[x]$, and let $\mathrm{lc}(g)^\ell f = gq + r$ be as in **a**. Prove that $\gcd(f, g) = \gcd(g, r')$, where r' is the primitive part of r; i.e. $r = \alpha r'$, $\alpha \in R$ and r' primitive.

e. Use **d** to give an algorithm for computing $\gcd(f, g)$.

f. Use **e** to compute $\gcd(f, g)$ for the examples in **c**.

g. Use the above to give an algorithm for computing $\gcd(f, g)$, where $f, g \in R[x_1, \dots, x_n]$, R a UFD.

h. Use **g** to compute $\gcd(f, g)$, where $f = (y^2+y)x^3 + (-y^3-y+1)x^2 + (-y^3+y^2-2)x + (-y^2+2y)$, $g = (y^2+2y+1)x^2 + (y^3+3y^2-2)x + (y^3-y^2-2y) \in \mathbb{Z}[x, y]$.

1.4. Term Orders. It was important in the last two sections to specify an order on the power products. In the linear case, we computed with x first, then y, etc. The particular order used was unimportant but did have to be specified. In the one variable case we used the highest degree term first and this was required by the procedures used. In more than one variable we need an order analogous to the ones used in these two special cases and this will be the focus of this section.

First recall that the set of power products is denoted by

$$\mathbb{T}^n = \left\{x_1^{\beta_1} \cdots x_n^{\beta_n} \mid \beta_i \in \mathbb{N}, i = 1, \ldots, n\right\}.$$

Sometimes we will denote $x_1^{\beta_1} \cdots x_n^{\beta_n}$ by $\boldsymbol{x}^{\boldsymbol{\beta}}$, where $\boldsymbol{\beta} = (\beta_1, \ldots, \beta_n) \in \mathbb{N}^n$.

We would like to emphasize that, throughout this book, "power product" will always refer to a product of the x_i variables, and "term" will always refer to a coefficient times a power product. So every power product is a term (with coefficient 1) but a term is not necessarily a power product. We will also always assume that the different terms in a polynomial have different power products (so we never write $3x^2y$ as $2x^2y + x^2y$).

There are many ways to order $\mathbb{T}^n$. However, we already know some properties that a desirable order must satisfy. For example, the orders in the linear and one variable cases were used to define a division (or reduction) algorithm, thus the order had to extend divisibility relations (see the discussion at the end of the previous section). That is, if $\boldsymbol{x}^{\boldsymbol{\alpha}}$ divides $\boldsymbol{x}^{\boldsymbol{\beta}}$, then we should have $\boldsymbol{x}^{\boldsymbol{\alpha}} \leq \boldsymbol{x}^{\boldsymbol{\beta}}$, or equivalently, if $\alpha_i \leq \beta_i$ for all $i = 1, \ldots, n$, then $\boldsymbol{x}^{\boldsymbol{\alpha}} \leq \boldsymbol{x}^{\boldsymbol{\beta}}$. Also, in the divisions described in Sections 1.2 and 1.3, we arranged the terms of the polynomials in increasing or decreasing order, and hence we must be able to compare any two power products. Thus the order must be a *total order*, that is, given any $\boldsymbol{x}^{\boldsymbol{\alpha}}, \boldsymbol{x}^{\boldsymbol{\beta}} \in \mathbb{T}^n$, exactly one of the following three relations must hold:

$$\boldsymbol{x}^{\boldsymbol{\alpha}} < \boldsymbol{x}^{\boldsymbol{\beta}},\ \boldsymbol{x}^{\boldsymbol{\alpha}} = \boldsymbol{x}^{\boldsymbol{\beta}} \text{ or } \boldsymbol{x}^{\boldsymbol{\alpha}} > \boldsymbol{x}^{\boldsymbol{\beta}}. \tag{1.4.1}$$

Moreover, the reduction $\longrightarrow_+$ described in Sections 1.2 and 1.3 must stop after a finite number of steps. Recall that whenever we had $f \xrightarrow{g}_+ r$, the polynomial r was such that its leading power product was less than the leading power product of g: in Section 1.2, that meant that the reduced polynomial r was obtained by eliminating a leading variable using g; in Section 1.3, that meant that the remainder polynomial r had degree less than that of g. Therefore, for the reduction to be finite, we need that the order be a *well-ordering*, that is, there is no infinite descending chain $\boldsymbol{x}^{\boldsymbol{\alpha}_1} > \boldsymbol{x}^{\boldsymbol{\alpha}_2} > \boldsymbol{x}^{\boldsymbol{\alpha}_3} > \cdots$ in $\mathbb{T}^n$. An order that satisfies all these conditions is called a *term order*, and it turns out that these conditions are captured in the following definition (this will be justified in Proposition 1.4.5 and Theorem 1.4.6).

DEFINITION 1.4.1. *By a* term order *on* $\mathbb{T}^n$ *we mean a total order* $<$ *on* $\mathbb{T}^n$ *satisfying the following two conditions:*

(i) $1 < \boldsymbol{x}^{\boldsymbol{\beta}}$ *for all* $\boldsymbol{x}^{\boldsymbol{\beta}} \in \mathbb{T}^n$, $\boldsymbol{x}^{\boldsymbol{\beta}} \neq 1$;

(ii) *If* $\boldsymbol{x}^{\boldsymbol{\alpha}} < \boldsymbol{x}^{\boldsymbol{\beta}}$, *then* $\boldsymbol{x}^{\boldsymbol{\alpha}}\boldsymbol{x}^{\boldsymbol{\gamma}} < \boldsymbol{x}^{\boldsymbol{\beta}}\boldsymbol{x}^{\boldsymbol{\gamma}}$, *for all* $\boldsymbol{x}^{\boldsymbol{\gamma}} \in \mathbb{T}^n$.

First recall that by a total order $<$ on $\mathbb{T}^n$ we mean a relation satisfying 1.4.1 (total) which is also transitive ($\boldsymbol{x}^{\boldsymbol{\alpha}} < \boldsymbol{x}^{\boldsymbol{\beta}}$ and $\boldsymbol{x}^{\boldsymbol{\beta}} < \boldsymbol{x}^{\boldsymbol{\gamma}}$ implies that $\boldsymbol{x}^{\boldsymbol{\alpha}} < \boldsymbol{x}^{\boldsymbol{\gamma}}$) and cannot have both $\boldsymbol{x}^{\boldsymbol{\alpha}} < \boldsymbol{x}^{\boldsymbol{\beta}}$ and $\boldsymbol{x}^{\boldsymbol{\beta}} < \boldsymbol{x}^{\boldsymbol{\alpha}}$. The latter implies that the relation $\leq$ (i.e. $\boldsymbol{x}^{\boldsymbol{\alpha}} < \boldsymbol{x}^{\boldsymbol{\beta}}$ or $\boldsymbol{x}^{\boldsymbol{\alpha}} = \boldsymbol{x}^{\boldsymbol{\beta}}$) is antisymmetric (i.e. $\boldsymbol{x}^{\boldsymbol{\alpha}} \leq \boldsymbol{x}^{\boldsymbol{\beta}}$ and $\boldsymbol{x}^{\boldsymbol{\beta}} \leq \boldsymbol{x}^{\boldsymbol{\alpha}}$ implies that $\boldsymbol{x}^{\boldsymbol{\alpha}} = \boldsymbol{x}^{\boldsymbol{\beta}}$). See also Appendix B.

Before we prove that the basic properties we discussed above follow from the conditions in Definition 1.4.1, we give three examples of term orders. The easy verification that they are term orders will be left to the exercises (Exercise 1.4.3).

DEFINITION 1.4.2. *We define the* lexicographical order *on* $\mathbb{T}^n$ *with* $x_1 > x_2 > \cdots > x_n$ *as follows: For*

$$\boldsymbol{\alpha} = (\alpha_1, \dots, \alpha_n), \boldsymbol{\beta} = (\beta_1, \dots, \beta_n) \in \mathbb{N}^n$$

we define

$$\boldsymbol{x}^{\boldsymbol{\alpha}} < \boldsymbol{x}^{\boldsymbol{\beta}} \iff \begin{cases} \textit{the first coordinates } \alpha_i \textit{ and } \beta_i \textit{ in } \boldsymbol{\alpha} \textit{ and } \boldsymbol{\beta} \\ \textit{from the left, which are different, satisfy } \alpha_i < \beta_i. \end{cases}$$

So, in the case of two variables x_1 and x_2, we have

$$1 < x_2 < x_2^2 < x_2^3 < \cdots < x_1 < x_2x_1 < x_2^2x_1 < \cdots < x_1^2 < \cdots .$$

As noted before, when we do examples in a small number of variables, we will usually use x, y, or z instead of the subscripted variables above. It is important to note that we need to specify the order on the variables. For example, if we use the lexicographic order with $x < y$, then we have

$$1 < x < x^2 < x^3 < \cdots < y < xy < x^2y < \cdots < y^2 < \cdots .$$

(We deliberately altered the order of x and y from what was probably expected to emphasize the point that an order on the variables must be specified.)

Note that, for this order, x_1^{μ} is always greater than x_2^{ν}, for all non-zero $\mu, \nu \in \mathbb{N}$. This will be of importance later on (see Section 2.3). We will always denote this order by "lex". We emphasize again that we always need to specify the order on the variables.

DEFINITION 1.4.3. *We define the* degree lexicographical order *on* $\mathbb{T}^n$ *with* $x_1 > x_2 > \cdots > x_n$ *as follows: For*

$$\boldsymbol{\alpha} = (\alpha_1, \dots, \alpha_n), \boldsymbol{\beta} = (\beta_1, \dots, \beta_n) \in \mathbb{N}^n$$

we define

$$\boldsymbol{x}^{\boldsymbol{\alpha}} < \boldsymbol{x}^{\boldsymbol{\beta}} \iff \begin{cases} \sum_{i=1}^n \alpha_i < \sum_{i=1}^n \beta_i \\ \textit{or} \\ \sum_{i=1}^n \alpha_i = \sum_{i=1}^n \beta_i \textit{ and } \boldsymbol{x}^{\boldsymbol{\alpha}} < \boldsymbol{x}^{\boldsymbol{\beta}} \\ \textit{with respect to lex with } x_1 > x_2 > \cdots > x_n. \end{cases}$$

So, with this order, we first order by total degree and we break ties by the lex order. In the case of two variables x_1 and x_2, we have

$$1 < x_2 < x_1 < x_2^2 < x_1x_2 < x_1^2 < x_2^3 < x_1x_2^2 < x_1^2x_2 < x_1^3 < \cdots .$$

Or, using the degree lexicographic ordering in $k[x, y]$ with $x < y$, we have

$$1 < x < y < x^2 < xy < y^2 < x^3 < x^2y < xy^2 < y^3 < \cdots .$$

We will always denote this order by "deglex". Again, we always need to specify the order of the variables.

DEFINITION 1.4.4. *We define the* degree reverse lexicographical order *on* $\mathbb{T}^n$ *with* $x_1 > x_2 > \cdots > x_n$ *as follows: For*

$$\boldsymbol{\alpha} = (\alpha_1, \ldots, \alpha_n), \boldsymbol{\beta} = (\beta_1, \ldots, \beta_n) \in \mathbb{N}^n$$

we define

$$\boldsymbol{x}^{\boldsymbol{\alpha}} < \boldsymbol{x}^{\boldsymbol{\beta}} \iff \begin{cases} \sum_{i=1}^n \alpha_i < \sum_{i=1}^n \beta_i \\ \textit{or} \\ \sum_{i=1}^n \alpha_i = \sum_{i=1}^n \beta_i \textit{ and the first coordinates } \alpha_i \textit{ and } \beta_i \textit{ in} \\ \boldsymbol{\alpha} \textit{ and } \boldsymbol{\beta} \textit{ from the right, which are different, satisfy } \alpha_i > \beta_i. \end{cases}$$

We will always denote this order by "degrevlex". It is easy to see that in the case of two variables deglex and degrevlex are the same orders (Exercise 1.4.4). However, if there are at least 3 variables, this is not the case anymore, as the following example shows:

$$x_1^2x_2x_3 > x_1x_2^3 \text{ with respect to deglex with } x_1 > x_2 > x_3$$

but

$$x_1^2x_2x_3 < x_1x_2^3 \text{ with respect to degrevlex with } x_1 > x_2 > x_3.$$

This order turns out to be extremely good for certain types of computations. The important property that this order possesses is given in Exercise 1.4.9.

Note that the term "right" in Definition 1.4.4 refers to the smallest variable. That is, we have $x_1 > x_2 > \cdots > x_n$. This must be especially kept in mind when we consider examples involving x, y, z (see Exercise 1.4.1).

There are many other orders on $\mathbb{T}^n$ which we will see later in both the exercises and the text. The three examples given above are the ones we will use the most. We will see that each order has different properties and which order we use will depend on the problem we want to solve.

We now return to the general definition of a term order. We want to observe that a term order, as defined in Definition 1.4.1, has all the properties discussed before that definition. That is, we need to prove that any term order extends the divisibility relation and is a well-ordering.

PROPOSITION 1.4.5. *For* $\boldsymbol{x}^{\boldsymbol{\alpha}}, \boldsymbol{x}^{\boldsymbol{\beta}} \in \mathbb{T}^n$, *if* $\boldsymbol{x}^{\boldsymbol{\alpha}}$ *divides* $\boldsymbol{x}^{\boldsymbol{\beta}}$ *then* $\boldsymbol{x}^{\boldsymbol{\alpha}} \leq \boldsymbol{x}^{\boldsymbol{\beta}}$.

PROOF. By assumption there is an $\boldsymbol{x}^{\boldsymbol{\gamma}} \in \mathbb{T}^n$ such that $\boldsymbol{x}^{\boldsymbol{\beta}} = \boldsymbol{x}^{\boldsymbol{\alpha}}\boldsymbol{x}^{\boldsymbol{\gamma}}$. By Condition (i) in Definition 1.4.1 we have $\boldsymbol{x}^{\boldsymbol{\gamma}} \geq 1$ and so by Condition (ii) we have $\boldsymbol{x}^{\boldsymbol{\beta}} = \boldsymbol{x}^{\boldsymbol{\alpha}}\boldsymbol{x}^{\boldsymbol{\gamma}} \geq \boldsymbol{x}^{\boldsymbol{\alpha}}$, as desired. □

Now from the Hilbert Basis Theorem (Theorem 1.1.1) we can prove

THEOREM 1.4.6. *Every term order on $\mathbb{T}^n$ is a well-ordering; that is, for every subset A of $\mathbb{T}^n$, there exists $\boldsymbol{x}^{\boldsymbol{\alpha}} \in A$ such that for all $\boldsymbol{x}^{\boldsymbol{\beta}} \in A$, $\boldsymbol{x}^{\boldsymbol{\alpha}} \leq \boldsymbol{x}^{\boldsymbol{\beta}}$.*

PROOF. Suppose to the contrary that the given term order is not a well-ordering. Then there exist $\boldsymbol{x}^{\boldsymbol{\alpha}_i} \in \mathbb{T}^n, i = 1, 2, \ldots$ such that

$$\boldsymbol{x}^{\boldsymbol{\alpha}_1} > \boldsymbol{x}^{\boldsymbol{\alpha}_2} > \boldsymbol{x}^{\boldsymbol{\alpha}_3} > \cdots . \tag{1.4.2}$$

This defines a chain of ideals in $k[x_1, \ldots, x_n]$

$$\langle \boldsymbol{x}^{\boldsymbol{\alpha}_1} \rangle \subseteq \langle \boldsymbol{x}^{\boldsymbol{\alpha}_1}, \boldsymbol{x}^{\boldsymbol{\alpha}_2} \rangle \subseteq \langle \boldsymbol{x}^{\boldsymbol{\alpha}_1}, \boldsymbol{x}^{\boldsymbol{\alpha}_2}, \boldsymbol{x}^{\boldsymbol{\alpha}_3} \rangle \subseteq \cdots . \tag{1.4.3}$$

We first note that $\langle \boldsymbol{x}^{\boldsymbol{\alpha}_1}, \ldots, \boldsymbol{x}^{\boldsymbol{\alpha}_i} \rangle \neq \langle \boldsymbol{x}^{\boldsymbol{\alpha}_1}, \ldots, \boldsymbol{x}^{\boldsymbol{\alpha}_{i+1}} \rangle$, since if we had equality, then

$$\boldsymbol{x}^{\boldsymbol{\alpha}_{i+1}} = \sum_{j=1}^{i} u_j \boldsymbol{x}^{\boldsymbol{\alpha}_j}, \tag{1.4.4}$$

where u_j is a polynomial in $k[x_1, \ldots, x_n], j = 1, \ldots, i$. If we expand each u_j as a linear combination of power products, we see that each term in $u_j\boldsymbol{x}^{\boldsymbol{\alpha}_j}$ is divisible by $\boldsymbol{x}^{\boldsymbol{\alpha}_j}$. Thus every term of the right-hand side of Equation (1.4.4) is divisible by some $\boldsymbol{x}^{\boldsymbol{\alpha}_j}, 1 \leq j \leq i$. But $\boldsymbol{x}^{\boldsymbol{\alpha}_{i+1}}$ must appear as the power product of a term on the right-hand side of Equation (1.4.4). Therefore $\boldsymbol{x}^{\boldsymbol{\alpha}_{i+1}}$ is divisible by some $\boldsymbol{x}^{\boldsymbol{\alpha}_j}, 1 \leq j \leq i$, and hence $\boldsymbol{x}^{\boldsymbol{\alpha}_{i+1}} \geq \boldsymbol{x}^{\boldsymbol{\alpha}_j}$ for some $j, 1 \leq j \leq i$, by Proposition 1.4.5 and this contradicts (1.4.2). So if we go back to the chain of ideals (1.4.3), we now see that this chain is a strictly ascending chain of ideals in $k[x_1, \ldots, x_n]$. This is a contradiction to the Hilbert Basis Theorem (Theorem 1.1.1). □

Theorem 1.4.6 will be used throughout this book for many proofs in a manner described in Appendix B.

To finish this section, we fix some notation. First we choose a term order[4] on $k[x_1, \ldots, x_n]$. Then for all $f \in k[x_1, \ldots, x_n]$, with $f \neq 0$,we may write

$$f = a_1\boldsymbol{x}^{\boldsymbol{\alpha}_1} + a_2\boldsymbol{x}^{\boldsymbol{\alpha}_2} + \cdots + a_r\boldsymbol{x}^{\boldsymbol{\alpha}_r},$$

where $0 \neq a_i \in k$, $\boldsymbol{x}^{\boldsymbol{\alpha}_i} \in \mathbb{T}^n$, and $\boldsymbol{x}^{\boldsymbol{\alpha}_1} > \boldsymbol{x}^{\boldsymbol{\alpha}_2} > \cdots > \boldsymbol{x}^{\boldsymbol{\alpha}_r}$. We will always try to write our polynomials in this way. We define:

- $\mathrm{lp}(f) = \boldsymbol{x}^{\boldsymbol{\alpha}_1}$, the *leading power product* of f;
- $\mathrm{lc}(f) = a_1$, the *leading coefficient* of f;
- $\mathrm{lt}(f) = a_1\boldsymbol{x}^{\boldsymbol{\alpha}_1}$, the *leading term* of f.

[4]We will say that we have a term order on $k[x_1, \ldots, x_n]$ when we have a term order on $\mathbb{T}^n$.

We also define $\text{lp}(0) = \text{lc}(0) = \text{lt}(0) = 0$.

Note that lp, lc, and lt are multiplicative; that is, $\text{lp}(fg) = \text{lp}(f)\,\text{lp}(g)$, $\text{lc}(fg) = \text{lc}(f)\,\text{lc}(g)$, and $\text{lt}(fg) = \text{lt}(f)\,\text{lt}(g)$. Also, if we change the term order, then $\text{lp}(f)$, $\text{lc}(f)$, and $\text{lt}(f)$ may change. For example, let $f = 2x^2yz + 3xy^3 - 2x^3$:

- if the order is lex with $x > y > z$, then $\text{lp}(f) = x^3, \text{lc}(f) = -2$, and $\text{lt}(f) = -2x^3$;
- if the order is deglex with $x > y > z$, then $\text{lp}(f) = x^2yz, \text{lc}(f) = 2$, and $\text{lt}(f) = 2x^2yz$;
- if the order is degrevlex with $x > y > z$, then $\text{lp}(f) = xy^3, \text{lc}(f) = 3$, and $\text{lt}(f) = 3xy^3$.

Exercises

1.4.1. Consider the polynomial $f = 3x^4z - 2x^3y^4 + 7x^2y^2z^3 - 8xy^3z^3 \in \mathbb{Q}[x, y, z]$. Determine the leading term, leading coefficient, and leading power product of f with respect to deglex, lex, and degrevlex with $x > y > z$. Repeat the exercise with $x < y < z$.

1.4.2. In the polynomial ring in one variable, $k[x]$, let $<$ be a term order. Show that it must be the usual one, i.e. the one such that

$$1 < x < x^2 < x^3 < \cdots .$$

1.4.3. Show that lex, deglex and degrevlex are term orderings.

1.4.4. Show that in $k[x, y]$, deglex and degrevlex are the same orders.

1.4.5. Given polynomials $f_1, \dots , f_s$ and $u_1, \dots , u_s$ in $k[x_1, \dots , x_n]$, show that $\text{lp}(f_1u_1 + \cdots + f_su_s) \leq \max_{1\leq i\leq s}(\text{lp}(f_i)\,\text{lp}(u_i))$. Does equality necessarily hold? (Prove or disprove.)

1.4.6. Let $<$ be a total order on $\mathbb{T}^n$ satisfying condition (ii) in Definition 1.4.1, and assume that $<$ is also a well-ordering. Prove that for all $\boldsymbol{x}^{\boldsymbol{\alpha}} \neq 1$ in $\mathbb{T}^n$, we have $1 < \boldsymbol{x}^{\boldsymbol{\alpha}}$. (This is a partial converse of Theorem 1.4.6).

1.4.7. Let $x_1, \dots , x_n$ be variables, and let $m < n$. Prove that any term order on power products in the variables $x_1, \dots , x_m$ is the restriction of a term order on power products in the variables $x_1, \dots , x_n$. [Hint: Use the idea of lex, grouping the variables $x_1, \dots , x_m$ together, and using the given term order on them.]

1.4.8. Let $f \in k[x_1, \dots , x_n]$ and consider the lex order with $x_1 > x_2 > \cdots > x_n$. Let $i \in \{1, \dots , n\}$. Prove that $f \in k[x_i, \dots , x_n]$ if and only if $\text{lt}(f) \in k[x_i, \dots , x_n]$.

1.4.9. We call a polynomial $f \in k[x_1, \dots , x_n]$ *homogeneous* provided that the total degree of every term is the same (e.g. $x^2y^2z + xy^4 - z^5$ is homogeneous since every term has total degree 5, while $x^3y^2 - x^2yz^2 + y^2z$ is not homogeneous; the latter polynomial is the sum of the two homogeneous polynomials $x^3y^2 - x^2yz^2$ and y^2z, called the *homogeneous components* of $x^3y^2 - x^2yz^2 + y^2z$). Let f be a homogeneous polynomial and let the term ordering be degrevlex with $x_1 > x_2 > \cdots > x_n$. Prove that x_n divides f if and only if x_n divides $\operatorname{lt}(f)$. Show more generally that $f \in \langle x_i, \ldots, x_n \rangle$ if and only if $\operatorname{lt}(f) \in \langle x_i, \ldots, x_n \rangle$.

1.4.10. The revlex ordering is defined as follows: For $\boldsymbol{\alpha} = (\alpha_1, \ldots, \alpha_n), \boldsymbol{\beta} = (\beta_1, \ldots, \beta_n) \in \mathbb{N}^n$ we define $\boldsymbol{x}^{\boldsymbol{\alpha}} < \boldsymbol{x}^{\boldsymbol{\beta}}$ if and only if the first coordinates α_i and β_i in $\boldsymbol{\alpha}$ and $\boldsymbol{\beta}$ from the right which are different satisfy $\alpha_i > \beta_i$. Show that revlex is not a term order on $k[x_1, \ldots, x_n]$.

1.4.11. Let $I \subseteq k[x_1, \ldots, x_n]$ be an ideal generated by (possibly infinitely many) power products (such an ideal is called a *monomial ideal*). Prove that there exist $\boldsymbol{\alpha}_1, \ldots, \boldsymbol{\alpha}_m \in \mathbb{N}^n$ such that $I = \langle \boldsymbol{x}^{\boldsymbol{\alpha}_1}, \ldots, \boldsymbol{x}^{\boldsymbol{\alpha}_m} \rangle$. [Hint: First show that a polynomial $f \in I$ if and only if each term of f is in I.]

1.4.12. (Dickson's Lemma) Prove that the result of Exercise 1.4.11 is equivalent to the following statement: Given any $A \subseteq \mathbb{N}^n$, there exist $\boldsymbol{\alpha}_1, \ldots, \boldsymbol{\alpha}_m \in A$ such that

$$A \subseteq \bigcup_{i=1}^{m} (\boldsymbol{\alpha}_i + \mathbb{N}^n).$$

(By $\boldsymbol{\alpha} + \mathbb{N}^n$ we mean $\{\boldsymbol{\alpha} + \boldsymbol{\gamma} \mid \boldsymbol{\gamma} \in \mathbb{N}^n\}$.)

1.4.13. Prove that every monomial ideal I (see Exercise 1.4.11) contains a unique minimal generating set. That is, prove there is a subset $G \subseteq I$ such that $I = \langle G \rangle$ and for all subsets $F \subseteq I$ with $I = \langle F \rangle$ we have $G \subseteq F$. [Hint: Prove first that if $I = \langle \boldsymbol{x}^{\boldsymbol{\alpha}_1}, \ldots, \boldsymbol{x}^{\boldsymbol{\alpha}_m} \rangle$ then for $\boldsymbol{\beta} \in \mathbb{N}^n$ we have $\boldsymbol{x}^{\boldsymbol{\beta}} \in I$ if and only if there is an i such that $\boldsymbol{x}^{\boldsymbol{\alpha}_i}$ divides $\boldsymbol{x}^{\boldsymbol{\beta}}$.]

1.4.14. (Mora-Robbiano [**MoRo**]) Let

$$\boldsymbol{u}_1 = (u_{11}, u_{12}, \ldots, u_{1n}), \ldots, \boldsymbol{u}_m = (u_{m1}, u_{m2}, \ldots, u_{mn}) \in \mathbb{Q}^n.$$

We define an order in $\mathbb{Q}^m$ as follows: $(\alpha_1, \ldots, \alpha_m) < (\beta_1, \ldots, \beta_m)$ if and only if the first α_i, β_i from the left which are different satisfy $\alpha_i < \beta_i$. (Note that this is just lex on $\mathbb{Q}^m$.) Now we define an order $<_u$ in $k[x_1, \ldots, x_n]$ as follows: for $\boldsymbol{\alpha}, \boldsymbol{\beta} \in \mathbb{N}^n$,

$$\boldsymbol{x}^{\boldsymbol{\alpha}} <_u \boldsymbol{x}^{\boldsymbol{\beta}} \iff (\boldsymbol{\alpha} \cdot \boldsymbol{u}_1, \ldots, \boldsymbol{\alpha} \cdot \boldsymbol{u}_m) < (\boldsymbol{\beta} \cdot \boldsymbol{u}_1, \ldots, \boldsymbol{\beta} \cdot \boldsymbol{u}_m),$$

where $\boldsymbol{\alpha} \cdot \boldsymbol{u}_i$ is the usual dot product in $\mathbb{Q}^n$.

a. Prove that $<_u$ is a transitive relation.

b. Prove that $\boldsymbol{x}^{\boldsymbol{\alpha}} <_u \boldsymbol{x}^{\boldsymbol{\beta}}$ implies $\boldsymbol{x}^{\boldsymbol{\alpha}} \boldsymbol{x}^{\gamma} <_u \boldsymbol{x}^{\boldsymbol{\beta}} \boldsymbol{x}^{\gamma}$ for all $\boldsymbol{\alpha}, \boldsymbol{\beta}, \boldsymbol{\gamma} \in \mathbb{N}^n$.

c. Prove that if the vectors $\boldsymbol{u}_1, \ldots, \boldsymbol{u}_m$ span $\mathbb{Q}^n$, then the order, $<_u$, is a total order.

d. Prove that if the vectors $\boldsymbol{u}_1, \ldots, \boldsymbol{u}_m$ span $\mathbb{Q}^n$, then $<_u$ is a term order if and only if for all i, the first $\boldsymbol{u}_j$ such that $u_{ji} \neq 0$ satisfies $u_{ji} > 0$.

e. Let $\boldsymbol{u}_1, \ldots, \boldsymbol{u}_m$ be vectors satisfying: for all i, the first $\boldsymbol{u}_j$ such that $u_{ji} \neq 0$ satisfies $u_{ji} > 0$. Show that the partial $<_u$ can be extended to a term order, $<_{u'}$, that is, $\boldsymbol{x}^{\boldsymbol{\alpha}} <_u \boldsymbol{x}^{\boldsymbol{\beta}}$ implies $\boldsymbol{x}^{\boldsymbol{\alpha}} <_{u'} \boldsymbol{x}^{\boldsymbol{\beta}}$.

1.4.15. What vectors $\boldsymbol{u}_1, \dots, \boldsymbol{u}_n \in \mathbb{Q}^n$ in Exercise 1.4.14 give rise to the lex, to the deglex, and to the degrevlex term orderings with $x_1 < x_2 < \cdots < x_n$? Same question with $x_1 > x_2 > \cdots > x_n$.

1.4.16. Let $f = 2x^4y^5 + 3x^5y^2 + x^3y^9 \in \mathbb{Q}[x, y]$. Show that there is no term ordering on $\mathbb{Q}[x, y]$ such that $\operatorname{lp}(f) = x^4y^5$.

1.4.17. (*) Let $X_i = x_1^{\alpha_{i1}} \cdots x_n^{\alpha_{in}}$, $i = 1, \dots, r$, be power products in $k[x_1, \dots, x_n]$ and let $f = \sum_{i=1}^r c_i X_i$, where $c_i \in k - \{0\}$, for $i = 1, \dots, r$. Assume that there is a term order $<$ such that $\operatorname{lp}(f) = X_1$. Consider the vectors $\boldsymbol{\alpha}_i = (\alpha_{i1}, \dots, \alpha_{in}) \in \mathbb{N}^n$, $i = 1, \dots, r$. In this exercise we show that there exists a vector $\boldsymbol{u} = (u_1, \dots, u_n) \in \mathbb{Q}^n$ such that $u_i \geq 0$ for $i = 1, \dots, n$ and $\boldsymbol{\alpha}_1 \cdot \boldsymbol{u} = \sum_{j=1}^n \alpha_{1j} u_j > \sum_{j=1}^n \alpha_{\ell j} u_j = \boldsymbol{\alpha}_\ell \cdot \boldsymbol{u}$ for all $\ell = 2, \dots, r$ (compare with Exercise 1.4.14). We will use the following result from linear algebra (see, for example, [**Ga**]):

THEOREM. Let A be any $r \times n$ matrix with rational entries, then exactly one of the following two alternatives holds:

- there exists a row vector $\boldsymbol{v} \in \mathbb{Q}^r$ with non-negative coordinates such that the coordinates of the vector $\boldsymbol{v}A$ are all negative or zero;
- There exists a column vector $\boldsymbol{u} \in \mathbb{Q}^n$ with non-negative coordinates such that the coordinates of the vector $A\boldsymbol{u}$ are all positive.

a. Use the above result to show that there is a vector $\boldsymbol{u} \in \mathbb{Q}^n$ with non-negative coordinates such that $\boldsymbol{\alpha}_1 \cdot \boldsymbol{u} > \boldsymbol{\alpha}_\ell \cdot \boldsymbol{u}$ for $\ell = 2, \dots, r$. [Hint: Consider the matrix A whose rows are the vectors $\boldsymbol{\alpha}_1 - \boldsymbol{\alpha}_\ell$.]

There is a geometric way to view the linear algebra theorem used above. First we define the *convex hull* of the vectors $\boldsymbol{\alpha}_1, \dots, \boldsymbol{\alpha}_r$ as follows:

$$\operatorname{conv}(\boldsymbol{\alpha}_1, \dots, \boldsymbol{\alpha}_r) = \left\{ \sum_{i=1}^r c_i \boldsymbol{\alpha}_i \mid c_i \geq 0, i = 1, \dots, r, \text{ and } \sum_{i=1}^r c_i = 1 \right\}.$$

Also, let $\{\boldsymbol{e}_1, \dots, \boldsymbol{e}_n\}$ be the standard basis for $\mathbb{Q}^n$, that is, $\boldsymbol{e}_i$ is the vector in $\mathbb{Q}^n$ with all coordinates equal to 0 except the ith coordinate which is equal to 1.

b. Show that the first alternative in the linear algebra theorem above is equivalent to the condition that the zero vector is in the convex hull of the rows of A together with the vectors $\boldsymbol{e}_i$, $i = 1, \dots, n$. Note that the second alternative implies that there is a vector $\boldsymbol{u}$ which makes an acute angle with every row of A; i.e. the hyperplane, L, orthogonal to $\boldsymbol{u}$ has all the rows of A on one side, and so L has the convex hull of the rows of A and the $\boldsymbol{e}_i$'s on one side.

c. Conclude that X_1 is the leading term of f with respect to some term order if and only if the zero vector is not in the convex hull of the vectors $\boldsymbol{\alpha}_1 - \boldsymbol{\alpha}_i$, $i = 2, \ldots, r$ and $\mathbf{e}_j$, $j = 1, \ldots, n$. [Hint: See Exercise 1.4.14, part **e.**]

d. Use the above to determine all the possible leading terms of $f = 2x^4y^5 + 3x^5y^2 + x^3y^9 - x^7y$.

1.4.18. In this exercise we prove the Fundamental Theorem of Symmetric Polynomials. Recall that a polynomial $f \in k[x_1, \ldots, x_n]$ is called *symmetric* provided that when the variables of f are rearranged in any way, the resulting polynomial is still equal to f. For example, for $n = 3$, $x_1 + x_2 + x_3, x_1x_2 + x_1x_3 + x_2x_3$, and $x_1x_2x_3$ are symmetric. For general n, let $\sigma_1 = x_1 + x_2 + \cdots + x_n, \sigma_2 = x_1x_2 + x_1x_3 + \cdots + x_{n-1}x_n, \ldots, \sigma_n = x_1x_2 \cdots x_n$. These polynomials are called the *elementary symmetric polynomials.* The theorem states that every symmetric polynomial is a polynomial in the elementary symmetric polynomials. Fix the lex term ordering on $k[x_1, \ldots, x_n]$ with $x_1 > x_2 > \cdots > x_n$. Let $f \in k[x_1, \ldots, x_n]$ be a symmetric polynomial. We need to show the existence of a polynomial $h \in k[x_1, \ldots, x_n]$ such that $f = h(\sigma_1, \ldots, \sigma_n)$.

a. Let $\operatorname{lt}(f) = c\boldsymbol{x}^{\boldsymbol{\alpha}}$ where $\boldsymbol{\alpha} = (\alpha_1, \ldots, \alpha_n) \in \mathbb{N}^n$ and $c \in k$. Show that $\alpha_1 \geq \alpha_2 \geq \cdots \geq \alpha_n$.

b. Let

$$g = \sigma_1^{\alpha_1 - \alpha_2} \sigma_2^{\alpha_2 - \alpha_3} \cdots \sigma_{n-1}^{\alpha_{n-1} - \alpha_n} \sigma_n^{\alpha_n}.$$

Show that $\operatorname{lp}(g) = \boldsymbol{x}^{\boldsymbol{\alpha}}$.

c. Now observe that $\operatorname{lp}(f - cg) < \operatorname{lp}(f)$ and that $f - cg$ is a symmetric polynomial. Use the well-ordering property of term orders to complete the proof of the existence of h and so to prove the Fundamental Theorem of Symmetric Polynomials.

d. Note that the above proof yields an algorithm for computing h given the symmetric polynomial f. Use it in the case $n = 2$ to write $x_1^4 + x_2^4$ as a polynomial in $\sigma_1 = x_1 + x_2$ and $\sigma_2 = x_1x_2$.

1.5. Division Algorithm. In this section we study the second ingredient in our solution method for the problems mentioned in Section 1.1: a division algorithm in $k[x_1, \ldots, x_n]$. In Sections 1.2 and 1.3 we had a division algorithm, also referred to as a reduction process. We will define a division algorithm in $k[x_1, \ldots, x_n]$ that extends both of the algorithms seen in the previous sections.

The basic idea behind the algorithm is the same as for linear and one variable polynomials: when dividing f by $f_1, \ldots, f_s$, we want to cancel terms of f using the leading terms of the f_i's (so the new terms which are introduced are smaller than the canceled terms) and continue this process until it cannot be done anymore.

Let us first look at the special case of the division of f by g, where $f, g \in k[x_1, \ldots, x_n]$. We fix a term order on $k[x_1, \ldots, x_n]$.

DEFINITION 1.5.1. *Given f, g, h in $k[x_1, \ldots, x_n]$, with $g \neq 0$, we say that f* reduces *to h modulo g in one step, written*

$$f \xrightarrow{g} h,$$

if and only if $\text{lp}(g)$ *divides a non-zero term*[5] *X that appears in f and*

$$h = f - \frac{X}{\text{lt}(g)} g.$$

It must be strongly emphasized that in this definition we have subtracted from f the entire term X and we have replaced X by terms strictly smaller than X. (We observe that in the special cases presented in Sections 1.2 and 1.3 we considered only the case where $X = \text{lt}(f)$.)

For example, let $f = 6x^2y - x + 4y^3 - 1$ and $g = 2xy + y^3$ be polynomials in $\mathbb{Q}[x, y]$. If the term order is lex with $x > y$, then $f \xrightarrow{g} h$, where $h = -3xy^3 - x + 4y^3 - 1$, since, in this case $X = 6x^2y$ is the term of f we have canceled using $\text{lt}(g) = 2xy$; in fact $X = \text{lt}(f)$. (We are *not* allowed to cancel, say, only $4x^2y$. Another way of saying this is that we are not allowed to write $f = 4x^2y + 2x^2y - x + 4y^3 - 1$ and just cancel $4x^2y$). We now consider the term order deglex with $x > y$ so that now $\text{lt}(g) = y^3$ and so $f \xrightarrow{g} h$, where now $h = 6x^2y - 8xy - x - 1$. We note that in this latter case we canceled the term $X = 4y^3$ from f which is not the leading term of f.

We can think of h in the definition as the remainder of a one step division of f by g similar to the one seen in Section 1.3. We can continue this process and subtract off all terms in f that are divisible by $\text{lt}(g)$.

EXAMPLE 1.5.2. Let $f = y^2x + 4yx - 3x^2, g = 2y + x + 1 \in \mathbb{Q}[x, y]$. Also, let the order be deglex with $y > x$. Then

$$f \xrightarrow{g} -\frac{1}{2}yx^2 + \frac{7}{2}yx - 3x^2 \xrightarrow{g} \frac{1}{4}x^3 + \frac{7}{2}yx - \frac{11}{4}x^2 \xrightarrow{g} \frac{1}{4}x^3 - \frac{9}{2}x^2 - \frac{7}{4}x.$$

Note that in the last polynomial, namely $\frac{1}{4}x^3 - \frac{9}{2}x^2 - \frac{7}{4}x$, no term is divisible by $\text{lp}(g) = y$ and so this procedure cannot continue. We could write this reduction process in long division format as

[5] From now on we will use capital letters, usually X, Y or Z, to denote power products or terms instead of the more cumbersome $\boldsymbol{x}^{\boldsymbol{\alpha}}$ or $a\boldsymbol{x}^{\boldsymbol{\alpha}}$ unless we need to make an explicit reference to the exponent $\boldsymbol{\alpha}$. We will also say $X > Y$, for term X, Y, provided that $\text{lp}(X) > \text{lp}(Y)$.

$$
\begin{array}{r|l}
 & \frac{1}{2}yx - \frac{1}{4}x^2 + \frac{7}{4}x \\
\hline
2y + x + 1 & y^2x + 4yx - 3x^2 \\
 & y^2x + \frac{1}{2}yx^2 + \frac{1}{2}yx \\
\cline{2-2}
 & -\frac{1}{2}yx^2 + \frac{7}{2}yx - 3x^2 \\
 & -\frac{1}{2}yx^2 - \frac{1}{4}x^3 - \frac{1}{4}x^2 \\
\cline{2-2}
 & \frac{1}{4}x^3 + \frac{7}{2}yx - \frac{11}{4}x^2 \\
 & \frac{7}{2}yx + \frac{7}{4}x^2 + \frac{7}{4}x \\
\cline{2-2}
 & \frac{1}{4}x^3 \qquad -\frac{9}{2}x^2 - \frac{7}{4}x
\end{array}
$$

In the multivariable case we may have to divide by more than one polynomial at a time, and so we extend the process of reduction defined above to include this more general setting.

DEFINITION 1.5.3. *Let f, h, and $f_1, \dots, f_s$ be polynomials in $k[x_1, \dots, x_n]$, with $f_i \neq 0$ $(1 \leq i \leq s)$, and let $F = \{f_1, \dots, f_s\}$. We say that f* reduces *to h modulo F, denoted*

$$f \xrightarrow{F}_+ h,$$

if and only if there exist a sequence of indices $i_1, i_2, \dots, i_t \in \{1, \dots, s\}$ and a sequence of polynomials $h_1, \dots, h_{t-1} \in k[x_1, \dots, x_n]$ such that

$$f \xrightarrow{f_{i_1}} h_1 \xrightarrow{f_{i_2}} h_2 \xrightarrow{f_{i_3}} \cdots \xrightarrow{f_{i_{t-1}}} h_{t-1} \xrightarrow{f_{i_t}} h.$$

EXAMPLE 1.5.4. Let $f_1 = yx - y, f_2 = y^2 - x \in \mathbb{Q}[x, y]$. Let the order be deglex with $y > x$. Let $F = \{f_1, f_2\}, f = y^2x$. Then

$$f \xrightarrow{F}_+ x,$$

since

$$y^2x \xrightarrow{f_1} y^2 \xrightarrow{f_2} x.$$

DEFINITION 1.5.5. *A polynomial r is called* reduced *with respect to a set of non-zero polynomials $F = \{f_1, \dots, f_s\}$ if $r = 0$ or no power product that appears in r is divisible by any one of the $\operatorname{lp}(f_i), i = 1, \dots, s$. In other words, r cannot be reduced modulo F.*

DEFINITION 1.5.6. *If $f \xrightarrow{F}_+ r$ and r is reduced with respect to F, then we call r a* remainder *for f with respect to F.*

The reduction process allows us to define a division algorithm that mimics the Division Algorithm in one variable. Given $f, f_1, \dots, f_s \in k[x_1, \dots, x_n]$ with

$f_i \neq 0$ $(1 \leq i \leq s)$, this algorithm returns quotients $u_1, \ldots, u_s \in k[x_1, \ldots, x_n]$, and a remainder $r \in k[x_1, \ldots, x_n]$, such that

$$f = u_1 f_1 + \cdots + u_s f_s + r.$$

This algorithm is given as Algorithm 1.5.1.

INPUT: $f, f_1, \ldots, f_s \in k[x_1, \ldots, x_n]$ with $f_i \neq 0$ $(1 \leq i \leq s)$

OUTPUT: $u_1, \ldots, u_s, r$ such that $f = u_1 f_1 + \cdots + u_s f_s + r$ and
r is reduced with respect to $\{f_1, \ldots, f_s\}$ and
$\max(\mathrm{lp}(u_1)\,\mathrm{lp}(f_1), \ldots, \mathrm{lp}(u_s)\,\mathrm{lp}(f_s), \mathrm{lp}(r)) = \mathrm{lp}(f)$.

INITIALIZATION: $u_1 := 0, u_2 := 0, \ldots, u_s := 0, r := 0, h := f$

WHILE $h \neq 0$ **DO**

IF there exists i such that $\mathrm{lp}(f_i)$ divides $\mathrm{lp}(h)$ **THEN**

choose i least such that $\mathrm{lp}(f_i)$ divides $\mathrm{lp}(h)$

$$u_i := u_i + \frac{\mathrm{lt}(h)}{\mathrm{lt}(f_i)}$$

$$h := h - \frac{\mathrm{lt}(h)}{\mathrm{lt}(f_i)} f_i$$

ELSE

$r := r + \mathrm{lt}(h)$

$h := h - \mathrm{lt}(h)$

ALGORITHM 1.5.1. *Multivariable Division Algorithm*

Note that in Algorithm 1.5.1 we have, in effect, assumed an ordering among the polynomials in the set $\{f_1, \ldots, f_s\}$ when we chose i to be least such that $\mathrm{lp}(f_i)$ divides $\mathrm{lp}(h)$. This is an important point and will be illustrated in Example 1.5.10.

It is informative to consider the similarities between Algorithm 1.3.1, the one variable Division Algorithm in Section 1.3, and Algorithm 1.5.1, the multivariable Division Algorithm. The quotients $u_1, \ldots, u_s$ in Algorithm 1.5.1 correspond to the single quotient q in Algorithm 1.3.1; we have s different quotients in Algorithm 1.5.1 because we are dividing f by s different polynomials $f_1, \ldots, f_s$ as opposed to dividing f by a single polynomial g in Algorithm 1.3.1. The remainders, denoted by r in both algorithms, have the same definition: no term of r is divisible by the leading term of any divisor. In Algorithm 1.3.1, once the leading term of r is not divisible by $\mathrm{lt}(g)$, we also know that no other term of r

is divisible by $\mathrm{lt}(g)$, and we have obtained the remainder. So in Algorithm 1.3.1 we start with $r = f$ and subtract off multiples of g until this occurs. This simple property is not true in the multivariable case, necessitating the introduction of the extra polynomial h in Algorithm 1.5.1. So we start with $h = f$ and $r = 0$ and subtract off the leading term of h when we can or add the leading term of h into r when we cannot, and so build up the remainder.

EXAMPLE 1.5.7. We recompute Example 1.5.2 but now we follow Algorithm 1.5.1. Let $F = \{f_1\}$, where $f_1 = 2y + x + 1 \in \mathbb{Q}[x, y]$. The order is deglex with $y > x$. Let $f = y^2x + 4yx - 3x^2$.

INITIALIZATION: $u_1 := 0, r := 0, h := y^2x + 4yx - 3x^2$

First pass through the WHILE loop:

$$y = \mathrm{lp}(f_1) \text{ divides } \mathrm{lp}(h) = y^2x$$
$$u_1 := u_1 + \tfrac{y^2x}{2y} = \tfrac{1}{2}yx$$
$$\begin{aligned} h &:= h - \tfrac{\mathrm{lt}(h)}{\mathrm{lt}(f_1)} f_1 \\ &= (y^2x + 4yx - 3x^2) - \tfrac{y^2x}{2y}(2y + x + 1) \\ &= -\tfrac{1}{2}yx^2 + \tfrac{7}{2}yx - 3x^2 \end{aligned}$$

Second pass through the WHILE loop:

$$y = \mathrm{lp}(f_1) \text{ divides } \mathrm{lp}(h) = yx^2$$
$$u_1 := u_1 + \tfrac{-\frac{1}{2}yx^2}{2y} = \tfrac{1}{2}yx - \tfrac{1}{4}x^2$$
$$\begin{aligned} h &:= h - \tfrac{\mathrm{lt}(h)}{\mathrm{lt}(f_1)} f_1 \\ &= (-\tfrac{1}{2}yx^2 + \tfrac{7}{2}yx - 3x^2) - \tfrac{-\frac{1}{2}yx^2}{2y}(2y + x + 1) \\ &= \tfrac{1}{4}x^3 + \tfrac{7}{2}yx - \tfrac{11}{4}x^2 \end{aligned}$$

Third pass through the WHILE loop:

$$y = \mathrm{lp}(f_1) \text{ does not divide } \mathrm{lp}(h) = x^3$$
$$r := r + \mathrm{lt}(h) = \tfrac{1}{4}x^3$$
$$h := h - \mathrm{lt}(h) = \tfrac{7}{2}yx - \tfrac{11}{4}x^2$$

Fourth pass through the WHILE loop:

$$y = \mathrm{lp}(f_1) \text{ divides } \mathrm{lp}(h) = yx$$
$$u_1 := u_1 + \tfrac{\frac{7}{2}yx}{2y} = \tfrac{1}{2}yx - \tfrac{1}{4}x^2 + \tfrac{7}{4}x$$
$$\begin{aligned} h &:= h - \tfrac{\mathrm{lt}(h)}{\mathrm{lt}(f_1)} f_1 \\ &= (\tfrac{7}{2}yx - \tfrac{11}{4}x^2) - \tfrac{\frac{7}{2}yx}{2y}(2y + x + 1) \\ &= -\tfrac{9}{2}x^2 - \tfrac{7}{4}x \end{aligned}$$

Fifth pass through the WHILE loop:

$$y = \mathrm{lp}(f_1) \text{ does not divide } \mathrm{lp}(h) = x^2$$
$$r := r + \mathrm{lt}(h) = \tfrac{1}{4}x^3 - \tfrac{9}{2}x^2$$
$$h := h - \mathrm{lt}(h) = -\tfrac{7}{4}x$$

Sixth pass through the WHILE loop:

$$y = \mathrm{lp}(f_1) \text{ does not divide } \mathrm{lp}(h) = x$$
$$r := r + \mathrm{lt}(h) = \tfrac{1}{4}x^3 - \tfrac{9}{2}x^2 - \tfrac{7}{4}x$$
$$h := h - \mathrm{lt}(h) = 0$$

The WHILE loop stops, and we have

$$f \xrightarrow{F}_+ \frac{1}{4}x^3 - \frac{9}{2}x^2 - \frac{7}{4}x$$

and

$$f = (\frac{1}{2}yx - \frac{1}{4}x^2 + \frac{7}{4}x)(2y + x + 1) + (\frac{1}{4}x^3 - \frac{9}{2}x^2 - \frac{7}{4}x).$$

Note that these are the same steps we used in Example 1.5.2.

EXAMPLE 1.5.8. Let $F = \{f_1, f_2\}$, where $f_1 = yx - y, f_2 = y^2 - x \in \mathbb{Q}[x, y]$. The order is deglex with $y > x$. Let $f = y^2x$.

INITIALIZATION: $u_1 := 0, u_2 := 0, r := 0, h := y^2x$

First pass through the WHILE loop:

$yx = \mathrm{lp}(f_1)$ divides $\mathrm{lp}(h) = y^2x$

$u_1 := u_1 + \frac{\mathrm{lt}(h)}{\mathrm{lt}(f_1)} = y$

$h := h - \frac{\mathrm{lt}(h)}{\mathrm{lt}(f_1)} f_1 = y^2x - \frac{y^2x}{yx}(yx - y) = y^2$

Second pass through the WHILE loop:

$yx = \mathrm{lp}(f_1)$ does not divide $\mathrm{lp}(h) = y^2$

$y^2 = \mathrm{lp}(f_2)$ divides $\mathrm{lp}(h) = y^2$

$u_2 := u_2 + \frac{\mathrm{lt}(h)}{\mathrm{lt}(f_2)} = 1$

$h := h - \frac{\mathrm{lt}(h)}{\mathrm{lt}(f_2)} f_2 = y^2 - \frac{y^2}{y^2}(y^2 - x) = x$

Third pass through the WHILE loop:

$yx = \mathrm{lp}(f_1)$ does not divide $\mathrm{lp}(h) = x$

$y^2 = \mathrm{lp}(f_2)$ does not divide $\mathrm{lp}(h) = x$

$r := r + \mathrm{lt}(h) = x$

$h := h - \mathrm{lt}(h) = 0$

The WHILE loop stops, and we get

$$f \xrightarrow{F}_+ x$$

and

$$f = yf_1 + f_2 + x.$$

THEOREM 1.5.9. *Given a set of non-zero polynomials* $F = \{f_1, \ldots, f_s\}$ *and* f *in* $k[x_1, \ldots, x_n]$, *the Division Algorithm (Algorithm* 1.5.1*) produces polynomials* $u_1, \ldots, u_s, r \in k[x_1, \ldots, x_n]$ *such that*

$$f = u_1 f_1 + \cdots + u_s f_s + r,$$

with r *reduced with respect to* F *and*

$$\mathrm{lp}(f) = \max(\max_{1 \le i \le s}(\mathrm{lp}(u_i)\,\mathrm{lp}(f_i)), \mathrm{lp}(r)).$$

PROOF. We first observe that the algorithm terminates. At each stage of the algorithm, the leading term of h is subtracted off until this can no longer be done. That is, we get a sequence $h_1, h_2, \ldots$ of the h's in the algorithm, where h_{i+1} is obtained from h_i by subtracting off $\operatorname{lt}(h_i)$ and possibly some smaller terms: $h_{i+1} = h_i - (\operatorname{lt}(h_i) +$ lower terms). This is because we compute h_{i+1} from h_i by subtracting off $\frac{\operatorname{lt}(h_i)}{\operatorname{lt}(f_j)} f_j = \operatorname{lt}(h_i) +$ lower terms (in case some $\operatorname{lp}(f_j)$ divides $\operatorname{lp}(h_i)$) or by subtracting off $\operatorname{lt}(h_i)$ (in case no $\operatorname{lp}(f_j)$ divides $\operatorname{lp}(h_i)$). So we have that for all i, $\operatorname{lp}(h_{i+1}) < \operatorname{lp}(h_i)$. Thus, since the term order is a well-ordering (Theorem 1.4.6), the list of the h_i's must stop.

To prove the second part, we note that from what we did above, and since $h = f$ at the beginning of the algorithm, we have at any stage in the algorithm $\operatorname{lp}(h) \leq \operatorname{lp}(f)$. Now, for each i, we obtain u_i by adding terms $\frac{\operatorname{lt}(h)}{\operatorname{lt}(f_i)}$, where $\frac{\operatorname{lt}(h)}{\operatorname{lt}(f_i)} f_i$ cancels the leading term of h. It is then immediate that $\operatorname{lp}(u_i)\operatorname{lp}(f_i) \leq \operatorname{lp}(f)$. Moreover, r is obtained by adding in terms $\operatorname{lt}(h)$ and so $\operatorname{lp}(r) \leq \operatorname{lp}(f)$, as well. □

With f written as in Theorem 1.5.9, we have $f - r \in \langle f_1, \ldots, f_s \rangle$. Therefore, if $r = 0$, then f is in $\langle f_1, \ldots, f_s \rangle$. However, the converse is not necessarily true; that is, f may be in the ideal $\langle f_1, \ldots, f_s \rangle$, but the remainder of the division of f by $f_1, \ldots, f_s$ may not be zero as the following example shows.

EXAMPLE 1.5.10. Consider the polynomial $f = y^2x - x \in \mathbb{Q}[x, y]$, and the ideal $I = \langle f_1, f_2 \rangle \subseteq \mathbb{Q}[x, y]$, where $f_1 = yx - y$, $f_2 = y^2 - x$. Set $F = \{f_1, f_2\}$. Using the deglex term order with $y > x$ and the Division Algorithm, we see that $f \xrightarrow{f_1} y^2 - x \xrightarrow{f_2} 0$, that is, $f \xrightarrow{F}_+ 0$ and indeed, $f = yf_1 + f_2$, and hence $f \in I$. However if we reverse the order of f_1 and f_2 (that is, we use f_2 first in the Division Algorithm) then $f \xrightarrow{f_2} x^2 - x$, and $x^2 - x$ is reduced with respect to F. So the remainder of the division of f by F is non-zero, but f is in the ideal $\langle f_1, f_2 \rangle$.

This difficulty already occurred in the one variable case. For example, if $f = x$, $f_1 = x^2$ and $f_2 = x^2 - x$, then f is reduced with respect to $\{f_1, f_2\}$, whereas $f = f_1 - f_2 \in \langle f_1, f_2 \rangle$. The difficulty was resolved by finding a better generating set for $\langle f_1, f_2 \rangle$, namely $x = \gcd(x^2, x^2 - x)$. To do this in the multivariable case is the subject of the next section.

Exercises

1.5.1. Let $f = x^3y^3 + 2y^2$, $f_1 = 2xy^2 + 3x + 4y^2$, $f_2 = y^2 - 2y - 2 \in \mathbb{Q}[x, y]$. Using lex with $x > y$, divide f by f_1, f_2 to obtain a remainder r and an expression as in Theorem 1.5.9. Repeat this exercise reversing the role of f_1 and f_2.

1.5.2. Let $f = x^2y^2 - w^2$, $f_1 = x - y^2w$, $f_2 = y - zw$, $f_3 = z - w^3$, $f_4 = w^3 - w \in \mathbb{Q}[x, y, z, w]$. Using lex with $x > y > z > w$, divide f by f_1, f_2, f_3, f_4 to obtain a remainder r and an expression as in Theorem 1.5.9. Repeat this exercise reversing the role of f_1, f_2, f_3, f_4, i.e. using f_4, f_3, f_2, f_1.

1.5.3. Prove that given a set of non-zero polynomials $F \subseteq k[x_1, \ldots, x_n]$, there can be no infinite chain $g_1 \xrightarrow{F} g_2 \xrightarrow{F} g_3 \xrightarrow{F} \cdots$. [Hint: The new point here that did not occur in Theorem 1.5.9 is that we may not be subtracting off leading terms in $g_i \xrightarrow{F} g_{i+1}$.]

1.5.4. Show that for any polynomials $f, g \in k[x_1, \dots, x_n]$, for any finite set of non-zero polynomials $F \subseteq k[x_1, \dots, x_n]$, and for any power product $X \in \mathbb{T}^n$, we have

a. If $f \in F$, then $fg \xrightarrow{F}_+ 0$.

b. If $f \xrightarrow{F}_+ g$, then $Xf \xrightarrow{F}_+ Xg$.

1.5.5. Let $f, g, h, r, s \in k[x_1, \dots, x_n]$ and let F be a collection of non-zero polynomials in $k[x_1, \dots, x_n]$. Disprove the following:

a. If $f \xrightarrow{F}_+ r$ and $g \xrightarrow{F}_+ s$, then $f + g \xrightarrow{F}_+ r + s$.

b. If $f \xrightarrow{F}_+ r$ and $g \xrightarrow{F}_+ s$, then $fg \xrightarrow{F}_+ rs$.

c. If $f+g \xrightarrow{F}_+ h$, $f \xrightarrow{F}_+ r$, and $g \xrightarrow{F}_+ s$, where h, r, s are reduced with respect to F, then $r + s = h$.

1.5.6. Let $F = \{f_1, \dots, f_s\} \subseteq k[x_1, \dots, x_n]$, with $f_i \neq 0$ $(1 \leq i \leq s)$, and let $f \in k[x_1, \dots, x_n]$ such that $f = \sum_{i=1}^{s} u_i f_i$ with $\text{lp}(f) = \max_{1 \leq i \leq s} \text{lp}(u_i f_i)$. Give an example that shows this does not imply that $f \xrightarrow{F}_+ 0$. [Compare with Theorem 1.6.2 part (iii).]

1.6. Gröbner Bases. In this section we finally define the fundamental object of this book, namely, a Gröbner basis.

DEFINITION 1.6.1. *A set of non-zero polynomials $G = \{g_1, \dots, g_t\}$ contained in an ideal I, is called a* Gröbner basis[6] *for I if and only if for all $f \in I$ such that $f \neq 0$, there exists $i \in \{1, \dots, t\}$ such that* $\text{lp}(g_i)$ *divides* $\text{lp}(f)$.

In other words, if G is a Gröbner basis for I, then there are no non-zero polynomials in I reduced with respect to G. We note that it is not clear from this definition that Gröbner bases exist. We will prove this in Corollary 1.6.5.

We first present three other characterizations of a Gröbner basis. In order to do this we need to make the following definition. For a subset S of $k[x_1, \dots, x_n]$, we define the *leading term ideal* of S to be the ideal

$$\text{Lt}(S) = \langle \text{lt}(s) \mid s \in S \rangle.$$

THEOREM 1.6.2. *Let I be a non-zero ideal of $k[x_1, \dots, x_n]$. The following statements are equivalent for a set of non-zero polynomials $G = \{g_1, \dots, g_t\} \subseteq I$.*

(i) *G is a Gröbner basis for I.*

(ii) *$f \in I$ if and only if $f \xrightarrow{G}_+ 0$.*

(iii) *$f \in I$ if and only if $f = \sum_{i=1}^{t} h_i g_i$ with* $\text{lp}(f) = \max_{1 \leq i \leq t}(\text{lp}(h_i)\,\text{lp}(g_i))$.

(iv) $\text{Lt}(G) = \text{Lt}(I)$.

---

[6] Another term which is commonly used in the literature is *standard basis*.

PROOF. (i) $\Longrightarrow$ (ii). Let $f \in k[x_1, \ldots, x_n]$. Then, by Theorem 1.5.9, there exists $r \in k[x_1, \ldots, x_n]$, reduced with respect to G, such that $f \xrightarrow{G}_+ r$. Thus $f - r \in I$ and so $f \in I$ if and only if $r \in I$. Clearly, if $r = 0$ (that is, $f \xrightarrow{G}_+ 0$), then $f \in I$. Conversely, if $f \in I$ and $r \neq 0$ then $r \in I$ and by (i), there exists $i \in \{1, \ldots, t\}$ such that $\text{lp}(g_i)$ divides $\text{lp}(r)$. This is a contradiction to the fact that r is reduced with respect to G. Thus $r = 0$ and $f \xrightarrow{G}_+ 0$.

(ii) $\Longrightarrow$ (iii). For $f \in I$, we know by hypothesis that $f \xrightarrow{G}_+ 0$, and since the process of reduction is exactly the same as the Division Algorithm, we see that (iii) follows from Theorem 1.5.9.

(iii) $\Longrightarrow$ (iv).Clearly, $\text{Lt}(G) \subseteq \text{Lt}(I)$. For the reverse inclusion it suffices to show that for all $f \in I$, $\text{lt}(f) \in \text{Lt}(G)$, since the $\text{lt}(f)$'s generate $\text{Lt}(I)$. Writing f as in the hypothesis, it immediately follows that

$$\text{lt}(f) = \sum_i \text{lt}(h_i)\,\text{lt}(g_i),$$

where the sum is over all i such that $\text{lp}(f) = \text{lp}(h_i)\,\text{lp}(g_i)$. The result follows immediately.

(iv) $\Longrightarrow$ (i). Let $f \in I$. Then $\text{lt}(f)$ is in $\text{Lt}(G)$, and hence

$$\text{lt}(f) = \sum_{i=1}^{t} h_i\,\text{lt}(g_i), \tag{1.6.1}$$

for some $h_i \in k[x_1, \ldots, x_n]$. If we expand the right-hand side of Equation (1.6.1), we see that each term is divisible by some $\text{lp}(g_i)$. Thus $\text{lt}(f)$, the only term in the left-hand side, is also divisible by some $\text{lp}(g_i)$, as desired. □

COROLLARY 1.6.3. *If $G = \{g_1, \ldots, g_t\}$ is a Gröbner basis for the ideal I, then $I = \langle g_1, \ldots, g_t\rangle$.*

PROOF. Clearly $\langle g_1, \ldots, g_t\rangle \subseteq I$, since each g_i is in I. For the reverse inclusion, let $f \in I$. By Theorem 1.6.2, $f \xrightarrow{G}_+ 0$, and hence $f \in \langle g_1, \ldots, g_t\rangle$. □

For the next corollary we first need some information about the special nature of ideals generated by terms.

LEMMA 1.6.4. *Let I be an ideal generated by a set S of non-zero terms, and let $f \in k[x_1, \ldots, x_n]$. Then f is in I if and only if for every term X appearing in f there exists $Y \in S$ such that Y divides X. Moreover, there exists a finite subset S_0 of S such that $I = \langle S_0\rangle$.*

PROOF. If $f \in I$, then

$$f = \sum_{i=1}^{\ell} h_i X_i, \tag{1.6.2}$$

where $h_i \in k[x_1, \ldots, x_n]$ and $X_i \in S$, for $i = 1, \ldots, \ell$. If we expand the right-hand side of Equation (1.6.2), we see that every term is divisible by some term

X_i in S, and hence every term of the left-hand side must also be divisible by some term $X_i \in S$.

Conversely, if for every term X appearing in f there exists a term $Y \in S$ such that Y divides X, then each such X is in $I = \langle S \rangle$, and hence f is in I.

In order to prove the last statement we note that, by the Hilbert Basis Theorem (Theorem 1.1.1), I has a finite generating set. By the first part of the lemma each term in each member of this generating set is divisible by an element of S. The finite set, S_0, of such divisors is clearly a generating set for I. □

COROLLARY 1.6.5. *Every non-zero ideal I of $k[x_1, \ldots, x_n]$ has a Gröbner basis.*

PROOF. By Lemma 1.6.4 the leading term ideal $\mathrm{Lt}(I)$ has a finite generating set which can be assumed to be of the form $\{\mathrm{lt}(g_1), \ldots, \mathrm{lt}(g_t)\}$ with $g_1, \ldots, g_t \in I$. If we let $G = \{g_1, \ldots, g_t\}$, then we have $\mathrm{Lt}(G) = \mathrm{Lt}(I)$ and hence G is a Gröbner basis for I by Theorem 1.6.2. □

We now give a fifth characterization of a Gröbner basis. We will expand our terminology a little.

DEFINITION 1.6.6. *We say that a subset $G = \{g_1, \ldots, g_t\}$ of $k[x_1, \ldots, x_n]$ is a* Gröbner basis *if and only if it is a Gröbner basis for the ideal $\langle G \rangle$ it generates.*

THEOREM 1.6.7. *Let $G = \{g_1, \ldots, g_t\}$ be a set of non-zero polynomials in $k[x_1, \ldots, x_n]$. Then G is a Gröbner basis if and only if for all $f \in k[x_1, \ldots, x_n]$, the remainder of the division of f by G is unique.*

PROOF. We first assume that G is a Gröbner basis. Let $f \xrightarrow{G}_+ r_1$ and $f \xrightarrow{G}_+ r_2$, with r_1 and r_2 reduced with respect to G. Since $f - r_1$ and $f - r_2$ are both in $\langle G \rangle = \langle g_1, \ldots, g_t \rangle$, so is $r_1 - r_2$. Moreover $r_1 - r_2$ is reduced with respect to G. But then $r_1 - r_2 = 0$, by Theorem 1.6.2 (ii).

Conversely, assume that remainders upon division by G are unique. We will prove condition (ii) in Theorem 1.6.2. So let $f \in \langle G \rangle$. Suppose that $f \xrightarrow{G}_+ r$ such that r is reduced. We must show that $r = 0$. (Of course, we know, by hypothesis, that r is unique.)

CLAIM: If $c \in k$ is non-zero, $X \in \mathbb{T}^n$ is a power product, and $g \in k[x_1, \ldots, x_n]$ is such that $g \xrightarrow{G}_+ r$, where r is reduced, then, for each $i \in \{1, \ldots, t\}$, $g - cXg_i \xrightarrow{G}_+ r$. (Note that we have *not* assumed that $cX\,\mathrm{lt}(g_i)$ actually cancels a term in g.)

We note that if the claim is true we are done. To see this, since $f \in I$, we can write $f = \sum_{\nu=1}^{\ell} c_\nu X_\nu g_{i_\nu}$, where c_ν is in k and is non-zero and $X_\nu \in \mathbb{T}^n$ and each $i_\nu \in \{1, \ldots, t\}$ (this can be done by writing $f = \sum_{i=1}^{t} h_i g_i$ and writing each h_i as a sum of terms). Then, applying the claim to $g = f$, we see that $f - c_1 X_1 g_{i_1} \xrightarrow{G}_+ r$. So now we can apply the claim to $g = f - c_1 X_1 g_{i_1}$ to obtain $f - c_1 X_1 g_{i_1} - c_2 X_2 g_{i_2} \xrightarrow{G}_+ r$. Thus, using induction, we see that $0 =$

$f - \sum_{\nu=1}^{\ell} c_\nu X_\nu g_{i_\nu} \xrightarrow{G}_+ r$. That is, $0 \xrightarrow{G}_+ r$ which immediately implies that $r = 0$, as desired.

PROOF OF THE CLAIM: Define d by letting $d\,\mathrm{lc}(g_i)$ be the coefficient of $X\,\mathrm{lp}(g_i)$ in g. We will consider three cases.

- Case 1. $d = 0$. Then the coefficient of $X\,\mathrm{lp}(g_i)$ in $g - cXg_i$ is $-c\,\mathrm{lc}(g_i)$ which is non-zero and so $g - cXg_i \xrightarrow{g_i} g \xrightarrow{G}_+ r$ which is the desired result.
- Case 2. $d = c$. Let r_1 be reduced and assume that $g - cXg_i \xrightarrow{G}_+ r_1$. Then, since $d = c \neq 0$ we see that $g \xrightarrow{g_i} g - cXg_i \xrightarrow{G}_+ r_1$. Thus, since we know $g \xrightarrow{G}_+ r$ also, we see $r = r_1$, as desired (by the assumption that the remainder is unique).
- Case 3. $d \neq 0$ and $d \neq c$. Set $h = g - dXg_i$. Then the coefficient of $X\,\mathrm{lp}(g_i)$ in h is 0. Since $d \neq 0$ we have $g \xrightarrow{g_i} h$. Also, since $d \neq c$ we have $g - cXg_i \xrightarrow{g_i} h$. So if $h \xrightarrow{G}_+ r_2$, such that r_2 is reduced, we get $g \xrightarrow{g_i} h \xrightarrow{G}_+ r_2$ and so $r_2 = r$, since the remainder is unique. And so $g - cXg_i \xrightarrow{g_i} h \xrightarrow{G}_+ r$, as desired.

The theorem is now proved. □

Although we have in Theorem 1.6.7 that remainders are unique for division by a Gröbner basis, we saw in Exercise 1.5.2 that the quotients are not necessarily unique (we will see in Exercise 1.6.2 that the polynomials in Exercise 1.5.2 do form a Gröbner basis).

EXAMPLE 1.6.8. We continue Example 1.5.10. So let $f = y^2x - x$, $f_1 = yx - x$ and $f_2 = y^2 - x$. Let $F = \{f_1, f_2\}$. We use deglex with $y > x$. We showed in Example 1.5.10 that $f \xrightarrow{F}_+ 0$ and $f \xrightarrow{F}_+ x^2 - x$, the latter being reduced with respect to F. Thus by Theorem 1.6.7, F is not a Gröbner basis. We can see this in another way. Namely, since $f = yf_1 + f_2 \in \langle f_1, f_2\rangle$ and $f \xrightarrow{F}_+ x^2 - x$ we have $x^2 - x \in \langle f_1, f_2\rangle$. But $x^2 = \mathrm{lp}(x^2 - x)$ is not divisible by either $\mathrm{lp}(f_1) = xy$ or $\mathrm{lp}(f_2) = y^2$. Thus by the definition of a Gröbner basis (Definition 1.6.1), F is not a Gröbner basis.

EXAMPLE 1.6.9. Consider the polynomials $g_1 = z + x, g_2 = y - x \in \mathbb{Q}[x, y, z]$. Let $G = \{g_1, g_2\}$, $I = \langle g_1, g_2\rangle$. We use the lex term order on $\mathbb{Q}[x, y, z]$ with $x < y < z$. We will prove that G is a Gröbner basis for I. Suppose to the contrary that there exists $f \in I$ such that $\mathrm{lt}(f) \notin \langle \mathrm{lt}(g_1), \mathrm{lt}(g_2)\rangle = \langle z, y\rangle$. Then, z does not divide $\mathrm{lt}(f)$, and y does not divide $\mathrm{lt}(f)$. Thus, because of the lex term order, z and y do not appear in any term of f, and so $f \in \mathbb{Q}[x]$. Let $f = (z + x)h_1 + (y - x)h_2$, where $h_1, h_2 \in \mathbb{Q}[x, y, z]$. Since y does not appear in f, we may set $y = x$, and we have $f = (z + x)h_1(x, x, z)$, and hence $z + x$ divides f, a contradiction to the fact that the only variable occurring in f is x.

We will give a systematic way of proving that a set of polynomials is a Gröbner basis in the next section.

We observe that if we have a Gröbner basis $G = \{g_1, \ldots, g_t\}$ for an ideal I, then we can solve some of the problems posed in Section 1.1 in a fashion similar to what we did in the one variable case. To decide whether a polynomial f is in I, we use the Division Algorithm and divide f by G. The remainder of the division is zero if and only if f is in I. Also, by Theorem 1.6.7, the representative of the element $f + I$ in the quotient ring $k[x_1, \ldots, x_n]/I$ is $r + I$, where r is the remainder of the division of f by G. Also, a basis for the k-vector space $k[x_1, \ldots, x_n]/I$ is the set of all cosets of power products that are not divisible by some $\text{lt}(g_i)$ (Exercise 1.6.10). All of these applications will be discussed fully in Chapter 2.

We note that a Gröbner basis with respect to one term order may not be a Gröbner basis with respect to a different term order. For example, if we use the lex term order with $x > y > z$ in Example 1.6.9, then $\{g_1, g_2\}$ is not a Gröbner basis for I (Exercise 1.6.3).

Clearly, the question now is how do we compute a Gröbner basis for an ideal I? The results in this section only prove existence, and the proofs of these results do not indicate any method for finding Gröbner bases. We will give Buchberger's Algorithm for their computation in the next section.

However, we have already computed (without knowing it!) Gröbner bases for two special cases. In the linear case, the polynomials obtained from row reducing the matrix of the original linear polynomials to row echelon form constitute a Gröbner basis for the ideal generated by these original polynomials, the variables being ordered according to the position of their column in the matrix of the system of equations (Exercise 1.6.5). In the one variable case, $G = \{\gcd(f_1, \ldots, f_s)\}$ is a Gröbner basis for the ideal $I = \langle f_1, \ldots, f_s \rangle$, by Theorem 1.6.2(ii) (Exercise 1.6.6). In both cases we do have an algorithm for computing the Gröbner basis.

Exercises

1.6.1. Show that the polynomials $f_1 = 2xy^2 + 3x + 4y^2, f_2 = y^2 - 2y - 2 \in \mathbb{Q}[x, y]$, with lex with $x > y$ do not form a Gröbner basis for the ideal they generate. (See Exercise 1.5.1.)

1.6.2. Show that the polynomials $f_1 = x - y^2w, f_2 = y - zw, f_3 = z - w^3, f_4 = w^3 - w \in \mathbb{Q}[x, y, z, w]$ in Exercise 1.5.2 form a Gröbner basis for the ideal they generate, with respect to lex with $x > y > z > w$. [Hint: Follow Example 1.6.9.] Show that they do not form a Gröbner basis with respect to lex with $w > x > y > z$.

1.6.3. Show that the polynomials g_1, g_2 in Example 1.6.9 do not form a Gröbner basis with respect to lex with $x > y > z$.

1.6.4. Let $<$ be any term order in $k[x, y, z]$ with $x > y > z$. Show that the polynomials f_1, f_2, f_3 in Example 1.2.2 do not form a Gröbner basis for I, whereas $f_1, f_2, -17z$ do.

1.6.5. Let $f_1, \ldots, f_m$ be non-zero linear polynomials in $k[x_1, \ldots, x_n]$ which are in row echelon form. Show that they form a Gröbner basis for the ideal they generate with respect to any order for which the variables are ordered according to the corresponding columns in the matrix.

1.6.6. In the polynomial ring in one variable, $k[x]$, consider a set of non-zero polynomials $F = \{f_1, \ldots, f_s\} \subseteq k[x]$. Let $d = \gcd(f_1, \ldots, f_s)$. Prove that F is a Gröbner basis if and only if $cd \in F$, for some $c \in k$, $c \neq 0$.

1.6.7. Generalize Exercise 1.6.6 to principal ideals in $k[x_1, \ldots, x_n]$. That is, show that if $I = \langle d \rangle$ is a principal ideal in $k[x_1, \ldots, x_n]$, then $F \subseteq I$ is a Gröbner basis for I if and only if $cd \in F$, for some $c \in k$, $c \neq 0$.

1.6.8. Let I be an ideal in $k[x_1, \ldots, x_n]$. Prove that $\mathrm{Lt}(I)$ is the k-vector space spanned by $\{\mathrm{lp}(f) \mid f \in I\}$.

1.6.9. Let $I \subseteq k[x_1, \ldots, x_n]$ be an ideal generated by a set $G = \{X_1, \ldots, X_s\}$ of non-zero terms. Prove that G is a Gröbner basis for I.

1.6.10. Let I be an ideal of $k[x_1, \ldots, x_n]$ and let $G = \{g_1, \ldots, g_t\}$ be a Gröbner basis for I. Prove that a basis for the k-vector space $k[x_1, \ldots, x_n]/I$ is $\{X + I \mid X \in \mathbb{T}^n \text{ and } \mathrm{lp}(g_i) \text{ does not divide } X \text{ for all } i = 1, \ldots, t\}$.

1.6.11. In this exercise we give another equivalent definition of a Gröbner basis. Let $I \subseteq k[x_1, \ldots, x_n]$ be an ideal. For a subset $S \subseteq k[x_1, \ldots, x_n]$ set $\mathcal{L}p(S) = \{\mathrm{lp}(f) \mid f \in S\}$. (Note that we have just taken the *set* of all $\mathrm{lp}(f)$ not the ideal generated by the $\mathrm{lp}(f)$'s.) Set $I^* = I - \{0\}$.

a. Show that $\mathbb{T}^n$ is a *monoid*; that is, $\mathbb{T}^n$ is closed under multiplication.

b. Show that $\mathcal{L}p(I^*)$ is a *monoideal* of $\mathbb{T}^n$; that is, show that for all $X \in \mathcal{L}p(I^*)$ and $Y \in \mathbb{T}^n$ we have $XY \in \mathcal{L}p(I^*)$. (Note: this is just Exercise 1.7.6.)

c. Prove that $F \subseteq I^*$ is a Gröbner basis for I if and only if $\mathcal{L}p(F)$ generates $\mathcal{L}p(I^*)$ as a monoideal. (We say that $\mathcal{L}p(F)$ generates $\mathcal{L}p(I^*)$ as a monoideal if and only if for all $X \in \mathcal{L}p(I^*)$ there exists $Y \in \mathbb{T}^n$ and $Z \in \mathcal{L}p(F)$ such that $X = YZ$.)

1.6.12. In this exercise we give another equivalent definition of a Gröbner basis. Let $G \subseteq k[x_1, \ldots, x_n]$ consist of non-zero polynomials. We call the reduction relation "$\xrightarrow{G}_+$" *confluent* provided that for all $f, g, h \in k[x_1, \ldots, x_n]$ such that $f \xrightarrow{G}_+ g$ and $f \xrightarrow{G}_+ h$, there exists an $r \in k[x_1, \ldots, x_n]$ such that $h \xrightarrow{G}_+ r$ and $g \xrightarrow{G}_+ r$. Prove that G is a Gröbner basis if and only if "$\xrightarrow{G}_+$" is confluent. [Hint: Use Theorem 1.6.7.]

1.6.13. Let $\{g_1, \ldots, g_t\} \subseteq k[x_1, \ldots, x_n]$ and let $0 \neq h \in k[x_1, \ldots, x_n]$. Prove that $\{g_1, \ldots, g_t\}$ is a Gröbner basis if and only if $\{hg_1, \ldots, hg_t\}$ is a Gröbner basis.

1.6.14. Let G be a Gröbner basis for an ideal I of $k[x_1, \ldots, x_n]$ and let K be an extension field of k. Let J be the ideal of $K[x_1, \ldots, x_n]$ generated by I. Prove that G is also a Gröbner basis for J.

1.6.15. Let G be a Gröbner basis for an ideal I and let $r, f \in k[x_1, \ldots, x_n]$, where r is reduced with respect to G. Prove that if $f - r \in I$, then $f \xrightarrow{G}_+ r$.

1.6.16. Let G and G' be two Gröbner bases for an ideal $I \subseteq k[x_1, \ldots, x_n]$ with respect to a single term order. Let $f \in k[x_1, \ldots, x_n]$. Assume that $f \xrightarrow{G}_+ r$ and $f \xrightarrow{G'}_+ r'$ where r is reduced with respect to G and r' is reduced with respect to G'. Prove that $r = r'$.

1.6.17. Let I be an ideal of $k[x_1, \ldots, x_n]$. Assume that we are given two term orderings, say $<_1$ and $<_2$. Let $\{g_1, \ldots, g_t\}$ be a Gröbner basis for I with respect to $<_1$. Assume that $\mathrm{lt}_{<_1}(g_i) = \mathrm{lt}_{<_2}(g_i)$, for $i = 1, \ldots, t$. Prove that $\{g_1, \ldots, g_t\}$ is also a Gröbner basis for I with respect to $<_2$.

1.6.18. Let I be an ideal in $k[x_1, \ldots, x_n]$ and let $G = \{g_1, \ldots, g_t\}$ and $G' = \{g'_1, \ldots, g'_{t'}\}$ be subsets of I of non-zero polynomials where we assume that G is a Gröbner basis for I. Assume that for all $g_i \in G$ we have $g_i = \sum_{j=1}^{t'} a_{ij} g'_j$ where $\mathrm{lp}(g_i) = \max_{1 \le i \le t'}(\mathrm{lp}(a_{ij})\,\mathrm{lp}(g'_j))$. Prove that G' is also a Gröbner basis for I, with respect to the same term order.

1.6.19. Let $f \in k[x_1, \ldots, x_n]$ have total degree d, and let w be a new variable. We define the *homogenization* of f to be $f^h = w^d f(\frac{x_1}{w}, \ldots, \frac{x_n}{w}) \in k[x_1, \ldots, x_n, w]$. Note that f^h is homogeneous (see Exercise 1.4.9). For an ideal I of $k[x_1, \ldots, x_n]$, we define I^h to be the ideal of $k[x_1, \ldots, x_n, w]$ defined by $I^h = \langle f^h \mid f \in I \rangle$. Also, for $g \in k[x_1, \ldots, x_n, w]$ we define $g_h = g(x_1, \ldots, x_n, 1) \in k[x_1, \ldots, x_n]$.

a. Give an example that shows that there is an ideal $I = \langle f_1, \ldots, f_s \rangle$ of $k[x_1, \ldots, x_n]$ such that I^h is strictly larger than $\langle f_1^h, \ldots, f_s^h \rangle \subseteq k[x_1, \ldots, x_n, w]$.

b. Let $<$ be the deglex or degrevlex order in $k[x_1, \ldots, x_n]$. Let $<_h$ be the order defined by extending $<$ to $k[x_1, \ldots, x_n, w]$ as follows:

$$x_1^{\nu_1} \cdots x_n^{\nu_n} w^{\nu} <_h x_1^{\nu'_1} \cdots x_n^{\nu'_n} w^{\nu'} \text{ if and only if}$$

$$\begin{cases} x_1^{\nu_1} \cdots x_n^{\nu_n} < x_1^{\nu'_1} \cdots x_n^{\nu'_n} \text{ or} \\ x_1^{\nu_1} \cdots x_n^{\nu_n} = x_1^{\nu'_1} \cdots x_n^{\nu'_n} \text{ and } \nu < \nu'. \end{cases}$$

Prove that $<_h$ is a term order in $k[x_1, \ldots, x_n, w]$ and that $\mathrm{lt}_{<_h}(f^h) = \mathrm{lt}(f)$.

c. Let I be an ideal of $k[x_1, \ldots, x_n]$ and let $G = \{g_1, \ldots, g_t\}$ be a Gröbner basis for I with respect to deglex or degrevlex. Prove that $G^h = \{g_1^h, \ldots, g_t^h\}$ is a Gröbner basis for I^h with respect to $<_h$. [Hint: It suffices to show that $\mathrm{Lt}_{<_h}(I^h) = \langle \mathrm{lt}_{<_h}(g_1^h), \ldots, \mathrm{lt}_{<_h}(g_t^h) \rangle$. If $f \in I^h$, then we may assume that f is homogeneous. Prove that $f = w^{\nu}(f_h)^h$, for some $\nu \ge 0$ and that $f_h \in I$.]

d. Use **c** to compute a set of generators for I^h, where $I = \langle yx - x, y^2 - x \rangle$ by first showing that $G = \{yx - x, y^2 - x, x^2 - x\}$ is a Gröbner basis for I with respect to deglex with $x > y$. [Hint: Show that if f is reduced with respect to G and in I, then $f = ax + by$ for some $a, b \in k$, and $f = h_1(yx - x) + h_2(y^2 - x) + h_3(x^2 - x)$. Then successively set $x = 0$ then $y = 1$.]

1.6.20. This material is taken from Robbiano and Sweedler [**RoSw**]. By a *k-subalgebra* $A \subseteq k[x_1, \dots, x_n]$ we mean a subring which is also a k-vector space. For a subset $F = \{f_1, \dots, f_s\} \subseteq k[x_1, \dots, x_n]$ we denote by $k[F]$ the k-subalgebra of $k[x_1, \dots, x_n]$ generated by F, that is $k[F] = \{\sum_\nu c_\nu f_1^{\nu_1} \cdots f_s^{\nu_s} \mid c_\nu \in k$ and $\nu = (\nu_1, \dots, \nu_s) \in \mathbb{N}^s$ and only finitely many c_ν's are non-zero$\}$. Fix a term order on $k[x_1, \dots, x_n]$. We will define a reduction procedure that answers the "algebra membership problem" for k-subalgebras of $k[x_1, \dots, x_n]$. This is the problem of determining, for $f \in k[x_1, \dots, x_n]$, whether $f \in k[F]$. (This problem will be solved using Gröbner bases in Section 2.4.)

a. For $F \subseteq k[x_1, \dots, x_n]$, let $\mathcal{L}p(F) = \{\mathrm{lp}(f) \mid f \in F\}$. (Note that we have only taken the *set* of all $\mathrm{lp}(f)$.) For a k-subalgebra $A \subseteq k[x_1, \dots, x_n]$ show that $\mathcal{L}p(A)$ is a multiplicative submonoid of $\mathbb{T}^n$. We call F a SAGBI basis for $A = k[F]$ provided $\mathcal{L}p(F)$ generates $\mathcal{L}p(A)$ as a monoid. (SAGBI stands for Subalgebra Analog to Gröbner Bases for Ideals. In Robbiano and Sweedler [**RoSw**], an algorithm for computing SAGBI bases is given.)

b. For $F = \{f_1, \dots, f_s\} \subseteq k[x_1, \dots, x_n]$ and $\boldsymbol{\nu} = (\nu_1, \dots, \nu_s) \in \mathbb{N}^s$ denote by $F^{\boldsymbol{\nu}} = f_1^{\nu_1} \cdots f_s^{\nu_s}$. For $g, h \in k[x_1, \dots, x_n]$ we write $g \xrightarrow{F} h$ to mean there is a $c \in k$ ($c \neq 0$) and $\boldsymbol{\nu} \in \mathbb{N}^s$ such that $\mathrm{lt}(cF^{\boldsymbol{\nu}})$ is a term in g and $h = g - cF^{\boldsymbol{\nu}}$. If we have $g \xrightarrow{F} g_1 \xrightarrow{F} g_2 \xrightarrow{F} \cdots \xrightarrow{F} h$ we write $g \xrightarrow{F}_+ h$. Show that $g \xrightarrow{F}_+ h$ implies that $g - h \in k[F]$.

c. Show that $F \subseteq k[x_1, \dots, x_n]$ is a SAGBI basis for $k[F]$ if and only if for all $f \in k[F]$ we have $f \xrightarrow{F}_+ 0$.

d. Show that if F consists entirely of terms then F is a SAGBI basis for $k[F]$.

e. Prove that $F = \{x^2, y^2, xy + y, xy^2\}$ is a SAGBI basis for $\mathbb{Q}[F] \subseteq \mathbb{Q}[x, y]$ for deglex with $y > x$. [Hint: For $f = \sum c_\nu (x^2)^{\nu_1} (y^2)^{\nu_2} (xy + y)^{\nu_3} (xy^2)^{\nu_4} \in \mathbb{Q}[F]$, expand out the $xy + y$ term and show that it suffices to prove that $x^n y^m$ is in the monoid generated by $\{x^2, y^2, xy, xy^2\}$ for $n \leq m$.]

1.7. S-Polynomials and Buchberger's Algorithm. In this section we first lay the theoretical foundation for the algorithm for computing Gröbner bases by presenting Buchberger's Theorem [**Bu65, Bu85**]. This result is given in Theorem 1.7.4. We then present his algorithm.

Let $I = \langle f_1, \dots, f_s \rangle$ be an ideal of $k[x_1, \dots, x_n]$, and let $F = \{f_1, \dots, f_s\}$, where we assume that $f_i \neq 0$ $(1 \leq i \leq s)$. In the previous section we defined F to be a Gröbner basis if and only if for all $f \in I$, there exists $i \in \{1, \dots, s\}$ such that $\mathrm{lp}(f_i)$ divides $\mathrm{lp}(f)$. So a difficulty arises with elements of I whose leading power products are not divisible by any $\mathrm{lp}(f_i)$. But if f is in I, $f = \sum_{i=1}^s h_i f_i$, for some $h_i \in k[x_1, \dots, x_n]$. Hence the difficulty occurs when the largest of the $\mathrm{lp}(h_i f_i) = \mathrm{lp}(h_i)\,\mathrm{lp}(f_i)$'s cancel. The simplest way for this to occur is in the following.

DEFINITION 1.7.1. *Let* $0 \neq f, g \in k[x_1, \dots, x_n]$. *Let*[7] $L = \text{lcm}(\text{lp}(f), \text{lp}(g))$. *The polynomial*

$$S(f,g) = \frac{L}{\text{lt}(f)} f - \frac{L}{\text{lt}(g)} g$$

is called the S-polynomial *of* f *and* g.

EXAMPLE 1.7.2. Let $f = 2yx - y, g = 3y^2 - x \in \mathbb{Q}[x,y]$, with the deglex term ordering with $y > x$. Then $L = y^2x$, and $S(f,g) = \frac{y^2x}{2yx} f - \frac{y^2x}{3y^2} g = \frac{1}{2}yf - \frac{1}{3}xg = -\frac{1}{2}y^2 + \frac{1}{3}x^2$. Moreover $\text{lp}(\frac{1}{2}yf) = y^2x = \text{lp}(\frac{1}{3}xg)$ have canceled in $S(f,g)$.

There is another way of viewing S-polynomials. Namely, in the division of f by $f_1, \dots, f_s$, it may happen that some term X appearing in f is divisible by both $\text{lp}(f_i)$ and $\text{lp}(f_j)$ for $i \neq j$ (hence X is divisible by $L = \text{lcm}(\text{lp}(f_i), \text{lp}(f_j))$. If we reduce f using f_i, we get the polynomial $h_1 = f - \frac{X}{\text{lt}(f_i)} f_i$, and if we reduce f using f_j, we get $h_2 = f - \frac{X}{\text{lt}(f_j)} f_j$. The ambiguity that is introduced is $h_2 - h_1 = \frac{X}{\text{lt}(f_i)} f_i - \frac{X}{\text{lt}(f_j)} f_j = \frac{X}{L} S(f_i, f_j)$.

EXAMPLE 1.7.3. Let $f = y^2x + 1, f_1 = yx - y, f_2 = y^2 - x \in \mathbb{Q}[x,y]$ with the deglex term ordering with $y > x$. We consider the term $X = y^2x$ in f. We have that $f \xrightarrow{f_1} y^2 + 1 = f - yf_1$, and $f \xrightarrow{f_2} x^2 + 1 = f - xf_2$. Note that $X = L = \text{lcm}(\text{lp}(f_1), \text{lp}(f_2)) = y^2x$, and that the ambiguity introduced is $-y^2 + x^2 = yf_1 - xf_2 = S(f_1, f_2)$. Also, note that $S(f_1, f_2) \in \langle f_1, f_2 \rangle$, and that it can be reduced: $S(f_1, f_2) \xrightarrow{f_2} x^2 - x$. The polynomial $x^2 - x$ is now reduced with respect to $\{f_1, f_2\}$, but is not zero.

Now that we have introduced S-polynomials as a way to "cancel" leading terms and to account for the ambiguity in the Division Algorithm, we can go ahead with a strategy for computing Gröbner bases. It turns out that the S-polynomials account for all ambiguities we need to be concerned about as the next result shows.

THEOREM 1.7.4 (BUCHBERGER). *Let* $G = \{g_1, \dots, g_t\}$ *be a set of non-zero polynomials in* $k[x_1, \dots, x_n]$. *Then* G *is a Gröbner basis for the ideal* $I = \langle g_1, \dots, g_t \rangle$ *if and only if for all* $i \neq j$,

$$S(g_i, g_j) \xrightarrow{G}_+ 0.$$

Before we can prove this result, we need one preliminary lemma.

LEMMA 1.7.5. *Let* $f_1, \dots, f_s \in k[x_1, \dots, x_n]$ *be such that* $\text{lp}(f_i) = X \neq 0$ *for all* $i = 1, \dots, s$. *Let* $f = \sum_{i=1}^{s} c_i f_i$ *with* $c_i \in k, i = 1, \dots, s$. *If* $\text{lp}(f) < X$, *then* f *is a linear combination, with coefficients in* k, *of* $S(f_i, f_j), 1 \leq i < j \leq s$.

[7] Recall that the least common multiple of two power products X, Y is the power product L such that X divides L, Y divides L and if Z is another power product such that X divides Z and Y divides Z then L divides Z. We denote L by $\text{lcm}(X, Y)$.

PROOF. Write $f_i = a_iX$+lower terms, $a_i \in k$. Then the hypothesis says that $\sum_{i=1}^{s} c_i a_i = 0$, since the c_i's are in k. Now, by definition, $S(f_i, f_j) = \frac{1}{a_i} f_i - \frac{1}{a_j} f_j$, since $\operatorname{lp}(f_i) = \operatorname{lp}(f_j) = X$. Thus

$$\begin{aligned} f &= c_1 f_1 + \cdots + c_s f_s \\ &= c_1 a_1 (\tfrac{1}{a_1} f_1) + \cdots + c_s a_s (\tfrac{1}{a_s} f_s) \\ &= c_1 a_1 (\tfrac{1}{a_1} f_1 - \tfrac{1}{a_2} f_2) + (c_1 a_1 + c_2 a_2)(\tfrac{1}{a_2} f_2 - \tfrac{1}{a_3} f_3) + \cdots \\ &\quad + (c_1 a_1 + \cdots + c_{s-1} a_{s-1})(\tfrac{1}{a_{s-1}} f_{s-1} - \tfrac{1}{a_s} f_s) + (c_1 a_1 + \cdots + c_s a_s) \tfrac{1}{a_s} f_s \\ &= c_1 a_1 S(f_1, f_2) + (c_1 a_1 + c_2 a_2) S(f_2, f_3) + \cdots \\ &\quad + (c_1 a_1 + \cdots + c_{s-1} a_{s-1}) S(f_{s-1}, f_s), \end{aligned}$$

since $c_1 a_1 + \cdots + c_s a_s = 0$. □

We are now ready to prove Buchberger's Theorem.

PROOF OF THEOREM 1.7.4. If $G = \{g_1, \ldots, g_t\}$ is a Gröbner basis for $I = \langle g_1, \ldots, g_t \rangle$, then $S(g_i, g_j) \xrightarrow{G}_+ 0$ for all $i \neq j$ by Theorem 1.6.2, since $S(g_i, g_j) \in I$.

Conversely, let us assume that $S(g_i, g_j) \xrightarrow{G}_+ 0$ for all $i \neq j$. We will use Theorem 1.6.2(iii) to prove that G is a Gröbner basis for I. Let $f \in I$. Then f can be written in many ways as a linear combination of the g_i's. We choose to write $f = \sum_{i=1}^{t} h_i g_i$, with

$$X = \max_{1 \leq i \leq t} (\operatorname{lp}(h_i) \operatorname{lp}(g_i))$$

least (here we use the well-ordering property of the term order). If $X = \operatorname{lp}(f)$, we are done. Otherwise, $\operatorname{lp}(f) < X$. We will find a representation of f with a smaller X, and this will be a contradiction. Let $S = \{i \mid \operatorname{lp}(h_i) \operatorname{lp}(g_i) = X\}$. For $i \in S$, write $h_i = c_i X_i$ + lower terms. Set $g = \sum_{i \in S} c_i X_i g_i$. Then, $\operatorname{lp}(X_i g_i) = X$, for all $i \in S$, but $\operatorname{lp}(g) < X$. By Lemma 1.7.5, there exist $d_{ij} \in k$ such that

$$g = \sum_{i,j \in S, i \neq j} d_{ij} S(X_i g_i, X_j g_j).$$

Now, $X = \operatorname{lcm}(\operatorname{lp}(X_i g_i), \operatorname{lp}(X_j g_j))$, so

$$\begin{aligned} S(X_i g_i, X_j g_j) &= \frac{X}{\operatorname{lt}(X_i g_i)} X_i g_i - \frac{X}{\operatorname{lt}(X_j g_j)} X_j g_j \\ &= \frac{X}{\operatorname{lt}(g_i)} g_i - \frac{X}{\operatorname{lt}(g_j)} g_j = \frac{X}{X_{ij}} S(g_i, g_j), \end{aligned}$$

where $X_{ij} = \operatorname{lcm}(\operatorname{lp}(g_i), \operatorname{lp}(g_j))$. By hypothesis, $S(g_i, g_j) \xrightarrow{G}_+ 0$, and so we see from this last equation that $S(X_i g_i, X_j g_j) \xrightarrow{G}_+ 0$ (See Exercise 1.5.4). This gives a representation

$$S(X_i g_i, X_j g_j) = \sum_{\nu=1}^{t} h_{ij\nu} g_\nu,$$

where, by Theorem 1.5.9,

$$\begin{aligned}\max_{1\le\nu\le t}(\mathrm{lp}(h_{ij\nu})\,\mathrm{lp}(g_\nu)) &= \mathrm{lp}(S(X_ig_i, X_jg_j))\\ &< \max(\mathrm{lp}(X_ig_i), \mathrm{lp}(X_jg_j)) = X.\end{aligned}$$

Substituting these expressions into g above, and g into f, we get $f = \sum_{i=1}^t h'_ig_i$, with $\max_{1\le i\le t}(\mathrm{lp}(h'_i)\,\mathrm{lp}(g_i)) < X$. This is a contradiction. □

We have as an immediate Corollary of the proof of Theorem 1.7.4, the following additional equivalent condition for a subset G of $k[x_1,\ldots,x_n]$ to be a Gröbner basis.

COROLLARY 1.7.6. *Let $G = \{g_1,\ldots,g_t\}$ with $g_i \neq 0$ $(1 \le i \le t)$. Then G is a Gröbner basis if and only if for all $i \neq j$ $(1 \le i,j \le t)$, we have*

$$S(g_i, g_j) = \sum_{\nu=1}^t h_{ij\nu}g_\nu, \text{ where } \mathrm{lp}(S(g_i,g_j)) = \max_{1\le\nu\le t}(\mathrm{lp}(h_{ij\nu})\,\mathrm{lp}(g_\nu)).$$

We note that Buchberger's Theorem (Theorem 1.7.4) gives a strategy for computing Gröbner bases: reduce the S-polynomials and if a remainder is non-zero, add this remainder to the list of polynomials in the generating set; do this until there are "enough" polynomials to make all S-polynomials reduce to zero. Let us first look at an example.

EXAMPLE 1.7.7. Let $f_1 = xy - x$, $f_2 = x^2 - y \in \mathbb{Q}[x,y]$ with the deglex term order with $x < y$. Let $F = \{f_1, f_2\}$. Then $S(f_1,f_2) = xf_1 - yf_2 = y^2 - x^2 \xrightarrow{F} y^2 - y$, and $f_3 = y^2 - y$ is reduced with respect to F. So we add f_3 to F, and let $F' = \{f_1, f_2, f_3\}$. Then $S(f_1, f_2) \xrightarrow{F'} 0$. Now $S(f_1, f_3) = yf_1 - xf_3 = 0$, and $S(f_2, f_3) = y^2f_2 - x^2f_3 = -y^3 + x^2y \xrightarrow{F'} x^2y - y^2 \xrightarrow{F'} 0$. Thus $\{f_1, f_2, f_3\}$ is a Gröbner basis.

We give Buchberger's Algorithm to compute Gröbner bases as Algorithm 1.7.1.

THEOREM 1.7.8. *Given $F = \{f_1,\ldots,f_s\}$ with $f_i \neq 0$ $(1 \le i \le s)$, Buchberger's Algorithm (Algorithm* 1.7.1*) will produce a Gröbner basis for the ideal $I = \langle f_1,\ldots,f_s\rangle$.*

PROOF. We first need to show that this algorithm terminates. Suppose to the contrary that the algorithm does not terminate. Then, as the algorithm progresses, we construct a set G_i strictly larger than G_{i-1} and obtain a strictly increasing infinite sequence

$$G_1 \subsetneq G_2 \subsetneq G_3 \subsetneq \cdots.$$

Each G_i is obtained from G_{i-1} by adding some $h \in I$ to G_{i-1}, where h is the non-zero reduction, with respect to G_{i-1}, of an S-polynomial of two elements of

INPUT: $F = \{f_1, \dots, f_s\} \subseteq k[x_1, \dots, x_n]$ with $f_i \neq 0$ $(1 \leq i \leq s)$

OUTPUT: $G = \{g_1, \dots, g_t\}$, a Gröbner basis for $\langle f_1, \dots, f_s \rangle$

INITIALIZATION: $G := F$, $\mathcal{G} := \{\{f_i, f_j\} \mid f_i \neq f_j \in G\}$

WHILE $\mathcal{G} \neq \emptyset$ **DO**

Choose any $\{f, g\} \in \mathcal{G}$

$\mathcal{G} := \mathcal{G} - \{\{f, g\}\}$

$S(f, g) \xrightarrow{G}_+ h$, where h is reduced with respect to G

IF $h \neq 0$ **THEN**

$\mathcal{G} := \mathcal{G} \cup \{\{u, h\} \mid$ for all $u \in G\}$

$G := G \cup \{h\}$

ALGORITHM 1.7.1. *Buchberger's Algorithm for Computing Gröbner Bases*

G_{i-1}. Since h is reduced with respect to G_{i-1}, we have that $\operatorname{lt}(h) \notin \operatorname{Lt}(G_{i-1})$. Thus we get

$$\operatorname{Lt}(G_1) \subsetneq \operatorname{Lt}(G_2) \subsetneq \operatorname{Lt}(G_3) \subsetneq \cdots .$$

This is a strictly ascending chain of ideals which contradicts the Hilbert Basis Theorem (Theorem 1.1.1).

Now we have $F \subseteq G \subseteq I$, and hence $I = \langle f_1, \dots, f_s \rangle \subseteq \langle g_1, \dots, g_t \rangle \subseteq I$. Thus G is a generating set for the ideal I. Moreover, if g_i, g_j are polynomials in G, then $S(g_i, g_j) \xrightarrow{G}_+ 0$ by construction. Therefore G is a Gröbner basis for I by Theorem 1.7.4. □

EXAMPLE 1.7.9. Let $f_1 = xy - x, f_2 = -y + x^2 \in \mathbb{Q}[x, y]$ ordered by the lex term ordering with $x < y$.

INITIALIZATION: $G := \{f_1, f_2\}, \mathcal{G} := \{\{f_1, f_2\}\}$

First pass through the WHILE loop

$\mathcal{G} := \emptyset$

$S(f_1, f_2) \xrightarrow{G}_+ x^3 - x = h$ (reduced with respect to G)

Since $h \neq 0$, let $f_3 := x^3 - x$

$\mathcal{G} := \{\{f_1, f_3\}, \{f_2, f_3\}\}$

$G := \{f_1, f_2, f_3\}$

Second pass through the WHILE loop

$\mathcal{G} := \{\{f_2, f_3\}\}$

$S(f_1, f_3) \xrightarrow{G}_+ 0 = h$

Third pass through the WHILE loop

$\mathcal{G} := \emptyset$
$S(f_2, f_3) \xrightarrow{G}_+ 0 = h$
The WHILE loop stops, since $\mathcal{G} = \emptyset$.

Thus $\{f_1, f_2, f_3\}$ is a Gröbner basis for the ideal $\langle f_1, f_2 \rangle$.

We will conclude this section by giving two more simple examples which illustrate Buchberger's Algorithm.

EXAMPLE 1.7.10. Let $f_1 = y^2 + yx + x^2$, $f_2 = y + x$, and $f_3 = y \in \mathbb{Q}[x, y]$. Let us use the lex term order with $y > x$ to compute a Gröbner basis for $I = \langle f_1, f_2, f_3 \rangle$.

INITIALIZATION: $G := \{f_1, f_2, f_3\}$
$\mathcal{G} := \{\{f_1, f_2\}, \{f_1, f_3\}, \{f_2, f_3\}\}$
First pass through the WHILE loop
$\mathcal{G} := \{\{f_1, f_3\}, \{f_2, f_3\}\}$
$S(f_1, f_2) = x^2$ (reduced with respect to $\{f_1, f_2, f_3\}$)
let $f_4 := x^2$
$\mathcal{G} := \{\{f_1, f_3\}, \{f_2, f_3\}, \{f_1, f_4\}, \{f_2, f_4\}, \{f_3, f_4\}\}$
$G := \{f_1, f_2, f_3, f_4\}$
Second pass through the WHILE loop
$\mathcal{G} := \{\{f_2, f_3\}, \{f_1, f_4\}, \{f_2, f_4\}, \{f_3, f_4\}\}$
$S(f_1, f_3) \xrightarrow{G}_+ 0$
Third pass through the WHILE loop
$\mathcal{G} := \{\{f_1, f_4\}, \{f_2, f_4\}, \{f_3, f_4\}\}$
$S(f_2, f_3) = x$ (reduced with respect to $\{f_1, f_2, f_3, f_4\}$)
$f_5 := x$
$\mathcal{G} := \{\{f_1, f_4\}, \{f_2, f_4\}, \{f_3, f_4\}, \{f_1, f_5\}, \{f_2, f_5\}$
$\{f_3, f_5\}, \{f_4, f_5\}\}$
$G := \{f_1, f_2, f_3, f_4, f_5\}$
Fourth pass through the WHILE loop
$\mathcal{G} := \{\{f_2, f_4\}, \{f_3, f_4\}, \{f_1, f_5\}, \{f_2, f_5\}, \{f_3, f_5\}, \{f_4, f_5\}\}$
$S(f_1, f_4) \xrightarrow{G}_+ 0$
Fifth pass through the WHILE loop
$\mathcal{G} := \{\{f_3, f_4\}, \{f_1, f_5\}, \{f_2, f_5\}, \{f_3, f_5\}, \{f_4, f_5\}\}$
$S(f_2, f_4) \xrightarrow{G}_+ 0.$

The sixth through tenth executions of the WHILE loop will also give S-polynomials which reduce to zero (Exercise 1.7.4) and thus $\{f_1, f_2, f_3, f_4, f_5\}$ is a Gröbner basis for $\langle f_1, f_2, f_3 \rangle$.

So far, in our examples, we have used the field $\mathbb{Q}$. The theory developed so far is valid for any field k. To illustrate this point, in our next example we compute a Gröbner basis in the case when k is a finite field.

EXAMPLE 1.7.11. In this example we consider the field $k = \mathbb{Z}_5 = \mathbb{Z}/5\mathbb{Z}$. Let $f_1 = x^2 + y^2 + 1$ and $f_2 = x^2y + 2xy + x$ be in $\mathbb{Z}_5[x, y]$. We use the lex term ordering with $x > y$ to compute a Gröbner basis for $I = \langle f_1, f_2 \rangle \subseteq \mathbb{Z}_5[x, y]$.

INITIALIZATION: $G := \{f_1, f_2\}$, $\mathcal{G} := \{\{f_1, f_2\}\}$
First pass through the WHILE loop
$\mathcal{G} := \emptyset$
$S(f_1, f_2) = yf_1 - f_2 = 3xy + 4x + y^3 + y$
(reduced with respect to G)
Let $f_3 := 3xy + 4x + y^3 + y$
$\mathcal{G} := \{\{f_1, f_3\}, \{f_2, f_3\}\}$
$G := \{f_1, f_2, f_3\}$
Second pass through the WHILE loop
$\mathcal{G} := \{\{f_2, f_3\}\}$
$S(f_1, f_3) = yf_1 - 2xf_3 \longrightarrow_+ 4y^5 + 3y^4 + y^2 + y + 3$
(reduced with respect to G)
Let $f_4 := 4y^5 + 3y^4 + y^2 + y + 3$
$\mathcal{G} : \{\{f_2, f_3\}, \{f_1, f_4\}, \{f_2, f_4\}, \{f_3, f_4\}\}$
$G := \{f_1, f_2, f_3, f_4\}$.

The third through the sixth executions of the WHILE loop give S-polynomials that reduce to zero (Exercise 1.7.4) and thus $G = \{f_1, f_2, f_3, f_4\}$ forms a Gröbner basis for $\langle f_1, f_2\rangle \subseteq \mathbb{Z}_5[x, y]$.

Exercises

1.7.1. Compute the S-polynomials of the following pairs in $\mathbb{Q}[x, y, z]$ with respect to the lex, deglex, and degrevlex orderings with $x > y > z$:
a. $f = 3x^2yz - y^3z^3$ and $g = xy^2 + z^2$.
b. $f = 3x^2yz - xy^3$ and $g = xy^2 + z^2$.
c. $f = 3x^2y - yz$ and $g = xy^2 + z^4$.

1.7.2. Use Theorem 1.7.4 to show that the polynomials given in Exercise 1.6.2 do form a Gröbner basis with respect to lex with $x > y > z > w$.

1.7.3. You should do the following exercises without a Computer Algebra System.
a. Find a Gröbner basis for $\langle x^2y + z, xz + y\rangle \subseteq \mathbb{Q}[x, y, z]$ with respect to deglex with $x > y > z$. [Answer: $x^2y + z, xz + y, xy^2 - z^2, y^3 + z^3$.]
b. Find a Gröbner basis for $\langle x^2y + z, xz + y\rangle \subseteq \mathbb{Q}[x, y, z]$ with respect to lex with $x < y < z$. [Answer: $z + x^2y, xz + y, x^3y - y$.]

1.7.4. Finish the computation in Examples 1.7.10 and 1.7.11 without a Computer Algebra System.

1.7.5. In Example 1.7.11 we obtained G using arithmetic modulo 5 throughout the computation. The reader might think that G could also be obtained by first computing a Gröbner basis G' for $I = \langle f_1, f_2\rangle$ viewed as an ideal in $\mathbb{Q}[x, y]$, where we assume that the polynomials in G' have relatively prime integer coefficients, and then reducing this basis modulo 5. This is not the case as we will see in this exercise.
a. Compute the Gröbner basis G' for $I = \langle f_1, f_2\rangle \subseteq \mathbb{Q}[x, y]$ with respect

to the lex ordering with $x > y$.

b. Reduce G' modulo 5 to obtain G'_5 and compare with Example 1.7.11.

1.7.6. Assume that we have fixed a term order on $\mathbb{T}^n$. For $0 \neq f \in k[x_1, \ldots, x_n]$ define the *multidegree* of f by $\deg(f) = \boldsymbol{\alpha}$ where $\text{lp}(f) = \boldsymbol{x}^{\boldsymbol{\alpha}}$ with $\boldsymbol{\alpha} \in \mathbb{N}^n$. Of course, this definition of deg depends on the term order in use. Define $\deg(S) = \{\deg(f) \mid f \in S\}$ for subsets $S \subseteq k[x_1, \ldots, x_n]$ not containing 0.

a. Let I be an ideal in $k[x_1, \ldots, x_n]$. Let $G = \{g_1, \ldots, g_t\}$ be a Gröbner basis for I. Assume that $\deg(g_i) = \boldsymbol{\alpha}_i$ for $1 \leq i \leq t$. Prove that

$$\deg(I^*) = \bigcup_{i=1}^{t} (\boldsymbol{\alpha}_i + \mathbb{N}^n)$$

where $I^* = I - \{0\}$ (see Exercise 1.4.12).

b. In $\mathbb{Q}[x, y]$ let $I = \langle x^2y - y + x, xy^2 - x \rangle$. Show that, with respect to the deglex ordering with $x < y$, $\deg(I^*)$ is represented by the shaded region in the diagram given below:

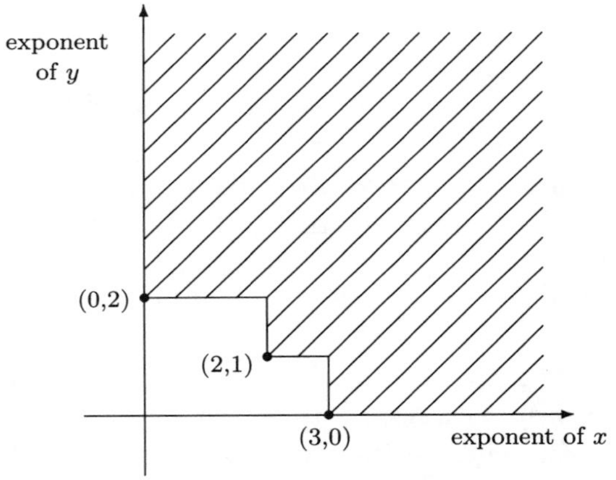

c. Draw the region which represents $\deg(I^*)$ if we use lex with $x < y$.

1.7.7. Show how the steps in the Euclidean Algorithm (Algorithm 1.3.2) parallel the steps in Buchberger's Algorithm (Algorithm 1.7.1).

1.7.8. Show how the steps in Gaussian Elimination (see Section 1.2) parallel the steps in Buchberger's Algorithm (Algorithm 1.7.1).

1.7.9. Assume that $F = \{f_1, \ldots, f_s\} \subseteq k[x_1, \ldots, x_n]$ and each f_j is a difference of two power products. Prove that, with respect to any term order, $\langle F \rangle$ has a Gröbner basis consisting of differences of power products.

1.8. Reduced Gröbner Bases. In the last section we saw how to compute Gröbner bases. However, the Gröbner basis obtained from Buchberger's Algorithm might not be unique. In this section we show that by putting certain conditions on the polynomials in the Gröbner basis, we obtain uniqueness.

In Buchberger's Algorithm there are two places where choices are made. First, there is the order in which the polynomials are inputed and this affects the application of the Division Algorithm. Second, in the WHILE loop of Buchberger's Algorithm where we compute S-polynomials, we choose $\{f, g\} \in \mathcal{G}$ at random. So, if we were to change either of these choices, we might end up with a different Gröbner basis.

For example, in Example 1.7.10 if we had computed $S(f_2, f_3) = x$ first, the S-polynomial $S(f_1, f_2)$ would have reduced to zero, and would not have appeared in the Gröbner basis. So we would have obtained a different Gröbner basis. Note that even after we have computed the Gröbner basis $G = \{f_1, f_2, f_3, f_4, f_5\}$ for I, we can observe that $f_4 = x^2$ can be removed from G, that is, $\{f_1, f_2, f_3, f_5\}$ is also a Gröbner basis for $\langle f_1, f_2, f_3\rangle$. This is because any term divisible by $\mathrm{lt}(f_4) = x^2$ is also divisible by $\mathrm{lt}(f_5) = x$. The set $\{f_1, f_2, f_3, f_5\}$ is the Gröbner basis for I we would have obtained had we computed $S(f_2, f_3) = x$ before $S(f_1, f_2)$.

This leads to the following definition.

DEFINITION 1.8.1. *A Gröbner basis* $G = \{g_1, \dots, g_t\}$ *is called* minimal *if for all* i, $\mathrm{lc}(g_i) = 1$ *and for all* $i \neq j$, $\mathrm{lp}(g_i)$ *does not divide* $\mathrm{lp}(g_j)$.

LEMMA 1.8.2. *Let* $G = \{g_1, \dots, g_t\}$ *be a Gröbner basis for the ideal* I. *If* $\mathrm{lp}(g_2)$ *divides* $\mathrm{lp}(g_1)$, *then* $\{g_2, \dots, g_t\}$ *is also a Gröbner basis for* I.

PROOF. Clearly, if a polynomial f is such that $\mathrm{lp}(f)$ is divisible by $\mathrm{lp}(g_1)$, then it is also divisible by $\mathrm{lp}(g_2)$. Therefore, using Definition 1.6.1, $\{g_2, \dots, g_t\}$ is a Gröbner basis for I. $\square$

As a direct consequence of this lemma, we now see how a minimal Gröbner basis can be obtained from a Gröbner basis.

COROLLARY 1.8.3. *Let* $G = \{g_1, \dots, g_t\}$ *be a Gröbner basis for the ideal* I. *To obtain a minimal Gröbner basis from* G, *eliminate all* g_i *for which there exists* $j \neq i$ *such that* $\mathrm{lp}(g_j)$ *divides* $\mathrm{lp}(g_i)$ *and divide each remaining* g_i *by* $\mathrm{lc}(g_i)$.

In Example 1.7.10 above, a minimal Gröbner basis for I can be obtained from $\{f_1, f_2, f_3, f_4, f_5\}$, by removing f_1, f_2, and f_4. We could also remove f_1, f_3, and f_4. So minimal Gröbner bases are not unique, but, as the following proposition shows, all minimal Gröbner bases for an ideal I have the same number of elements, and the same leading terms.

PROPOSITION 1.8.4. *If* $G = \{g_1, \dots, g_t\}$ *and* $F = \{f_1, \dots, f_s\}$ *are minimal Gröbner bases for an ideal* I, *then* $s = t$, *and after renumbering if necessary,* $\mathrm{lt}(f_i) = \mathrm{lt}(g_i)$ *for all* $i = 1, \dots, t$.

PROOF. Since f_1 is in I and since G is a Gröbner basis for I, there exists i such that $\mathrm{lp}(g_i)$ divides $\mathrm{lp}(f_1)$. After renumbering if necessary, we may assume that $i = 1$. Now g_1 is also in I, and hence, since F is a Gröbner basis for I, there exists j such that $\text{lp}(f_j)$ divides $\text{lp}(g_1)$. Therefore $\text{lp}(f_j)$ divides $\text{lp}(f_1)$, and hence $j = 1$, since F is a minimal Gröbner basis. Thus $\text{lp}(f_1) = \text{lp}(g_1)$.

Now f_2 is in I, and hence there exists i such that $\text{lp}(g_i)$ divides $\text{lp}(f_2)$, since G is a Gröbner basis. The minimality of F and the fact that $\text{lp}(g_1) = \text{lp}(f_1)$ imply that $i \neq 1$, and, after renumbering if necessary, we may assume that $i = 2$. As above we get that $\text{lp}(g_2) = \text{lp}(f_2)$. This process continues until all f's and g's are used up. Thus $s = t$ and after renumbering $\text{lp}(f_i) = \text{lp}(g_i)$ for all $i = 1, \ldots, t$. □

As we mentioned after Corollary 1.8.3, minimal Gröbner bases are not unique. To get uniqueness, we need to add a stronger condition on the polynomials in the Gröbner basis.

DEFINITION 1.8.5. *A Gröbner basis $G = \{g_1, \ldots, g_t\}$ is called a* reduced *Gröbner basis if, for all i, $\text{lc}(g_i) = 1$ and g_i is reduced with respect to $G - \{g_i\}$. That is, for all i, no non-zero term in g_i is divisible by any $\text{lp}(g_j)$ for any $j \neq i$.*

Note that a reduced Gröbner basis is also minimal. We now prove that reduced Gröbner bases exist.

COROLLARY 1.8.6. *Let $G = \{g_1, \ldots, g_t\}$ be a minimal Gröbner basis for the ideal I. Consider the following reduction process:*

$g_1 \xrightarrow{H_1}_+ h_1$, *where h_1 is reduced with respect to $H_1 = \{g_2, \ldots, g_t\}$*

$g_2 \xrightarrow{H_2}_+ h_2$, *where h_2 is reduced with respect to $H_2 = \{h_1, g_3, \ldots, g_t\}$*

$g_3 \xrightarrow{H_3}_+ h_3$, *where h_3 is reduced with respect to $H_3 = \{h_1, h_2, g_4, \ldots, g_t\}$*

$\vdots$

$g_t \xrightarrow{H_t}_+ h_t$, *where h_t is reduced with respect to $H_t = \{h_1, h_2, \ldots, h_{t-1}\}$.*

Then $H = \{h_1, \ldots, h_t\}$ is a reduced Gröbner basis for I.

PROOF. Note that, since G is a minimal Gröbner basis, we have that $\text{lp}(h_i) = \text{lp}(g_i)$ for each $i = 1, \ldots, t$. Therefore, H is also a Gröbner basis for I (in fact, it is a minimal Gröbner basis). Since the division of g_i by $h_1, \ldots, h_{i-1}, g_{i+1}, \ldots, g_t$ is done by eliminating terms of g_i using $\text{lp}(h_1), \ldots, \text{lp}(h_{i-1}), \text{lp}(g_{i+1}), \ldots, \text{lp}(g_t)$, and since $\text{lp}(h_j) = \text{lp}(g_j)$, for all j, H is a reduced Gröbner basis. □

THEOREM 1.8.7 (BUCHBERGER). *Fix a term order. Then every non-zero ideal I has a unique reduced Gröbner basis with respect to this term order.*

PROOF. We proved in the previous result that every ideal has a reduced Gröbner basis. Thus we only need to prove uniqueness. Let $G = \{g_1, \ldots, g_t\}$ and $H = \{h_1, \ldots, h_t\}$ be reduced Gröbner bases for I. We note that by Proposition 1.8.4, since a reduced Gröbner basis is minimal, both G and H have the same number of elements and we may assume that, for each i, $\text{lt}(g_i) = \text{lt}(h_i)$. Let i be given, $1 \leq i \leq t$. If $g_i \neq h_i$, then $g_i - h_i \in I$ implies that there exists j such that $\text{lp}(h_j)$ divides $\text{lp}(g_i - h_i)$. Since $\text{lp}(g_i - h_i) < \text{lp}(h_i)$, we see that $j \neq i$. But then

$\text{lp}(h_j) = \text{lp}(g_j)$ divides a term of g_i or h_i. This contradicts the fact that G and H are reduced Gröbner bases. So $g_i = h_i$. $\square$

EXAMPLE 1.8.8. Let us go back to Example 1.7.10. We have seen right after Corollary 1.8.3 that $\{f_3, f_5\}$ and $\{f_2, f_5\}$ are both minimal Gröbner bases for the ideal I. Since $f_3 = y$ and $f_5 = x$, we see that $\{f_3, f_5\}$ is a reduced Gröbner basis for I. Since $f_2 = y + x$, we see that f_2 can be reduced to y using f_5. So $\{f_2, f_5\}$ is not a reduced Gröbner basis. Of course, the reduced Gröbner basis obtained from $\{f_2, f_5\}$ is in fact $\{f_3, f_5\}$, since the reduced Gröbner basis is unique.

EXAMPLE 1.8.9. We go back to Example 1.7.11. There we showed that a Gröbner basis for $I = \langle x^2 + y^2 + 1, x^2y + 2xy + x\rangle \subseteq \mathbb{Z}_5[x, y]$ with respect to the lex ordering with $x > y$ is $\{x^2+y^2+1, x^2y+2xy+x, 3xy+4x+y^3+y, 4y^5+3y^4+y^2+y+3\}$. By Corollary 1.8.3, $\{x^2+y^2+1, xy+3x+2y^3+2y, y^5+2y^4+4y^2+4y+2\}$ is a minimal Gröbner basis for I. In fact it is easy to see that it is the reduced Gröbner basis for I.

Exercises

1.8.1. Compute the reduced Gröbner basis for the ideal in Example 1.7.9.

1.8.2. Compute the reduced Gröbner basis for the ideals in Exercise 1.7.3.

1.8.3. Let $I \subseteq k[x_1, \ldots, x_n]$ be an ideal. We call I a *homogeneous ideal* provided that $I = \langle f_1, \ldots, f_s\rangle$ where each f_i is homogeneous (see Exercise 1.4.9). Fix an arbitrary term order on $k[x_1, \ldots, x_n]$.

a. Show that I is homogeneous if and only if for all $f \in I$, each homogeneous component of f is also in I.

b. Show that any homogeneous ideal has a Gröbner basis consisting of homogeneous polynomials.

c. Prove that I is a homogeneous ideal if and only if the reduced Gröbner basis for I consists of homogeneous polynomials.

d. Prove that a subset G of a homogeneous ideal I is a reduced Gröbner basis with respect to the lex ordering if and only if G is a reduced Gröbner basis with respect to the deglex ordering.

1.8.4. Let $F \subseteq k[x_1, \ldots, x_n]$ and let $I = \langle F\rangle$. Find an algorithm that will determine a subset $F' \subseteq F$ such that $I = \langle F'\rangle$ and for which no proper subset of F' generates I.

1.8.5. We use the same notation as in Exercise 1.6.19. Let $<$ be any order on $k[x_1, \ldots, x_n]$. We let $<_h$ be as in Exercise 1.6.19. Let $I = \langle f_1, \ldots, f_s\rangle$ be an ideal of $k[x_1, \ldots, x_n]$. Let $G = \{g_1, \ldots, g_t\}$ be a Gröbner basis for $\langle f_1^h, \ldots, f_s^h\rangle$ with respect to $<_h$. We may assume that the polynomials in G are all homogeneous. Prove that $G_h = \{(g_1)_h, \ldots, (g_t)_h\}$ is a Gröbner basis for I. [Hint: First note that even without the assumption that G is a Gröbner basis we have $\langle G_h\rangle = \langle f_h \mid f \in \langle f_1^h, \ldots, f_s^h\rangle\rangle = I$. Then note that $\text{lt}_{<_h}(f)_h = \text{lt}(f_h)$ for every homogeneous polynomial in $k[x_1, \ldots, x_n, w]$. Prove that $S((g_i)_h, (g_j)_h) = S(g_i, g_j)_h$. Prove that given

$f, g \in k[x_1, \ldots, x_n, w]$, if $f \xrightarrow{G}_+ g$, then $f_h \xrightarrow{G_h}_+ g_h$.]

1.8.6. (*) As mentioned before, a Gröbner basis with respect to one term ordering might not be a Gröbner basis with respect to another term ordering. In this exercise we show that for a given ideal there are only finitely many possible reduced Gröbner bases. Let I be an ideal of $k[x_1, \ldots, x_n]$.

a. Let $\mathcal{T}$ be the (infinite) set of all possible term orderings on $k[x_1, \ldots, x_n]$. Let $\mathcal{R} = \{$reduced Gröbner bases for I with respect to the term orders in $\mathcal{T}\}$, and let $\mathcal{L} = \{$leading term ideals of I with respect to the term orders in $\mathcal{T}\}$. Prove that there is a one to one correspondence between $\mathcal{R}$ and $\mathcal{L}$.

b. Prove that $\mathcal{L}$ is finite. [Sketch of the proof ([**MoRo**]): Suppose to the contrary that $\mathcal{L}$ is infinite. For each leading term ideal in $\mathcal{L}$, choose a term ordering which gives this leading term ideal. Let $\mathcal{T}_0 \subseteq \mathcal{T}$ be the infinite set of these chosen term orderings. Also, let I be generated by $\{f_1, \ldots, f_s\}$. Since there are only finitely many terms which appear in the f_i's, there exist terms $m_1, \ldots, m_s$ and an infinite set $\mathcal{T}_1 \subseteq \mathcal{T}_0$ such that $\operatorname{lt}(f_i) = m_i$ for each term order in $\mathcal{T}_1$. Consider the two possible cases: either $\langle m_1, \ldots, m_s \rangle$ is the leading term ideal for I with respect to one term order in $\mathcal{T}_1$, or it is not. In the first case, use Exercise 1.6.17 and in the second add a polynomial to $f_1, \ldots, f_s$ and repeat the argument.]

c. Conclude that there are only finitely many reduced Gröbner bases for a given ideal.

d. A set F which is a Gröbner basis for an ideal I with respect to every term order is called a *universal Gröbner basis*. Use **c** to show that every ideal has a universal Gröbner basis. An example of such a basis is given in Exercise 1.8.7.

1.8.7. Find a universal Gröbner basis for the ideal $\langle x - y^2, xy - x \rangle \subseteq \mathbb{Q}[x, y]$. [Hint: At each stage of Buchberger's Algorithm, consider all possible choices of leading terms.] (Answer: $\{x - y^2, xy - x, y^3 - y^2, x^2 - x\}$.)

1.9. Summary. We conclude this chapter by giving a summary of the most important results that we have seen so far. The first theorem lists all the equivalent conditions that we now have for a set $G = \{g_1, \ldots, g_t\}$ to be a Gröbner basis.

THEOREM 1.9.1. *The following statements are equivalent for a set of non-zero polynomials*

$$G = \{g_1, \ldots, g_t\} \text{ and } I = \langle G \rangle.$$

(i) *For all* $f \in I$, *there exists* i *such that* $\operatorname{lp}(g_i)$ *divides* $\operatorname{lp}(f)$, *that is,* G *is a Gröbner basis.*

(ii) $\operatorname{Lt}(G) = \operatorname{Lt}(I)$.

(iii) $f \in I$ *if and only if* $f \xrightarrow{G}_+ 0$.

(iv) *For all* $f \in k[x_1, \ldots, x_n]$, *if* $f \xrightarrow{G}_+ r_1$, $f \xrightarrow{G}_+ r_2$, *and* r_1, r_2 *are reduced with respect to* G, *then* $r_1 = r_2$.

(v) *For all* $i \neq j$, $S(g_i, g_j) \xrightarrow{G}_+ 0$.

(vi) *For all* $f \in I$, *there exists* $h_1, \ldots, h_t \in k[x_1, \ldots, x_n]$ *such that* $f = h_1 g_1 + \cdots + h_t g_t$ *and* $\operatorname{lp}(f) = \max_{1 \le \nu \le t}(\operatorname{lp}(h_\nu)\operatorname{lp}(g_\nu))$.

(vii) *For all* $i \neq j$ $(1 \le i, j \le t)$, *we have* $S(g_i, g_j) = h_{ij1} g_1 + \cdots + h_{ijt} g_t$ *such that* $\operatorname{lp}(S(g_i, g_j)) = \max_{1 \le \nu \le t}(\operatorname{lp}(h_{ij\nu})\operatorname{lp}(g_\nu))$.

The proofs that all of these conditions are equivalent are contained in Theorems 1.6.2, 1.6.7, and 1.7.4, and Corollary 1.7.6.

THEOREM 1.9.2. *Fix a term order on* $k[x_1, \ldots, x_n]$. *Then every ideal* I *has a reduced Gröbner basis with respect to this term order. This Gröbner basis is effectively computable once* I *has been given as generated by a finite set of polynomials. Moreover this reduced Gröbner basis for* I *(with respect to the given term order) is unique.*

Chapter 2. Applications of Gröbner Bases

This chapter is devoted to giving a number of applications of the theory developed in Chapter 1 to computations in polynomial rings. We also give some applications to computations which use polynomial rings. In Section 2.1 we give methods for doing basic computations in $k[x_1, \ldots, x_n]$ and $k[x_1, \ldots, x_n]/I$, e.g. determining if one polynomial is a member of an explicitly given ideal. In Section 2.2 we introduce the Hilbert Nullstellensatz and use it to connect Gröbner bases to some elementary questions in algebraic geometry. In the next section we give a method for eliminating variables in systems of polynomial equations and use it, for example, to compute generators for the intersection of ideals. In Section 2.4 we study homomorphisms between polynomial rings. In particular we determine generators for the kernel of such a homomorphism and we give a method to determine whether it is onto. We then generalize these results to the case of polynomial rings modulo an ideal (affine algebras). In the next section we give more applications to algebraic geometry, e.g. we show how to find the ideal corresponding to the projection of a variety, and to a parametrically given variety. The last three sections present applications of Gröbner bases to problems in computational mathematics: determining minimal polynomials of elements in field extensions, determining whether graphs can be colored by three colors, and finding solutions to integer programming problems.

2.1. Elementary Applications of Gröbner Bases. Let $I = \langle f_1, \ldots, f_s \rangle$ be an ideal of $k[x_1, \ldots, x_n]$. In this section we want to show how to perform effectively[1] the following tasks:

(i) Given $f \in k[x_1, \ldots, x_n]$, determine whether f is in I (this is the ideal membership problem), and if so, find $v_1, \ldots, v_s \in k[x_1, \ldots, x_n]$ such that $f = v_1 f_1 + \cdots + v_s f_s$;

(ii) Determine whether two ideals I, J of $k[x_1, \ldots, x_n]$ are equal;

(iii) Find coset representatives for every element of $k[x_1, \ldots, x_n]/I$;

(iv) Find a basis of the k-vector space $k[x_1, \ldots, x_n]/I$;

[1]We remind the reader that by "perform effectively" or by "determine" we mean that one can give an algorithm than can be programmed on a computer.

(v) Determine the operations in $k[x_1, \dots, x_n]/I$;

(vi) Find inverses in $k[x_1, \dots, x_n]/I$ when they exist.

We begin with Task (i). Let $F = \{f_1, \dots, f_s\}$ and let $G = \{g_1, \dots, g_t\}$ be a Gröbner basis for $I = \langle f_1, \dots, f_s \rangle$ with respect to a fixed term ordering. We have already seen in Theorem 1.9.1 that

$$f \in I \iff f \xrightarrow{G}_+ 0.$$

So the ideal membership question is answered. Moreover, applying the Division Algorithm to $f \in I$ yields $u_1, \dots, u_t$ such that

$$f = u_1 g_1 + \cdots + u_t g_t. \tag{2.1.1}$$

Also, Buchberger's Algorithm can be implemented so as to keep track of the linear combinations of the f_i's that give rise to the g_j's. This can be seen as follows: during Buchberger's Algorithm (Algorithm 1.7.1) for the computation of a Gröbner basis, a new polynomial g is added to the basis if it is the non-zero remainder of the division of an S-polynomial by the current basis, say $\{h_1, \dots, h_\ell\}$. That is,

$$g = S(h_\nu, h_\mu) - \sum_{i=1}^{\ell} w_i h_i,$$

for some $\nu, \mu \in \{1, 2, \dots, \ell\}$ and some polynomials w_i which are explicitly computed in the Division Algorithm. This procedure is illustrated in Example 2.1.1 below. So we can obtain as an output of Buchberger's Algorithm not only the Gröbner basis $\{g_1, \dots, g_t\}$ but also a $t \times s$ matrix M with polynomial entries such that

$$\begin{bmatrix} g_1 \\ g_2 \\ \vdots \\ g_t \end{bmatrix} = M \begin{bmatrix} f_1 \\ f_2 \\ \vdots \\ f_s \end{bmatrix}. \tag{2.1.2}$$

Thus Equation (2.1.1) can be transformed to give the polynomial f as a linear combination of the original polynomials $f_1, \dots, f_s$:

$$f = v_1 f_1 + \cdots + v_s f_s.$$

EXAMPLE 2.1.1. In this example we consider $k = \mathbb{Q}$. Let $f_1 = x^2y - y + x$, $f_2 = xy^2 - x$, and $I = \langle f_1, f_2 \rangle$. We use the deglex term ordering with $x < y$. We follow Algorithm 1.7.1 given in Section 1.7, but we keep track of the linear combinations that give rise to the new polynomials in the generating set.

INITIALIZATION: $G := \{f_1, f_2\}, \mathcal{G} := \{\{f_1, f_2\}\}$.

First pass through the WHILE loop

$\mathcal{G} := \emptyset$

$S(f_1, f_2) = yf_1 - xf_2 = -y^2 + xy + x^2$

(reduced with respect to G)

Let $f_3 := -y^2 + xy + x^2$
Note that $f_3 = yf_1 - xf_2$
$\mathcal{G} := \{\{f_1, f_3\}, \{f_2, f_3\}\}$
$G := \{f_1, f_2, f_3\}$

Second pass through the WHILE loop
$\mathcal{G} := \{\{f_2, f_3\}\}$
$S(f_1, f_3) = yf_1 + x^2 f_3 = x^3y + x^4 - y^2 + xy$
$\xrightarrow{f_1} x^4 - y^2 + 2xy - x^2$
$\xrightarrow{f_3} x^4 + xy - 2x^2$
(reduced with respect to G)
Let $f_4 := x^4 + xy - 2x^2$
Note that $f_4 = (yf_1 + x^2 f_3) - xf_1 - f_3$
$= (x^2y - x)f_1 + (-x^3 + x)f_2$
$\mathcal{G} := \{\{f_2, f_3\}, \{f_1, f_4\}, \{f_2, f_4\}, \{f_3, f_4\}\}$
$G := \{f_1, f_2, f_3, f_4\}$

Third pass through the WHILE loop
$\mathcal{G} := \{\{f_1, f_4\}, \{f_2, f_4\}, \{f_3, f_4\}\}$
$S(f_2, f_3) = f_2 + xf_3 = x^2y + x^3 - x \xrightarrow{f_1} x^3 + y - 2x$
(reduced with respect to G)
Let $f_5 := x^3 + y - 2x$
Note that $f_5 = (f_2 + xf_3) - f_1 = (xy - 1)f_1 + (-x^2 + 1)f_2$
$\mathcal{G} := \{\{f_1, f_4\}, \{f_2, f_4\}, \{f_3, f_4\}, \{f_1, f_5\},$
$\{f_2, f_5\}, \{f_3, f_5\}, \{f_4, f_5\}\}$
$G := \{f_1, f_2, f_3, f_4, f_5\}$

The reader can verify that all the remaining S-polynomials reduce to zero, and hence $G = \{f_1, f_2, f_3, f_4, f_5\}$ is a Gröbner basis for I. It is also easy to see that $\{f_1, f_3, f_5\}$ is a Gröbner basis for I, since $\operatorname{lt}(f_2)$ is divisible by $\operatorname{lt}(f_3)$ and $\operatorname{lt}(f_4)$ is divisible by $\operatorname{lt}(f_5)$. In fact $\{f_1, f_3, f_5\}$ is the reduced Gröbner basis for I with respect to the deglex term ordering with $x < y$. Moreover, in the above computation we kept track of the linear combinations of f_1 and f_2 giving rise to f_3 and f_5 and this gives us the following:

$$\begin{bmatrix} f_1 \\ f_3 \\ f_5 \end{bmatrix} = \begin{bmatrix} 1 & 0 \\ y & -x \\ xy-1 & -x^2+1 \end{bmatrix} \begin{bmatrix} f_1 \\ f_2 \end{bmatrix}. \tag{2.1.3}$$

Now consider the polynomial

$$f = x^4y - 2x^5 + 2x^2y^2 - 2x^3y - 2x^4 - 2y^3 + 4xy^2 - 3x^2y + 2x^3 - y + 2x.$$

We show[2] that $f \in I$:

$$
\begin{aligned}
f &\xrightarrow{x^2,f_1} -2x^5+2x^2y^2-2x^3y-2x^4-2y^3+4xy^2-2x^2y+x^3-y+2x\\
&\xrightarrow{-2x^2,f_5} 2x^2y^2-2x^3y-2x^4-2y^3+4xy^2-3x^3-y+2x\\
&\xrightarrow{2y,f_1} -2x^3y-2x^4-2y^3+4xy^2-3x^3+2y^2-2xy-y+2x\\
&\xrightarrow{-2y,f_5} -2x^4-2y^3+4xy^2-3x^3+4y^2-6xy-y+2x\\
&\xrightarrow{-2x,f_5} -2y^3+4xy^2-3x^3+4y^2-4xy-4x^2-y+2x\\
&\xrightarrow{2y,f_3} 2xy^2-2x^2y-3x^3+4y^2-4xy-4x^2-y+2x\\
&\xrightarrow{-2x,f_3} -x^3+4y^2-4xy-4x^2-y+2x\\
&\xrightarrow{-1,f_5} 4y^2-4xy-4x^2\\
&\xrightarrow{-4,f_3} 0.
\end{aligned}
$$

So we see that

$$
\begin{aligned}
f &= x^2f_1-2x^2f_5+2yf_1-2yf_5-2xf_5+2yf_3-2xf_3-f_5-4f_3\\
&= (x^2+2y)f_1+(2y-2x-4)f_3+(-2x^2-2y-2x-1)f_5.
\end{aligned}
$$

Using Equation (2.1.3) we have

$$
\begin{aligned}
f &= (x^2+2y)f_1+(2y-2x-4)(yf_1-xf_2)\\
&\quad +(-2x^2-2y-2x-1)((xy-1)f_1+(-x^2+1)f_2)\\
&= (-2x^3y-2xy^2-2x^2y+2y^2-3xy+3x^2+2x+1)f_1\\
&\quad +(2x^4+2x^2y+2x^3-2xy+x^2-2y+2x-1)f_2.
\end{aligned}
$$

EXAMPLE 2.1.2. We give another illustration of this for $k=\mathbb{Z}_5$. We go back to Example 1.8.9. Recall that the reduced Gröbner basis for $I=\langle f_1,f_2\rangle\subseteq\mathbb{Z}_5[x,y]$, where $f_1=x^2+y^2+1$ and $f_2=x^2y+2xy+x$, with respect to the lex ordering with $x>y$, is $\{g_1,g_2,g_3\}$, where $g_1=f_1$, $g_2=2f_3=xy+3x+2y^3+2y$, and $g_3=4f_4=y^5+2y^4+4y^2+4y+2$. It is easy to keep track of how g_2 and g_3 are generated during the algorithm, and we get

$$
\begin{aligned}
g_2 &= 2yf_1+3f_2\\
g_3 &= 4(y+3)g_1+2(3x+4y^2+3y)g_2\\
&= (2xy+y^3+2y^2+4y+2)f_1+(3x+4y^2+3y)f_2,
\end{aligned}
$$

[2]It is convenient, when we are trying to keep track of the linear combinations in the reduction process, to include in the notation the term by which we multiply the polynomial we are using for reduction. That is, if $f,g,h\in k[x_1,\ldots,x_n]$, and X is a term, then $f\xrightarrow{X,g}h$ means that $h=f-Xg$.

and so

$$\begin{bmatrix} g_1 \\ g_2 \\ g_3 \end{bmatrix} = \begin{bmatrix} 1 & 0 \\ 2y & 3 \\ 2xy + y^3 + 2y^2 + 4y + 2 & 3x + 4y^2 + 3y \end{bmatrix} \begin{bmatrix} f_1 \\ f_2 \end{bmatrix}.$$

The second task, determining whether two ideals I, J are equal, is a consequence of Theorem 1.8.7. That is, $I = J$ if and only if I and J have the same reduced Gröbner basis. In particular, we note that for a given ideal I, we have that $I = k[x_1, \ldots, x_n]$ if and only if the reduced Gröbner basis for I is $\{1\}$. Alternatively, $I = \langle f_1, \ldots, f_s\rangle \subseteq J$ if and only if $f_1, \ldots, f_s \in J$, and we know how to determine whether this is true; so, to determine if $I = J$, we may simply check whether $I \subseteq J$ and $J \subseteq I$.

We now consider Task (iii), that is, finding coset representatives for every element of $k[x_1, \ldots, x_n]/I$. We keep the notation from the beginning of the section: $I = \langle G\rangle$, where $G = \{g_1, \ldots, g_t\}$ is a Gröbner basis for I. We know that for all $f \in k[x_1, \ldots, x_n]$ there exists a unique element $r \in k[x_1, \ldots, x_n]$, reduced with respect to G, such that $f \xrightarrow{G}_+ r$ (Theorem 1.6.7).

DEFINITION 2.1.3. *The element r above is called the* normal form *of f with respect to G, and is denoted $N_G(f)$.*

PROPOSITION 2.1.4. *Let $f, g \in k[x_1, \ldots, x_n]$. Then*

$$f \equiv g \pmod I \text{ if and only if } N_G(f) = N_G(g).$$

Therefore $\{N_G(f) \mid f \in k[x_1, \ldots, x_n]\}$ is a set of coset representatives for $k[x_1, \ldots, x_n]/I$. Moreover, the map $N_G \colon k[x_1, \ldots, x_n] \longrightarrow k[x_1, \ldots, x_n]$ is k-linear.

PROOF. From the Division Algorithm, there exists $q \in I$ such that $f = q + N_G(f)$, so that $f - N_G(f) \in I$. Thus $f + I = N_G(f) + I$ in $k[x_1, \ldots, x_n]/I$.

Also, for any $c_1, c_2 \in k$, and for any $f_1, f_2 \in k[x_1, \ldots, x_n]$, $c_1f_1 + c_2f_2 - (c_1N_G(f_1) + c_2N_G(f_2)) \in I$ and $c_1N_G(f_1) + c_2N_G(f_2)$ is reduced with respect to G. Therefore $N_G(c_1f_1 + c_2f_2) = c_1N_G(f_1) + c_2N_G(f_2)$ (see Exercise 1.6.15) and so the map $N_G \colon k[x_1, \ldots, x_n] \longrightarrow k[x_1, \ldots, x_n]$ is k-linear.

Now $f \equiv g \pmod I$ if and only if there exists $q \in I$ such that $f = q + g$. Thus $N_G(f) = N_G(q) + N_G(g)$. But $N_G(q) = 0$, since $q \in I$, so $N_G(f) = N_G(g)$. Conversely, if $N_G(f) = N_G(g)$, then $f - g = (f - N_G(f)) - (g - N_G(g)) \in I$ and hence $f \equiv g \pmod I$. □

EXAMPLE 2.1.5. We go back to Example 2.1.1. We note that

$$x^3 \xrightarrow{f_5} -y + 2x.$$

Since $-y + 2x$ is reduced, we have $N_G(x^3) = -y + 2x$. Also,

$$x^2y + y \xrightarrow{f_1} 2y - x.$$

Since $2y-x$ is reduced, we have $N_G(x^2y+y) = 2y-x$. Moreover, since $N_G(x^3) \neq N_G(x^2y+y)$, we see that $x^3 \not\equiv x^2y+y \pmod{I}$.

The next task we want to consider is Task (iv), that is, we wish to find a basis of the k-vector space $k[x_1,\dots,x_n]/I$. We keep the same notation as above.

PROPOSITION 2.1.6. *A basis for the k-vector space $k[x_1,\dots,x_n]/I$ consists of the cosets of all the power products $X \in \mathbb{T}^n$ such that* $\text{lp}(g_i)$ *does not divide X for all $i = 1, 2, \dots, t$.*

PROOF. We have seen that for any $f \in k[x_1,\dots,x_n]$, $f + I = N_G(f) + I$ in $k[x_1,\dots,x_n]/I$. Since $N_G(f)$ is reduced with respect to G, it is, by the definition of reduced, a k-linear combination of power products $X \in \mathbb{T}^n$ such that $\text{lp}(g_i)$ does not divide X for all $i = 1, 2, \dots, t$. Finally, the cosets of such power products are linearly independent by the uniqueness of the normal form. □

EXAMPLE 2.1.7. Again, we go back to Example 2.1.1. A Gröbner basis for I with respect to deglex with $x < y$ is $G = \{x^2y - y + x, -y^2 + xy + x^2, x^3 + y - 2x\}$. So a basis for $\mathbb{Q}[x,y]/I$ consists of the cosets of $1, x, y, x^2, xy$ and so $\dim_{\mathbb{Q}}(\mathbb{Q}[x,y]/I) = 5$.

We are now able to complete Task (v), that is we can now give a multiplication table for $k[x_1,\dots,x_n]/I$. The representative of the coset of f times the coset of g will be the normal form of fg.

EXAMPLE 2.1.8. We go back to Example 2.1.7 and give a multiplication table for $\mathbb{Q}[x,y]/I$. The representative of the coset $y + I$ times the coset $xy + I$ is the normal form of xy^2. Since $xy^2 \xrightarrow{G}_+ x$, and x is reduced with respect to G, we have $N_G(xy^2) = x$ and so $(y + I)(xy + I) = x + I$. The other products are computed in a similar fashion and we obtain the following multiplication table for the representatives $1, x, y, x^2, xy$ of the $\mathbb{Q}$-basis $\{1+I, x+I, y+I, x^2+I, xy+I\}$ for $\mathbb{Q}[x,y]/I$.

$\times$	1	x	y	x^2	xy
1	1	x	y	x^2	xy
x	x	x^2	xy	$-y+2x$	$y-x$
y	y	xy	$xy+x^2$	$y-x$	x
x^2	x^2	$-y+2x$	$y-x$	$-xy+2x^2$	$xy-x^2$
xy	xy	$y-x$	x	$xy-x^2$	x^2

So, for example, $(2x^2+y)(3xy-5) = 6x^3y - 10x^2 + 3xy^2 - 5y \equiv 6(xy-x^2) - 10x^2 + 3x - 5y = 6xy - 16x^2 - 5y + 3x \pmod{I}$ and so $(2x^2+y+I)(3xy-5+I) = 6xy - 16x^2 - 5y + 3x + I$.

EXAMPLE 2.1.9. We go back to Example 1.8.9. Recall that the reduced Gröbner basis for $I = \langle f_1, f_2\rangle \subseteq \mathbb{Z}_5[x,y]$, where $f_1 = x^2 + y^2 + 1$, and $f_2 = x^2y + 2xy + x$, with respect to the lex ordering with $x > y$, is $\{g_1, g_2, g_3\}$, where $g_1 = f_1$, $g_2 = xy + 3x + 2y^3 + 2y$, and $g_3 = y^5 + 2y^4 + 4y^2 + 4y + 2$. So a basis for $\mathbb{Z}_5[x,y]/I$ consists of the cosets of $1, x, y, y^2, y^3, y^4$, and so $\dim_{\mathbb{Z}_5}(\mathbb{Z}_5[x,y]/I) = 6$.

To conclude this section, we consider Task (vi), that is, we want to determine whether an element $f+I$ of $k[x_1,\ldots,x_n]/I$ has an inverse and, in the case when $f+I$ has an inverse, we want to compute that inverse. Of course, given a k-basis and the multiplication table, this problem translates into an exercise in linear algebra provided the k-basis is finite (see Theorem 2.2.7). We illustrate this in the following example.

EXAMPLE 2.1.10. Using Example 2.1.1 again, we would like to determine whether $y+x+1+I$ is invertible, and, if so, determine its inverse. So we need to find $a,b,c,d,e\in\mathbb{Q}$ such that

$$(axy+bx^2+cy+dx+e)(y+x+1)\equiv 1 \pmod{I}.$$

Now,

$$\begin{aligned}
&(axy+bx^2+cy+dx+e)(y+x+1)\\
&\quad= axy^2+ax^2y+axy+bx^2y+bx^3+bx^2+cy^2+cxy\\
&\qquad +cy+dxy+dx^2+dx+ey+ex+e\\
&\quad\equiv ax+a(y-x)+axy+b(y-x)+b(-y+2x)+bx^2\\
&\qquad +c(xy+x^2)+cxy+cy+dxy+dx^2+dx+ey+ex+e \pmod{I}\\
&\quad= (a+2c+d)xy+(b+c+d)x^2+(a+c+e)y+(b+d+e)x+e.
\end{aligned}$$

So $(axy+bx^2+cy+dx+e)(y+x+1)\equiv 1 \pmod{I}$ if and only if

$$\left\{\begin{array}{ccccccccccc}
a & + & & + & 2c & + & d & & & = & 0\\
& & b & + & c & + & d & & & = & 0\\
a & & & + & c & & & + & e & = & 0\\
& & b & & & + & d & + & e & = & 0\\
& & & & & & & & e & = & 1,
\end{array}\right.$$

since the cosets of $1,x,y,x^2,xy$ form a basis of the $\mathbb{Q}$-vector space $\mathbb{Q}[x,y]/I$. These equations are easily solved to yield $a=-2, b=-1, c=1, d=0$, and $e=1$. Hence $(-2xy-x^2+y+1)+I$ is an inverse of $y+x+1+I$ in $\mathbb{Q}[x,y]/I$. Of course if we had started with an element of $\mathbb{Q}[x,y]/I$ that did not have an inverse, these equations would have had no solution.

An alternative approach to the method used in Example 2.1.10, which does not suffer from the defect that $k[x_1,\ldots,x_n]/I$ must have a finite k-basis, is to recognize that $f+I$ has an inverse in $k[x_1,\ldots,x_n]/I$ if and only if the ideal $\langle I,f\rangle$ is, in fact, all of $k[x_1,\ldots,x_n]$, since $fg-1\in I$ if and only if $1\in\langle I,f\rangle$. Thus, given an ideal $I=\langle f_1,\ldots,f_s\rangle$ and a polynomial $f\in k[x_1,\ldots,x_n]$, to determine if $f+I$ has an inverse in $k[x_1,\ldots,x_n]/I$ and to compute that inverse, we first find a reduced Gröbner basis H for the ideal $\langle f_1,\ldots,f_s,f\rangle$. If $H\neq\{1\}$, then $f+I$ does not have an inverse in $k[x_1,\ldots,x_n]/I$. If $H=\{1\}$, then, as in the solution to Task (i), we can express 1 as a linear combination of $f_1,\ldots,f_s,f$,

$$1=h_1f_1+\cdots+h_sf_s+gf.$$

The polynomial g is then the inverse of f modulo I.

EXAMPLE 2.1.11. We go back to Example 2.1.10. We first compute a Gröbner basis for the ideal $\langle f_1, f_3, f_5, y+x+1\rangle$ with respect to the deglex order with $x < y$, keeping track of the multipliers as we did in Example 2.1.1. Letting $f_6 = y+x+1$ we compute that $S(f_1, f_6) \longrightarrow_+ -x^2-x = f_7$, $S(f_3, f_6) \longrightarrow_+ -2x-1 = f_8$. The S-polynomials $S(f_5, f_6)$, $S(f_1, f_7)$, $S(f_5, f_7)$, $S(f_6, f_7)$, $S(f_1, f_8)$, and $S(f_3, f_8)$ all reduce to zero. Finally, $S(f_5, f_8) \longrightarrow_+ \frac{1}{4}$. At this point we stop, since all other S-polynomials must reduce to zero using the polynomial $\frac{1}{4}$. Working backwards we compute

$$\begin{aligned} 1 &= 4f_5 - 4f_6 - 2f_7 + (2x^2-5)f_8 \\ &= (-2x^2+3)f_1 + (2x^2-5)f_3 + (-2x^2+7)f_5 \\ &\quad +(2x^4+2x^2y-4x^3-5x^2-5y+10x+1)f_6, \end{aligned}$$

giving us the inverse $(2x^4 + 2x^2y - 4x^3 - 5x^2 - 5y + 10x + 1) + I$. Using the multiplication table of Example 2.1.8 we readily see that this is the same answer we obtained in Example 2.1.10.

Exercises

2.1.1. You should do this exercise without the aid of a Computer Algebra System. Let $f = xy^4 + 2x^3y^2 - xy^2 + 2x^2y - x^3 - y$. In Example 2.1.1 show that $f \in I$ and write f as a linear combination of f_1, f_3, f_5 and also as a linear combination of f_1, f_2.

2.1.2. Compute the multiplication table for Example 2.1.9.

2.1.3. Consider Example 2.1.8. Let $f = -1+x^2+xy$. Show that for all $g \in k[x, y]$ such that $g(0) = 0$ we have $fg \in I$. [Hint: Note that it suffices to show that $xf, yf \in I$.]

2.1.4. Show that $\dim_{\mathbb{Q}}(\mathbb{Q}[x, y, z]/\langle y^4 + 3y^2z + z^2, x^2 + z, xy + y^2 + z\rangle) = \infty$.

2.1.5. In $\mathbb{Q}[x, y, z]$, let $I = \langle x^2 + z, xy + y^2 + z, xz - y^3 - 2yz, y^4 + 3y^2z + z^2\rangle$ and $J = \langle x^2 + z, xy + y^2 + z, x^3 - yz\rangle$. Determine which of the following (if any) are true: $I \subset J$, $J \subset I$, or $I = J$.

2.1.6. In Example 2.1.9 determine which of the cosets, $y^2+I, x+I, 2+x+y^2+I$ has an inverse. For those that do have an inverse find it.

2.1.7. Rationalize the denominator of $\dfrac{1}{x+\sqrt{3}+\sqrt[3]{25}}$. [Hint: Consider the ideal $I = \langle y_1^2-3, y_2^3-5\rangle \subseteq \mathbb{Q}(x)[y_1, y_2]$. Note that $\{y_1^2-3, y_2^3-5\}$ is a Gröbner basis for I. Follow the technique used in Example 2.1.10, keeping in mind that the field is $\mathbb{Q}(x)$.]

2.1.8. Show that in $\mathbb{Q}[x, y]/I$, where $I = \langle x^2+y, y^2+x\rangle$, the coset $xy+y+a+I$, for $a \in \mathbb{Q}$, has an inverse if and only if $a \neq 0$.

2.1.9. Let $I \subseteq k[x_1, \ldots, x_n]$ be an ideal.

a. Devise a method similar to that used in the first solution of Task (vi) for determining whether for $f \in k[x_1, \ldots, x_n]$, $f+I$ is a zero divisor in

$k[x_1, \ldots, x_n]/I$. (Recall that in a commutative ring A, $\alpha \in A$ is called a *zero divisor* provided that $\alpha \neq 0$ and there is a $\beta \neq 0$ in A such that $\alpha\beta = 0$.)

b. Show in Example 2.1.8 that $xy + I$ is a zero divisor.

c. Show in Example 2.1.8 that if $J = \{g + I \in k[x_1, \ldots, x_n]/I \mid (g + I)(x^2 + I) = 0\}$, then J is the set of all multiples of $-1 + x^2 + xy + I$ by elements of k.

2.1.10. In $\mathbb{Q}[x, y, z]$, let $I = \langle x + y^2, x^2y + z\rangle$. Show that I is a *prime ideal* (that is, $\mathbb{Q}[x, y, z]/I$ contains no zero divisors). [Hint: Note that for lex and $z > x > y$, the given generators for I form a Gröbner basis.]

2.2. Hilbert Nullstellensatz. In Section 1.1, we saw that there was a correspondence between subsets of $k[x_1, \ldots, x_n]$ and subsets of k^n. The purpose of this section is to analyze this correspondence further. We need to expand somewhat the notions given there.

Let K be an extension field of k, that is, K is a field such that $k \subseteq K$. Given a subset $S \subseteq k[x_1, \ldots, x_n]$, we define the *variety*, $V_K(S)$, in K^n by

$$V_K(S) = \{(a_1, \ldots, a_n) \in K^n \mid f(a_1, \ldots, a_n) = 0 \text{ for all } f \in S\}.$$

We note that, as in Section 1.1, if $I = \langle f_1, \ldots, f_s\rangle \subseteq k[x_1, \ldots, x_n]$ then

$$V_K(I) = \{(a_1, \ldots, a_n) \in K^n \mid f_i(a_1, \ldots, a_n) = 0, 1 \leq i \leq s\} = V_K(f_1, \ldots, f_s).$$

We emphasize that the variety is in K^n and the ideal is in $k[x_1, \ldots, x_n]$. (It makes sense to evaluate $f \in k[x_1, \ldots, x_n]$ at a point $(a_1, \ldots, a_n) \in K^n$ since $k \subseteq K$.) Also, given a subset $V \subseteq K^n$ we define the ideal, $I(V)$, in $k[x_1, \ldots, x_n]$ by

$$I(V) = \{f \in k[x_1, \ldots, x_n] \mid f(a_1, \ldots, a_n) = 0 \text{ for all } (a_1, \ldots, a_n) \in V\}.$$

So now we have the correspondences

$$\begin{array}{ccc} \{\text{ Subsets of } k[x_1, \ldots, x_n]\} & \longrightarrow & \{\text{ Subsets of } K^n\} \\ S & \longmapsto & V_K(S) \end{array} \tag{2.2.1}$$

and

$$\begin{array}{ccc} \{\text{ Subsets of } K^n\} & \longrightarrow & \{\text{ Ideals of } k[x_1, \ldots, x_n]\} \\ V & \longmapsto & I(V). \end{array} \tag{2.2.2}$$

The reason for introducing this extended notion of a variety is that the set of solutions of a system of equations depends on the field K. That is, the field K will affect the properties of the maps above. This is illustrated in the following two examples.

EXAMPLE 2.2.1. For $K = \mathbb{R}$ we have $V_{\mathbb{R}}(x^2 + y^2) = V_{\mathbb{R}}(x, y) = \{(0, 0)\} \subseteq \mathbb{R}^2$. On the other hand for $K = \mathbb{C}$ we have that $V_{\mathbb{C}}(x^2 + y^2)$ is the union of the two lines $y = \pm ix$, where $i = \sqrt{-1}$, while $V_{\mathbb{C}}(x, y)$ is still $\{(0, 0)\}$.

EXAMPLE 2.2.2. Consider the polynomial $f = x^2 + y^2 + 1$. Then for $K = \mathbb{R}$ we see that $V_{\mathbb{R}}(f) = \emptyset$, whereas $V_{\mathbb{C}}(f)$ has an infinite number of points. The situation is similar for $f = x^4 + y^4 + 1$. In fact there are infinitely many ideals of $\mathbb{R}[x, y]$ whose corresponding variety in $\mathbb{R}^2$ is empty.

Examples 2.2.1 and 2.2.2 show that a system of equations may have "too few" solutions in k^n to give us insight into the algebraic and geometric properties of I and $V_K(I)$. In Example 2.2.1, the ideals $\langle x^2 + y^2 \rangle$ and $\langle x, y \rangle$ give rise to the same variety over $\mathbb{R}$, but are "essentially different". The same situation occurs in Example 2.2.2. By enlarging the field $\mathbb{R}$ to $\mathbb{C}$, "essentially different" ideals will give rise to different varieties. This will be clarified later and the key result for this is the Hilbert Nullstellensatz[3] given below.

In order to state the Nullstellensatz, we consider the *algebraic closure* of the field k, denoted $\overline{k}$. Recall that a field K is *algebraically closed* if for every polynomial $f \in K[x]$ in one variable, the equation $f = 0$ has a solution in K. Every field k is contained in a field $\overline{k}$ which is algebraically closed and such that every element of $\overline{k}$ is the root of a non-zero polynomial in one variable with coefficients in k. This field is unique up to isomorphism and is called the algebraic closure of k (see [**Hun, Lan**]). For example, the algebraic closure of $\mathbb{R}$ is $\mathbb{C}$. For the remainder of this section we will consider the correspondences (2.2.1) and (2.2.2) with $K = \overline{k}$.

The Hilbert Nullstellensatz has many forms and we will present two of them below in Theorems 2.2.3 and 2.2.5. We will not include the proofs of these theorems. The interested reader can find them in [**AtMD, Hun**].

THEOREM 2.2.3 (WEAK HILBERT NULLSTELLENSATZ). *Let I be an ideal contained in $k[x_1, \dots, x_n]$. Then $V_{\overline{k}}(I) = \emptyset$ if and only if $I = k[x_1, \dots, x_n]$.*

Note that the result is clear for one variable since the field $\overline{k}$ is algebraically closed. Before we go to the next form of the Nullstellensatz, we need a definition.

DEFINITION 2.2.4. *For an ideal I of $k[x_1, \dots, x_n]$ we define the* radical *of I, denoted $\sqrt{I}$, by*

$$\sqrt{I} = \{f \in k[x_1, \dots x_n] \mid \text{ there exists } e \in \mathbb{N} \text{ such that } f^e \in I\}.$$

It is easily checked that $\sqrt{I}$ is an ideal in $k[x_1, \dots, x_n]$. Moreover, we have that I and $\sqrt{I}$ give rise to the same variety; that is, for all fields $K \supseteq k$,

$$V_K(I) = V_K(\sqrt{I}).$$

THEOREM 2.2.5 (STRONG HILBERT NULLSTELLENSATZ). *$I(V_{\overline{k}}(I)) = \sqrt{I}$ for all ideals I of $k[x_1, \dots, x_n]$.*

[3]The word "Nullstellensatz" is a German word for "zero point theorem". The theorem is given this name because, as we see in Theorem 2.2.3, it gives information about the zero set, i.e. the variety, of an ideal.

Theorem 2.2.5 implies that two ideals I and J correspond to the same variety, i.e. $V_{\overline{k}}(I) = V_{\overline{k}}(J)$, if and only if their radicals are equal, i.e. $\sqrt{I} = \sqrt{J}$. This allows us to make more precise what we meant earlier by "essentially different" ideals. Two ideals are "essentially different" if and only if they have different radicals. In Example 2.2.1, note that $\sqrt{\langle x, y\rangle} = \langle x, y\rangle$ and $\sqrt{\langle x^2 + y^2\rangle} = \langle x^2 + y^2\rangle$ and so are "essentially different" and correspond to different varieties in $\mathbb{C}^2$.

We now consider some applications of the above results. Let $I = \langle f_1, \dots, f_s\rangle$ be an ideal of $k[x_1, \dots, x_n]$, and let $G = \{g_1, \dots, g_t\}$ be the reduced Gröbner basis for I with respect to a term ordering.

THEOREM 2.2.6. *$V_{\overline{k}}(I) = \emptyset$ if and only if $1 \in G$. (i.e., given polynomials $f_1, \dots, f_s$, then there are no solutions to the system $f_1 = 0, f_2 = 0, \dots, f_s = 0$ in $\overline{k}^n$ if and only if $G = \{1\}$.)*

PROOF. By Theorem 2.2.3, $V_{\overline{k}}(I) = \emptyset$ if and only if $1 \in I$. But the last condition is equivalent to $G = \{1\}$, since G is the reduced Gröbner basis. □

THEOREM 2.2.7. *The following statements are equivalent.*

(i) *The variety $V_{\overline{k}}(I)$ is finite.*

(ii) *For each $i = 1, \dots, n$, there exists $j \in \{1, \dots, t\}$ such that $\operatorname{lp}(g_j) = x_i^\nu$ for some $\nu \in \mathbb{N}$.*

(iii) *The dimension of the k-vector space $k[x_1, \dots, x_n]/I$ is finite.*

PROOF. (i) $\Longrightarrow$ (ii). Let $V_{\overline{k}}(I)$ be finite. If $V_{\overline{k}}(I)$ is empty, then, by Theorem 2.2.3, $I = k[x_1, \dots, x_n]$ and hence $G = \{1\}$ and (ii) is trivially satisfied. So we may assume that $V_{\overline{k}}(I)$ is not empty. Fix $i \in \{1, \dots, n\}$. Let $a_{ij}, j = 1, \dots, \ell$ be the distinct ith coordinates of the points in $V_{\overline{k}}(I)$. For each $j, 1 \leq j \leq \ell$, let $0 \neq f_j \in k[x_i]$ be such that $f_j(a_{ij}) = 0$ (this can be done by the definition of $\overline{k}$). Let $f = f_1 f_2 \cdots f_\ell \in k[x_i] \subseteq k[x_1, \dots, x_n]$. Then we see that $f \in I(V_{\overline{k}}(I))$, and hence, by Theorem 2.2.5, there exists e such that $f^e \in I$. Since $\operatorname{lp}(f^e) = x_i^{em}$ for some natural number m, and since the leading power product of every element of I is divisible by the leading power product of some element of G, there exists a polynomial in G whose leading power product is a power of x_i alone. This is true for every $i = 1, \dots, n$.

(ii) $\Longrightarrow$(iii). We saw in Section 2.1 that a k-basis of $k[x_1, \dots, x_n]/I$ is the set of cosets of power products reduced with respect to G. Since for every $i \in \{1, \dots, n\}$ a power of x_i is a leading power product of some g_j, there are only finitely many power products which are reduced with respect to G, and hence $\dim_k k[x_1, \dots, x_n]/I$ is finite.

(iii)$\Longrightarrow$(i). We will show that for any $i = 1, \dots, n$, there are only finitely many distinct values for the ith coordinate of points in $V_{\overline{k}}(I)$. Fix $i \in \{1, \dots, n\}$. Since, by assumption, $k[x_1, \dots, x_n]/I$ is a finite dimensional k-vector space, the powers $1, x_i, x_i^2, \dots$ of x_i are linearly dependent modulo I. Therefore there is an integer

m and constants $c_j \in k, 0 \leq j \leq m$, not all zero, such that

$$\sum_{j=0}^{m} c_j x_i^j \in I.$$

Since the above polynomial can only have finitely many roots in $\overline{k}$, there are only finitely many values for the ith coordinates of the points of $V_{\overline{k}}(I)$. □

An ideal $I \neq k[x_1, \ldots, x_n]$ that satisfies any one of the equivalent conditions in Theorem 2.2.7 is called *zero-dimensional*. This terminology is adopted because $V_{\overline{k}}(I)$ consists of only finitely many points.

EXAMPLE 2.2.8. In Example 2.1.1 we saw that a Gröbner basis for the ideal $I = \langle x^2y - y + x, xy^2 - x\rangle$ in $\mathbb{Q}[x, y]$ with respect to the deglex term ordering with $x < y$ is $G = \{x^2y - y + x, -y^2 + xy + x^2, x^3 + y - 2x\}$. We see that x^3 and y^2 appear as a leading power product of elements of G, and hence $V_{\overline{k}}(I)$ is finite. In fact it is easy to solve the equations to get $V_{\overline{\mathbb{Q}}}(I) = \{(0, 0), (\alpha, -1), (-\alpha, 1), (\alpha', -1), (-\alpha', 1)\}$, where α and α' are the roots of the equation $z^2 - z - 1 = 0$.

We note that a Gröbner basis for I with respect to the lex term ordering with $x < y$ is $G' = \{x^5 - 3x^3 + x, y + x^3 - 2x\}$, and again we have that some power of x and some power of y appear as leading power products of elements of G'. These equations may easily be solved to yield the same answer as above.

EXAMPLE 2.2.9. We go back to Example 2.1.9. Recall that the reduced Gröbner basis for $I = \langle f_1, f_2\rangle \subseteq \mathbb{Z}_5[x, y]$, where $f_1 = x^2 + y^2 + 1$ and $f_2 = x^2y + 2xy + x$, with respect to the lex ordering with $x > y$, is $\{g_1, g_2, g_3\}$, where $g_1 = f_1$, $g_2 = xy + 3x + 2y^3 + 2y$, and $g_3 = y^5 + 2y^4 + 4y^2 + 4y + 2$. We see that x^2 and y^5 appear as leading power products and hence $V_{\overline{\mathbb{Z}}_5}(I)$ is finite. We note that not all of the solutions are in $\mathbb{Z}_5$; some are in the algebraic closure $\overline{\mathbb{Z}}_5$ of $\mathbb{Z}_5$.

EXAMPLE 2.2.10. As a third example we again let $k = \mathbb{Q}$ and consider the intersection of the circle $f_1 = (x - 1)^2 + y^2 - 1 = 0$ and the ellipse $f_2 = 4(x - 1)^2 + y^2 + xy - 2 = 0$. Using the lex term ordering with $x > y$ we see that the Gröbner basis for the ideal $\langle f_1, f_2\rangle$ is $\{g_1, g_2\}$, where $g_1 = 5y^4 - 3y^3 - 6y^2 + 2y + 2$ and $g_2 = x - 5y^3 + 3y^2 + 3y - 2$. Since $\text{lp}(g_1) = y^4$ and $\text{lp}(g_2) = x$, we see that Theorem 2.2.7 implies that the number of points in the intersection is finite. Also, clearly $g_1 = 0$ has at most four solutions and for each solution of $g_1 = 0$ we get precisely one solution of $g_2 = 0$. Thus we see in this case the geometrically obvious fact that the intersection of a circle and an ellipse can consist of at most four points.

We note that in the last example the form of the Gröbner basis was particularly convenient for determining the points in the variety. That is, the first polynomial contained only the y variable, and the leading power product of the second polynomial was a power of x. We will now show that, in the case of zero-dimensional ideals, this type of structure in the Gröbner basis is always present when the lex term ordering is used.

COROLLARY 2.2.11. *Let I be a zero-dimensional ideal and G be the reduced Gröbner basis for I with respect to the lex term order with $x_1 < x_2 < \cdots < x_n$. Then we can order $g_1, \ldots, g_t$ such that g_1 contains only the variable x_1, g_2 contains only the variables x_1 and x_2 and $\operatorname{lp}(g_2)$ is a power of x_2, g_3 contains only the variables x_1, x_2 and x_3 and $\operatorname{lp}(g_3)$ is a power of x_3, and so forth until g_n.*

PROOF. This follows immediately from Part (ii) in Theorem 2.2.7. That is, we may reorder the g_j such that $\operatorname{lp}(g_j)$ is a power of x_j. It then follows, because of the lex ordering, that the only variables that may appear in g_j are $x_1, x_2, \ldots, x_j$. □

We see that the Gröbner basis for a zero-dimensional ideal I is in "triangular" form (this is similar to the row echelon form in the linear case). Thus, in order to solve the system of equations determined by a zero-dimensional ideal I, it suffices to have an algorithm to find the roots of polynomials in one variable. That is, we first solve the equation in one variable $g_1 = 0$. For each solution α of $g_1 = 0$, we solve the equation $g_2(\alpha, x_2) = 0$. We continue in this manner all the way until $g_n = 0$. The solutions obtained in this way are the only possible solutions. We still have to test them in the equations $g_{n+1} = 0, \ldots, g_t = 0$ (in the case when $t > n$) in order to obtain the set of solutions of the full system of equations. The techniques for finding the roots of polynomials in one variable are not part of the theory of Gröbner bases. The interested reader should consult [**Coh**]. These ideas will be illustrated in the following example.

EXAMPLE 2.2.12. Consider the ideal $I = \langle z^2y+z^2, x^3y+x+y+1, z+x^2+y^3\rangle$ in $\mathbb{Q}[x,y,z]$. We compute[4] the reduced Gröbner basis G for I with respect to the lex ordering with $x > y > z$. We get $G = \{z^4 - z^3, y^{11} + 3y^8z - 2y^7 - 4y^4z + y^3 + y^2 + 2y + z^3 - z^2 + z + 1, x^2 + y^3 + z, yz^2 + z^2, xy + x + y^7 + 2y^4z - y^3 - z^2 - z, xz + y^{10} - y^9 + y^8 + 3y^7z - y^7 - 2y^6z - y^6 + 2y^5z + y^5 - 2y^4z - y^4 - 2y^3z + y^3 + y^2z - yz + y - z^3 + 5z^2 + z + 1\}$. So using the notation of Corollary 2.2.11, we have $g_1 = z^4 - z^3$ is a polynomial in z alone. Also, $g_2 = y^{11} + 3y^8z - 2y^7 - 4y^4z + y^3 + y^2 + 2y + z^3 - z^2 + z + 1$ is a polynomial in y and z alone whose leading power product is $\operatorname{lp}(g_2) = y^{11}$. Finally $g_3 = x^2 + y^3 + z$ is a polynomial in x, y, z whose leading power product is $\operatorname{lp}(g_3) = x^2$. So to find the solutions of the original set of equations, we first note that $z = 0$ or $z = 1$. Then in order to find the corresponding y values, we would have to solve the 11th degree equations $g_2(y, 0) = 0$ and $g_2(y, 1) = 0$. We continue this way as described above.

[4]For the remainder of this chapter and following chapters, Gröbner basis computations will, most often, not be done explicitly in the text and will often require the use of a Computer Algebra System. The reader who wants to verify the computations stated in the text should avail themselves of such a system. The authors usually used CoCoA, but other systems could have been used for some of the computations. This is discussed in the Appendix.

We have seen in Theorem 2.2.5 the importance of computing $\sqrt{I}$. For zero-dimensional ideals we will show how to do this in Exercises 2.3.23 and 2.3.24. But this is a difficult task, in general, which is beyond the scope of this book. The interested reader should consult [**EHV, GTZ**]. However we can now give an easy criterion for membership in $\sqrt{I}$.

THEOREM 2.2.13. *Let* $I = \langle f_1, \dots, f_s \rangle$ *be an ideal in* $k[x_1, \dots, x_n]$. *Then* $f \in \sqrt{I}$ *if and only if* $1 \in \langle f_1, \dots, f_s, 1 - wf \rangle \subseteq k[x_1, \dots, x_n, w]$, *where* w *is a new indeterminate.*

PROOF. By Theorem 2.2.5, $\sqrt{I} = I(V_{\bar{k}}(I))$, and hence $f \in \sqrt{I}$ if and only if $f(a_1, \dots, a_n) = 0$ for all $(a_1, \dots, a_n) \in V_{\bar{k}}(I)$. Let $f \in \sqrt{I}$. If $(a_1, \dots, a_n, b) \in V_{\bar{k}}(\langle f_1, \dots, f_s, 1 - wf \rangle)$, then

$$f_i(a_1, \dots, a_n) = 0 \text{ for all } i = 1, 2, \dots, s \text{ and } 1 - bf(a_1, \dots, a_n) = 0.$$

But then $(a_1, \dots, a_n) \in V_{\bar{k}}(I)$, and hence $f(a_1, \dots, a_n) = 0$, which is a contradiction. Therefore $V_{\bar{k}}(\langle f_1, \dots, f_s, 1 - wf \rangle) = \emptyset$, and by Theorem 2.2.3, we have $1 \in \langle f_1, \dots, f_s, 1 - wf \rangle$. Conversely, let $1 \in \langle f_1, \dots, f_s, 1 - wf \rangle$. Then

$$1 = \sum_{i=1}^{s} h_i f_i + h(1 - wf),$$

for some $h_i, h \in k[x_1, \dots, x_n, w]$. Then for every $(a_1, \dots, a_n) \in V_{\bar{k}}(I)$, we have

$$1 = (1 - wf(a_1, \dots, a_n))h(a_1, \dots, a_n, w).$$

Note that the right-hand side is a polynomial in w. If $f(a_1, \dots, a_n) \neq 0$, then we can set $w = \frac{1}{f(a_1, \dots, a_n)}$ to obtain a contradiction. Therefore $f(a_1, \dots, a_n) = 0$, and so $f \in \sqrt{I}$. □

So the radical membership question can be answered by deciding whether 1 is in an ideal. Thus, as we showed in Section 2.1, to decide whether f is in $\sqrt{I}$, we first compute a reduced Gröbner basis G for the ideal $\langle f_1, \dots, f_s, 1 - wf \rangle$. If $1 \in G$, then $f \in \sqrt{I}$, otherwise, $f \notin \sqrt{I}$.

EXAMPLE 2.2.14. Let $I = \langle xy^2 + 2y^2, x^4 - 2x^2 + 1 \rangle$ be an ideal of $\mathbb{Q}[x, y]$. We would like to determine whether $f = y - x^2 + 1$ is in $\sqrt{I}$. So let us consider the ideal $\langle xy^2 + 2y^2, x^4 - 2x^2 + 1, 1 - w(y - x^2 + 1) \rangle$ in the ring $\mathbb{Q}[x, y, w]$. A Gröbner basis for this ideal with respect to the deglex term ordering with $x < y < w$ can be computed to be $\{1\}$, so that f is indeed in $\sqrt{I}$. Since $f \in \sqrt{I}$, we know that $f^e \in I$ for some e, and we may want to determine the smallest such e. To do this we first compute a Gröbner basis for I. For example, with respect to the deglex term ordering with $x < y$, we have $G = \{y^2, x^4 - 2x^2 + 1\}$. We use this Gröbner basis to compute the normal form of f^i for $i = 1, 2, \dots$ until the first time that normal form is zero. For example, we can compute that $N_G(f) = f \neq 0$, $N_G(f^2) = -2yx^2 + 2y \neq 0$, but $N_G(f^3) = 0$, so that $f^3 \in I$.

Theorem 2.2.13 gives a method for determining whether two ideals I and J have the same radical and therefore correspond to the same variety in $\overline{k}^n$. Let $I = \langle f_1, \dots, f_s \rangle$ and $J = \langle g_1, \dots, g_t \rangle$. Using Theorem 2.2.13 we can decide whether each f_i is in $\sqrt{J}$. If so, then $I \subseteq \sqrt{J}$ and hence $\sqrt{I} \subseteq \sqrt{J}$. The reverse inclusion is checked similarly.

EXAMPLE 2.2.15. Let $I = \langle x^2z^2+x^3, xz^4+2x^2z^2+x^3, y^2z-2yz^2+z^3, x^2y+y^3 \rangle$ and $J = \langle xz^2+x^2, yz^2-z^3, x^2y-x^2z, y^4-x^3, x^4z-x^3z, z^6+x^4, x^5-x^4 \rangle$ be ideals in $\mathbb{Q}[x,y,z]$. The reader can easily verify by the method above that $\sqrt{I} = \sqrt{J}$.

Exercises

2.2.1. Consider the system of equations over $\mathbb{C}$

$$\begin{cases} x^{10} - 22x^6 + 51x^4 - 48x^2 + 18 & = & -18y \\ x^{10} - 22x^6 + 51x^4 - 30x^2 + 18 & = & 18z \\ x^{12} - 9x^{10} + 32x^8 - 57x^6 + 51x^4 - 18x^2 & = & 0. \end{cases}$$

Obtain an explicit solution involving ζ and ζ^2 where $1, \zeta$, and ζ^2 are the three cube roots of unity. [Hint: Use lex with $x > y > z$. There are 11 solutions.]

2.2.2. Use Lagrange Multipliers to maximize the function $f = x^2 + y^2 + xy$ subject to the constraint $x^2 + 2y^2 = 1$. (At least find explicitly the 4 points where the maximum could occur.)

2.2.3. Show (using a Computer Algebra System) that the function $f(x,y,z) = (x^2+y^2)(x^2+y^2-1)z + z^3 + x + y$ has no real critical points (i.e. places where the three partial derivatives vanish simultaneously).

2.2.4. In $\mathbb{Q}[x,y,z]$, let $I = \langle x^4y^2 + z^2 - 4xy^3z - 2y^5z, x^2 + 2xy^2 + y^4 \rangle$. Let $f = yz - x^3$. First show that $f \in \sqrt{I}$. Then find the least power of f which lies in I.

2.2.5. Verify the assertions made in Example 2.2.15.

2.2.6. Show that the following are equivalent for an ideal $I \subseteq k[x_1, \dots, x_n]$.

a. I is zero-dimensional.

b. For all i, $1 \leq i \leq n$, there is a polynomial $f \in I$ such that f contains only the variable x_i.

2.2.7. Let I be a zero-dimensional ideal of $k[x_1, \dots, x_n]$. Corollary 2.2.11 gives one way to compute the monic generator of $I \cap k[x_i]$. In this exercise we present a more efficient way to compute this polynomial, for $i = 1, \dots, n$. We will assume that we have a Gröbner basis G for I with respect to some term order (any order will do). This method will not require any new Gröbner basis computation, instead it will use simple techniques of linear algebra applied to the vector space $k[x_1, \dots, x_n]/I$.

a. Let m be least such that $\{1+I, x_i+I, \dots, x_i^m+I\}$ is a set of linearly dependent vectors in $k[x_1, \dots, x_n]/I$. Let $\sum_{\nu=0}^m a_\nu x_i^\nu \equiv 0 \pmod{I}$. Prove that $f = \frac{1}{a_m}(\sum_{\nu=0}^m a_\nu x_i^\nu)$ is the monic generator of $I \cap k[x_i]$.

In view of this, we need a method for determining whether $\{1+I, x_i+I, \ldots, x_i^m+I\}$ is linearly dependent in $k[x_1,\ldots,x_n]/I$ and, if so, for finding a linear combination $\sum_{\nu=0}^m a_\nu x_i^\nu$ in I. Consider $m+1$ new variables $y_0, y_1, \ldots, y_m$ and the polynomial $g = \sum_{\nu=0}^m y_\nu N_G(x_i^\nu) \in k[x_1,\ldots,x_n,y_0,\ldots,y_m]$. We view g as a polynomial in $k[x_1,\ldots,x_n]$ with coefficients in $k[y_0,\ldots,y_m]$ (which we note are all linear), and we let J be the ideal in $k[y_0,\ldots,y_m]$ generated by the coefficients of g.

b. Prove that $\{1+I, x_i+I, \ldots, x_i^m+I\}$ is a linearly dependent subset of $k[x_1,\ldots,x_n]/I$ if and only if $V_k(J) \neq \{\mathbf{0}\}$.

c. Prove that if $(a_0,\ldots,a_m) \in V_k(J)$, then $f = \sum_{\nu=0}^m a_\nu x_i^\nu \in I$.

d. Use the above to give an algorithm that inputs a Gröbner basis G for I and outputs the monic generator of $I \cap k[x_i]$.

e. Use the algorithm above to find $I \cap k[x]$ and $I \cap k[y]$ in Examples 2.2.8, 2.2.9, and 2.2.10. Verify your answers by computing the appropriate Gröbner bases.

2.2.8. (Faugère, Gianni, Lazard, Mora [**FGLM**]) Let I be a zero-dimensional ideal of $k[x_1,\ldots,x_n]$. Let G_1 be a Gröbner basis for I with respect to a term order $<_1$, and let $<_2$ be another term order. The technique presented in Exercise 2.2.7 can be used to compute a Gröbner basis, G_2, for I with respect to $<_2$ using only linear algebra.

a. Let $X_1,\ldots,X_r$ be power products in $k[x_1,\ldots,x_n]$. Use the technique of Exercise 2.2.7 to give a method for determining whether $\{X_1+I,\ldots,X_r+I\}$ is linearly dependent or not.

b. Assume that all the the power products in $k[x_1,\ldots,x_n]$ are ordered using $<_2$ as follows

$$1 <_2 X_1 <_2 X_2 <_2 X_3 <_2 \cdots .$$

Modify **a** to give a method for deciding whether there exists a polynomial f in I whose leading term with respect to $<_2$ is X_r.

c. Use the above to find an algorithm that inputs a Gröbner basis, G_1, for I with respect to $<_1$ and outputs a Gröbner basis, G_2, for I with respect to $<_2$. (Note that any power product $X \in k[x_1,\ldots,x_n]$ is either reduced with respect to G_2, a leading power product of some polynomial in G_2, or a multiple of the leading power product of some polynomial in G_2. Moreover, eventually in the algorithm, every power product not yet examined will be a multiple of a leading power product already generated.)

d. Why does the method of this exercise not work for ideals which are not zero-dimensional? [Hint: Think about the stopping condition in **c**.]

e. Use this algorithm to compute a Gröbner basis for I in Example 2.2.8 with respect to degrevlex with $x < y$.

2.2.9. Let I be a zero-dimensional ideal in $k[x_1, \ldots, x_n]$. For $i = 1, \ldots, n$, let μ_i be the degree of the monic generator of $I \cap k[x_i]$.
a. Prove that $V_{\overline{k}}(I)$ has at most $\mu_1 \mu_2 \cdots \mu_n$ elements.
b. Prove that $\dim_k(k[x_1, \ldots, x_n]/I) \leq \mu_1 \mu_2 \cdots \mu_n$.

2.3. Elimination. In the previous section we saw the advantage of computing Gröbner bases with respect to the lex term order. In this section we present a far reaching generalization of this idea.

Consider two sets of variables $\{x_1, \ldots, x_n\}$ and $\{y_1, \ldots, y_m\}$. Assume that the power products in the x variables and the power products in the y variables are ordered by term orders $<_x, <_y$ respectively. We define a term order $<$ on the power products in the x, y variables as follows.

DEFINITION 2.3.1. *For X_1, X_2 power products in the x variables and Y_1, Y_2 power products in the y variables, we define*

$$X_1 Y_1 < X_2 Y_2 \iff \begin{cases} X_1 <_x X_2 \\ \text{or} \\ X_1 = X_2 \text{ and } Y_1 <_y Y_2. \end{cases}$$

This term order is called an elimination order *with the x variables larger than the y variables.*

Elimination orders have the following fundamental property whose proof we leave to the exercises (Exercise 2.3.2).

LEMMA 2.3.2. *The elimination order defined in Definition 2.3.1 is a term order. Moreover, if Y is a power product in the y variables and Z is a power product in the x, y variables such that one of the x_i appears to a positive power in Z, then $Y < Z$.*

EXAMPLE 2.3.3. If the orders $<_x$ and $<_y$ are lex term orderings, then the elimination order defined in Definition 2.3.1 is the lex term ordering on all the variables with the y variables smaller than the x variables (Exercise 2.3.3).

The elimination order is "like" a lexicographic term ordering between the x and y variables. The advantage of this order is that when one is interested in properties that the lexicographic term ordering between the two sets of variables is advantageous for, the order within the two sets is unimportant. It is a fact that computations using the lexicographic term ordering are slow (see, for example, [**Bu83, GMNRT**]) and it is better to have as "little lexicographic ordering as possible". There is a great advantage to the elimination order as the following result shows.

THEOREM 2.3.4. *Let I be a non-zero ideal of $k[y_1, \ldots, y_m, x_1, \ldots, x_n]$, and let $<$ be an elimination order with the x variables larger than the y variables. Let $G = \{g_1, \ldots, g_t\}$ be a Gröbner basis for this ideal. Then $G \cap k[y_1, \ldots, y_m]$ is a Gröbner basis for the ideal $I \cap k[y_1, \ldots, y_m]$.*

The ideal $I \cap k[y_1, \ldots, y_m]$ is called an *elimination ideal*, since the x variables have been "eliminated".

PROOF. Clearly $G \cap k[y_1, \ldots, y_m]$ is contained in $I \cap k[y_1, \ldots, y_m]$. Now let $0 \neq f(y_1, \ldots, y_m) \in I \cap k[y_1, \ldots, y_m]$. Since G is a Gröbner basis for I, there exists i such that $\text{lp}(g_i)$ divides $\text{lp}(f)$. Moreover, since f has only y variables, we see that $\text{lp}(g_i)$ involves only the y variables and so, from Lemma 2.3.2, every term in g_i involves only y variables, i.e. $g_i \in G \cap k[y_1, \ldots, y_m]$. Thus, for every $f \in I \cap k[y_1, \ldots, y_m]$, there exists $g_i \in G \cap k[y_1, \ldots, y_m]$ such that $\text{lp}(g_i)$ divides $\text{lp}(f)$, and hence $G \cap k[y_1, \ldots, y_m]$ is a Gröbner basis for $I \cap k[y_1, \ldots, y_m]$. □

As a first application of Theorem 2.3.4 we now present a method for finding generators for the intersection of two ideals.

PROPOSITION 2.3.5. *Let I, J be ideals in $k[x_1, \ldots, x_n]$, and let w be a new variable. Consider the ideal $\langle wI, (1-w)J \rangle$ in $k[x_1, \ldots, x_n, w]$. Then*

$$I \cap J = \langle wI, (1-w)J \rangle \cap k[x_1, \ldots, x_n].$$

REMARK: If $I = \langle f_1, \ldots, f_s \rangle$, and $J = \langle f_1', \ldots, f_p' \rangle$, then a set of generators for the ideal $\langle wI, (1-w)J \rangle$ is $\{wf_1, \ldots, wf_s, (1-w)f_1', \ldots, (1-w)f_p'\}$.

PROOF. Let $f \in I \cap J$. Since

$$f = wf + (1-w)f,$$

we have $f \in \langle wI, (1-w)J \rangle \cap k[x_1, \ldots, x_n]$. Conversely, suppose that $f \in \langle wI, (1-w)J \rangle \cap k[x_1, \ldots, x_n]$. Then, since $f \in \langle wI, (1-w)J \rangle \subseteq k[x_1, \ldots, x_n, w]$, we have

$$\begin{aligned} f(x_1, \ldots, x_n) = &\sum_{i=1}^{s} wf_i(x_1, \ldots, x_n)h_i(x_1, \ldots, x_n, w) \\ &+ \sum_{j=1}^{p} (1-w)f_j'(x_1, \ldots, x_n)h_j'(x_1, \ldots, x_n, w). \end{aligned}$$

Since w does not appear in $f(x_1, \ldots, x_n)$, we can let $w = 1$ and get $f \in I$, and then let $w = 0$ and get $f \in J$. □

As a consequence of the above result we obtain a method for computing generators for the ideal $I \cap J$. First we compute a Gröbner basis G for the ideal $\langle wI, (1-w)J \rangle \subseteq k[x_1, \ldots, x_n, w]$ using an elimination order with $x_1, \ldots, x_n$ smaller than w. We then obtain a Gröbner basis for $I \cap J$ by computing $G \cap k[x_1, \ldots, x_n]$, which is done simply by inspection. A similar technique can be used to compute the intersection of more than two ideals (see Exercise 2.3.8).

EXAMPLE 2.3.6. Consider the following ideals in $\mathbb{Q}[x, y]$:

$$I = \langle x^2 + y^3 - 1, x - yx + 3\rangle \text{ and } J = \langle x^2y - 1\rangle.$$

We wish to compute $I \cap J$. We compute a Gröbner basis G for the ideal

$$\langle w(x^2 + y^3 - 1), w(x - yx + 3), (1 - w)(x^2y - 1)\rangle \subseteq \mathbb{Q}[x, y, w]$$

using the deglex term ordering on the variables x and y with $x > y$ and an elimination order with w greater than x, y. We get

$$G = \{x^3y^2 - x^3y - 3x^2y - xy + x + 3, x^2y^4 + x^4y - x^2y - y^3 - x^2 + 1,$$

$$12853w + 118x^4y + 9x^2y^3 - 357x^3y - 972x^2y^2 + 2152x^2y - 118x^2 - 9y^2 + 357x$$

$$+972y - 2152, x^5y + 3x^2y^3 + 3x^2y^2 - x^3 + 3x^2y - 3y^2 - 3y - 3\}.$$

So a Gröbner basis for the ideal $I \cap J$ is

$$\{x^3y^2 - x^3y - 3x^2y - xy + x + 3, x^2y^4 + x^4y - x^2y - y^3 - x^2 + 1,$$

$$x^5y + 3x^2y^3 + 3x^2y^2 - x^3 + 3x^2y - 3y^2 - 3y - 3\}.$$

Fix a term order on $k[x_1, \ldots, x_n]$. For $f, g \in k[x_1, \ldots, x_n]$, both non-zero, we define the greatest common divisor of f and g, denoted $\gcd(f, g)$, to be the polynomial d such that (i) d divides both f and g; (ii) if h divides both f and g then h divides d; (iii) $\mathrm{lc}(d) = 1$. Dually, we define the least common multiple of f and g, denoted by $\mathrm{lcm}(f, g)$, to be the polynomial ℓ such that (i) f and g both divide ℓ; (ii) if f and g both divide a polynomial h, then ℓ divides h; (iii) $\mathrm{lc}(\ell) = \mathrm{lc}(f)\,\mathrm{lc}(g)$. It may be shown that $\mathrm{lcm}(f, g)$ and $\gcd(f, g)$ exist[5] and that

$$fg = \mathrm{lcm}(f, g)\gcd(f, g).$$

We now show that Proposition 2.3.5 can be used to compute least common multiples and greatest common divisors of polynomials in $k[x_1, \ldots, x_n]$.

LEMMA 2.3.7. *For $f, g \in k[x_1, \ldots, x_n]$, both non-zero, we have*

$$\langle f\rangle \cap \langle g\rangle = \langle \mathrm{lcm}(f, g)\rangle.$$

PROOF. Let $\ell = \mathrm{lcm}(f, g)$. Then $\ell \in \langle f\rangle \cap \langle g\rangle$, by the definition of ℓ. Conversely, if $h \in \langle f\rangle \cap \langle g\rangle$, then $h = af = bg$ for some $a, b \in k[x_1, \ldots, x_n]$. Hence f divides h and g divides h, and thus ℓ divides h by the definition of $\mathrm{lcm}(f, g)$, so that $h \in \langle \ell\rangle$. □

[5]The reader should recognize that this is due to the fact that $k[x_1, \ldots, x_n]$ is a unique factorization domain (UFD) (see [**Hun**].)

Therefore to compute the lcm and then the gcd of two non-zero polynomials f and g in $k[x_1, \ldots, x_n]$, we first compute the reduced Gröbner basis G for the ideal $\langle wf, (1-w)g\rangle$ with respect to an elimination order with $x_1, \ldots, x_n$ smaller than w. Then $\operatorname{lcm}(f,g)$ is the polynomial in G in which w does not appear. To obtain $\gcd(f,g)$, we use the Division Algorithm to compute $\gcd(f,g) = \dfrac{fg}{\operatorname{lcm}(f,g)}$.

To compute the lcm and gcd of more than 2 polynomials, we use the above method repeatedly and the fact that

$$\operatorname{lcm}(f_1, f_2, f_3) = \operatorname{lcm}(f_1, \operatorname{lcm}(f_2, f_3)), \text{ and}$$

$$\gcd(f_1, f_2, f_3) = \gcd(f_1, \gcd(f_2, f_3)).$$

(Alternatively we could note that $\langle f_1\rangle \cap \langle f_2\rangle \cap \langle f_3\rangle = \langle \operatorname{lcm}(f_1, f_2, f_3)\rangle$ and apply Exercise 2.3.8).

EXAMPLE 2.3.8. Let $f = x^2y^2 - y^2 + x^2 - 1$ and $g = xy^2 - y^2 - x + 1$ be polynomials in $\mathbb{Q}[x, y]$. To compute $\operatorname{lcm}(f, g)$, we first compute the reduced Gröbner basis G for the ideal $\langle wf, (1-w)g\rangle = \langle w(x^2y^2 - y^2 + x^2 - 1), (1 - w)(xy^2 - y^2 - x + 1)\rangle \subseteq \mathbb{Q}[x, y, w]$ using the lex term ordering with $w > x > y$ to get $G = \{x^2y^4 - x^2 - y^4 + 1, -wxy^2 + wx + wy^2 - w + xy^2 - x - y^2 + 1, 2wx^2 - 2w + x^2y^2 - x^2 - y^2 + 1\}$. Therefore $\operatorname{lcm}(f, g) = 1 - x^2 - y^4 + x^2y^4$. To compute $\gcd(f, g)$, we use the Division Algorithm to divide fg by $\operatorname{lcm}(f, g)$ to get a quotient equal to $x - 1$, and this is $\gcd(f, g)$.

Proposition 2.3.5 has another application: the computation of ideal quotients. We will discuss the geometric significance of these ideals after Proposition 2.5.1.

DEFINITION 2.3.9. *Let I and J be ideals in $k[x_1, \ldots, x_n]$. The* ideal quotient *$J\colon I$ is defined to be*

$$J\colon I = \{g \in k[x_1, \ldots, x_n] \mid gI \subseteq J\}.$$

LEMMA 2.3.10. *Let $I = \langle f_1, \ldots, f_s\rangle$ and J be ideals in $k[x_1, \ldots, x_n]$. Then*

$$J\colon I = \bigcap_{i=1}^{s} J\colon \langle f_i\rangle.$$

PROOF. If $g \in J\colon I$, then $gI \subseteq J$, so in particular $gf_i \in J$ for $i = 1, \ldots, s$, and hence $g \in \bigcap_{i=1}^{s} J\colon \langle f_i\rangle$. Conversely, if $g \in \bigcap_{i=1}^{s} J\colon \langle f_i\rangle$, then $g\langle f_i\rangle \subseteq J$ for $i = 1, \ldots, s$, and hence $gI \subseteq J$, so that $g \in J\colon I$. □

We have seen a method for computing intersections of ideals, so, in view of the above lemma, we only need to concentrate on computing $J\colon \langle f\rangle$ for a single polynomial f.

LEMMA 2.3.11. *Let J be an ideal and $f \neq 0$ be a polynomial in $k[x_1, \ldots, x_n]$. Then*

$$J\colon \langle f\rangle = \frac{1}{f}(J \cap \langle f\rangle).$$

PROOF. If $g \in \frac{1}{f}(J \cap \langle f \rangle)$, then $gf \in J$, and hence $g \in J\colon \langle f \rangle$. Conversely, if $g \in J\colon \langle f \rangle$, then $gf \in J$, and hence $gf \in J \cap \langle f \rangle$, so that $g \in \frac{1}{f}(J \cap \langle f \rangle)$. □

EXAMPLE 2.3.12. Let $g_1 = x(x+y)^2$, $g_2 = y$, $f_1 = x^2$, and $f_2 = x+y$ in $\mathbb{Q}[x,y]$. Consider the ideals $I = \langle f_1, f_2 \rangle$ and $J = \langle g_1, g_2 \rangle$. We wish to compute $J\colon I$. By Lemma 2.3.10 we have

$$J\colon I = (J\colon \langle f_1 \rangle) \bigcap (J\colon \langle f_2 \rangle),$$

and so by Lemma 2.3.11

$$J\colon I = \frac{1}{f_1}(J \cap \langle f_1 \rangle) \bigcap \frac{1}{f_2}(J \cap \langle f_2 \rangle).$$

First we compute $J \cap \langle f_1 \rangle$ by computing a Gröbner basis G_1 for the ideal $\langle wg_1, wg_2, (1-w)f_1 \rangle \subseteq \mathbb{Q}[x,y,w]$ with respect to the lex term ordering with $w > x > y$ to obtain

$$G_1 = \{x^2w - x^2, wy, x^3, x^2y\},$$

so that $\frac{1}{f_1}(J \cap \langle f_1 \rangle) = \langle x, y \rangle$. Second we compute $J \cap \langle f_2 \rangle$ by computing a Gröbner basis G_2 for the ideal $\langle wg_1, wg_2, (1-w)f_2 \rangle \subseteq \mathbb{Q}[x,y,w]$ using the same order as above, and we obtain

$$G_2 = \{wx - x - y, wy, x^3 + y^3, xy + y^2\},$$

so that $\frac{1}{f_2}(J \cap \langle f_2 \rangle) = \langle x^2 - xy + y^2, y \rangle$. Finally we compute $\langle x, y \rangle \cap \langle x^2 - xy + y^2, y \rangle$ by computing a Gröbner basis G for the ideal $\langle wx, wy, (1-w)(x^2 - xy + y^2), (1-w)y \rangle \subseteq \mathbb{Q}[x,y,w]$ with respect to the lex ordering with $w > x > y$, to obtain

$$G = \{wx, x^2, y\}.$$

Therefore $J\colon I = \langle x^2, y \rangle$.

Exercises

2.3.1. Assume that we have the lex ordering on x, y with $x > y$ and the degrevlex ordering on u, v, w with $u > v > w$. Use the elimination order with the variables x, y larger than the variables u, v, w to write the polynomial $f = xu^2vw + 3y^3u^2vw - 6xuvw - 100x^2yuv^2 + 2xu^3w - x^2yu^2w + 9xuw^2 \in k[x,y,u,v,w]$ in order of descending terms.

2.3.2. Prove Lemma 2.3.2.

2.3.3. Verify the assertion in Example 2.3.3.

2.3.4. In this exercise we generalize the concept of elimination orders. Let S and T be subsets of $\{x_1, \ldots, x_n\}$ such that $S \cup T = \{x_1, \ldots, x_n\}$ (S and T may overlap). For a power product $X = x_1^{\alpha_1} \cdots x_n^{\alpha_n} \in k[x_1, \ldots, x_n]$ we define X_S (resp. X_T) to be $\prod_{x_i \in S} x_i^{\alpha_i}$ (resp. $\prod_{x_i \in T} x_i^{\alpha_i}$). Let $<_S$ and $<_T$ be term orders on the variables in S and T respectively. We define a new order $<$ as follows: for power products X and Y in $k[x_1, \ldots, x_n]$,

$$X < Y \iff X_S <_S Y_S \text{ or } (X_S = Y_S \text{ and } X_T <_T Y_T).$$

a. Prove that $<$ is a term order.

b. Let $n = 3$ and the variables be x, y, z. Let $S = \{x, y\}$ and $T = \{x, y, z\}$. Let $<_S$ be deglex with $x < y$, and let $<_T$ be lex with $x > y > z$. Order the following power products according to $<$:

$$x^2y^2z, xy^3z, xy^3z^4, x^2y^2, x^2y^2z^3, xyz, z^5.$$

2.3.5. Let $\leq_p$ be a relation on $\mathbb{T}^n$ satisfying

- for all $X, Y \in \mathbb{T}^n$, $X \leq_p Y$ or $Y \leq_p X$;
- $\leq_p$ is reflexive and transitive;
- $1 \leq_p X$ for all $X \in \mathbb{T}^n$;
- $X \leq_p Y$ implies $XZ \leq_p YZ$, for all $X, Y, Z, \in \mathbb{T}^n$.

a. Prove that there is a term order $<$ which extends $\leq_p$. That is, if $X < Y$, then $X \leq_p Y$. [Hint: Use the idea of an elimination order to combine $\leq_p$ with any term order.]

b. Ler $\boldsymbol{u} = (u_1, \ldots, u_n) \in \mathbb{R}^n$ be a vector with non-negative coordinates. Define $\leq_p$ on $\mathbb{T}^n$ as follows. For $X = x_1^{\alpha_1} \cdots x_n^{\alpha_n}$, and $Y = x_1^{\beta_1} \cdots x_n^{\beta_n} \in \mathbb{T}^n$, $X \leq_p Y$ if and only if $\sum_{i=1}^n u_i\alpha_i \leq \sum_{i=1}^n u_i\beta_i$. Show that $\leq_p$ extends to a term order on $\mathbb{T}^n$.

c. Construct the deglex ordering using **b**.

2.3.6. Without the use of a Computer Algebra System, compute $\langle x, y\rangle \cap \langle x - 1, y\rangle \subseteq \mathbb{Q}[x, y]$.

2.3.7. Compute the intersection, $I \cap J$, where $I = \langle x^2y - z - 1, xy + y + 1\rangle$ and $J = \langle x - y, z^2 - x\rangle \subseteq \mathbb{Q}[x, y, z]$.

2.3.8. In this exercise, we extend Proposition 2.3.5. Let $I_1, I_2, \ldots, I_m$ be ideals of $k[x_1, \ldots, x_n]$. For each $i = 1, \ldots, m$, consider a new variable w_i. Let $J = \langle 1 - (w_1 + w_2 + \cdots + w_m), w_1I_1, \ldots, w_mI_m\rangle \subseteq k[x_1, \ldots, x_n, w_1, \ldots, w_m]$.

a. Prove that $I_1 \cap I_2 \cap \cdots \cap I_m = J \cap k[x_1, \ldots, x_n]$.

b. Use **a** to give a method for computing generators of $I_1 \cap I_2 \cap \cdots \cap I_m$.

c. Use **b** to compute generators for the ideal $\langle x, y\rangle \cap \langle x-1, y\rangle \cap \langle x-2, y-1\rangle \subseteq \mathbb{Q}[x, y]$.

2.3.9. Verify the assertions in Example 2.3.8.

2.3.10. In $\mathbb{Q}[x, y]$ compute $\gcd(x^3 - x^2y - 3x^2 + xy - y^2 - 3y, x^3y + xy^2 + 2x^2 + 2y)$. [Answer: $x^2 + y$.]

2.3.11. In $\mathbb{Q}[x,y]$, compute generators for $I\colon \langle 1-x^2-xy\rangle$, where I is the ideal defined in Example 2.1.8. (See Exercise 2.1.3.)

2.3.12. In $\mathbb{Q}[x,y,z]$ compute generators for $I\colon J$ and $J\colon I$ where $I = \langle x^2y-y+x, xy^2-x\rangle$ and $J = \langle x^2+y^2, x^3+y\rangle$. [Answer: $I\colon J = \langle y^2-1, x^2+xy-1\rangle$.]

2.3.13. Let $g_1x_n, \dots, g_\nu x_n, g_{\nu+1}, \dots, g_t$ be homogeneous polynomials (that is, every power product occurring in each g_i has the same total degree) where x_n does not divide any of $g_{\nu+1}, \dots, g_t$. Consider degrevlex with $x_1 > x_2 > \cdots > x_n$. Prove that if $g_1x_n, \dots, g_\nu x_n, g_{\nu+1}, \dots, g_t$ is a Gröbner basis for the ideal I, then $g_1, \dots, g_t$ is a Gröbner basis for the ideal $I\colon \langle x_n\rangle$. [Hint: See Exercise 1.4.9.]

2.3.14. Let I be an ideal of $k[x_1, \dots, x_n]$, and let $f \in k[x_1, \dots, x_n]$. We define

$$I\colon f^\infty = \bigcup_{i=1}^{\infty} I\colon \langle f^i\rangle.$$

(The ideal $I\colon f^\infty$ can be thought of as the ideal defining the points in $V(I)$ which do not lie on the hyper-surface defined by $f=0$; see the discussion following Proposition 2.5.1.)

a. Show that $I = I\colon\langle 1\rangle \subseteq I\colon\langle f\rangle \subseteq I\colon\langle f^2\rangle \subseteq \cdots$. Conclude that $I\colon f^\infty$ is an ideal of $k[x_1, \dots, x_n]$.

b. Prove that $I\colon f^\infty = I\colon\langle f^m\rangle$ for some m.

c. Let w be a new variable, and let $J = \langle I, 1-wf\rangle \subseteq k[x_1, \dots, x_n, w]$. Prove that $I\colon f^\infty = J \cap k[x_1, \dots, x_n]$. (Compare with Theorem 2.2.13.)

d. Let $\{f_1, \dots, f_s\}$ be a generating set for I and $\{g_1, \dots, g_t\}$ be a generating set for $J \cap k[x_1, \dots, x_n]$. Now write $g_i = (1-wf)h_i + \sum_{j=1}^{s} u_{ij}f_j$ for $i = 1, \dots, t$. Define $m = \max_{ij}(\deg_w(u_{ij}))$. Prove that $I\colon f^\infty = I\colon\langle f^m\rangle$.

e. Compute generators for $I\colon f^\infty$, where I and f are as in Exercise 2.2.4.

2.3.15. Let I be an ideal of $k[x_1, \dots, x_n]$. Prove that if $\{x_n^{a_1}g_1, \dots, x_n^{a_t}g_t\}$ is a Gröbner basis for I with respect to the degrevlex ordering with $x_1 > x_2 > \cdots > x_n$ with $g_1, \dots, g_t$ homogeneous and with no g_i divisible by x_n, then $\{g_1, \dots, g_t\}$ is a Gröbner basis for $I\colon x_n^\infty$. [Hint: Recall Exercise 1.4.9.]

2.3.16. In this exercise we show how to compute the generator for the radical of a principal ideal.

a. For a polynomial $f \in k[x_1, \dots, x_n]$ write $f = p_1^{\nu_1}p_2^{\nu_2}\cdots p_m^{\nu_m}$ where the polynomials $p_1, p_2, \dots, p_m$ are irreducible. Set $f^* = p_1p_2\cdots p_m$, called the *square free part* of f. Show that $\sqrt{\langle f\rangle} = \langle f^*\rangle$.

b. Let k be a field of characteristic zero. Show that in $k[x_1, \dots, x_n]$

$$f^* = \frac{f}{\gcd(f, \frac{\partial f}{\partial x_1}, \dots, \frac{\partial f}{\partial x_n})}.$$

c. Find the radical of the ideal $\langle -x^2y^2 + x^3y^2 + 2x^4y^2 - 2x^5y^2 - x^6y^2 + x^7y^2 - 2xy^3 + 4x^2y^3 - 4x^4y^3 + 2x^5y^3 - y^4 + 3xy^4 - 3x^2y^4 + x^3y^4\rangle \subseteq \mathbb{Q}[x,y]$. [Answer: $\langle x^3y - y^2 + xy^2 - xy\rangle$.]

2.3.17. In this exercise, we compute the solutions to a system of congruence equations. More precisely, let $I_1, \dots, I_m$ be distinct ideals in $k[x_1, \dots, x_n]$, and let $f_1, \dots, f_m \in k[x_1, \dots, x_n]$. We wish to compute the set of solutions $f \in k[x_1, \dots, x_n]$ of the system

$$f \equiv f_i \pmod{I_i},\ 1 \le i \le m. \tag{2.3.1}$$

As in Exercise 2.3.8, let $w_1, \dots, w_m$ be new variables, $J = \langle 1 - (w_1 + \cdots + w_m), w_1 I_1, \dots, w_m I_m \rangle$, and G be a Gröbner basis for J with respect to an elimination order with the w variables larger than the x variables. Finally, let $g = \sum_{i=1}^m w_i f_i$.

a. Prove that System (2.3.1) has a solution if and only if $g \xrightarrow{G}_+ h$, where $h \in k[x_1, \dots, x_n]$.

b. Prove that if $h = N_G(g)$ and $h \in k[x_1, \dots, x_n]$, then the set of solutions of System (2.3.1) is given by $h + \bigcap_{i=1}^m I_i$. In particular, prove that $u \in k[x_1, \dots, x_n]$ is a solution of System (2.3.1) if and only if $h = N_{G \cap k[x_1, \dots, x_n]}(u)$.

c. Show that the following system of congruence equations has a solution and compute the set of solutions.

$$\begin{aligned} f &\equiv x - 1 \pmod{\langle x, y \rangle} \\ f &\equiv x \pmod{\langle x - 1, y \rangle} \\ f &\equiv y \pmod{\langle x - 2, y - 1 \rangle}. \end{aligned}$$

Prove that if we replace $\langle x, y \rangle$ by $\langle x^2, y \rangle$, then the system above has no solution.

2.3.18. In this exercise we use the Chinese Remainder Theorem (see [**Hun**]) which states that if all the ideals I_i are maximal then System (2.3.1) always has a solution. Let $\boldsymbol{a}_1, \dots, \boldsymbol{a}_m \in k^n$ be distinct, and let $\alpha_1, \dots, \alpha_m \in k$ be given.

a. Prove that there exists $f \in k[x_1, \dots, x_n]$ such that $f(\boldsymbol{a}_i) = \alpha_i$ for $i = 1, \dots, m$. [Hint: Use Exercise 2.3.17.]

b. Give a method for computing such an f.

c. Find a polynomial f in $\mathbb{Q}[x, y]$ which satisfies $f(0, 0) = -1$, $f(1, 0) = 1$, and $f(2, 1) = 1$.

2.3.19. Let $f \in k[x_1, \dots, x_n]$ be a homogeneous polynomial. Consider the ideal $I \subseteq k[x_1, \dots, x_n]$ generated by all power products X which appear in f and such that $\operatorname{lp}(f) = X$ with respect to some term order. Prove that $f \in \sqrt{I}$. [Hint: First show that if X and Y have the same total degree, and if X divides Y, then $X = Y$. Then show that if x_i^ν appears as a term in f, then any term in f with x_i is in $\sqrt{I}$. So we may assume that no power of any single variable x_i appears in f. Use the same argument for terms in f of the form $x_i x_j$. Use Exercise 2.3.4.]

2.3.20. (Lakshman [**Lak**]) Let Q_i be a zero-dimensional ideal of $k[x_1, \dots, x_n]$ ($1 \le i \le r$) and let $<$ be a term order on $k[x_1, \dots, x_n]$. Let G_i be a Gröbner basis for Q_i, $i = 1, \dots, r$. Let $I = Q_1 \cap Q_2 \cap \dots \cap Q_r$. We will compute a Gröbner basis for I with respect to $<$ using linear algebra techniques only.

a. As in Exercise 2.2.8, write all the power products in $k[x_1, \dots, x_n]$ ordered by $<$: $1 < X_1 < X_2 < \cdots$. Modify the method of Exercise 2.2.8 to give a method for deciding whether there exists a polynomial f in I whose leading term with respect to $<$ is X_r. [Hint: $\{1 + I, X_1 + I, \dots, X_r + I\}$ is linearly dependent if and only if $\{1 + Q_i, X_1 + Q_i, \dots, X_r + Q_i\}$ is linearly dependent for each i.]

b. Use **a** to give an algorithm that inputs a Gröbner basis G_i for each Q_i with respect to $<$ and outputs a Gröbner basis for I with respect to $<$.

c. Use **b** to compute generators for the intersection $Q_1 \cap Q_2$ in the following cases:

(i) $Q_1 = \langle x, y \rangle$, $Q_2 = \langle x - 1, y \rangle \subseteq \mathbb{Q}[x, y]$.

(ii) $Q_1 = \langle xy + 1, x^2 - 1 \rangle$, $Q_2 = \langle x^2 + y, y + x \rangle \subseteq \mathbb{Q}[x, y]$. [Answer: For lex with $x > y$, $\langle y^3 - y, x + y \rangle$.]

2.3.21. (Lakshman [**Lak**]) Let I be a zero-dimensional ideal of $k[x_1, \dots, x_n]$, and let G be a Gröbner basis for I with respect to some term order $<$. Let $J = \langle h_1, \dots, h_s \rangle$ be an ideal of $k[x_1, \dots, x_n]$. We wish to compute a Gröbner basis for $I \colon J$ using linear algebra techniques only.

a. As in Exercise 2.2.8, write all the power products in $k[x_1, \dots, x_n]$ ordered by $<$: $1 < X_1 < X_2 < \cdots$. Modify the method of Exercise 2.2.8 to give a method for deciding whether there exists a polynomial f in $I \colon J$ whose leading term with respect to $<$ is X_r. [Hint: $\{1 + (I \colon J), X_1 + (I \colon J), \dots, X_r + (I \colon J)\}$ is linearly dependent if and only if $\{h_i + I, X_1 h_i + I, \dots, X_r h_i + I\}$ is linearly dependent for each i.]

b. Use **a** to give an algorithm that inputs a Gröbner basis G for I with respect to $<$ and a generating set for J and outputs a Gröbner basis for $I \colon J$ with respect to $<$.

c. Use **b** on the following example: $I = \langle 2xy - 2x - y^2 + y, y^3 - 3y^2 + 2y, x^3 - 2x^2 - x - y^2 + 3y \rangle$, $J = \langle y - x^2, x^2 - y^2 \rangle \subseteq \mathbb{Q}[x, y]$. [Answer: for lex $x > y$, $I \colon J = \langle x + y - 3, y^2 - 3y + 2 \rangle$.]

2.3.22. (Lakshman [**Lak**]) Let

$$\begin{array}{rccc} \phi\colon & k[x_1, \dots, x_n] & \longrightarrow & k[y_1, \dots, y_n] \\ & x_i & \longmapsto & \sum_{j=1}^{n} a_{ij} y_j, \end{array}$$

where $a_{ij} \in k$ be a non-singular linear transformation. For an ideal I of $k[x_1, \dots, x_n]$, note that $\phi(I)$ is an ideal of $k[y_1, \dots, y_n]$. We will consider a zero-dimensional ideal I. Let G_ϕ be a Gröbner basis for $\phi(I)$. We wish to compute a Gröbner basis for I with respect to some term order $<$ using linear algebra techniques only.

a. As in Exercise 2.2.8, write all the power products in $k[x_1, \ldots, x_n]$ ordered by $<$: $1 < X_1 < X_2 < \cdots$. Modify the method of Exercise 2.2.8 to give a method for deciding whether there exists a polynomial f in I whose leading term with respect to $<$ is X_r. [Hint: $\{1+I, X_1+I, \ldots, X_r+I\}$ is linearly dependent if and only if $\{\phi(1)+\phi(I), \phi(X_1)+\phi(I), \ldots, \phi(X_r)+\phi(I)\}$ is linearly dependent.]

b. Use **a** to give an algorithm that inputs a Gröbner basis G for $\phi(I)$ and outputs a Gröbner basis for I with respect to $<$.

c. Use **b** on the following example: Consider the map

$$\begin{array}{rccc} \phi\colon & \mathbb{Q}[x,y] & \longrightarrow & \mathbb{Q}[u,v] \\ & x & \longmapsto & 2u-v \\ & y & \longmapsto & -u+v \end{array}$$

and let $I = \langle x^3+3x^2y+3xy^2+y^3+x^2+3xy+2y^2, x^2+4xy+4y^2+2x+3y\rangle$. Use deglex with $x > y$.

2.3.23. (*)(Seidenberg [**Se**]) Let $\mathrm{char}(k) = 0$ and let I be a zero-dimensional ideal of $k[x_1, \ldots, x_n]$. Assume that for each $i = 1, \ldots, n$, I contains a square-free polynomial $g_i \in k[x_i]$.

a. Prove that $I = \sqrt{I}$. [Sketch of the proof: Use induction on n. For $n > 1$, let $g_1 = p_1 \cdots p_r$, with $p_j \in k[x_1]$ irreducible and pairwise non-associate. Prove that $I = \bigcap_{j=1}^{r}\langle I, p_j\rangle$, and so it suffices to show that $\sqrt{\langle I, p_j\rangle} = \langle I, p_j\rangle$ for $j = 1, \ldots, r$. Consider the map

$$\phi_j\colon (k[x_1])[x_2, \ldots, x_n] \longrightarrow (k[x_1]/\langle p_j\rangle)[x_2, \ldots, x_n].$$

Use Exercise 2.3.16 to observe that a polynomial f is square-free if and only if $\gcd(f, f') = 1$ and the fact that gcd's are invariant under field extension to conclude that the images of g_i, $i = 2, \ldots, n$ are square-free in $(k[x_1]/\langle p_j\rangle)[x_2, \ldots, x_n]$. By induction hypothesis, $\sqrt{\phi_j(\langle I, p_j\rangle)} = \phi_j(\langle I, p_j\rangle)$. Conclude that $\sqrt{\langle I, p_j\rangle} = \langle I, p_j\rangle$.]

b. Conclude that a zero-dimensional ideal I of $k[x_1, \ldots, x_n]$ satisfies $\sqrt{I} = I$ if and only if it contains a univariate square-free polynomial in each of the variables.

c. Use this to give an algorithm which inputs a zero-dimensional ideal I and outputs TRUE if $\sqrt{I} = I$ and FALSE otherwise.

d. Give an example that shows that part **a** is not necessarily true if $\mathrm{char}(k) \neq 0$.

e. Use the above to determine whether $I = \sqrt{I}$ where $I = \langle x^2y - x - y, xy + x + z - yz, z^2 - y\rangle \subseteq \mathbb{Q}[x, y, z]$.

2.3.24. Let $\mathrm{char}(k) = 0$ and let I be a zero-dimensional ideal of $k[x_1, \ldots, x_n]$. For $i = 1, \ldots, n$, let g_i be the monic generator of $I \cap k[x_i]$, and let g_i^* be its square-free part.

a. Use Exercise 2.3.23 to prove that $\sqrt{I} = \langle I, g_1^*, g_2^*, \dots, g_n^* \rangle$.

b. Use **a** to give an algorithm for computing generators for $\sqrt{I}$, where I is a zero-dimensional ideal.

c. Compute generators for the radical of $I = \langle x^2y - x - y, xy + x + z - xy, z^4 - 2yz^2 + yz + y^2 - xy + z - x \rangle \subseteq \mathbb{Q}[x, y, z]$.

2.3.25. Let $\text{char}(k) = 0$ and let I be a zero-dimensional ideal of $k[x_1, \dots, x_n]$. Assume that $\sqrt{I} = I$. Also, let K be an extension field of k. Let J be the ideal of $K[x_1, \dots, x_n]$ generated by I. Prove that $\sqrt{J} = J$. Give an example that shows that this is not necessarily the case if $\text{char}(k) \neq 0$. [Hint: Use Exercises 2.3.16 and 2.3.23.]

2.3.26. (*) Let $\text{char}(k) = 0$ and let I be a zero-dimensional ideal of $k[x_1, \dots, x_n]$.

a. Prove that the number of points in $V_{\overline{k}}(I)$ is less than or equal to $\dim_k(k[x_1, \dots, x_n]/I)$. [Sketch of the proof: Let G be a Gröbner basis for I and let J be the ideal of $\overline{k}[x_1, \dots, x_n]$ generated by G. Use Exercise 1.6.14 to prove that J is zero-dimensional and that

$$\dim_k(k[x_1, \dots, x_n]/I) = \dim_{\overline{k}}(\overline{k}[x_1, \dots, x_n]/J).$$

For each point $\boldsymbol{a} = (a_1, \dots, a_n) \in V_{\overline{k}}(I)$, consider the ideal $J_{\boldsymbol{a}} = \langle x_1 - a_1, \dots, x_n - a_n \rangle \subseteq \overline{k}[x_1, \dots, x_n]$. Prove that $f + J_{\boldsymbol{a}} = f(\boldsymbol{a}) + J_{\boldsymbol{a}}$ for all $f \in k[x_1, \dots, x_n]$. Consider the map

$$\begin{array}{rcl} \phi\colon \overline{k}[x_1, \dots, x_n] & \longrightarrow & \prod_{\boldsymbol{a} \in V_{\overline{k}}} \overline{k}[x_1, \dots, x_n]/J_{\boldsymbol{a}} \\ f & \longmapsto & (f(\boldsymbol{a}) + J_{\boldsymbol{a}} \mid \boldsymbol{a} \in V_{\overline{k}}). \end{array}$$

Using Exercise 2.3.18 show that ϕ is surjective. show that ϕ induces a map

$$\psi\colon \overline{k}[x_1, \dots, x_n]/J \longrightarrow \prod_{\boldsymbol{a} \in V_{\overline{k}}} \overline{k}[x_1, \dots, x_n]/J_{\boldsymbol{a}}.$$

Since $\overline{k}[x_1, \dots, x_n]/J_{\boldsymbol{a}} \cong \overline{k}$, we see that ψ is a $\overline{k}$-vector space homomorphism.]

b. Prove that if $\sqrt{I} = I$ and $\text{char}(k) = 0$, then equality holds in **a**. [Hint: If $\sqrt{I} = I$, then $\sqrt{J} = J$ by Exercise 2.3.25. Prove that $J = \bigcap_{\boldsymbol{a} \in V_{\overline{k}}(J)} J_{\boldsymbol{a}}$, and conclude that ψ above is a bijection.]

c. Compute the dimension of $\mathbb{Q}[x, y]/I$ and the number of points in $V_{\mathbb{C}}(I)$ where $I = \langle x^2y - x - y, xy + x \rangle$. [Answer: dim = 3 and there are 3 points.]

d. Compute the dimension of $\mathbb{Q}[x, y]/I$ and the number of points in $V_{\mathbb{C}}(I)$ where $I = \langle x^2y - x - y, x^2 + xy + 2x^2 + y + y^2 \rangle$. [Answer: dim = 6 and there are 3 points.]

2.4. Polynomial Maps. In this section we are interested in k-algebra homomorphisms between the polynomial rings $k[y_1, \dots, y_m]$ and $k[x_1, \dots, x_n]$. We recall that such a *k-algebra homomorphism* is a ring homomorphism

$$\phi\colon k[y_1, \dots, y_m] \longrightarrow k[x_1, \dots, x_n]$$

which is also a k-vector space linear transformation. Such a map is uniquely determined by

$$\phi\colon y_i \longmapsto f_i, \tag{2.4.1}$$

where $f_i \in k[x_1, \dots, x_n], 1 \leq i \leq m$. That is, if we let $h \in k[y_1, \dots, y_m]$, say $h = \sum_\nu c_\nu y_1^{\nu_1} \cdots y_m^{\nu_m}$, where $c_\nu \in k$, $\nu = (\nu_1, \dots, \nu_m) \in \mathbb{N}^m$, and only finitely many c_ν's are non-zero, then we have

$$\phi(h) = \sum_\nu c_\nu f_1^{\nu_1} \cdots f_m^{\nu_m} = h(f_1, \dots, f_m) \in k[x_1, \dots, x_n]. \tag{2.4.2}$$

Also, conversely, given any such assignment (2.4.1), we get a k-algebra homomorphism from formula (2.4.2).

Recall that the *kernel* of ϕ is the ideal

$$\ker(\phi) = \{h \in k[y_1, \dots, y_m] \mid \phi(h) = 0\},$$

and the *image* of ϕ is the k-subalgebra of $k[x_1, \dots, x_n]$,

$$\operatorname{im}(\phi) = \{f \in k[x_1, \dots, x_n] \mid \text{ there exists } h \in k[y_1, \dots, y_m] \text{ with } f = \phi(h)\}.$$

This subalgebra is denoted by $k[f_1, \dots, f_m]$. We know from the theory of abelian groups that

$$k[y_1, \dots, y_m]/\ker(\phi) \cong k[f_1, \dots, f_m]$$

as abelian groups under the map

$$k[y_1, \dots, y_m]/\ker(\phi) \longrightarrow k[f_1, \dots, f_m]$$

defined by

$$g + \ker(\phi) \longmapsto \phi(g).$$

This map is, in fact, a k-algebra homomorphism, as is easily seen, and thus is a k-algebra isomorphism. This is called the *First Isomorphism Theorem* for k-algebras. Another way to think of $\ker(\phi)$ is that $h \in \ker(\phi)$ if and only if $h(f_1, \dots, f_m) = 0$ and so $\ker(\phi)$ is often called the *ideal of relations* among the polynomials $f_1, \dots, f_m$.

We will use the theory of elimination presented in Section 2.3 to determine the following:

(i) The kernel of ϕ or more precisely, a Gröbner basis for the kernel of ϕ;

(ii) The image of ϕ or more precisely, an algorithm to decide whether a polynomial f is in the image of ϕ and an algorithm to decide whether ϕ is onto.

Before we give a characterization of the kernel of the map ϕ, we need a technical lemma.

LEMMA 2.4.1. *Let $a_1, a_2, \dots, a_n, b_1, b_2, \dots, b_n$ be elements of a commutative ring R. Then the element $a_1 a_2 \cdots a_n - b_1 b_2 \cdots b_n$ is in the ideal $\langle a_1 - b_1, a_2 - b_2, \dots, a_n - b_n\rangle$.*

PROOF. The proof is easily done by induction using the fact that

$$a_1a_2\cdots a_n - b_1b_2\cdots b_n = a_1(a_2\cdots a_n - b_2\cdots b_n) + b_2\cdots b_n(a_1 - b_1).$$

□

THEOREM 2.4.2. *Let* $K = \langle y_1 - f_1, \dots, y_m - f_m\rangle \subseteq k[y_1, \dots, y_m, x_1, \dots, x_n]$. *Then* $\ker(\phi) = K \cap k[y_1, \dots, y_m]$.

PROOF. Let $g \in K \cap k[y_1, \dots, y_m]$. Then

$$g(y_1, \dots, y_m) = \sum_{i=1}^{m}(y_i - f_i(x_1, \dots, x_n))h_i(y_1, \dots, y_m, x_1, \dots, x_n),$$

where $h_i \in k[y_1, \dots, y_m, x_1, \dots, x_n]$. Therefore g is zero when evaluated at $(y_1, \dots, y_m) = (f_1, \dots, f_m)$ and hence $g \in \ker(\phi)$.

Conversely, let $g \in \ker(\phi)$. We can write

$$g = \sum_{\nu} c_\nu y_1^{\nu_1}\cdots y_m^{\nu_m},$$

where $c_\nu \in k$, $\nu = (\nu_1, \dots, \nu_m) \in \mathbb{N}^m$, and only finitely many c_ν's are non-zero. Therefore, since $g(f_1, \dots, f_m) = 0$, we have

$$g = g - g(f_1, \dots, f_m) = \sum_{\nu} c_\nu(y_1^{\nu_1}\cdots y_m^{\nu_m} - f_1^{\nu_1}\cdots f_m^{\nu_m}).$$

By Lemma 2.4.1, each term in the sum above is in the ideal K, and hence $g \in K \cap k[y_1, \dots, y_m]$. □

We now have an algorithm for computing a Gröbner basis for the kernel of ϕ. We first compute a Gröbner basis G for the ideal $K = \langle y_1 - f_1, \dots, y_m - f_m\rangle$ in $k[y_1, \dots, y_m, x_1, \dots, x_n]$ with respect to an elimination order in which the x variables are larger than the y variables. The polynomials in G without any x variables form a Gröbner basis for the kernel of ϕ.

EXAMPLE 2.4.3. Let $\phi\colon \mathbb{Q}[r, u, v, w] \longrightarrow \mathbb{Q}[x, y]$ be the map defined by

$$\begin{array}{ccc} r & \longmapsto & x^4 \\ u & \longmapsto & x^3y \\ v & \longmapsto & xy^3 \\ w & \longmapsto & y^4. \end{array}$$

We first compute a Gröbner basis G for the ideal

$$K = \langle r - x^4, u - x^3y, v - xy^3, w - y^4\rangle \subseteq \mathbb{Q}[r, u, v, w, x, y]$$

with respect to the deglex term ordering on the x, y variables with $y > x$ and the degrevlex term ordering on the r, u, v, w variables with $r > u > v > w$, with an elimination order between them with the x, y variables larger than the r, u, v, w variables. We get $G = \{x^4 - r, x^3y - u, xy^3 - v, y^4 - w, yv - xw, yr - xu, y^2u -$

$x^2v, x^2y^2w - v^2, uv - rw, v^3 - uw^2, rv^2 - u^2w, yuw - xv^2, u^3 - r^2v, yu^2 - xrv\}$. Therefore a Gröbner basis for $\ker(\phi)$ is

$$G \cap k[r,u,v,w] = \{uv - rw, v^3 - uw^2, rv^2 - u^2w, u^3 - r^2v\}.$$

We now turn our attention to the second question posed at the beginning of the section, namely, the question of finding an algorithm to determine whether an element $f \in k[x_1, \ldots, x_n]$ is in the image of the map ϕ and an algorithm to determine whether ϕ is onto. This material has been adapted from D. Shannon and M. Sweedler [**ShSw**].

THEOREM 2.4.4. *Let $K = \langle y_1 - f_1, \ldots, y_m - f_m \rangle \subseteq k[y_1, \ldots, y_m, x_1, \ldots, x_n]$ be the ideal considered in Theorem 2.4.2, and let G be the reduced Gröbner basis for K with respect to an elimination order with the x variables larger than the y variables. Then $f \in k[x_1, \ldots, x_n]$ is in the image of ϕ if and only if there exists $h \in k[y_1, \ldots, y_m]$ such that $f \xrightarrow{G}_+ h$. In this case, $f = \phi(h) = h(f_1, \ldots, f_m)$.*

PROOF. Let $f \in k[x_1, \ldots, x_n]$ be in $\text{im}(\phi)$. Then $f = g(f_1, \ldots, f_m)$ for some $g \in k[y_1, \ldots, y_m]$. Consider the polynomial

$$f(x_1, \ldots, x_n) - g(y_1, \ldots, y_m) \in k[y_1, \ldots, y_m, x_1, \ldots, x_n].$$

Note that $f(x_1, \ldots, x_n) - g(y_1, \ldots, y_m) = g(f_1, \ldots, f_m) - g(y_1, \ldots, y_m)$, and hence, using Lemma 2.4.1, we see that $f(x_1, \ldots, x_n) - g(y_1, \ldots, y_m)$ is in K. Therefore, by Proposition 2.1.4, $g \xrightarrow{G}_+ h$, and $f \xrightarrow{G}_+ h$, where $h = N_G(g) = N_G(f)$. But, since $g \in k[y_1, \ldots, y_m]$, g can only be reduced by polynomials in G which have leading terms in the y variables alone. Since the x variables are larger than the y variables in our elimination order, the polynomials used to reduce g are in $k[y_1, \ldots, y_m]$. Therefore $h \in k[y_1, \ldots, y_m]$.

Conversely, let $f \xrightarrow{G}_+ h$, where $h \in k[y_1, \ldots, y_m]$. Then $f - h \in K$, so

$$\begin{aligned} f(x_1, \ldots, x_n) &- h(y_1, \ldots, y_m) \\ &= \sum_{i=1}^{m} g_i(y_1, \ldots, y_m, x_1, \ldots, x_n)(y_i - f_i(x_1, \ldots, x_n)). \end{aligned}$$

If we substitute f_i for y_i, we see that $f = h(f_1, \ldots, f_m) = \phi(h)$, and f is in the image of ϕ. □

The following Corollary shows that the result of the preceding theorem gives an algorithmic method for determining whether f is in the image of ϕ.

COROLLARY 2.4.5. *With the notation of Theorem 2.4.4, $f \in k[x_1, \ldots, x_n]$ is in the image of ϕ if and only if $N_G(f) \in k[y_1, \ldots, y_m]$.*

PROOF. If $N_G(f) \in k[y_1, \ldots, y_m]$, then, since $f \xrightarrow{G}_+ N_G(f)$, we have $f \in \text{im}(\phi)$ by the Theorem. Conversely, if $f \in \text{im}(\phi)$, then by the Theorem $f \xrightarrow{G}_+ h$

with $h \in k[y_1, \ldots, y_m]$. By Proposition 2.1.4, $N_G(f) = N_G(h)$ and as in the proof above we see that $h \in k[y_1, \ldots, y_m]$ implies $N_G(h) \in k[y_1, \ldots, y_m]$. □

EXAMPLE 2.4.6. Let ϕ: $\mathbb{Q}[u, v] \longrightarrow \mathbb{Q}[x]$ be the map defined by

$$\begin{array}{ccc} u & \longmapsto & x^4 + x \\ v & \longmapsto & x^3. \end{array}$$

We would like to determine if x^5 is in the image of ϕ. We first compute a Gröbner basis G for the ideal

$$K = \langle u - x^4 - x, v - x^3 \rangle \subseteq \mathbb{Q}[u, v, x]$$

with respect to the lex term ordering with $x > u > v$ to get

$$G = \{u^3 - v^4 - 3v^3 - 3v^2 - v, xv + x - u, xu^2 - v^3 - 2v^2 - v, x^2u - v^2 - v, x^3 - v\}.$$

We now reduce x^5 using G to get

$$x^5 \xrightarrow{x^3 - v} x^2v \xrightarrow{xv + x - u} -x^2 + xu,$$

and $-x^2 + xu$ is reduced with respect to G. Since $N_G(x^5) = -x^2 + xu \notin \mathbb{Q}[u, v]$, we have, by Corollary 2.4.5, that x^5 is not in the image of ϕ.

Now that we have an algorithm to determine whether a polynomial f is in the image of ϕ, we can determine whether ϕ is onto. We simply check whether $x_1, \ldots, x_n \in \operatorname{im}(\phi)$. In the next result we see that we can do this simply by inspecting the Gröbner basis.

THEOREM 2.4.7. *Let $K = \langle y_1 - f_1, \ldots, y_m - f_m \rangle \subseteq k[y_1, \ldots, y_m, x_1, \ldots, x_n]$ be the ideal considered in Theorem 2.4.2, and let G be the reduced Gröbner basis for K with respect to an elimination order with the x variables larger than the y variables. Then ϕ is onto if and only if for each $i = 1, \ldots, n$, there exists $g_i \in G$ such that $g_i = x_i - h_i$, where $h_i \in k[y_1, \ldots, y_m]$. Moreover, in this case, $x_i = h_i(f_1, \ldots, f_m)$.*

PROOF. Let us first assume that ϕ is onto. Also, without loss of generality, let us assume that the order is such that $x_1 < \cdots < x_n$. Then by Theorem 2.4.4, since x_1 is in the image of ϕ, there exists $h_1' \in k[y_1, \ldots, y_m]$ such that $x_1 \xrightarrow{G}_+ h_1'$. Therefore $x_1 - h_1' \in K$, and hence there exists $g_1 \in G$ such that $\operatorname{lp}(g_1)$ divides $\operatorname{lp}(x_1 - h_1') = x_1$. Therefore, since the only terms strickly smaller than x_1 are terms in the y variables alone, $g_1 = x_1 - h_1$ for some $h_1 \in k[y_1, \ldots, y_m]$. Similarly, since x_2 is in the image of ϕ, there exists $h_2' \in k[y_1, \ldots, y_m]$ such that $x_2 \xrightarrow{G}_+ h_2'$, and hence there exists $g_2 \in G$ such that $\operatorname{lp}(g_2)$ divides $\operatorname{lp}(x_2 - h_2') = x_2$. Since the only terms strictly smaller than x_2 are terms involving x_1 and the y variables only, and since G is the reduced Gröbner basis and any term involving x_1 could be reduced using $g_1 = x_1 - h_1$, we must have $g_2 = x_2 - h_2$ for some $h_2 \in k[y_1, \ldots, y_m]$. We proceed in a similar fashion for the remaining x_i's.

For the converse we first note that ϕ is onto if and only if $x_i \in \text{im}(\phi)$ for $1 \leq i \leq n$. Since $x_i - h_i \in G$, we have $x_i \xrightarrow{G}_+ h_i$. Since h_i is a polynomial in the y variables alone, we have that x_i is in the image of ϕ by Theorem 2.4.4, and hence ϕ is onto. Again by Theorem 2.4.4, we see that $x_i = h_i(f_1, \ldots, f_m)$. □

The above result gives an algorithm for determining whether the map ϕ is onto or not. We first compute the reduced Gröbner basis G for the ideal K, and, by inspection, we check whether there exists $g_i = x_i - h_i \in G$ for each $i = 1, \ldots, n$, with $h_i \in k[y_1, \ldots, y_m]$.

EXAMPLE 2.4.8. We have seen in Example 2.4.6 that the map ϕ was not onto, since x^5 was not in the image of ϕ. Also, the Gröbner basis G did not have a polynomial of the form $x - h(u, v)$. Now consider

$$\begin{array}{rccc} \phi^*\colon & \mathbb{Q}[u,v,w] & \longrightarrow & \mathbb{Q}[x] \\ & u & \longmapsto & x^4 + x \\ & v & \longmapsto & x^3 \\ & w & \longmapsto & x^5. \end{array}$$

We want to determine whether ϕ^* is onto. We first compute a Gröbner basis G^* for the ideal

$$K^* = \langle u - x^4 - x, v - x^3, w - x^5 \rangle \subseteq \mathbb{Q}[u, v, w, x],$$

with respect to the lex term ordering with $x > u > v > w$ to get

$$G^* = \{x - uv^2 + uv - u + w^2, v^5 - w^3, -uw + v^3 + v^2, -uv^3 + vw^2 + w^2,$$

$$-u^2v + v^2w + 2vw + w, u^3 - v^4 - 3v^3 - 3v^2 - v\}$$

Since we have $x - uv^2 + uv - u + w^2 \in G^*$, the map ϕ^* is onto. In fact we have $x = \phi^*(uv^2 - uv + u - w^2) = (x^4 + x)(x^3)^2 - (x^4 + x)x^3 + x^4 + x - (x^5)^2$.

We now extend the preceding results to quotient rings of polynomial rings.

DEFINITION 2.4.9. *An k-algebra is called an* affine k-algebra *if it is isomorphic as a k-algebra to $k[x_1, \ldots, x_n]/I$ for some ideal I of $k[x_1, \ldots, x_n]$.*

Clearly the polynomial ring $k[x_1, \ldots, x_n]$ is an affine k-algebra. Moreover, if $f_1, \ldots, f_m \in k[x_1, \ldots, x_n]$, then, as we saw at the beginning of this section, the image, $k[f_1, \ldots, f_m]$, of the map

$$\phi\colon k[y_1, \ldots, y_m] \longrightarrow k[x_1, \ldots, x_n]$$

which sends y_i to f_i is isomorphic to $k[y_1, \ldots, y_m]/\ker(\phi)$ and hence is an affine k-algebra.

We now want to study k-algebra homomorphisms between affine k-algebras. Let J be an ideal of $k[y_1, \ldots, y_m]$ and let I be an ideal of $k[x_1, \ldots, x_n]$. Consider a k-algebra homomorphism

$$\phi\colon k[y_1, \ldots, y_m]/J \longrightarrow k[x_1, \ldots, x_n]/I.$$

Let us assume that

$$\phi\colon y_i + J \longmapsto f_i + I.$$

We note that the map ϕ is well-defined if and only if the following condition is satisfied:

$$\text{if } J = \langle g_1, \ldots, g_t\rangle, \text{ then, for } i = 1, \ldots, t,\ g_i(f_1, \ldots, f_m) \in I.$$

This condition can be checked easily using a Gröbner basis for I and Theorem 1.6.2.

Generalizing Theorem 2.4.2 we have

THEOREM 2.4.10. *Let K be the ideal of $k[y_1, \ldots, y_m, x_1, \ldots, x_n]$ whose generators are those of I together with the polynomials $y_i - f_i, 1 \leq i \leq m$, that is, $K = \langle I, y_1 - f_1, \ldots, y_m - f_m\rangle$. Then $\ker(\phi) = K \cap k[y_1, \ldots, y_m]$ (mod J). That is, if $K \cap k[y_1, \ldots, y_m] = \langle f_1', \ldots, f_p'\rangle$, then $\ker(\phi) = \langle f_1' + J, \ldots, f_p' + J\rangle$.*

PROOF. Let $f' \in K \cap k[y_1, \ldots, y_m]$. Then we can write

$$f'(y_1, \ldots, y_m) = \sum_{i=1}^{m}(y_i - f_i(x_1, \ldots, x_n))h_i(y_1, \ldots, y_m, x_1, \ldots, x_n)$$
$$+w(y_1, \ldots, y_m, x_1, \ldots, x_n),$$

where

$$w(y_1, \ldots, y_m, x_1, \ldots, x_n) = \sum_{\nu} u_\nu(y_1, \ldots, y_m, x_1, \ldots, x_n)v_\nu(x_1, \ldots, x_n)$$

with $v_\nu \in I$, and $h_i, u_\nu \in k[y_1, \ldots, y_m, x_1, \ldots, x_n]$. Then

$$\phi(f' + J) = f'(f_1, \ldots, f_m) + I = w(f_1, \ldots, f_m, x_1, \ldots, x_n) + I = 0,$$

since

$$w(f_1, \ldots, f_m, x_1, \ldots, x_n) = \sum_{\nu} u_\nu(f_1, \ldots, f_m, x_1, \ldots, x_n)v_\nu(x_1, \ldots, x_n) \in I,$$

since each $v_\nu \in I$.

Conversely, let $f' \in k[y_1, \ldots, y_m]$ with $\phi(f'+J) = 0$. Then $f'(f_1, \ldots, f_m) \in I$. Let $f'(y_1, \ldots, y_m) = \sum_\nu c_\nu y_1^{\nu_1} \cdots y_m^{\nu_m}$, where $\nu = (\nu_1, \ldots, \nu_m) \in \mathbb{N}^m$, $c_\nu \in k$, and only finitely many c_ν are non-zero. Then

$$f'(y_1, \ldots, y_m) = (f'(y_1, \ldots, y_m) - f'(f_1, \ldots, f_m)) + f'(f_1, \ldots, f_m)$$
$$= \sum_{\nu} c_\nu(y_1^{\nu_1} \cdots y_m^{\nu_m} - f_1^{\nu_1} \cdots f_m^{\nu_m}) + f'(f_1, \ldots, f_m).$$

By Lemma 2.4.1,

$$\sum_{\nu} c_\nu(y_1^{\nu_1} \cdots y_m^{\nu_m} - f_1^{\nu_1} \cdots f_m^{\nu_m})$$

is in the ideal $\langle y_1 - f_1, \ldots, y_m - f_m\rangle$ and hence

$$f'(y_1, \ldots, y_m) \in \langle I, y_1 - f_1, \ldots, y_m - f_m\rangle = K,$$

since $f'(f_1, \ldots, f_m) \in I$. Therefore $f'(y_1, \ldots, y_m) \in K \cap k[y_1, \ldots, y_m]$. $\square$

We now prove the analog of Theorem 2.4.4.

THEOREM 2.4.11. *Let $K = \langle I, y_1 - f_1, \ldots, y_m - f_m \rangle$ be the ideal as in Theorem 2.4.10, and let G be a Gröbner basis for K with respect to an elimination order with the x variables larger than the y variables. Then $f + I \in k[x_1, \ldots, x_n]/I$ is in the image of ϕ if and only if there exists $h \in k[y_1, \ldots, y_m]$ such that $f \xrightarrow{G}_+ h$. In this case $f + I = \phi(h + J) = h(f_1, \ldots, f_m) + I$.*

PROOF. Let $f + I$ be in the image of ϕ. Then there exists $g \in k[y_1, \ldots, y_m]$ such that $f - g(f_1, \ldots, f_m) \in I$. We consider the polynomial $f(x_1, \ldots, x_n) - g(y_1, \ldots, y_m) \in k[y_1, \ldots, y_m, x_1, \ldots, x_n]$. Since $f(x_1, \ldots, x_n) - g(y_1, \ldots, y_m) = g(f_1, \ldots, f_m) - g(y_1, \ldots, y_m) + (f(x_1, \ldots, x_n) - g(f_1, \ldots, f_m))$, we have, using Lemma 2.4.1, that $f(x_1, \ldots, x_n) - g(y_1, \ldots, y_m)$ is in K. The argument proceeds as in the proof of Theorem 2.4.4 (Exercise 2.4.8).

Conversely, let $f \in k[x_1, \ldots, x_n]$ be such that $f \xrightarrow{G}_+ h$ with $h \in k[y_1, \ldots, y_m]$. Then $f - h \in K$, and hence

$$f(x_1, \ldots, x_n) - h(y_1, \ldots, y_m) =$$

$$\sum_{i=1}^{m} g_i(y_1, \ldots, y_m, x_1, \ldots, x_n)(y_i - f_i(x_1, \ldots, x_n)) + w(y_1, \ldots, y_m, x_1, \ldots, x_n),$$

where

$$w(y_1, \ldots, y_m, x_1, \ldots, x_n) = \sum_{\nu} u_\nu(y_1, \ldots, y_m, x_1, \ldots, x_n) v_\nu(x_1, \ldots, x_n)$$

with $v_\nu \in I$ and where $g_i, u_\nu \in k[y_1, \ldots, y_m, x_1, \ldots, x_n]$. If we substitute f_i for y_i, we see that $f - h(f_1, \ldots, f_m) \in I$, and hence $f + I = \phi(h + J)$. $\square$

As before we have

COROLLARY 2.4.12. *Continuing the notation of Theorem 2.4.11, we have that $f + I \in k[x_1, \ldots, x_n]/I$ is in the image of ϕ if and only if $N_G(f) \in k[y_1, \ldots, y_m]$.*

We finally determine whether the map ϕ is onto, again in a fashion similar to Theorem 2.4.7.

THEOREM 2.4.13. *Let K be the ideal as in Theorem 2.4.10, and let G be the reduced Gröbner basis for K with respect to an elimination order with the x variables larger than the y variables. Then ϕ is onto if and only if for each $i = 1, \ldots, n$ there exists a polynomial $g_i = x_i - h_i \in G$, where $h_i \in k[y_1, \ldots, y_m]$.*

PROOF. The proof is similar to the one of Theorem 2.4.7 and is left to the reader (Exercise 2.4.9). $\square$

EXAMPLE 2.4.14. Consider the following $\mathbb{Q}$-algebra homomorphism

$$\begin{array}{rccc} \phi\colon & \mathbb{Q}[u,v,w]/J & \longrightarrow & \mathbb{Q}[x,y]/I \\ & u+J & \longmapsto & x+y+I \\ & v+J & \longmapsto & x-y+I \\ & w+J & \longmapsto & 2xy+I, \end{array}$$

where $J = \langle uv - w, u^2 - v^2 - w^2 - 2w\rangle \subseteq \mathbb{Q}[u,v,w]$, and $I = \langle x^2 - 2xy, y^2\rangle \subseteq \mathbb{Q}[x,y]$. We first note that the map ϕ is well-defined, since

$$\begin{aligned} \phi(uv - w + J) &= (x+y)(x-y) - 2xy + I \\ &= x^2 - y^2 - 2xy + I = 0, \end{aligned}$$

and

$$\begin{aligned} \phi(u^2 - v^2 - w^2 - 2w + J) &= (x+y)^2 - (x-y)^2 - 4x^2y^2 - 4xy + I \\ &= -4x^2y^2 + I = 0. \end{aligned}$$

We now compute the kernel of ϕ as in Theorem 2.4.10. So let

$$K = \langle x^2 - 2xy, y^2, u - (x+y), v - (x-y), w - 2xy\rangle \subseteq \mathbb{Q}[x,y,u,v,w].$$

We compute the reduced Gröbner basis G for K with respect to the lex order with $x > y > u > v > w$ to get

$$G = \left\{u^2 - 2w, v^2, y - \frac{1}{2}u + \frac{1}{2}v, x - \frac{1}{2}u - \frac{1}{2}v, uv - w, vw, uw, w^2\right\}.$$

By Theorem 2.4.10, we have

$$\begin{aligned} \ker(\phi) &= \langle u^2 - 2w + J, v^2 + J, uv - w + J, vw + J, uw + J, w^2 + J\rangle \\ &= \langle u^2 - 2w + J, v^2 + J, vw + J, uw + J, w^2 + J\rangle = L/J, \end{aligned}$$

where $L = \langle u^2 - 2w, v^2, vw, uw, w^2, uv - w, u^2 - v^2 - w^2 - 2w\rangle$ (recall that $uv - w \in J$). Also, by Theorem 2.4.13, we see that the map ϕ is onto, since $x - \frac{1}{2}u - \frac{1}{2}v$ and $y - \frac{1}{2}u + \frac{1}{2}v$ are in G. In fact we have

$$x + I = \frac{1}{2}\phi(u + v + J) \text{ and } y + I = \frac{1}{2}\phi(u - v + J).$$

To conclude, note that, by the First Isomorphism Theorem, we have

$$(\mathbb{Q}[u,v,w]/J)/(L/J) \cong \mathbb{Q}[u,v,w]/L,$$

and hence

$$\mathbb{Q}[u,v,w]/L \cong \mathbb{Q}[x,y]/I,$$

where the isomorphism is obtained from ϕ.

Exercises

2.4.1. Without using a Computer Algebra System, use the method of Example 2.4.3 to find $\ker(\phi)$, where $\phi\colon \mathbb{Q}[u,v] \longrightarrow \mathbb{Q}[x]$ is defined by $\phi\colon u \longmapsto x^2$ and $\phi\colon v \longmapsto x^3$.

2.4.2. Compute generators for $\ker(\phi)$, where $\phi\colon \mathbb{Q}[u,v,w] \longrightarrow \mathbb{Q}[x,y]$ is defined by $\phi\colon u \longmapsto x^2+y$, $\phi\colon v \longmapsto x+y$ and $\phi\colon w \longmapsto x-y^2$. Is ϕ onto?

2.4.3. Determine whether the map ϕ is onto, where $\phi\colon \mathbb{Q}[r,u,v,w] \longrightarrow \mathbb{Q}[x,y]$ is defined by $\phi\colon r \longmapsto x^2+y$, $\phi\colon u \longmapsto x+y$, $\phi\colon v \longmapsto x-y^2$, and $\phi\colon w \longmapsto x^2+y^2$.

2.4.4. (Shannon and Sweedler [**ShSw**]) This exercise assumes that the reader is familiar with the definition of the minimal polynomial of an algebraic element over a field. Let $f, f_1, \ldots, f_m \in k[x_1,\ldots,x_n]$. Consider the ideal $K = \langle y-f, y_1-f_1, \ldots, y_m-f_m\rangle$ in $k[y,y_1,\ldots,y_m,x_1,\ldots,x_n]$. Let G be the reduced Gröbner basis for K with respect to the lex ordering with $x_1 > \cdots > x_n > y > y_1 > \cdots > y_m$. Let G_0 be the set of all polynomials in G involving only the variables $y, y_1, \ldots, y_m$ and in which y actually appears.

a. Show that f is algebraic over $k(f_1,\ldots,f_m)$ if and only if $G_0 \neq \emptyset$. [Hint: In the case where we assume that f is algebraic over $k(f_1,\ldots,f_m)$, we can find $h = \sum_\nu h_\nu(y_1,\ldots,y_m)y^\nu$ such that $h(f,f_1,\ldots,f_m)=0$ and $h_0(f_1,\ldots,f_m) \neq 0$. We have $h \xrightarrow{G}_+ 0$. Analyze this reduction.]

b. In the case that $G_0 \neq \emptyset$ let $g_0 \in G_0$ be such that $\operatorname{lp}(g_0)$ is least. Show that $g_0(y,f_1,\ldots,f_m)$ is a minimal polynomial for f over $k(f_1,\ldots,f_m)$.

2.4.5. A polynomial $f \in k[x_1,\ldots,x_n]$ is called *symmetric* if

$$f(x_1,\ldots,x_n) = f(x_{\sigma(1)},\ldots,x_{\sigma(n)}),$$

for all permutations σ of $\{1,\ldots,n\}$. In Exercise 1.4.18 we saw that the set of symmetric functions is a k-algebra generated by the following n functions:

$$\begin{aligned}
\sigma_1 &= x_1+x_2+\cdots+x_n \\
\sigma_2 &= x_1x_2+x_1x_3+\cdots+x_1x_n+x_2x_3+\cdots+x_{n-1}x_n \\
&\vdots \\
\sigma_n &= x_1x_2\cdots x_n.
\end{aligned}$$

Use Theorem 2.4.4 to give a method for deciding whether a given function $f \in k[x_1,\ldots,x_n]$ is a symmetric function. Use this method to check your answer in Exercise 1.4.18, part **d**.

2.4.6. (Shannon and Sweedler [**ShSw**]) In this exercise we extend the results of this section to maps from subrings of $k(y_1,\ldots,y_m)$ to $k(x_1,\ldots,x_n)$, where $k(y_1,\ldots,y_m)$ and $k(x_1,\ldots,x_n)$ are the fields of fractions of $k[y_1,\ldots,y_m]$ and $k[x_1,\ldots,x_n]$ respectively.

a. We consider the following map

$$\begin{array}{cccc} \phi\colon & k[y_1,\dots,y_m] & \longrightarrow & k[x_1,\dots,x_n] \\ & y_i & \longmapsto & f_i. \end{array}$$

Let $P=\ker(\phi)$. Prove that ϕ can be extended to a k-algebra homomorphism

$$\psi\colon k[y_1,\dots,y_m]_P \longrightarrow k(x_1,\dots,x_n),$$

where $k[y_1,\dots,y_m]_P = \{\frac{f}{g} \mid f,g \in k[y_1,\dots,y_m] \text{ and } g\notin P\}$. Note that $k[y_1,\dots,y_m]_P$ is a subring of $k(y_1,\dots,y_m)$. It is called the *localization* of $k[y_1,\dots,y_m]$ at P.

b. Give a method for computing generators of $\ker(\psi)$.

c. Prove the following analog of Theorem 2.4.7. Let K be the ideal of $k[y_1,\dots,y_m,x_1,\dots,x_n]$ generated by y_i-f_i, $i=1,\dots,m$. Let G be a Gröbner basis for K with respect to the lex term order with $x_1>x_2>\cdots>x_n>y_1>\cdots>y_m$. Then ψ is onto if and only if for each $i=1,\dots,n$ there exists $g_i\in G$ such that $g_i=\alpha_i x_i-\beta_i$, where $\alpha_i\notin P$, $\alpha_i\in k[y_1,\dots,y_m]$ and $\beta_i\in k[x_{i+1},\dots,x_n,y_1,\dots,y_m]$. [Hint: If x_i is in the image of ψ, then $x_i=\frac{\delta_i(f_1,\dots,f_m)}{\gamma_i(f_1,\dots,f_m)}$ for some $\delta_i,\gamma_i\in k[y_1,\dots,y_m]$ such that $\gamma_i\notin P$. Choose γ_i such that $\mathrm{lp}(\gamma_i)$ is the smallest possible. Consider $t_i=\gamma_i(y_1,\dots,y_m)x_i-\delta_i(y_1,\dots,y_m)$. Prove that $t_i\in K$. Prove that there exists a polynomial $g_i\in G$ such that $\mathrm{lp}(g_i)=y_1^{\nu_1}\cdots y_m^{\nu_m}x_i$, and so conclude that g_i is of the required form. For the converse, first prove that x_n is in the image, then x_{n-1}, etc.]

d. Consider the map $\phi\colon k[u,v]\longrightarrow k[x]$ defined by $u\longmapsto x^4+x$ and $v\longmapsto x^3$. Compute generators of P and determine whether ψ is onto.

e. Consider the map $\phi\colon \mathbb{Q}[u,v,w]\longrightarrow \mathbb{Q}[x,y]$ defined by $u\longmapsto x^2+y$, and $v\longmapsto x+y$ and $w\longmapsto x-y^2$. Compute generators of P and determine whether ψ is onto.

2.4.7. Let $\phi\colon \mathbb{Q}[u,v]\longrightarrow \mathbb{Q}[x]$ be defined by $\phi\colon u\longmapsto x^4+x^2+x$ and $\phi\colon v\longmapsto x^3-x$.

a. Using Theorem 2.4.7 show that ϕ is not onto.

b. Show that x^3 is not in the image of ϕ.

c. Show that the map ψ, corresponding to the one given in Exercise 2.4.6, is onto.

2.4.8. Complete the proof of Theorem 2.4.11.

2.4.9. Prove Theorem 2.4.13.

2.4.10. Let $\phi\colon \mathbb{Q}[u,v,w]/J\longrightarrow \mathbb{Q}[x,y]/I$ be defined by $\phi\colon u+J\longmapsto x^2+y+I$, $\phi\colon v+J\longmapsto x+y+I$, and $\phi\colon w+J\longmapsto x^3-xy^2+I$ and where $J=\langle uv-w\rangle$ and $I=\langle xy+y\rangle$.

a. Prove that ϕ is well-defined.

b. Find the kernel of ϕ.

c. Is ϕ onto?

2.4.11. For those who have the appropriate algebra skills, generalize Exercise 2.4.6 to the case of

$$\begin{array}{rccc} \phi\colon & k[y_1,\dots,y_m]/J & \longrightarrow & k[x_1,\dots,x_n]/I \\ & y_i + J & \longmapsto & f_i + I, \end{array}$$

where we assume that J, I are ideals in the appropriate rings and I is a prime ideal.

2.5. Some Applications to Algebraic Geometry. In this section we will apply the results of the previous sections to study maps between varieties. Throughout this section, for I an ideal in $k[x_1,\dots,x_n]$, we will consider the variety $V_{\overline{k}}(I) \subseteq \overline{k}^{\,n}$ as we did in Section 2.2. We will abbreviate this variety more simply by $V(I)$.

We begin by considering *projection maps*

$$\text{(2.5.1)} \qquad \begin{array}{rccc} \pi\colon & \overline{k}^{\,m+n} & \longrightarrow & \overline{k}^{\,m} \\ & (a_1,\dots,a_m,b_1,\dots,b_n) & \longmapsto & (a_1,\dots,a_m). \end{array}$$

If we apply this map to a variety V, we may not obtain a variety. For example, the variety $V(xy-1)$ projects onto the x-axis minus the origin, and this is not a variety. We are interested in finding the smallest variety containing $\pi(V)$. Before we do this we give the following general proposition.

PROPOSITION 2.5.1. *If $S \subset \overline{k}^{\,n}$, then $V(I(S))$ is the smallest variety containing S. That is, if W is any variety containing S, then $V(I(S)) \subseteq W$. This set is called the* Zariski closure *of S.*

PROOF. Let $W = V(J) \subseteq \overline{k}^{\,n}$ be a variety containing S, where J is an ideal in $k[x_1,\dots,x_n]$. Then $I(W) \subset I(S)$ and $V(I(S)) \subset V(I(W))$. But $V(I(W)) = V(\sqrt{J}) = V(J) = W$, by Theorem 2.2.5. Therefore $V(I(S)) \subseteq W$. □

As a simple example of the above proposition, consider two varieties V and W contained in $\overline{k}^{\,n}$. Then $V - W$ need not be a variety, and its Zariski closure is $V(I(V-W))$. We note that $I(V-W) = I(V)\colon I(W)$ (Exercise 2.5.2). Recall that we showed how to compute the ideal quotient in Lemmas 2.3.10 and 2.3.11.

EXAMPLE 2.5.2. Consider the varieties $V = V(x(y-z), y(x-z))$ and $W = V(y-z)$ in $\mathbb{C}^3$. Then V consists of the four lines $y = z = 0$, $x = z = 0$, $x = y = 0$, and $x = y = z$. Moreover W is the plane $y = z$ which contains just two of these lines, namely $y = z = 0$ and $x = y = z$. Thus $V - W$ consists of the union of the two lines $x = z = 0$ and $x = y = 0$ excluding the origin. We use the above method to compute the smallest variety containing $V - W$, namely $V(I(V-W))$ (although it is geometrically obvious that this variety is the union of the two lines including the origin). By the above we have $I(V-W) = I(V)\colon I(W)$. Also, it is easy to see that $I(V) = \langle x(y-z), y(x-z)\rangle$ and $I(W) = \langle y-z\rangle$, since $\sqrt{\langle x(y-z), y(x-z)\rangle} = \langle x(y-z), y(x-z)\rangle$ and $\sqrt{\langle y-z\rangle} = \langle y-z\rangle$. We compute $I(V): I(W)$ using Lemma 2.3.11 to get

$$\langle x(y-z), y(x-z)\rangle : \langle y-z\rangle = \langle x, yz\rangle.$$

Therefore the smallest variety containing $V - W$ is, as we observed above, the union of the two lines $x = z = 0$ and $x = y = 0$.

We now return to the projection map (2.5.1).

THEOREM 2.5.3. *Let I be an ideal in $k[y_1, \dots, y_m, x_1, \dots, x_n]$. The Zariski closure of $\pi(V(I))$ is $V(I \cap k[y_1, \dots, y_m])$.*

PROOF. Let $V = V(I)$ and $I_y = I \cap k[y_1, \dots, y_m]$. Let us first prove that $\pi(V) \subseteq V(I_y)$. Let $(a_1, \dots, a_m, b_1, \dots, b_n) \in V$, so that $(a_1, \dots, a_m) \in \pi(V)$. If $f \in I_y$, then $f(a_1, \dots, a_m, b_1, \dots, b_n) = 0$, since $f \in I$, and thus $f(a_1, \dots, a_m) = 0$, since f contains only y variables. Therefore $\pi(V) \subseteq V(I_y)$. In view of Proposition 2.5.1, to complete the proof of the theorem we need to show that $V(I_y) \subseteq V(I(\pi(V)))$. We first show that $I(\pi(V)) \subseteq \sqrt{I_y}$. Let $f \in I(\pi(V))$, so that $f(a_1, \dots, a_m) = 0$ for all $(a_1, \dots, a_m) \in \pi(V)$. If we view f as an element of $k[y_1, \dots, y_m, x_1, \dots, x_n]$, then $f(a_1, \dots, a_m, b_1, \dots, b_n) = 0$ for all $(a_1, \dots, a_m, b_1, \dots, b_n) \in V$. By Theorem 2.2.5, there exists an e such that $f^e \in I$. But, since f involves only $y_1, \dots, y_m$, $f^e \in I_y$, and hence $f \in \sqrt{I_y}$. Now we have $V(I_y) = V(\sqrt{I_y}) \subseteq V(I(\pi(V)))$. This complete the proof of the theorem. □

We now turn our attention to an application of Theorems 2.3.4 and 2.5.3.

We consider a map

$$\varphi \colon \overline{k}^{\,n} \longrightarrow \overline{k}^{\,m},$$

given by $\varphi(x_1, \dots, x_n) = (f_1(x_1, \dots, x_n), \dots, f_m(x_1, \dots, x_n))$, where the f_i's are in $k[x_1, \dots, x_n]$. One can think of this as a subset of $\overline{k}^{\,m}$ parametrized by $f_1, \dots, f_m$:

$$\begin{aligned} y_1 &= f_1(x_1, \dots, x_n) \\ y_2 &= f_2(x_1, \dots, x_n) \\ &\vdots \\ y_m &= f_m(x_1, \dots, x_n). \end{aligned}$$

These equations define a variety in $\overline{k}^{\,m+n}$, namely

$$V = V(y_1 - f_1, \dots, y_m - f_m).$$

(We note that V is, in fact, the graph of φ.) We want to convert the parametric equations into polynomial equations in the y variables alone. This process is called *implicitization*. One has to be careful, since parametric equations do not always define a variety and so we will find the Zariski closure of $\mathrm{im}(\varphi)$.

Consider the projection

$$\begin{array}{rccc} \pi: & \overline{k}^{\,m+n} & \longrightarrow & \overline{k}^{\,m} \\ & (y_1, \ldots, y_m, x_1, \ldots, x_n) & \longmapsto & (y_1, \ldots, y_m). \end{array}$$

Then we can see that $\pi(V)$ is the image of φ, and so we apply Theorem 2.5.3. That is we let

$$I = \langle y_1 - f_1, \ldots, y_m - f_m \rangle \subseteq k[y_1, \ldots, y_m, x_1, \ldots, x_n].$$

Then the Zariski closure of $\pi(V)$ is $V(I \cap k[y_1, \ldots, y_m])$. Note from Theorem 2.4.2 that the Zariski closure of the image of φ is defined precisely by the ideal of relations of the polynomials $f_1, \ldots, f_m$.

Therefore, to find a set of defining equations for that variety, one computes a Gröbner basis G for I with respect to an elimination order with the x variables larger than the y variables. The polynomials in G which are in the y variables only are the desired polynomials.

EXAMPLE 2.5.4. Consider the map

$$\begin{array}{rccc} \varphi: & \mathbb{C}^2 & \longrightarrow & \mathbb{C}^4 \\ & (x, y) & \longmapsto & (x^4, x^3y, xy^3, y^4) \end{array}$$

so that the parametrization is given by:

$$\begin{array}{rcl} r & = & x^4 \\ u & = & x^3y \\ v & = & xy^3 \\ w & = & y^4. \end{array} \tag{2.5.2}$$

The Gröbner basis for the ideal $I = \langle r - x^4, u - x^3y, v - xy^3, w - y^4 \rangle$ with respect to the lex term ordering with $x > y > r > u > v > w$ is $G = \{uw^2 - v^3, rw - uv, -rv^2 + u^2w, -r^2v + u^3, -y^4 + w, -xw + yv, -xv^2 + yuw, -xu + yr, xrv - yu^2, -xy^3 + v, -x^2v + y^2u, -x^2y^2r + u^2, -x^3y + u, -x^4 + r\}$. The polynomials that do not involve x or y are those that determine the smallest variety containing the solutions of the parametric equations (2.5.2):

$$uw^2 - v^3, rw - uv, -rv^2 + u^2w, -r^2v + u^3.$$

More generally, we consider maps between two varieties $V \subseteq \overline{k}^{\,n}$ and $W \subseteq \overline{k}^{\,m}$ given by polynomials; i.e.,

$$\begin{array}{rccc} \alpha: & V & \longrightarrow & W \\ & (a_1, \ldots, a_n) & \longmapsto & (f_1(a_1, \ldots, a_n), \ldots, f_m(a_1, \ldots, a_n)), \end{array}$$

where $f_1, \ldots, f_m \in k[x_1, \ldots, x_n]$.

Such a map α gives rise to a k-algebra homomorphism α^* between the affine k-algebras $k[y_1, \ldots, y_m]/I(W)$ and $k[x_1, \ldots, x_n]/I(V)$ as follows:

$$\begin{array}{rccc} \alpha^*: & k[y_1, \ldots, y_m]/I(W) & \longrightarrow & k[x_1, \ldots, x_n]/I(V) \\ & y_i + I(W) & \longmapsto & f_i + I(V). \end{array}$$

To see that α^* is well-defined, we need to show that for all $g \in I(W)$, we have $g(f_1, \ldots, f_m) \in I(V)$. But for all $(a_1, \ldots, a_n) \in V$, we have $\alpha(a_1, \ldots, a_n) \in W$, and hence

$$0 = g(\alpha(a_1, \ldots, a_n)) = g(f_1, \ldots, f_m)(a_1, \ldots, a_n),$$

as desired.

We have immediately then that α^* is a k-algebra homomorphism. Note also that if the map α is the identity map of the variety V onto itself, then the corresponding map α^* is the identity map of the affine algebra $k[x_1, \ldots, x_n]/I(V)$ onto itself.

Thus the study of the map α between varieties might be done by studying the corresponding map α^* between the corresponding affine k-algebras. We will give two examples illustrating this idea: determining the image of a variety and determining whether a given map is a variety isomorphism.

Suppose that we have a variety V in $\overline{k}^{\,n}$ and a map α into $\overline{k}^{\,m}$ given by polynomials $f_1, \ldots, f_m \in k[x_1, \ldots, x_n]$:

$$\begin{array}{cccc} \alpha\colon & V & \longrightarrow & \overline{k}^{\,m} \\ & (a_1, \ldots, a_n) & \longmapsto & (f_1(a_1, \ldots, a_n), \ldots, f_m(a_1, \ldots, a_n)). \end{array}$$

We would like to determine the Zariski closure of the image of the map α. In the case when $V = \overline{k}^{\,n}$ we did this at the begining of this section. We can find $I(\mathrm{im}(\alpha))$ by considering the corresponding map

$$\begin{array}{cccc} \alpha^*\colon & k[y_1, \ldots, y_m] & \longrightarrow & k[x_1, \ldots, x_n]/I(V) \\ & y_i & \longmapsto & f_i + I(V). \end{array}$$

PROPOSITION 2.5.5. *A polynomial* $g \in k[y_1, \ldots, y_m]$ *is in* $I(\mathrm{im}(\alpha))$ *if and only if* $g \in \ker(\alpha^*)$.

PROOF. Let $g \in I(\mathrm{im}(\alpha))$. Then for any $(a_1, \ldots, a_n) \in V$, $g(\alpha(a_1, \ldots, a_n)) = 0$, and hence $g(\alpha(x_1, \ldots, x_n)) \in I(V)$, so that $\alpha^*(g) = 0$ and $g \in \ker(\alpha^*)$. The argument is clearly reversible. □

This proposition together with Theorem 2.4.10 gives us an algorithm for computing the ideal $I(\mathrm{im}(\alpha))$, and hence for determining the smallest variety containing $\mathrm{im}(\alpha)$.

EXAMPLE 2.5.6. Let V be the variety in $\mathbb{C}^2$ defined by $x^2 + y^2 - 1$ (a circle in the x, y plane). Consider the map α given by the polynomials $f_1 = x^2$, $f_2 = y^2$, and $f_3 = xy$; i.e.

$$\begin{array}{cccc} \alpha\colon & V & \longrightarrow & \mathbb{C}^3 \\ & (x, y) & \longmapsto & (x^2, y^2, xy). \end{array}$$

The corresponding map α^* is

$$\begin{array}{rccc} \alpha^*\colon & \mathbb{C}[u,v,w] & \longrightarrow & \mathbb{C}[x,y]/\langle x^2+y^2-1\rangle \\ & u & \longmapsto & x^2+\langle x^2+y^2-1\rangle \\ & v & \longmapsto & y^2+\langle x^2+y^2-1\rangle \\ & w & \longmapsto & xy+\langle x^2+y^2-1\rangle. \end{array}$$

To find $I(\mathrm{im}(\alpha))$, we consider the ideal $K = \langle x^2+y^2-1, u-x^2, v-y^2, w-xy\rangle$ and find a Gröbner basis G for K with respect to the lex term ordering with $x > y > u > v > w$ to get $G = \{x^2+v-1, xy-w, xv-yw, xw+yv-y, y^2-v, u+v-1, v^2-v+w^2\}$. Therefore

$$I(\mathrm{im}(\alpha)) = \ker(\alpha^*) = K \cap \mathbb{C}[u,v,w] = \langle u+v-1, v^2-v+w^2\rangle.$$

Geometrically, the equation $u+v-1=0$ is the equation of a plane parallel to the w axis. The equation $v^2-v+w^2=0$ is the equation of a cylinder whose axis is parallel to the u axis. The intersection of these two surfaces is an ellipse.

The second example illustrates how α^* can be used for determining whether two varieties are isomorphic.

DEFINITION 2.5.7. *Two varieties $V \subseteq \overline{k}^n$ and $W \subseteq \overline{k}^m$ are said to be* isomorphic *over k if there are maps $\alpha\colon V \longrightarrow W$ and $\beta\colon W \longrightarrow V$ given by polynomials with coefficients in k such that $\alpha\circ\beta = id_W$ and $\beta\circ\alpha = id_V$, where id_V and id_W are the identity maps of V and W respectively.*

THEOREM 2.5.8. *The varieties $V \subseteq \overline{k}^n$ and $W \subseteq \overline{k}^m$ are isomorphic over k if and only if there exists a k-algebra isomorphism of the affine k-algebras*

$$k[y_1,\ldots,y_m]/I(W) \text{ and } k[x_1,\ldots,x_n]/I(V).$$

PROOF. First let α and β be inverse polynomial maps. Suppose that

$$\begin{array}{rccc} \alpha\colon & V & \longrightarrow & W \\ & (a_1,\ldots,a_n) & \longmapsto & (f_1(a_1,\ldots,a_n),\ldots,f_m(a_1,\ldots,a_n)), \end{array}$$

where $f_1,\ldots,f_m \in k[x_1,\ldots,x_n]$, and suppose that

$$\begin{array}{rccc} \beta\colon & W & \longrightarrow & V \\ & (b_1,\ldots,b_m) & \longmapsto & (g_1(b_1,\ldots,b_m),\ldots,g_n(b_1,\ldots,b_m)), \end{array}$$

where $g_1,\ldots,g_n \in k[y_1,\ldots,y_m]$. Note that the map $(\beta\circ\alpha)^*$ is defined by $x_i \longmapsto g_i(f_1,\ldots,f_m)$ while the map $\alpha^*\circ\beta^*$ is defined by

$$x_i \longmapsto g_i \longmapsto g_i(f_1,\ldots,f_m),$$

and so, $(\beta\circ\alpha)^* = \alpha^*\circ\beta^*$. Now, since $\beta\circ\alpha = id_V$, $(\beta\circ\alpha)^* = id_{k[x_1,\ldots,x_n]/I(V)}$ and hence $\alpha^*\circ\beta^*$ is also the identity of $k[x_1,\ldots,x_n]/I(V)$ onto itself. Similarly, we have that $\beta^*\circ\alpha^*$ is the identity of $k[y_1,\ldots,y_m]/I(W)$ onto itself. Therefore α^* is a k-algebra isomorphism.

For the converse, we assume that we have a k-algebra isomorphism

$$\Theta\colon k[y_1,\dots,y_m]/I(W) \longrightarrow k[x_1,\dots,x_n]/I(V).$$

We will see that $\Theta = \alpha^*$ for a map α between the varieties V and W which is given by polynomials and such that α^{-1} exists and is also given by polynomials.

Let $\Theta(y_i + I(W)) = f_i + I(V)$, where $f_i \in k[x_1,\dots,x_n]$ for $i = 1,\dots,m$, and let $\Theta^{-1}(x_j + I(V)) = g_j + I(W)$, where $g_j \in k[y_1,\dots,y_m]$ for $j = 1,\dots,n$. Consider the maps

$$\begin{array}{rccl} \alpha\colon & V & \longrightarrow & W \\ & (a_1,\dots,a_n) & \longmapsto & (f_1(a_1,\dots,a_n),\dots,f_m(a_1,\dots,a_n)), \end{array}$$

and

$$\begin{array}{rccl} \beta\colon & W & \longrightarrow & V \\ & (b_1,\dots,b_m) & \longmapsto & (g_1(b_1,\dots,b_m),\dots,g_n(b_1,\dots,b_m)). \end{array}$$

It is readily seen that α maps V into W, β maps W into V, and that α and β are inverse maps. □

Therefore, to determine variety isomorphism, we need to check whether α^* is a k-algebra isomorphism. We have seen in Theorem 2.4.10 how to compute the kernel of α^* and in Theorem 2.4.13 how to determine whether α^* is onto.

EXAMPLE 2.5.9. Consider the variety $V \subseteq \mathbb{C}^2$ defined by the equation $x^2 - yx + 1$ in the x, y plane. Also, consider the variety $W \subseteq \mathbb{C}^2$ defined by the equation $u^4 + u^3 + 2u^2v + v^2 + uv + 1$ in the u, v plane. Finally consider the map

$$\begin{array}{rccl} \alpha\colon & V & \longrightarrow & W \\ & (x,y) & \longmapsto & (y, -y^2 - x). \end{array}$$

We will show that this gives an isomorphism of the varieties V and W.

First we show that α maps V into W. So let $(x, y) \in V$. Then if we replace u and v by y and $-y^2 - x$ respectively in the equation defining W we get

$$y^4 + y^3 + 2y^2(-y^2 - x) + (-y^2 - x)^2 + y(-y^2 - x) + 1 = x^2 - xy + 1 = 0,$$

since $(x, y) \in V$.

Now consider the corresponding map

$$\begin{array}{rccl} \alpha^*\colon & \mathbb{C}[u,v]/J & \longrightarrow & \mathbb{C}[x,y]/I \\ & f + J & \longmapsto & f(y, -y^2 - x) + I, \end{array}$$

where $J = \langle u^4 + u^3 + 2u^2v + v^2 + uv + 1\rangle$ and $I = \langle x^2 - yx + 1\rangle$ (see Exercise 2.3.16 part **a** to see why $I = I(V)$ and $J = J(W)$). Let $K = \langle x^2 - yx + 1, u - y, v + y^2 + x\rangle$ be the ideal in $\mathbb{C}[u, v, x, y]$ as in Theorem 2.4.10. We compute a Gröbner basis for K with respect to the lex term ordering with $x > y > u > v$ to get

$$G = \{x + u^2 + v, y - u, u^4 + u^3 + 2u^2v + uv + v^2 + 1\}.$$

Thus $K \cap k[u, v] = \langle u^4 + u^3 + 2u^2v + uv + v^2 + 1\rangle = J$, and hence $\ker(\alpha^*) = 0$ by Theorem 2.4.10 and so α^* is one to one. Also, since $x + u^2 + v$ and $y - u$

are in G, the map α^* is onto by Theorem 2.4.13. Therefore α^* is a $\mathbb{C}$-algebra isomorphism and, so by Theorem 2.5.8, α gives an isomorphism of the varieties V and W. Note that the inverse map is given by $\alpha^{-1}(u, v) = (-u^2 - v, u)$, for $(u, v) \in W$.

If the reader is interested in studying further the ideas presented in this section we strongly recommend the book of Cox, Little, and O'Shea [**CLOS**].

Exercises

2.5.1. Show that in $\mathbb{C}^3$, if the plane defined by $x = 1$ is removed from the variety $V = V(xy^2 + xz^2 - xy - y^2 - z^2 + y, x^2 + xy - 2x - y + 1)$, the Zariski closure of what remains is an ellipse. Conclude that V is the union of this ellipse and the given plane.

2.5.2. Let V and W be varieties in $\overline{k}^n$. Prove that $I(V - W) = I(V) : I(W)$.

2.5.3. **a.** Find the equation in $\mathbb{C}^2$ for the curve parametrized by $x = t^3, y = t^2 + 1$.

b. Find the equation in $\mathbb{C}^2$ for the curve parametrized by $x = t^3 + 1, y = t^2$.

c. Find the ideal for the intersection of these two varieties and then determine all points on this intersection.

d. Do part **c** by solving the equations directly.

2.5.4. Show by the method of this section that the variety in $\mathbb{C}^3$ parametrized by $x = u + uv + w, y = u + v^2 + w^2, z = u^2 + v$ is all of $\mathbb{C}^3$. [Hint: If you try to compute this example using lex, your computer may not be able to complete the computation. However, if you use deglex on the u, v, w variables and also on the x, y, z variables with an elimination order between them, you should encounter no difficulties.]

2.5.5. Consider the variety V parametrized by $x = t^3, y = t^4, z = t^5$ in $\mathbb{C}^3$.

a. Show that $I(V) = \langle y^5 - z^4, -y^2 + xz, xy^3 - z^3, x^2y - z^2, x^3 - yz\rangle$. [Hint: Use lex with $x > y > z$.]

b. Verify that also $I(V) = \langle xz - y^2, x^3 - yz, x^2y - z^2\rangle$.

c. Show that the tangent variety of V is parametrized by $x = t^3 + 3t^2u, y = t^4 + 4t^3u, z = t^5 + 5t^4u$. [Hint: The tangent variety is defined to be the union of all the tangent lines of V. So this exercise is done using elementary multivariable calculus.]

d. Compute generators for the ideal of the tangent variety of V. [Answer: $\langle 15x^4y^2 - 48y^5 - 16x^5z + 80xy^3z - 30x^2yz^2 - z^4\rangle$.]

2.5.6. Let V be the variety in $\mathbb{C}^3$ defined by $x^2 + y^2 - z^2 = 0$ and $x^3 + y = 0$. Define $\alpha\colon V \longrightarrow \mathbb{C}^4$ by $(a, b, c) \longmapsto (a^2, a + b, c^2 + a, c)$. Find the ideal of the image of α.

2.5.7. In $\mathbb{C}[x, y, z]$ let $J = \langle -2y - y^2 + 2z + z^2, 2x - yz - z^2\rangle$ and in $\mathbb{C}[u, v]$ let $I = \langle uv + v\rangle$. Define the map $\alpha\colon V(I) \longrightarrow V(J)$ by $(a, b) \longmapsto (a^2 + b, a + b, a - b)$. Prove that α defines an isomorphism between $V(I)$ and $V(J)$ (you may assume that $I(V(J)) = J$).

2.6. Minimal Polynomials of Elements in Field Extensions. In this section we will use the results of Section 2.4 to find the minimal polynomial of an element algebraic over a field k. We will assume that the reader is familiar with the most elementary facts about field extensions [**Go, He**]. The results of this section will not be used in the remainder of the book and may be skipped.

Let $k \subseteq K$ be a field extension. Recall that if $\alpha \in K$ is algebraic over k, then the *minimal polynomial* of α over k is defined to be the monic polynomial p in one variable, with coefficients in k, of smallest degree such $p(\alpha) = 0$. Alternatively, considering the k-algebra homomorphism $\phi\colon k[x] \longrightarrow k(\alpha)$ defined by $x \longmapsto \alpha$, we have that $\ker(\phi) = \langle p \rangle$. Moreover, the map ϕ is onto, since α is algebraic, and so

$$k[x]/\langle p \rangle \cong k(\alpha), \tag{2.6.1}$$

under the map defined by $x + \langle p \rangle \longmapsto \alpha$.

We first consider the case where $K = k(\alpha)$, with α algebraic over k, and our goal is to compute the minimal polynomial of any $\beta \in K$. We note that in order to compute in $k(\alpha)$ it suffices, by Equation (2.6.1), to compute in the affine k-algebra $k[x]/\langle p \rangle$. We assume that we know the minimal polynomial p of α.

THEOREM 2.6.1. *Let $k \subseteq K$ be a field extension, and let $\alpha \in K$ be algebraic over k. Let $p \in k[x]$ be the minimal polynomial of α over k. Let $0 \neq \beta \in k(\alpha)$, say*

$$\beta = \frac{a_0 + a_1\alpha + \cdots + a_n\alpha^n}{b_0 + b_1\alpha + \cdots + b_m\alpha^m},$$

where $a_i, b_j \in k, 0 \leq i \leq n, 0 \leq j \leq m$. Let $f(x) = a_0 + a_1x + \cdots + a_nx^n$ and $g(x) = b_0 + b_1x + \cdots + b_mx^m$ be the corresponding polynomials in $k[x]$. Consider the ideal $J = \langle p, gy - f \rangle$ of $k[x, y]$. Then the minimal polynomial of β over k is the monic polynomial that generates the ideal $J \cap k[y]$.

Note that $J \cap k[y]$ is generated by a single polynomial, since this is true for every ideal in $k[y]$ (i.e. $k[y]$ is a principal ideal domain).

PROOF. Note that since $k[x]/\langle p \rangle$ is a field, and $g(\alpha) \neq 0$ (it is the denominator of β), there is a polynomial $\ell \in k[x]$ such that $g\ell \equiv 1 \pmod{\langle p \rangle}$, that is $g\ell - 1 \in \langle p \rangle$. Let $h = f\ell$. Note that $h(\alpha) = \beta$. Now consider ϕ, the composition of the affine algebra homomorphisms

$$\begin{array}{rccccc} \phi\colon & k[y] & \longrightarrow & k[x]/\langle p \rangle & \xrightarrow{\cong} & k(\alpha) \\ & y & \longmapsto & h + \langle p \rangle & \longmapsto & \beta. \end{array}$$

Note that q is in the kernel of ϕ if and only if $q(\beta) = 0$. Therefore to find the minimal polynomial of β, we find the generator of the kernel of ϕ. By Theorem 2.4.10, the kernel of the map ϕ is $\langle p, y-h \rangle \cap k[y]$. Therefore it suffices to show that $\langle p, y - h \rangle = \langle p, gy - f \rangle$. First note that $y - h = y - f\ell \equiv \ell(gy - f) \pmod{\langle p \rangle}$, and hence $y - h \in \langle p, gy - f\rangle$ so that $\langle p, y - h\rangle \subseteq \langle p, gy - f\rangle$. Conversely, $gy - f \equiv g(y - f\ell) = g(y - h) \pmod{\langle p\rangle}$. Therefore $gy - f \in \langle p, y - h\rangle$, and $\langle p, gy - f\rangle \subseteq \langle p, y - h\rangle$. □

The above result gives an algorithm for finding the minimal polynomial of an element β in $k(\alpha)$: given α and β as in the theorem, we compute the reduced Gröbner basis G for the ideal $\langle p, gy - f\rangle$ of $k[x, y]$ with respect to the lex term ordering with $x > y$. The polynomial in G which is in y alone is the minimal polynomial of β.

EXAMPLE 2.6.2. Consider the extension field $\mathbb{Q}(\alpha)$ of $\mathbb{Q}$, where α is a root of the irreducible polynomial $x^5 - x - 2$. Now consider the element $\beta = \frac{1-\alpha-2\alpha^3}{\alpha} \in \mathbb{Q}(\alpha)$. We wish to find the minimal polynomial of β. We consider the ideal $J = \langle x^5 - x - 2, xy + 2x^3 + x - 1\rangle$ in $\mathbb{Q}[x, y]$. We compute the reduced Gröbner basis G for J with respect to the lex term ordering with $x > y$ to obtain

$$G = \left\{x - \frac{1438}{45887}y^4 - \frac{2183}{45887}y^3 + \frac{10599}{45887}y^2 - \frac{8465}{45887}y - \frac{101499}{45887},\right.$$

$$\left. y^5 + \frac{11}{2}y^4 + 4y^3 - 5y^2 + 95y + 259\right\}.$$

Therefore the minimal polynomial of β is $y^5 + \frac{11}{2}y^4 + 4y^3 - 5y^2 + 95y + 259$.

This technique can be extended to a more general setting of field extensions of the form $K = k(\alpha_1, \dots, \alpha_n)$. For this we need the following notation which we use for the remainder of this section. For $i = 2, \dots, n$ and $p \in k(\alpha_1, \dots, \alpha_{i-1})[x_i]$, we let $\overline{p}$ be any polynomial in $k[x_1, \dots, x_i]$ such that $\overline{p}(\alpha_1, \dots, \alpha_{i-1}, x_i) = p$. We note that $\overline{p}$ is not uniquely defined, but every application we will make of $\overline{p}$ will not depend on the particular choice we have made.

We now determine the minimal polynomial of any element β of $k(\alpha_1, \dots, \alpha_n)$ using the following result, which is similar to Theorem 2.6.1.

THEOREM 2.6.3. *Let $K = k(\alpha_1, \dots, \alpha_n)$ be an algebraic extension of k. For $i = 1, \dots, n$, let $p_i \in k(\alpha_1, \dots, \alpha_{i-1})[x_i]$ be the minimal polynomial of α_i over $k(\alpha_1, \dots, \alpha_{i-1})$. Let $\beta \in k(\alpha_1, \dots, \alpha_n)$, say*

$$\beta = \frac{f(\alpha_1, \dots, \alpha_n)}{g(\alpha_1, \dots, \alpha_n)},$$

where $f, g \in k[x_1, \dots, x_n]$. Consider the ideal $J = \langle \overline{p}_1, \dots, \overline{p}_n, gy - f\rangle$ contained in $k[x_1, \dots, x_n, y]$. Then the minimal polynomial of β over k is the monic polynomial that generates the ideal $J \cap k[y]$.

PROOF. We first show that

$$k[x_1, \dots, x_n]/\langle \overline{p}_1, \dots, \overline{p}_n\rangle \cong k(\alpha_1, \dots, \alpha_n)$$

using the map

$$\begin{array}{rccc}\phi_n\colon & k[x_1,\dots,x_n] & \longrightarrow & k(\alpha_1,\dots,\alpha_n)\\ & x_i & \longmapsto & \alpha_i.\end{array}$$

We note that ϕ_n is onto, since $\alpha_1,\dots,\alpha_n$ are algebraic over k. It remains to show that $\ker(\phi_n) = \langle \overline{p}_1,\dots,\overline{p}_n\rangle$. For this we use induction on n. The case $n=1$ is Equation (2.6.1). The fact that $\overline{p}_1,\dots,\overline{p}_n \in \ker(\phi_n)$ is immediate. Now let $f \in k[x_1,\dots,x_n]$ be such that $f(\alpha_1,\dots,\alpha_n) = 0$. Let

$$h(x_n) = f(\alpha_1,\dots,\alpha_{n-1},x_n) \in k(\alpha_1,\dots,\alpha_{n-1})[x_n],$$

and note that $h(\alpha_n) = 0$. Therefore p_n divides h, by definition of p_n. Say $h = p_n\ell_n$, for some $\ell_n \in k(\alpha_1,\dots,\alpha_{n-1})[x_n]$. Consider $f - \overline{p}_n\overline{\ell}_n \in k[x_1,\dots,x_n]$ and write

$$f - \overline{p}_n\overline{\ell}_n = \sum_\nu g_\nu(x_1,\dots,x_{n-1})x_n^\nu.$$

Then, since

$$(f - \overline{p}_n\overline{\ell}_n)(\alpha_1,\dots,\alpha_{n-1},x_n) = h - p_n\ell_n = 0,$$

we see that for all ν, $g_\nu(\alpha_1,\dots,\alpha_{n-1}) = 0$. Therefore $g_\nu(x_1,\dots,x_{n-1})$ is in the kernel of

$$\phi_{n-1}\colon k[x_1,\dots,x_{n-1}] \longrightarrow k(\alpha_1,\dots,\alpha_{n-1}),$$

and hence

$$g_\nu(x_1,\dots,x_{n-1}) \in \langle \overline{p}_1,\dots,\overline{p}_{n-1}\rangle,$$

by induction. Thus $f - \overline{p}_n\overline{\ell}_n \in \langle \overline{p}_1,\dots,\overline{p}_{n-1}\rangle$, and $f \in \langle \overline{p}_1,\dots,\overline{p}_n\rangle$. Thus $\ker(\phi_n) = \langle \overline{p}_1,\dots,\overline{p}_n\rangle$ as desired. The proof now proceeds as in Theorem 2.6.1. □

EXAMPLE 2.6.4. Consider the field extension $\mathbb{Q} \subseteq \mathbb{Q}(\sqrt{2},\sqrt[3]{5})$. The minimal polynomial of $\sqrt{2}$ over $\mathbb{Q}$ is $p_1 = x_1^2 - 2 \in \mathbb{Q}[x_1]$ and the minimal polynomial of $\sqrt[3]{5}$ over $\mathbb{Q}(\sqrt{2})$ is $p_2 = x_2^3 - 5 \in \mathbb{Q}(\sqrt{2})[x_2]$. We wish to find the minimal polynomial of $\sqrt{2}+\sqrt[3]{5}$. We compute the reduced Gröbner basis G for the ideal

$$J = \langle \overline{p}_1, \overline{p}_2, y - (x_1+x_2)\rangle = \langle x_1^2 - 2, x_2^3 - 5, y - (x_1+x_2)\rangle \subseteq \mathbb{Q}[x_1,x_2,y]$$

with respect to the lex term ordering wih $x_1 > x_2 > y$ to obtain

$$\begin{aligned} G = \{&1187x_1 - 48y^5 - 45y^4 + 320y^3 + 780y^2 - 735y + 1820,\\ &1187x_2 + 48y^5 + 45y^4 - 320y^3 - 780y^2 - 452y - 1820,\\ &y^6 - 6y^4 - 10y^3 + 12y^2 - 60y + 17\}.\end{aligned}$$

Therefore the minimal polynomial of $\sqrt{2}+\sqrt[3]{5}$ over $\mathbb{Q}$ is $y^6 - 6y^4 - 10y^3 + 12y^2 - 60y + 17$.

We also see that $\sqrt{2}+\sqrt[3]{5}$ has degree 6 over $\mathbb{Q}$ and hence $\mathbb{Q}(\sqrt{2}+\sqrt[3]{5}) = \mathbb{Q}(\sqrt{2},\sqrt[3]{5})$.

EXAMPLE 2.6.5. The minimal polynomial of $\sqrt[4]{2}$ over $\mathbb{Q}$ is $p_1 = x_1^4 - 2 \in \mathbb{Q}[x_1]$. All the roots of this polynomial generate the extension $\mathbb{Q}(\sqrt[4]{2}, i)$, where i the complex number such that $i^2 = -1$. We consider the element $\beta = 1 + \frac{i}{\sqrt[4]{2}}$. We wish to find the minimal polynomial of β over $\mathbb{Q}$. First, the minimal polynomial of i over $\mathbb{Q}(\sqrt[4]{2})$ is $p_2 = x_2^2 + 1$. Then $\overline{p}_1 = p_1$, and $\overline{p}_2 = p_2$. We consider the ideal $J = \langle x_1^4 - 2, x_2^2 + 1, x_1 y - (x_1 + x_2)\rangle$ of $\mathbb{Q}[x_1, x_2, y]$. The reduced Gröbner basis for J with respect to the lex term ordering with $x_1 > x_2 > y$ is computed to be

$$G = \left\{x_1 - 2y^3x_2 + 6y^2x_2 - 6yx_2 + 2x_2, x_2^2 + 1, y^4 - 4y^3 + 6y^2 - 4y + \frac{1}{2}\right\}.$$

Therefore the minimal polynomial of β is $y^4 - 4y^3 + 6y^2 - 4y + \frac{1}{2}$.

Alternatively, $\mathbb{Q}(\sqrt[4]{2}, i) = \mathbb{Q}(\sqrt[4]{2}, i\sqrt[4]{2})$. The minimal polynomial of $i\sqrt[4]{2}$ over $\mathbb{Q}(\sqrt[4]{2})$ is $p_2 = x_2^2 + \sqrt{2}$. Thus $\overline{p}_2 = x_2^2 + x_1^2$. We again want to compute the minimum polynomial of $\beta = 1 + \frac{i}{\sqrt[4]{2}}$. We consider the ideal $J = \langle x_1^4 - 2, x_2^2 + x_1^2, x_1^2 y - (x_1^2 + x_2)\rangle$. The reduced Gröbner basis for J with respect to the lex term ordering with $x_1 > x_2 > y$ is computed to be

$$G = \left\{x_1^2 + 2y^2 - 4y + 2, x_2 + 2y^3 - 6y^2 + 6y - 2, y^4 - 4y^3 + 6y^2 - 4y + \frac{1}{2}\right\}.$$

So we obtain the same result as before.

Since the degree of β over $\mathbb{Q}$ is 4, and since the degree of $\mathbb{Q}(\sqrt[4]{2}, i)$ over $\mathbb{Q}$ is 8, we see that $\mathbb{Q}(\beta)$ is a subfield of $\mathbb{Q}(\sqrt[4]{2}, i)$, but is not equal to it.

In the preceding two examples, we used a degree argument for deciding whether $k(\beta)$ is equal to $k(\alpha_1, \ldots, \alpha_n)$. We will give another method for determining this which has the added advantage of expressing the α_i's in terms of β. This algorithm is a consequence of the following theorem.

THEOREM 2.6.6. *Let $\alpha_1, \ldots, \alpha_n$ and $\overline{p}_1, \ldots, \overline{p}_n$ be as in Theorem 2.6.3. Let $\beta \in k(\alpha_1, \ldots, \alpha_n)$, say $\beta = \frac{f(\alpha_1, \ldots, \alpha_n)}{g(\alpha_1, \ldots, \alpha_n)}$, with $f, g \in k[x_1, \ldots, x_n]$. Let J be as in Theorem 2.6.3, and let G be the reduced Gröbner basis for J with respect to an elimination order with the x variables larger than y. Then $k(\alpha_1, \ldots, \alpha_n) = k(\beta)$ if and only if, for each $i = 1, \ldots, n$, there is a polynomial $g_i \in G$ such that $g_i = x_i - h_i$, for some $h_i \in k[y]$. Moreover, in this case, $\alpha_i = h_i(\beta)$.*

PROOF. Let $I = \langle \overline{p}_1, \ldots, \overline{p}_n\rangle$, and let $\ell \in k[x_1, \ldots, x_n]$ be such that $g\ell - 1 \in I$. Set $h = f\ell$, and note that $h(\alpha_1, \ldots, \alpha_n) = \beta$. Consider

$$\begin{array}{ccccc} k[y] & \xrightarrow{\phi} & k[x_1, \ldots, x_n]/I & \xrightarrow{\cong} & k(\alpha_1, \ldots, \alpha_n) \\ y & \longmapsto & h + I & \longmapsto & \beta. \end{array}$$

Then $k(\alpha_1, \ldots, \alpha_n) = k(\beta)$ if and only if ϕ is onto. We conclude by using Theorem 2.4.13 and the fact that $J = \langle I, y - h\rangle$. $\square$

EXAMPLE 2.6.7. In Example 2.6.4, the Gröbner basis G contained the following polynomials

$$1187x_1 - 48y^5 - 45y^4 + 320y^3 + 780y^2 - 735y + 1820,$$

$$1187x_2 + 48y^5 + 45y^4 - 320y^3 - 780y^2 - 452y - 1820.$$

This gives another proof that $\mathbb{Q}(\sqrt{2} + \sqrt[3]{5}) = \mathbb{Q}(\sqrt{2}, \sqrt[3]{5})$. Moreover, we have

(2.6.2)
$$\sqrt{2} = \frac{1}{1187}\Big(48(\sqrt{2} + \sqrt[3]{5})^5 + 45(\sqrt{2} + \sqrt[3]{5})^4 - 320(\sqrt{2} + \sqrt[3]{5})^3 \\ -780(\sqrt{2} + \sqrt[3]{5})^2 + 735(\sqrt{2} + \sqrt[3]{5}) - 1820\Big),$$

and

(2.6.3)
$$\sqrt[3]{5} = \frac{1}{1187}\Big(-48(\sqrt{2} + \sqrt[3]{5})^5 - 45(\sqrt{2} + \sqrt[3]{5})^4 + 320(\sqrt{2} + \sqrt[3]{5})^3 \\ +780(\sqrt{2} + \sqrt[3]{5})^2 + 452(\sqrt{2} + \sqrt[3]{5}) + 1820\Big).$$

However, in Example 2.6.5, no polynomial of the form $x_1 - h_1$ exists in G with $h_1 \in k[y]$, and this gives another proof that $\mathbb{Q}(\beta) \neq \mathbb{Q}(\sqrt[4]{2}, i)$.

Finally, we also note that in Example 2.6.2 we have $\mathbb{Q}\left(\frac{1-\alpha-\alpha^3}{\alpha}\right) = \mathbb{Q}(\alpha)$ and

$$\alpha = \frac{1438}{45887}\beta^4 + \frac{2183}{45887}\beta^3 - \frac{10599}{45887}\beta^2 + \frac{8465}{45887}\beta + \frac{101499}{45887},$$

where $\beta = \frac{1-\alpha-\alpha^3}{\alpha}$.

Exercises

2.6.1. Compute the minimal polynomial of the following over $\mathbb{Q}$.
  a. $\sqrt[3]{2} + \sqrt[3]{4} + 5$.
  b. $\dfrac{\sqrt{2} + 7}{\sqrt[4]{2} + 1}$.
  c. $\frac{\alpha^2+\alpha}{\alpha+3}$, where $\alpha^3 - \alpha - 1 = 0$.

2.6.2. Compute the minimal polynomial of the following over $\mathbb{Q}$.
  a. $\sqrt{2} + \sqrt[3]{2} + 5$.
  b. $\sqrt[3]{2} + \sqrt[4]{5} + 5$.
  c. $\sqrt{2} + \sqrt{3} + \sqrt{5}$.

2.6.3. Show that $\mathbb{Q}(\alpha_1, \alpha_2) = \mathbb{Q}(\beta)$ and express α_1 and α_2 in terms of β, where $\alpha_1^3 - \alpha_1 - 1 = 0$, $\alpha_2^2 = 5$ and $\beta = \dfrac{\alpha_1\alpha_2 + 1}{\alpha_1 + \alpha_2}$.

2.6.4. In Theorem 2.6.3 we required that $p_{i+1} \in k(\alpha_1, \dots, \alpha_i)[x_{i+1}]$ be the minimal polynomial of α_{i+1} over $k(\alpha_1, \dots, \alpha_i)$. The minimal polynomial of α_{i+1} over k would not do as the following example shows. Let $K =$

$\mathbb{Q}(\sqrt[4]{2}, i\sqrt[4]{2})$ and $\beta = \sqrt[4]{2} + i\sqrt[4]{2}$. We wish to find the minimal polynomial of β.

a. Use Theorem 2.6.3 to compute the minimal polynomial of β over k.

b. Instead of the minimal polynomial of $i\sqrt[4]{2}$ over $\mathbb{Q}(\sqrt[4]{2})$, use the minimal polynomial of $i\sqrt[4]{2}$ over $\mathbb{Q}$ in the method described in Theorem 2.6.3, and see that, now, the method does not give the correct minimal polynomial of β over $\mathbb{Q}$.

2.6.5. Find, explicitly, the quadratic subfield of $\mathbb{Q}(\zeta)$ where ζ is a primitive 7th root of unity. [Hint: Using the method of the text, note that the minimal polynomial of ζ is $\frac{x^7-1}{x-1}$ and that this quadratic subfield is the fixed field of the automorphism defined by $\zeta \longmapsto \zeta^2$.]

2.6.6. Complete the proof of Theorem 2.6.3.

2.7. The 3-Color Problem. In this section we want to illustrate how one can apply the technique of Gröbner bases to solve a well-known problem in graph theory: determining whether a given graph can be 3-colored. (The same technique would work for any coloring.) This material is based on a portion of D. Bayer's thesis [**Ba**]. The material in this section will not be used elsewhere in the text and may be skipped.

Let us first state the problem precisely. We are given a graph $\mathcal{G}$ with n vertices with at most one edge between any two vertices. We want to color the vertices in such a way that only 3 colors are used, and no two vertices connected by an edge are colored the same way. If $\mathcal{G}$ can be colored in this fashion, then $\mathcal{G}$ is called *3-colorable.* This can be seen to be the same as the 3-color problem for a map: the vertices represent the regions to be colored, and two vertices are connected by an edge if the two corresponding regions are adjacent.

First, we let $\xi = e^{\frac{2\pi i}{3}} \in \mathbb{C}$ be a cube root of unity (i.e. $\xi^3 = 1$). We represent the 3-colors by $1, \xi, \xi^2$, the 3 distinct cube roots of unity. Now, we let $x_1, \dots, x_n$ be variables representing the distinct vertices of the graph $\mathcal{G}$. Each vertex is to be assigned one of the 3 colors $1, \xi, \xi^2$. This can be represented by the following n equations

$$x_i^3 - 1 = 0, \ 1 \le i \le n. \tag{2.7.1}$$

Also, if the vertices x_i and x_j are connected by an edge, they need to have a different color. Since $x_i^3 = x_j^3$, we have $(x_i - x_j)(x_i^2 + x_i x_j + x_j^2) = 0$. Therefore x_i and x_j will have different colors if and only if

$$x_i^2 + x_i x_j + x_j^2 = 0. \tag{2.7.2}$$

Let I be the ideal of $\mathbb{C}[x_1, \dots, x_n]$ generated by the polynomials in Equation (2.7.1) and for each pair of vertices x_i, x_j which are connected by an edge by the polynomials in Equation (2.7.2). We will consider the variety $V(I)$ contained in $\mathbb{C}^n$. The following theorem is now immediate.

THEOREM 2.7.1. *The graph $\mathcal{G}$ is 3-colorable if and only if $V(I) \neq \emptyset$.*

We have seen in Section 2.2 that we can use Gröbner bases to determine if $V(I) = \emptyset$. We first compute a reduced Gröbner basis for I. If $1 \in G$, then $V(I) = \emptyset$ and otherwise $V(I) \neq \emptyset$ (see Theorem 2.2.6).

EXAMPLE 2.7.2. Consider the graph $\mathcal{G}$ of Figure 2.1.

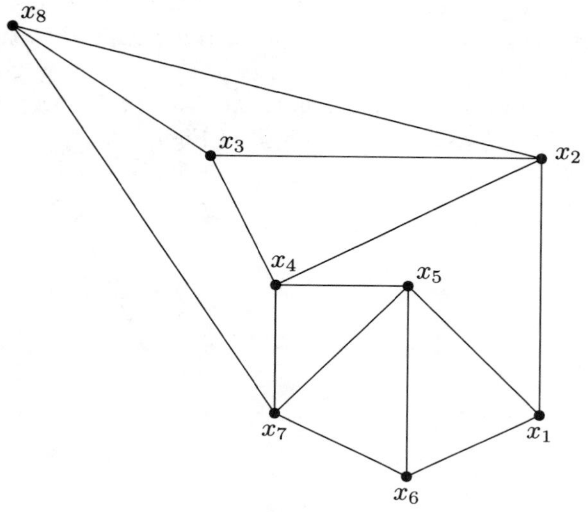

FIGURE 2.1. The graph $\mathcal{G}$

The polynomials corresponding to $\mathcal{G}$ are:

$$x_i^3 - 1, \text{ for } i = 1, \dots, 8$$

and

$$x_i^2 + x_i x_j + x_j^2, \text{ for the pairs } (i,j) \in \{(1,2),(1,5),(1,6),(2,3),(2,4),$$
$$(2,8),(3,4),(3,8),(4,5),(4,7),(5,6),(5,7),(6,7),(7,8)\}.$$

We compute a Gröbner basis G for the ideal I corresponding to the above polynomials. Keeping in mind Corollary 2.2.11, we use the lex term ordering with $x_1 > x_2 > \cdots > x_8$. We obtain

$$G = \{x_1 - x_7, x_2 + x_7 + x_8, x_3 - x_7, x_4 - x_8, x_5 + x_7 + x_8,$$
$$x_6 - x_8, x_7^2 + x_7 x_8 + x_8^2, x_8^3 - 1\}.$$

Since $1 \notin G$, we have that $V(I) \neq \emptyset$, and hence, by Theorem 2.7.1, $\mathcal{G}$ is 3-colorable. We can use the Gröbner basis G to give an explicit coloring, since the system of equations represented by G turns out to be easy to solve. Let us assume that the 3 colors we are using are blue, red, and green. We must first choose a color for x_8, say red, since the only polynomial in one variable in G is $x_8^3 - 1$. We then must choose a different color for x_7, say blue, because of the polynomial $x_7^2 + x_7 x_8 + x_8^2 \in G$. Then we have that x_1 and x_3 must be blue because of the polynomials $x_1 - x_7, x_3 - x_7 \in G$, and x_4, x_6 must be red because of the polynomials $x_4 - x_8$, $x_6 - x_8 \in G$. Finally x_2 and x_5 have the same color, which is a different color from the colors assigned to x_7 and x_8, so x_2 and x_5 are green; this is because the polynomials $x_2 + x_7 + x_8$, and $x_5 + x_7 + x_8$ are in G.

We note that it is evident from the Gröbner basis in Example 2.7.2 that there is only one way to color the graph $\mathcal{G}$, up to permuting the colors, and so it should not be too surprising that solving the equations determined by the Gröbner basis is easy. However, if there is more than one possible coloring, the Gröbner basis may look more complicated. This is illustrated in the following example.

EXAMPLE 2.7.3. Consider the graph $\mathcal{G}'$ in Figure 2.2.

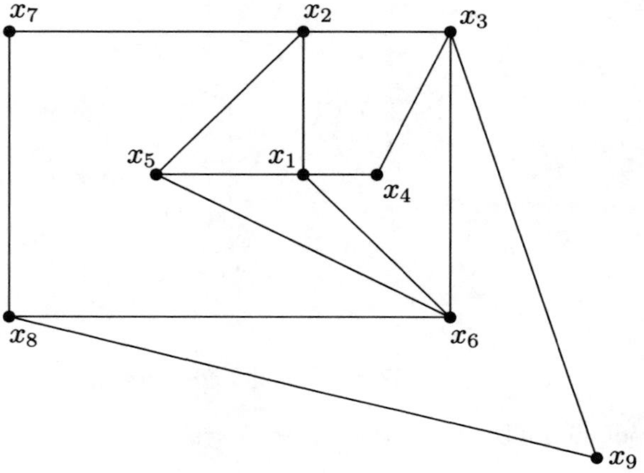

FIGURE 2.2. The graph $\mathcal{G}'$

The polynomials corresponding to $\mathcal{G}'$ are:

$$x_i^3 - 1, \text{ for } i = 1, \dots, 9$$

and

$$x_i^2 + x_i x_j + x_j^2, \text{ for the pairs } (i,j) \in \{(1,2), (1,4), (1,5), (1,6), (2,3), (2,5), (2,7), (3,4), (3,6), (3,9), (5,6), (6,8), (7,8), (8,9)\}.$$

We compute a Gröbner basis G' for the ideal I' corresponding to the above polynomials using the lex term ordering with $x_1 > x_2 > \cdots > x_8 > x_9$ to obtain

$$\begin{aligned} G' = \{&x_9^3 - 1, x_8^2 + x_8x_9 + x_9^2, (x_7 - x_9)(x_7 + x_8 + x_9), \\ &x_6 + x_7 + x_8, (-x_5 + x_7)(-x_5 + x_8), \\ &(x_5 - x_9)(x_4x_7 + x_4x_8 + x_7x_8 + x_4x_9 + x_7x_9 + x_8x_9), \\ &(x_4 - x_5)(x_4 + x_7 + x_8), x_3x_7 + x_3x_8 - x_7x_8 + x_3x_9 + x_9^2, \\ &x_3x_4 + x_4x_5 - x_4x_7 + x_5x_7 - x_4x_8 + x_5x_8 - x_3x_9 - x_9^2, \\ &x_3^2 + x_3x_9 + x_9^2, x_2 + x_7 + x_8, x_1 + x_5 - x_7 - x_8\}. \end{aligned}$$

Since $1 \notin G'$, we have that $V(I') \neq \emptyset$, and hence, by Theorem 2.7.1, $\mathcal{G}'$ is 3-colorable. This Gröbner basis looks much more complicated than the one in Example 2.7.2. This reflects the fact that there are many possible colorings of this graph. In fact, it is easy to 3-color the graph $\mathcal{G}'$ by trial and error.

Exercises

We note that we have tried to keep these problems small, without making them too trivial looking, but your Computer Algebra System may still have trouble doing the computations.

2.7.1. Show that if we add an edge between x_2 and x_5 in the graph $\mathcal{G}$ of Figure 2.1, the graph is no longer 3-colorable. (This can be done either by computing a Gröbner basis for the new ideal, or by observing what was done in Example 2.7.2.)

2.7.2. Show that if we add one edge between vertices x_1 and x_3 in the graph $\mathcal{G}'$ of Figure 2.2, then $\mathcal{G}'$ is still 3-colorable. Show that now the 3-coloring is unique except for the permutation of the colors.

2.7.3. Generalize the method given in this section to the case of determining whether graphs are 4-colorable.

2.7.4. Use the method of Exercise 2.7.3 to show that in the trivial example where there are 4 vertices, each pair of which is connected by an edge, the graph is 4-colorable and show that the equations imply that all 4 vertices must be colored a different color.

2.7.5. Show that in principle (by this we mean that the computations would probably be too lengthy to make the scheme practical) the method in this section could be generalized to giving a method to determine whether a graph is m-colorable for any positive integer m.

2.8. Integer Programming. The material in this section is taken from P. Conti and C. Traverso [**CoTr**]. No use of this section will be made elsewhere in the book.

The integer programming problem has the following form: let $a_{ij} \in \mathbb{Z}$, $b_i \in \mathbb{Z}$, and $c_j \in \mathbb{R}$, $i = 1, \dots, n$, $j = 1, \dots, m$; we wish to find a solution $(\sigma_1, \sigma_2, \dots, \sigma_m)$ in $\mathbb{N}^m$ of the system

$$\left\{\begin{array}{ccccccccc} a_{11}\sigma_1 & + & a_{12}\sigma_2 & + & \cdots & + & a_{1m}\sigma_m & = & b_1 \\ a_{21}\sigma_1 & + & a_{22}\sigma_2 & + & \cdots & + & a_{2m}\sigma_m & = & b_2 \\ & & & & \vdots & & & & \\ a_{n1}\sigma_1 & + & a_{n2}\sigma_2 & + & \cdots & + & a_{nm}\sigma_m & = & b_n, \end{array}\right. \tag{2.8.1}$$

which minimizes the "cost function"

$$c(\sigma_1, \sigma_2, \dots, \sigma_m) = \sum_{j=1}^{m} c_j\sigma_j. \tag{2.8.2}$$

This problem occurs often in scientific and engineering applications. There are many books on the subject which the reader may consult, see, for example, [**Schri**]. Our purpose here is to apply the results of Section 2.4 to indicate a solution method to this problem.

Our strategy is to:

(i) Translate the integer programming problem into a problem about polynomials;

(ii) Use the Gröbner bases techniques developed so far to solve the polynomial problem;

(iii) Translate the solution of the polynomial problem back into a solution of the integer programming problem.

In order to motivate the general technique presented below, we will first start with the special case when all a_{ij}'s and b_i's are non-negative integers. We will also concentrate first on solving System (2.8.1) without taking into account the cost function condition (Equation (2.8.2)).

We introduce a variable for each linear equation in (2.8.1), say $x_1, x_2, \ldots, x_n$, and a variable for each unknown σ_j, say $y_1, y_2, \ldots, y_m$. We then represent the equations in (2.8.1) as

$$x_i^{a_{i1}\sigma_1+\cdots+a_{im}\sigma_m} = x_i^{b_i},$$

for $i = 1, \ldots, n$. Then System (2.8.1) can be written as a single equation of power products

$$x_1^{a_{11}\sigma_1+\cdots+a_{1m}\sigma_m} \cdots x_n^{a_{n1}\sigma_1+\cdots+a_{nm}\sigma_m} = x_1^{b_1}x_2^{b_2}\cdots x_n^{b_n},$$

or equivalently,

$$(2.8.3)\qquad (x_1^{a_{11}}x_2^{a_{21}}\cdots x_n^{a_{n1}})^{\sigma_1}\cdots(x_1^{a_{1m}}x_2^{a_{2m}}\cdots x_n^{a_{nm}})^{\sigma_m} = x_1^{b_1}x_2^{b_2}\cdots x_n^{b_n}.$$

We note that the left-hand side power product in Equation (2.8.3) can be viewed as the image of the power product $y_1^{\sigma_1}y_2^{\sigma_2}\cdots y_m^{\sigma_m}$ under the polynomial map

$$\begin{array}{ccc} k[y_1, \ldots, y_m] & \stackrel{\phi}{\longrightarrow} & k[x_1, \ldots, x_n] \\ y_j & \longmapsto & x_1^{a_{1j}}x_2^{a_{2j}}\cdots x_n^{a_{nj}}. \end{array}$$

The following lemma is then clear.

LEMMA 2.8.1. *We use the notation set above, and we assume that all* a_{ij}*'s and* b_i*'s are non-negative. Then there exists a solution* $(\sigma_1, \sigma_2, \ldots, \sigma_m) \in \mathbb{N}^m$ *of System (2.8.1) if and only if the power product* $x_1^{b_1}x_2^{b_2}\cdots x_n^{b_n}$ *is the image under* ϕ *of a power product in* $k[y_1, \ldots, y_m]$. *Moreover if* $x_1^{b_1}x_2^{b_2}\cdots x_n^{b_n} = \phi(y_1^{\sigma_1}y_2^{\sigma_2}\cdots y_m^{\sigma_m})$, *then* $(\sigma_1, \sigma_2, \ldots, \sigma_m) \in \mathbb{N}^m$ *is a solution of System (2.8.1).*

We have presented in Section 2.4 an algorithmic method for determining whether an element of $k[x_1, \ldots, x_n]$ is in the image of a polynomial map such as ϕ (see Theorem 2.4.4). However the above lemma requires that the power product $x_1^{b_1}x_2^{b_2}\cdots x_n^{b_n}$ be the image of a power product, not a polynomial. But, because the map ϕ sends the variables y_j to power products in $k[x_1, \dots, x_n]$, we have

LEMMA 2.8.2. *We use the notation above, and we assume that all a_{ij}'s and b_i's are non-negative. If $x_1^{b_1} x_2^{b_2} \cdots x_n^{b_n}$ is in the image of ϕ, then it is the image of a power product $y_1^{\sigma_1} y_2^{\sigma_2} \cdots y_m^{\sigma_m} \in k[y_1, \dots, y_m]$.*

PROOF. Let $K = \langle y_j - x_1^{a_{1j}} x_2^{a_{2j}} \cdots x_n^{a_{nj}} \mid j = 1, \dots, m\rangle$ be the ideal considered in Theorem 2.4.4. Let G be a Gröbner basis for K with respect to an elimination order with the x variables larger than the y variables. Then, by Theorem 2.4.4

$$x_1^{b_1} x_2^{b_2} \cdots x_n^{b_n} \in \mathrm{im}(\phi) \iff x_1^{b_1} x_2^{b_2} \cdots x_n^{b_n} \xrightarrow{G}_+ h \text{ with } h \in k[y_1, \dots, y_m].$$

Moreover, if $x_1^{b_1} x_2^{b_2} \cdots x_n^{b_n} \xrightarrow{G}_+ h$ with $h \in k[y_1, \dots, y_m]$ then $x_1^{b_1} x_2^{b_2} \cdots x_n^{b_n} = \phi(h)$.

We first note that the polynomials that generate K are all differences of two power products. Therefore, during Buchberger's Algorithm to compute G (see Algorithm (1.7.1)), only polynomials which are differences of two power products will be generated. Indeed, the S-polynomial of two polynomials which are both differences of two power products is itself a difference of two power products, and the one step reduction of a polynomial which is a difference of two power products by another polynomial of the same form produces a polynomial which is itself a difference of two power products. Therefore the polynomials in G are all differences of two power products. Now if $x_1^{b_1} x_2^{b_2} \cdots x_n^{b_n}$ is in the image of ϕ, then it reduces to a polynomial $h \in k[y_1, \dots, y_m]$. But the one step reduction of a power product by a polynomial which is a difference of two power products produces a power product. Therefore h is a power product and we are done. □

The proof of Lemma 2.8.2 gives us a method for determining whether System (2.8.1) has a solution, and for finding a solution:

(i) Compute a Gröbner basis G for $K = \langle y_j - x_1^{a_{1j}} x_2^{a_{2j}} \cdots x_n^{a_{nj}} \mid j = 1, \dots, m\rangle$ with respect to an elimination order with the x variables larger than the y variables;

(ii) Find the remainder h of the division of the power product $x_1^{b_1} x_2^{b_2} \cdots x_n^{b_n}$ by G;

(iii) If $h \notin k[y_1, \dots, y_m]$, then System (2.8.1) does not have non-negative integer solutions. If $h = y_1^{\sigma_1} y_2^{\sigma_2} \cdots y_m^{\sigma_m}$, then $(\sigma_1, \sigma_2, \dots, \sigma_m)$ is a solution of System (2.8.1).

To illustrate the ideas presented so far, we consider a simple example.

EXAMPLE 2.8.3. Consider the system

$$\left\{ \begin{array}{ccccccccc} 3\sigma_1 & + & 2\sigma_2 & + & \sigma_3 & + & \sigma_4 & = & 10 \\ 4\sigma_1 & + & \sigma_2 & + & \sigma_3 & & & = & 5. \end{array} \right. \tag{2.8.4}$$

We have two x variables, x_1, x_2, one for each equation. We also have four y variables, y_1, y_2, y_3, y_4, one for each unknown. The corresponding polynomial map is

$$\begin{array}{ccc} \mathbb{Q}[y_1, y_2, y_3, y_4] & \xrightarrow{\phi} & \mathbb{Q}[x_1, x_2] \\ y_1 & \longmapsto & x_1^3 x_2^4 \\ y_2 & \longmapsto & x_1^2 x_2 \\ y_3 & \longmapsto & x_1 x_2 \\ y_4 & \longmapsto & x_1, \end{array}$$

and so $K = \langle y_1 - x_1^3x_2^4, y_2 - x_1^2x_2, y_3 - x_1x_2, y_4 - x_1\rangle \subseteq \mathbb{Q}[y_1, y_2, y_3, y_4, x_1, x_2]$. The Gröbner basis for K with respect to the lex order with $x_1 > x_2 > y_1 > y_2 > y_3 > y_4$ is $G = \{f_1, f_2, f_3, f_4, f_5\}$, where $f_1 = x_1 - y_4$, $f_2 = x_2y_4 - y_3$, $f_3 = x_2y_3^3 - y_1$, $f_4 = y_2 - y_3y_4$, $f_5 = y_1y_4 - y_3^4$. Then

$$x_1^{10}x_2^5 \xrightarrow{\{f_1, f_2\}} y_3^5y_4^5,$$

and $h = y_3^5y_4^5$ is reduced with respect to G. Using the exponents of h we get that $(0, 0, 5, 5)$ is a solution of System (2.8.4).

Now we turn our attention to the more general case where the a_{ij}'s and b_i's in (2.8.1) are any integers, not necessarily non-negative. We still focus our attention on determining whether System (2.8.1) has solutions and on finding solutions, that is, we still ignore the cost function condition (Equation (2.8.2)). We will proceed as before, except that we now have negative exponents on the x variables. Of course, this cannot be done in the polynomial ring $k[x_1, \ldots, x_n]$. Instead, we introduce a new variable w and we work in the affine ring $k[x_1, \ldots, x_n, w]/I$, where $I = \langle x_1x_2 \cdots x_nw - 1\rangle$. We may choose non-negative integers a'_{ij} and α_j, for each $j = 1, \ldots, m$ and $i = 1, \ldots, n$ such that for each $j = 1, \ldots, m$ we have

$$(a_{1j}, a_{2j}, \ldots, a_{nj}) = (a'_{1j}, a'_{2j}, \ldots, a'_{nj}) + \alpha_j(-1, -1, \ldots, -1).$$

For example, $(-3, 2, -5) = (2, 7, 0) + 5(-1, -1, -1)$. Then in the affine ring $k[x_1, \ldots, x_n, w]/I$ we can give meaning to the coset $x_1^{a_{1j}}x_2^{a_{2j}} \cdots x_n^{a_{nj}} + I$ by defining

$$x_1^{a_{1j}}x_2^{a_{2j}} \cdots x_n^{a_{nj}} + I = x_1^{a'_{1j}}x_2^{a'_{2j}} \cdots x_n^{a'_{nj}}w^{\alpha_j} + I.$$

Similarly, $(b_1, b_2, \ldots, b_n) = (b'_1, b'_2, \ldots, b'_n) + \beta(-1, -1, \ldots, -1)$, where b'_i and β are non-negative integers for $i = 1, \ldots, n$, and define

$$x_1^{b_1}x_2^{b_2} \cdots x_n^{b_n} + I = x_1^{b'_1}x_2^{b'_2} \cdots x_n^{b'_n}w^{\beta} + I.$$

Therefore we have the following equation which corresponds to Equation (2.8.3)

(2.8.5)
$$(x_1^{a'_{11}} \cdots x_n^{a'_{n1}}w^{\alpha_1})^{\sigma_1} \cdots (x_1^{a'_{1m}} \cdots x_n^{a'_{nm}}w^{\alpha_m})^{\sigma_m} + I = x_1^{b'_1} \cdots x_n^{b'_n}w^{\beta} + I.$$

We therefore proceed as before, and we note that the left-hand side of Equation (2.8.5) can be viewed as the image of the power product $y_1^{\sigma_1}y_2^{\sigma_2} \cdots y_m^{\sigma_m}$ under the algebra homomorphism

$$\begin{array}{ccc} k[y_1,\ldots,y_m] & \xrightarrow{\phi} & k[x_1,\ldots,x_n,w]/I \\ y_j & \longmapsto & x_1^{a'_{1j}}x_2^{a'_{2j}}\cdots x_n^{a'_{nj}}w^{\alpha_j}+I. \end{array}$$

As before we have

LEMMA 2.8.4. *We use the notation set above. Then there exists a solution* $(\sigma_1,\sigma_2,\ldots,\sigma_m)\in\mathbb{N}^m$ *of System (2.8.1) if and only if* $x_1^{b'_1}x_2^{b'_2}\cdots x_n^{b'_n}w^\beta+I$ *is the image under* ϕ *of a power product in* $k[y_1,\ldots,y_m]$*. Moreover if* $x_1^{b'_1}x_2^{b'_2}\cdots x_n^{b'_n}w^\beta+I=\phi(y_1^{\sigma_1}y_2^{\sigma_2}\cdots y_m^{\sigma_m})$*, then* $(\sigma_1,\sigma_2,\ldots,\sigma_m)$ *is a solution of System (2.8.1).*

We have presented in Section 2.4 an algorithmic method for determining whether an element of $k[x_1,\ldots,x_n,w]/I$ is in the image of an affine algebra homomorphism such as ϕ (see Theorem 2.4.13). As in the first case we considered, the above lemma requires that $x_1^{b'_1}x_2^{b'_2}\cdots x_n^{b'_n}w^\beta+I$ be the image of a power product, not a polynomial. As before we have

LEMMA 2.8.5. *We use the notation set above. If* $x_1^{b'_1}x_2^{b'_2}\cdots x_n^{b'_n}w^\beta+I$ *is in the image of* ϕ*, then it is the image of a power product* $y_1^{\sigma_1}y_2^{\sigma_2}\cdots y_m^{\sigma_m}\in k[y_1,\ldots,y_m]$*.*

PROOF. As in Theorem 2.4.13, let $K\subseteq k[y_1,\ldots,y_m,x_1,\ldots,x_n,w]$ be the ideal generated by $x_1x_2\cdots x_nw-1$ and $\{y_j-x_1^{a'_{1j}}x_2^{a'_{2j}}\cdots x_n^{a'_{nj}}w^{\alpha_j}\mid j=1,\ldots,m\}$. Let G be a Gröbner basis for K with respect to an elimination order with the x and w variables larger than the y variables. Then, by Theorem 2.4.13,

$$x_1^{b'_1}x_2^{b'_2}\cdots x_n^{b'_n}w^\beta+I\in\operatorname{im}(\phi)\iff x_1^{b'_1}x_2^{b'_2}\cdots x_n^{b'_n}w^\beta\xrightarrow{G}_+ h\in k[y_1,\ldots,y_m].$$

Moreover, if $x_1^{b'_1}x_2^{b'_2}\cdots x_n^{b'_n}w^\beta\xrightarrow{G}_+ h$ with $h\in k[y_1,\ldots,y_m]$ then

$$x_1^{b'_1}x_2^{b'_2}\cdots x_n^{b'_n}w^\beta+I=\phi(h).$$

As in Lemma 2.8.2, the polynomials that generate K are all differences of two power products, therefore, the argument used in the proof of Lemma 2.8.2 can again be applied. That is, the polynomials in G are all differences of two power products, and the reduction by G of a power product produces a power product. □

EXAMPLE 2.8.6. We consider the following system

$$\left\{\begin{array}{rcrcrcrcr} 3\sigma_1 & - & 2\sigma_2 & + & \sigma_3 & - & \sigma_4 & = & -1 \\ 4\sigma_1 & + & \sigma_2 & - & \sigma_3 & & & = & 5. \end{array}\right. \tag{2.8.6}$$

We have two x variables, x_1,x_2, one for each equation. We also have four y variables, y_1,y_2,y_3,y_4, one for each unknown. We consider the ideal $I=\langle x_1x_2w-1\rangle$

of $\mathbb{Q}[x_1, x_2, w]$ and the algebra homomorphism

$$\begin{array}{ccc} \mathbb{Q}[y_1, y_2, y_3, y_4] & \xrightarrow{\phi} & \mathbb{Q}[x_1, x_2, w]/I \\ y_1 & \longmapsto & x_1^3x_2^4 + I \\ y_2 & \longmapsto & x_2^3w^2 + I \\ y_3 & \longmapsto & x_1^2w + I \\ y_4 & \longmapsto & x_2w + I. \end{array}$$

Thus $K = \langle y_1 - x_1^3x_2^4, y_2 - x_2^3w^2, y_3 - x_1^2w, y_4 - x_2w, x_1x_2w - 1\rangle$. The Gröbner basis for K with respect to the lex order with $x_1 > x_2 > w > y_1 > y_2 > y_3 > y_4$ is $G = \{f_1, f_2, f_3, f_4, f_5, f_6, f_7, f_8, f_9\}$, where $f_1 = x_1 - y_1y_3^4y_4^6$, $f_2 = x_2 - y_1y_3^3y_4^6$, $f_3 = w - y_3y_4^2$, $f_4 = y_1y_3^4y_4^7 - 1$, $f_5 = y_1y_3^3y_4^8 - y_2$, $f_6 = y_1y_3^2y_4^9 - y_2^2$, $f_7 = y_1y_3y_4^{10} - y_2^3$, $f_8 = y_1y_4^{11} - y_2^4$, $f_9 = y_2y_3 - y_4$. We now reduce the power product x_2^6w by G (note that $x_1^{-1}x_2^5 + I = x_2^6w + I$)

$$\begin{array}{rcl} x_2^6w & \xrightarrow{\{f_2,f_3\}} & y_1^6y_3^{19}y_4^{38} \\ & \xrightarrow{f_4} & y_1^5y_3^{15}y_4^{31} \\ & \xrightarrow{f_4} & y_1^4y_3^{11}y_4^{24} \\ & \xrightarrow{f_4} & y_1^3y_3^7y_4^{17} \\ & \xrightarrow{f_4} & y_1^2y_3^3y_4^{10} \\ & \xrightarrow{f_5} & y_1y_2y_4^2, \end{array}$$

and $y_1y_2y_4^2$ is reduced with respect to G. Observing the exponents of the different power products obtained during the reduction, we have the following solutions of System (2.8.6)

$$(6, 0, 19, 38), (5, 0, 15, 31), (4, 0, 11, 24), (3, 0, 7, 17), (2, 0, 3, 10), (1, 1, 0, 2).$$

We return to the original problem. That is, we want to find solutions of System (2.8.1) that minimize the cost function $c(\sigma_1, \sigma_2, \ldots, \sigma_m) = \sum_{j=1}^m c_j\sigma_j$ (Equation (2.8.2)).

As we mentioned before, the only requirement on the term order in the method for obtaining solutions of System (2.8.1) described above, is that we have an elimination order between the x, w and the y variables with the x and w variables larger. Our strategy for minimizing the cost function is to use the c_j's to define such a term order.

DEFINITION 2.8.7. *A term order $<_c$ on the y variables is said to be* compatible *with the cost function c and the map ϕ if*

$$\left.\begin{array}{l} \phi(y_1^{\sigma_1}y_2^{\sigma_2}\cdots y_m^{\sigma_m}) = \phi(y_1^{\sigma_1'}y_2^{\sigma_2'}\cdots y_m^{\sigma_m'}) \\ \text{and} \\ c(\sigma_1, \ldots, \sigma_m) < c(\sigma_1', \ldots, \sigma_m') \end{array}\right\} \Longrightarrow y_1^{\sigma_1}y_2^{\sigma_2}\cdots y_m^{\sigma_m} <_c y_1^{\sigma_1'}y_2^{\sigma_2'}\cdots y_m^{\sigma_m'}.$$

Term orders on the y variables which are compatible with c and ϕ are exactly those orders which will give rise to solutions of System (2.8.1) with minimum cost as the next proposition shows.

PROPOSITION 2.8.8. *We use the notation set above. Let G be a Gröbner basis for K with respect to an elimination order with the x and w variables larger than the y variables, and an order $<_c$ on the y variables which is compatible with the cost function c and the map ϕ. If $x_1^{b_1'}x_2^{b_2'}\cdots x_n^{b_n'}w^\beta \xrightarrow{G}_+ y_1^{\sigma_1}\cdots y_m^{\sigma_m}$, where $y_1^{\sigma_1}\cdots y_m^{\sigma_m}$ is reduced with respect to G, then $(\sigma_1,\ldots,\sigma_m)$ is a solution of System (2.8.1) which minimizes the cost function c.*

PROOF. Let $x_1^{b_1'}x_2^{b_2'}\cdots x_n^{b_n'}w^\beta \xrightarrow{G}_+ y_1^{\sigma_1}\cdots y_m^{\sigma_m}$, with $y_1^{\sigma_1}\cdots y_m^{\sigma_m}$ reduced with respect to G. Then $(\sigma_1,\ldots,\sigma_m)$ is a solution of System (2.8.1) by Lemma 2.8.4. Now assume to the contrary that there exists a solution $(\sigma_1',\ldots,\sigma_m')$ of System (2.8.1) such that $\sum_{j=1}^m c_j\sigma_j' < \sum_{j=1}^m c_j\sigma_j$. Consider the corresponding power product $y_1^{\sigma_1'}\cdots y_m^{\sigma_m'}$. Note that $\phi(y_1^{\sigma_1}\cdots y_m^{\sigma_m}) = \phi(y_1^{\sigma_1'}\cdots y_m^{\sigma_m'}) = x_1^{b_1'}x_2^{b_2'}\cdots x_n^{b_n'}w^\beta + I$. Therefore $y_1^{\sigma_1}\cdots y_m^{\sigma_m} - y_1^{\sigma_1'}\cdots y_m^{\sigma_m'} \in \ker(\phi)$. By Theorem 2.4.10, $\ker(\phi)\subseteq K$, so that $y_1^{\sigma_1}\cdots y_m^{\sigma_m} - y_1^{\sigma_1'}\cdots y_m^{\sigma_m'} \in K$. Hence $y_1^{\sigma_1}\cdots y_m^{\sigma_m} - y_1^{\sigma_1'}\cdots y_m^{\sigma_m'} \xrightarrow{G}_+ 0$. Since $y_1^{\sigma_1}\cdots y_m^{\sigma_m} >_c y_1^{\sigma_1'}\cdots y_m^{\sigma_m'}$ by assumption, we have $\operatorname{lt}(y_1^{\sigma_1}\cdots y_m^{\sigma_m} - y_1^{\sigma_1'}\cdots y_m^{\sigma_m'}) = y_1^{\sigma_1}\cdots y_m^{\sigma_m}$. But $y_1^{\sigma_1}\cdots y_m^{\sigma_m}$ is reduced with respect to G, and therefore $y_1^{\sigma_1}\cdots y_m^{\sigma_m} - y_1^{\sigma_1'}\cdots y_m^{\sigma_m'}$ cannot reduce to 0 by G. □

We note that a different minimal solution may be obtained if we use a different order, as long as we have an elimination order with the x and w variables larger than the y variables, and as long as we use an order on the y variables compatible with the cost function c and the map ϕ.

For some cases, the term order $<_c$ is easy to obtain, however the general case is more involved. We refer the reader to the original paper [**CoTr**].

One particular simple case is when the cost function only involves positive coefficients, that is $c_j \geq 0$ for $j = 1,\ldots,m$. Then the following order is a term order compatible with the cost function and the map ϕ: first order power products using the cost function, and break ties by any other order (see Exercise 2.3.5). The following illustrates this idea.

EXAMPLE 2.8.9. We go back to Example 2.8.6 but we now consider the cost function

$$c(\sigma_1,\sigma_2,\sigma_3,\sigma_4) = 1000\sigma_1 + \sigma_2 + \sigma_3 + 100\sigma_4.$$

We use the lex order on w and the x variables with $x_1 > x_2 > w$. The power products in y are first ordered using the cost function and ties are broken by lex with $y_1 > y_2 > y_3 > y_4$. That is $y_1^{\sigma_1}y_2^{\sigma_2}y_3^{\sigma_3}y_4^{\sigma_4} < y_1^{\sigma_1'}y_2^{\sigma_2'}y_3^{\sigma_3'}y_4^{\sigma_4'}$ if and only if $1000\sigma_1+\sigma_2+\sigma_3+100\sigma_4 < 1000\sigma_1'+\sigma_2'+\sigma_3'+100\sigma_4'$ or $1000\sigma_1+\sigma_2+\sigma_3+100\sigma_4 = 1000\sigma_1'+\sigma_2'+\sigma_3'+100\sigma_4'$ and $y_1^{\sigma_1}y_2^{\sigma_2}y_3^{\sigma_3}y_4^{\sigma_4} <_{lex} y_1^{\sigma_1'}y_2^{\sigma_2'}y_3^{\sigma_3'}y_4^{\sigma_4'}$. Finally we use an elimination order with w and the x variables larger than the y variables. The reduced Gröbner basis for K is

$$G = \{w - y_2^2 y_4^3,\ y_4 - y_2 y_3,\ x_1 - y_1 y_2^6 y_3^{10},\ x_2 - y_1 y_2^6 y_3^9,\ y_1 y_2^7 y_3^{11} - 1\}.$$

We have $x_2^6 w \xrightarrow{G}_+ y_1 y_2^3 y_3^2$ which gives the solution $(1, 3, 2, 0)$. This solution is of minimal cost.

Exercises

2.8.1. Use the method of Lemma 2.8.1 to solve the following system of equations for non-negative integers.

$$\begin{cases} 3\sigma_1 + 2\sigma_2 + \sigma_3 = 10 \\ 4\sigma_1 + 3\sigma_2 + \sigma_3 = 12. \end{cases}$$

2.8.2. Use the method of Lemma 2.8.1 to solve the following system of equations for non-negative integers.

$$\begin{cases} 3\sigma_1 + 2\sigma_2 + \sigma_3 = 10 \\ 4\sigma_1 + 3\sigma_2 + \sigma_3 = 4. \end{cases}$$

2.8.3. Use the method of Lemma 2.8.1 to solve the following system of equations for non-negative integers.

$$\begin{cases} 3\sigma_1 + 2\sigma_2 + \sigma_3 + 2\sigma_4 = 10 \\ 4\sigma_1 + 3\sigma_2 + \sigma_3 = 12 \\ 2\sigma_1 + 4\sigma_2 + 2\sigma_3 + \sigma_4 = 25. \end{cases}$$

2.8.4. Use the method of Lemma 2.8.1 to solve the following system of equations for non-negative integers.

$$\begin{cases} 3\sigma_1 + 2\sigma_2 + \sigma_3 + 2\sigma_4 = 10 \\ 4\sigma_1 + 3\sigma_2 + \sigma_3 = 11 \\ 2\sigma_1 + 4\sigma_2 + 2\sigma_3 + \sigma_4 = 10. \end{cases}$$

2.8.5. Prove Lemma 2.8.4.

2.8.6. Use the method of Lemma 2.8.4 to solve the following system of equations for non-negative integers.

$$\begin{cases} 2\sigma_1 + \sigma_2 - 3\sigma_3 + \sigma_4 = 4 \\ -3\sigma_1 + 2\sigma_2 - 2\sigma_3 - \sigma_4 = -3. \end{cases}$$

2.8.7. In Example 2.8.9 replace the cost function by

$$c(\sigma_1, \sigma_2, \sigma_3, \sigma_4) = 1000\sigma_1 + \sigma_2 + \sigma_3 + \sigma_4.$$

2.8.8. In Example 2.8.9 replace the cost function by

$$c(\sigma_1, \sigma_2, \sigma_3, \sigma_4) = \sigma_1 + 1000\sigma_2 + \sigma_3 + \sigma_4.$$

Chapter 3. Modules and Gröbner Bases

Let k be a field. In this chapter we consider submodules of $k[x_1, \ldots, x_n]^m$ and their quotient modules. In Section 3.1 we briefly review the concepts from the theory of modules that we require (see [**Hun**]). Then in Section 3.2 we define the module of solutions of a homogeneous linear equation with polynomial coefficients (called the syzygy module) and use it to give yet another equivalent condition for a set to be a Gröbner basis for an ideal. In Section 3.3 we follow Buchberger [**Bu79**] and show how to use this new condition to significantly improve the computations of Gröbner bases. In Section 3.4 we show how to compute an explicit generating set for the syzygy module of a vector of polynomials. We then generalize the definition of Gröbner bases and the results of the previous two chapters to modules (Section 3.5) and show that the same type of applications we had for ideals are possible in the more general context of submodules of free modules (Section 3.6). In Section 3.7 we generalize the results of Section 3.4 to systems of linear equations of polynomials. In Section 3.8 we show how this theory can be applied to give more efficient methods for the computations of Chapter 2 that required elimination. In the next to last section we explicitly compute Hom. That is, we compute a presentation of $\mathrm{Hom}(M, N)$ given two explicitly presented modules M and N. Finally, in the last section we apply the previous material to give results on free resolutions and outline the computation of $\mathrm{Ext}(M, N)$.

3.1. Modules. In this section we let A be any commutative ring. The ring that will be of interest to us in later sections is $A = k[x_1, \ldots, x_n]$. We will consider the cartesian product

$$A^m = \left\{ \begin{bmatrix} a_1 \\ \vdots \\ a_m \end{bmatrix} \mid a_i \in A, i = 1, \ldots, m \right\}.$$

That is, A^m consists of all *column vectors* with coordinates in A of length m. Although we will *always* consider the elements of A^m as column vectors (which we will enclose in square brackets), in the interest of saving space in the book we will usually write the elements of A^m as row vectors enclosed in parentheses. In other words, we will often use

$$(a_1, \ldots, a_m) \text{ instead of } \begin{bmatrix} a_1 \\ \vdots \\ a_m \end{bmatrix}.$$

The use of column vectors is necessitated by the desire to have function composition match the usual notation used in matrix multiplication.

The set A^m is called a *free A-module*. A set M is called an *A-module* provided that M is an additive abelian group with a multiplication by elements of A (scalar multiplication) satisfying:

(i) for all $a \in A$ and for all $\boldsymbol{m} \in M$, $a\boldsymbol{m} \in M$;
(ii) for all $a \in A$ and for all $\boldsymbol{m}, \boldsymbol{m}' \in M$, $a(\boldsymbol{m} + \boldsymbol{m}') = a\boldsymbol{m} + a\boldsymbol{m}'$;
(iii) for all $a, a' \in A$ and for all $\boldsymbol{m} \in M$, $(a + a')\boldsymbol{m} = a\boldsymbol{m} + a'\boldsymbol{m}$;
(iv) for all $a, a' \in A$ and for all $\boldsymbol{m} \in M$, $a(a'\boldsymbol{m}) = (aa')\boldsymbol{m}$;
(v) for all $\boldsymbol{m} \in M$, $1\boldsymbol{m} = \boldsymbol{m}$.

Scalar multiplication in A^m is done componentwise; that is,

$$a \begin{bmatrix} a_1 \\ \vdots \\ a_m \end{bmatrix} = \begin{bmatrix} aa_1 \\ \vdots \\ aa_m \end{bmatrix},$$

or using our space saving notation, $a(a_1, \ldots, a_m) = (aa_1, \ldots, aa_m)$, for $a \in A$ and $(a_1, \ldots, a_m) \in A^m$. The module A^m is called free because it has a *basis*, that is a *generating set of linearly independent* vectors. For example,

$$\boldsymbol{e}_1 = (1, 0, \ldots, 0), \boldsymbol{e}_2 = (0, 1, 0, \ldots, 0), \ldots, \boldsymbol{e}_m = (0, 0, \ldots, 0, 1)$$

is a basis which we will call the *standard basis* for A^m. In other words, every element $\boldsymbol{a} = (a_1, \ldots, a_m)$ of A^m can be written in a unique fashion

$$\boldsymbol{a} = \sum_{i=1}^{m} a_i \boldsymbol{e}_i, \text{ where } a_i \in A.$$

Note that an ideal of A is an A-module, and in fact a submodule of the A-module $A = A^1$. A *submodule* of an A-module M is a subset of M which is an A-module in its own right. For example, if $\boldsymbol{a}_1, \ldots, \boldsymbol{a}_s$ are vectors in A^m, then

$$M = \{b_1\boldsymbol{a}_1 + \cdots + b_s\boldsymbol{a}_s \mid b_i \in A, i = 1, \ldots, s\} \subseteq A^m$$

is a submodule of A^m. We denote this submodule by $\langle \boldsymbol{a}_1, \ldots, \boldsymbol{a}_s \rangle \subseteq A^m$, and we call $\{\boldsymbol{a}_1, \ldots, \boldsymbol{a}_s\}$ a *generating set* of M.

The concept of an A-module is similar to the one of a vector space, except that the set of scalars in the module case is the ring A which is not necessarily a field. Submodules of A^m are used for linear algebra in A^m in the same way subspaces of k^m are used for linear algebra in k^m. For example, let $M = \langle \boldsymbol{a}_1, \ldots, \boldsymbol{a}_s \rangle \subseteq A^m$.

If we think of the vectors $\boldsymbol{a}_i$'s as the columns of an $m \times s$ matrix S, then M is the "column space" of the matrix S, that is,

$$M = \left\{ S\boldsymbol{b} \mid \boldsymbol{b} = \begin{bmatrix} b_1 \\ \vdots \\ b_s \end{bmatrix} \in A^s \right\}.$$

So, given $\boldsymbol{a} \in A^m$, the system of m linear equations (with unknown the coordinates of $\boldsymbol{b} \in A^s$) determined by the matrix equation

$$S\boldsymbol{b} = \boldsymbol{a} \tag{3.1.1}$$

can be translated into the question of whether the vector $\boldsymbol{a}$ is in M or not. In the case where $m = 1$, that is, M is an ideal of A, and System (3.1.1) has only one equation, then this problem is the ideal membership problem discussed in Section 2.1. Other linear algebra problems in A^m can also be translated this way into a module theoretic question. In this chapter we will develop algorithmic tools to answer these questions in the case where $A = k[x_1, \ldots, x_n]$.

We first go back to the general theory of modules. Since we are mainly interested in the ring $k[x_1, \ldots, x_n]$ which is Noetherian by the Hilbert Basis Theorem (Theorem 1.1.1), we will assume that the ring A we are considering is Noetherian. We have the following

THEOREM 3.1.1. *Every submodule M of A^m has a finite generating set.*

PROOF. Let M be a submodule of A^m. We use induction on m. If $m = 1$, then M is an ideal of A, and the result follows from the Hilbert Basis Theorem (Theorem 1.1.1). For $m > 1$, let

$$I = \{a \in A \mid a \text{ is the first coordinate of an element of } M\}.$$

Then I is an ideal of A, and hence, using the Hilbert Basis Theorem, I is finitely generated. Let

$$I = \langle a_1, \ldots, a_t \rangle.$$

Let $\boldsymbol{m}_1, \ldots, \boldsymbol{m}_t \in M$ be such that the first coordinate of $\boldsymbol{m}_i$ is a_i. Now consider

$$M' = \{(b_2, \ldots, b_m) \mid (0, b_2, \ldots, b_m) \in M\}.$$

Note that M' is a submodule of A^{m-1} and so, by induction, is finitely generated, say by $\boldsymbol{n}'_1, \ldots, \boldsymbol{n}'_\ell \in M' \subseteq A^{m-1}$. For $i = 1, \ldots, \ell$, let $\boldsymbol{n}_i$ be the element of A^m with 0 in the first coordinate and the coordinates of $\boldsymbol{n}'_i$ in the remaining $m - 1$ coordinates. Note that $\boldsymbol{n}_i \in M$. To conclude, we show that

$$M = \langle \boldsymbol{m}_1, \ldots, \boldsymbol{m}_t, \boldsymbol{n}_1, \ldots, \boldsymbol{n}_\ell \rangle.$$

Let $\boldsymbol{m} \in M$. Then the first coordinate m_1 of $\boldsymbol{m}$ can be written as $m_1 = \sum_{i=1}^{t} d_i a_i$. Now consider $\boldsymbol{m}' = \boldsymbol{m} - \sum_{i=1}^{t} d_i \boldsymbol{m}_i$. Then $\boldsymbol{m}' \in M$ and its first coordinate is zero. Therefore $\boldsymbol{m}' = \sum_{i=1}^{\ell} c_i \boldsymbol{n}_i$. So,

$$\boldsymbol{m} = \boldsymbol{m}' + \sum_{i=1}^{t} d_i \boldsymbol{m}_i = \sum_{i=1}^{\ell} c_i \boldsymbol{n}_i + \sum_{i=1}^{t} d_i \boldsymbol{m}_i,$$

as desired. □

DEFINITION 3.1.2. *An A-module M is called* Noetherian *if and only if every submodule of M is finitely generated.*

Thus Theorem 3.1.1 states that if A is a Noetherian ring then A^m is a Noetherian module for all $m \geq 1$ and, in fact, any submodule of A^m is a Noetherian module. Definition 3.1.2 is equivalent to saying that if

$$M_1 \subseteq M_2 \subseteq \cdots \subseteq M_n \subseteq \cdots$$

is an ascending chain of submodules of M, then there exists n_0 such that $M_n = M_{n_0}$ for all $n \geq n_0$. That these two statements are equivalent is proved in exactly the same way as Theorem 1.1.2 was proved.

Now for N any submodule of the A-module M, we define

$$M/N = \{\boldsymbol{m} + N \mid \boldsymbol{m} \in M\}.$$

M/N is the quotient abelian group with the usual addition of cosets. We make M/N into an A-module by defining

$$a \cdot (\boldsymbol{m} + N) = a\boldsymbol{m} + N, \text{ for all } a \in A, \boldsymbol{m} \in M.$$

It is an easy exercise to see that this multiplication is well-defined and gives M/N the structure of an A-module. We call M/N the *quotient module* of M by N.

For M and M' two A-modules, we call a function $\phi\colon M \longrightarrow M'$ an *A-module homomorphism* provided that it is an abelian group homomorphism, that is,

$$\phi(\boldsymbol{m} + \boldsymbol{m}') = \phi(\boldsymbol{m}) + \phi(\boldsymbol{m}') \text{ for all } \boldsymbol{m}, \boldsymbol{m}' \in M,$$

which satisfies

$$a\phi(\boldsymbol{m}) = \phi(a\boldsymbol{m}), \text{ for all } a \in A, \boldsymbol{m} \in M.$$

The homomorphism ϕ is called an *isomorphism* provided that ϕ is one to one and onto. In this case we write $M \cong M'$.

Let $N = \ker(\phi) = \{\boldsymbol{m} \in M \mid \phi(\boldsymbol{m}) = \boldsymbol{0}\}$. Then, it is easy to see that N is a submodule of M. Also, we note that $\phi(M)$ is a submodule of M'. We know from the theory of abelian groups that

$$M/N \cong \phi(M)$$

as abelian groups under the map

$$\begin{array}{rcl} M/N & \longrightarrow & \phi(M) \\ \boldsymbol{m}+N & \longmapsto & \phi(\boldsymbol{m}). \end{array}$$

This map is, in fact, an A-module homomorphism as is easily seen and thus is an A-module isomorphism. This fact is referred to as the *First Isomorphism Theorem* for modules.

As in the theory of abelian groups, the submodules of M/N are all of the form L/N, where L is a submodule of M containing N.

Now, let M be an A-module, and let $\boldsymbol{m}_1, \ldots, \boldsymbol{m}_s \in M$. Consider the map $\phi\colon A^s \longrightarrow M$ defined as follows:

$$\phi(a_1, \ldots, a_s) = \sum_{i=1}^{s} a_i \boldsymbol{m}_i.$$

It is easy to prove that ϕ is an A-module homomorphism. Moreover, the image of ϕ is the submodule of M generated by $\boldsymbol{m}_1, \ldots, \boldsymbol{m}_s$. Hence if $\boldsymbol{m}_1, \ldots, \boldsymbol{m}_s$ generate M, then ϕ is onto.

Letting $\boldsymbol{e}_1, \ldots, \boldsymbol{e}_s$ denote the standard basis elements in A^s, we note that ϕ is uniquely defined by specifying the image of each $\boldsymbol{e}_i \in A^s$, namely by specifying $\phi(\boldsymbol{e}_i) = \boldsymbol{m}_i$. We will often define homomorphisms ϕ from A^s by simply specifying $\phi(\boldsymbol{e}_i)$, for $i = 1, \ldots, s$.

Now given generators $\boldsymbol{m}_1, \ldots, \boldsymbol{m}_s$ of the A-module M, we define ϕ as above. Since $\boldsymbol{m}_1, \ldots, \boldsymbol{m}_s$ generates M we see that ϕ is onto. Let N be the kernel of ϕ. Then, by the First Isomorphism Theorem for modules, we have

$$M \cong A^s/N.$$

So we conclude

LEMMA 3.1.3. *Every finitely generated A-module M is isomorphic to A^s/N for some positive integer s and some submodule N of A^s.*

Our purpose in this chapter is to do explicit computations in finitely generated modules over A. So the first question we have to answer is what do we mean when we say that we have an explicitly given finitely generated A-module M? The first way is to be given $N = \langle \boldsymbol{a}_1, \ldots, \boldsymbol{a}_m \rangle$ for explict $\boldsymbol{a}_1, \ldots, \boldsymbol{a}_m \in A^s$ such that $M \cong A^s/N$ for some explicit isomorphism. Lemma 3.1.3 ensures the existence of such an s and N.

DEFINITION 3.1.4. *If $M \cong A^s/N$, then we call A^s/N a* presentation *of M.*

The second way to have an explicitly given module M, provided that M is a submodule of A^s, is to have explicit $\boldsymbol{m}_1, \ldots, \boldsymbol{m}_t \in A^s$ such that $M = \langle \boldsymbol{m}_1, \ldots, \boldsymbol{m}_t \rangle$. Or, more generally, if we have an explicitly given submodule N of A^s, the submodule $M = \langle \boldsymbol{m}_1 + N, \ldots, \boldsymbol{m}_t + N \rangle$ of A^s/N is explicitly given.

It is also useful sometimes to have a presentation of M in the last two cases, and we will show how we can obtain this in Section 3.8.

COROLLARY 3.1.5. *Every finitely generated A-module M is Noetherian.*

PROOF. By Lemma 3.1.3, $M \cong A^s/N$, for some s and some submodule N of A^s. The submodules of A^s/N are of the form L/N, where L is a submodule of A^s containing N. Since A^s is Noetherian (Theorem 3.1.1), we have that every submodule of A^s is finitely generated, and hence every submodule of A^s/N is finitely generated. Therefore A^s/N and hence M is Noetherian. □

3.2. Gröbner Bases and Syzygies. In this section we let A be the Noetherian ring $k[x_1, \ldots, x_n]$. Let $I = \langle f_1, \ldots, f_s \rangle$ be an ideal of A. We consider the A-module homomorphism ϕ defined in Section 3.1,

$$\phi\colon A^s \longrightarrow I$$

given by

$$(h_1, \ldots, h_s) \longmapsto \sum_{i=1}^{s} h_i f_i.$$

As we have seen in Section 3.1,

$$I \cong A^s/\ker(\phi), \text{ as } A\text{-modules.} \tag{3.2.1}$$

DEFINITION 3.2.1. *The kernel of the map ϕ is called the* syzygy module *of the $1 \times s$ matrix $\begin{bmatrix} f_1 & \cdots & f_s \end{bmatrix}$. It is denoted* $\mathrm{Syz}(f_1, \ldots, f_s)$. *An element $(h_1, \ldots, h_s)$ of* $\mathrm{Syz}(f_1, \ldots, f_s)$ *is called a* syzygy *of $\begin{bmatrix} f_1 & \cdots & f_s \end{bmatrix}$ and satisfies*

$$h_1 f_1 + \cdots + h_s f_s = 0.$$

Another way to say this is that $\mathrm{Syz}(f_1, \ldots, f_s)$ is the set of all solutions of the single linear equation with polynomial coefficients (the f_i's)

$$f_1\chi_1 + \cdots + f_s\chi_s = 0, \tag{3.2.2}$$

where the solutions χ_i are also to be polynomials in A.

We note that the map ϕ can also be viewed as matrix multiplication:

$$\phi(h_1, \ldots, h_s) = \begin{bmatrix} f_1 & \cdots & f_s \end{bmatrix} \begin{bmatrix} h_1 \\ \vdots \\ h_s \end{bmatrix} = \sum_{i=1}^{s} h_i f_i.$$

That is, if F is the $1 \times s$ matrix $\begin{bmatrix} f_1 & \cdots & f_s \end{bmatrix}$, and $\boldsymbol{h} = \begin{bmatrix} h_1 \\ \vdots \\ h_s \end{bmatrix} \in A^s$, then $\phi(h_1, \ldots, h_s) = F\boldsymbol{h}$ and $\mathrm{Syz}(f_1, \ldots, f_s)$ is the set of all solutions $\boldsymbol{h}$ of the linear equation $F\boldsymbol{h} = \mathbf{0}$.

EXAMPLE 3.2.2. Let $A = \mathbb{Q}[x, y, z, w]$, and

$$I = \langle x^2 - yw, xy - wz, y^2 - xz \rangle.$$

The map ϕ is

$$\phi\colon A^3 \longrightarrow I$$

given by

$$(h_1, h_2, h_3) \longmapsto h_1(x^2 - yw) + h_2(xy - wz) + h_3(y^2 - xz).$$

Then $(y, -x, w)$ and $(-z, y, -x)$ are both syzygies of

$$\left[\begin{array}{ccc} x^2 - yw & xy - wz & y^2 - xz \end{array} \right],$$

since

$$y(x^2 - yw) - x(xy - wz) + w(y^2 - xz) = 0$$

and

$$-z(x^2 - yw) + y(xy - wz) - x(y^2 - xz) = 0.$$

We will show later (see Example 3.4.2) that in fact these two syzygies generate $\operatorname{Syz}(x^2 - yw, xy - wz, y^2 - xz)$, that is,

$$\operatorname{Syz}(x^2 - yw, xy - wz, y^2 - xz) = \langle (y, -x, w), (-z, y, -x) \rangle \subseteq A^3.$$

Because of the isomorphism given in Equation (3.2.1), the ideal I can be described as a quotient of a free A-module and $\operatorname{Syz}(f_1, \dots, f_s)$. Moreover, we can view $\operatorname{Syz}(f_1, \dots, f_s)$ as the set of all linear relations among $f_1, \dots, f_s$. Also, homogeneous systems such as Equation (3.2.2) or, more generally, such systems in A^m play a central role in the theory of rings and modules, similar to the role they play in the usual linear algebra over fields. Finally $\operatorname{Syz}(f_1, \dots, f_s)$ will play a critical role in the theory of Gröbner bases; in particular, its use will lead to improvements of Buchberger's Algorithm (see Section 3.3).

For these reasons and others, $\operatorname{Syz}(f_1, \dots, f_s)$ is a very important object in commutative algebra.

We note that $\operatorname{Syz}(f_1, \dots, f_s)$ is finitely generated, since it is a submodule of A^s (Theorem 3.1.1). One of our goals is to compute generators for $\operatorname{Syz}(f_1, \dots, f_s)$. The next lemma shows how to compute these generators in a special case (the general case is presented in Section 3.4).

PROPOSITION 3.2.3. *Let* $c_1, \dots, c_s \in k - \{0\}$ *and let* $X_1, X_2, \dots, X_s$ *be power products in* A. *For* $i \neq j \in \{1, \dots, s\}$, *we define* $X_{ij} = \operatorname{lcm}(X_i, X_j)$. *Then the module* $\operatorname{Syz}(c_1X_1, \dots, c_sX_s)$ *is generated by*

$$\left\{ \frac{X_{ij}}{c_iX_i}\mathbf{e}_i - \frac{X_{ij}}{c_jX_j}\mathbf{e}_j \in A^s \mid 1 \leq i < j \leq s \right\},$$

where $\mathbf{e}_1, \dots, \mathbf{e}_s$ *form the standard basis for* A^s.

PROOF. First note that for all $i \neq j$ we have $\frac{X_{ij}}{c_iX_i}\mathbf{e}_i - \frac{X_{ij}}{c_jX_j}\mathbf{e}_j$ is a syzygy of $\left[\begin{array}{cccc} c_1X_1 & c_2X_2 & \cdots & c_sX_s \end{array}\right]$, since[1]

$$\left[\begin{array}{cccc} c_1X_1 & c_2X_2 & \cdots & c_sX_s \end{array}\right](0,\ldots,0,\underbrace{\frac{X_{ij}}{c_iX_i}}_{\substack{i\text{th}\\ \text{coord}}},0,\ldots,0,\underbrace{-\frac{X_{ij}}{c_jX_j}}_{\substack{j\text{th}\\ \text{coord}}},0,\ldots,0) = 0.$$

Therefore

$$\left\langle \frac{X_{ij}}{c_iX_i}\mathbf{e}_i - \frac{X_{ij}}{c_jX_j}\mathbf{e}_j \mid 1 \le i < j \le s \right\rangle \subseteq \mathrm{Syz}(c_1X_1,\ldots,c_sX_s).$$

To prove the converse, let $(h_1,\ldots,h_s)$ be a syzygy of $\left[\begin{array}{ccc} c_1X_1 & \cdots & c_sX_s \end{array}\right]$, that is,

$$h_1c_1X_1 + \cdots + h_sc_sX_s = 0.$$

Let X be any power product in $\mathbb{T}^n$. Then the coefficient of X in $h_1c_1X_1 + \cdots + h_sc_sX_s$ must be zero. Thus it suffices to consider the case for which $h_i = c_i'X_i'$, $i = 1,\ldots,s$, and where $c_i' = 0$ or $X_iX_i' = X$ for a fixed power product X. Let $c_{i_1}',\ldots,c_{i_t}'$, with $i_1 < i_2 < \cdots < i_t$, be the non-zero c_j''s. Then we have $c_1'c_1 + c_2'c_2 + \cdots + c_s'c_s = c_{i_1}'c_{i_1} + \cdots + c_{i_t}'c_{i_t} = 0$. Therefore, using the same technique as in Lemma 1.7.5, we have

$$\begin{aligned}
(h_1,\ldots,h_s) = {} & (c_1'X_1',\ldots,c_s'X_s') = c_{i_1}'X_{i_1}'\mathbf{e}_{i_1} + \cdots + c_{i_t}'X_{i_t}'\mathbf{e}_{i_t} \\
= {} & c_{i_1}'c_{i_1}\frac{X}{c_{i_1}X_{i_1}}\mathbf{e}_{i_1} + \cdots + c_{i_t}'c_{i_t}\frac{X}{c_{i_t}X_{i_t}}\mathbf{e}_{i_t} \\
= {} & c_{i_1}'c_{i_1}\frac{X}{X_{i_1i_2}}\left(\frac{X_{i_1i_2}}{c_{i_1}X_{i_1}}\mathbf{e}_{i_1} - \frac{X_{i_1i_2}}{c_{i_2}X_{i_2}}\mathbf{e}_{i_2}\right) \\
& + (c_{i_1}'c_{i_1} + c_{i_2}'c_{i_2})\frac{X}{X_{i_2i_3}}\left(\frac{X_{i_2i_3}}{c_{i_2}X_{i_2}}\mathbf{e}_{i_2} - \frac{X_{i_2i_3}}{c_{i_3}X_{i_3}}\mathbf{e}_{i_3}\right) + \cdots \\
& + (c_{i_1}'c_{i_1} + \cdots + c_{i_{t-1}}'c_{i_{t-1}})\frac{X}{X_{i_{t-1}i_t}}\left(\frac{X_{i_{t-1}i_t}}{c_{i_{t-1}}X_{i_{t-1}}}\mathbf{e}_{i_{t-1}} - \frac{X_{i_{t-1}i_t}}{c_{i_t}X_{i_t}}\mathbf{e}_{i_t}\right) \\
& + \underbrace{(c_{i_1}'c_{i_1} + \cdots + c_{i_t}'c_{i_t})}_{=0}\frac{X}{c_{i_t}X_{i_t}},
\end{aligned}$$

---

[1]Recall that $(a_1,\ldots,a_s)$ stands for the column vector $\left[\begin{array}{c} a_1 \\ \vdots \\ a_s \end{array}\right]$, and so

$$\left[\begin{array}{ccc} b_1 & \cdots & b_s \end{array}\right](a_1,\ldots,a_s) = \left[\begin{array}{ccc} b_1 & \cdots & b_s \end{array}\right]\left[\begin{array}{c} a_1 \\ \vdots \\ a_s \end{array}\right] = b_1a_1 + \cdots + b_sa_s,$$

as in the usual matrix multiplication. We have adopted this awkward looking convention instead of the usual "dot product" because it will be consistent with the notation of computing syzygies of column vectors of matrices later.

as desired. □

We observe that if $c_1X_1, \dots, c_sX_s$ are the leading terms of the polynomials $f_1, \dots, f_s$, and if $(h_1, \dots, h_s) \in \operatorname{Syz}(c_1X_1, \dots, c_sX_s)$, then $\sum_{i=1}^{s} h_if_i$ has a leading term strictly smaller than

$$\max_{1\le i\le s} \operatorname{lp}(h_i)\operatorname{lp}(f_i).$$

In particular the syzygy $\frac{X_{ij}}{c_iX_i}\boldsymbol{e}_i - \frac{X_{ij}}{c_jX_j}\boldsymbol{e}_j$ of $\begin{bmatrix} c_1X_1 & \cdots & c_sX_s \end{bmatrix}$ gives rise to the S-polynomial of f_i and f_j, since

$$\begin{bmatrix} f_1 & \cdots & f_s \end{bmatrix}\Big(\frac{X_{ij}}{c_iX_i}\boldsymbol{e}_i - \frac{X_{ij}}{c_jX_j}\boldsymbol{e}_j\Big) =$$

$$\begin{bmatrix} f_1 & \cdots & f_s \end{bmatrix}(0, \dots, 0, \underbrace{\frac{X_{ij}}{c_iX_i}}_{\substack{i\text{th}\\ \text{coord}}}, 0, \dots, 0, \underbrace{-\frac{X_{ij}}{c_jX_j}}_{\substack{j\text{th}\\ \text{coord}}}, 0, \dots, 0) =$$

$$\frac{X_{ij}}{c_iX_i}f_i - \frac{X_{ij}}{c_jX_j}f_j = S(f_i, f_j).$$

This last observation and Proposition 3.2.3 seem to indicate that the syzygy module of $\begin{bmatrix} \operatorname{lt}(f_1) & \cdots & \operatorname{lt}(f_s) \end{bmatrix}$ might be relevant to the computation of a Gröbner basis for $\langle f_1, \dots, f_s\rangle$. In order to implement this idea we need the following definition.

DEFINITION 3.2.4. *Let $X_1, \dots, X_s$ be power products and $c_1, \dots, c_s$ be in $k - \{0\}$. Then, for a power product X, we call a syzygy $\boldsymbol{h} = (h_1, \dots, h_s) \in$* Syz$(c_1X_1, \dots, c_sX_s)$ homogeneous *of degree X provided that each h_i is a term (that is, $\operatorname{lt}(h_i) = h_i$ for all i) and $X_i\operatorname{lp}(h_i) = X$ for all i such that $h_i \neq 0$. We say that $\boldsymbol{h} \in$* Syz$(c_1X_1, \dots, c_sX_s)$ *is* homogeneous *if it is homogeneous of degree X for some power product X.*

Note that the generating set given in Proposition 3.2.3 consists of a finite set of homogeneous syzygies. The next theorem presents another equivalent condition for a set to be a Gröbner basis.

THEOREM 3.2.5. *Let $G = \{g_1, \dots, g_t\}$ be a set of non-zero polynomials in A. Let $\mathcal{B}$ be a homogeneous generating set of* Syz$(\operatorname{lt}(g_1), \dots, \operatorname{lt}(g_t))$. *Then G is a Gröbner basis for the ideal $\langle g_1, \dots, g_t\rangle$ if and only if for all $(h_1, \dots, h_t) \in \mathcal{B}$, we have*

$$h_1g_1 + \cdots + h_tg_t \xrightarrow{G}_+ 0.$$

PROOF. If G is a Gröbner basis, then by Theorem 1.6.2,

$$h_1 g_1 + \cdots + h_t g_t \xrightarrow{G}_+ 0$$

for all $h_1, \ldots, h_t$.

Conversely, let $g \in \langle g_1, \ldots, g_t \rangle$, say

$$g = u_1 g_1 + \cdots + u_t g_t. \tag{3.2.3}$$

Choose a representation as in Equation (3.2.3) with

$$X = \max_{1 \leq i \leq t} (\operatorname{lp}(u_i) \operatorname{lp}(g_i))$$

least. Since, by Theorem 1.6.2, we need to show that g can be written as in Equation (3.2.3) with $X = \operatorname{lp}(g)$, we may assume that $\operatorname{lp}(g) < X$ and show that then we can obtain an expression such as (3.2.3) for g with a smaller value for X. Let $S = \{i \in \{1, \ldots, t\} \mid \operatorname{lp}(u_i) \operatorname{lp}(g_i) = X\}$. Then

$$\sum_{i \in S} \operatorname{lt}(u_i) \operatorname{lt}(g_i) = 0.$$

Let $\boldsymbol{h} = \sum_{i \in S} \operatorname{lt}(u_i) \boldsymbol{e}_i$ (where $\boldsymbol{e}_1, \ldots, \boldsymbol{e}_t$ is the standard basis for A^t). Then $\boldsymbol{h} \in \operatorname{Syz}(\operatorname{lt}(g_1), \ldots, \operatorname{lt}(g_t))$ and $\boldsymbol{h}$ is homogeneous. Now let $\mathcal{B} = \{\boldsymbol{h}_1, \ldots, \boldsymbol{h}_\ell\}$, where for each $j = 1, \ldots, \ell$, $\boldsymbol{h}_j = (h_{1j}, \ldots, h_{tj})$. So $\boldsymbol{h} = \sum_{j=1}^{\ell} a_j \boldsymbol{h}_j$, where, since $\boldsymbol{h}$ is a homogeneous syzygy, we may assume that the a_j's are terms such that $\operatorname{lp}(a_j) \operatorname{lp}(h_{ij}) \operatorname{lp}(g_i) = X$ for all i, j such that $a_j h_{ij} \neq 0$. By hypothesis, for each j, $\sum_{i=1}^{t} h_{ij} g_i \xrightarrow{G}_+ 0$. Thus by Theorem 1.5.9 we have for each $j = 1, \ldots, \ell$

$$\sum_{i=1}^{t} h_{ij} g_i = \sum_{i=1}^{t} v_{ij} g_i,$$

such that

$$\max_{1 \leq i \leq t} \operatorname{lp}(v_{ij}) \operatorname{lp}(g_i) = \operatorname{lp}(\sum_{i=1}^{t} h_{ij} g_i) < \max_{1 \leq i \leq t} \operatorname{lp}(h_{ij}) \operatorname{lp}(g_i).$$

The latter strict inequality is because $\sum_{i=1}^{t} h_{ij} \operatorname{lt}(g_i) = 0$.

Thus,

$$\begin{aligned}
g &= u_1 g_1 + \cdots + u_t g_t \\
&= \sum_{i \in S} \operatorname{lt}(u_i) g_i + \underbrace{\sum_{i \in S} (u_i - \operatorname{lt}(u_i)) g_i + \sum_{i \notin S} u_i g_i}_{\text{terms lower than } X} \\
&= \sum_{j=1}^{\ell} \sum_{i=1}^{t} a_j h_{ij} g_i + \text{ terms lower than } X \\
&= \sum_{j=1}^{\ell} \sum_{i=1}^{t} a_j v_{ij} g_i + \text{ terms lower than } X\ .
\end{aligned}$$

We have

$$\max_{i,j} \operatorname{lp}(a_j)\operatorname{lp}(v_{ij})\operatorname{lp}(g_i) < \max_{i,j} \operatorname{lp}(a_j)\operatorname{lp}(h_{ij})\operatorname{lp}(g_i) = X.$$

We have obtained a representation of g as a linear combination of the g_i's such that the maximum leading power product of any summand is less than X. Thus the theorem is proved. □

In Exercise 3.2.1 we will give an example which shows that the hypothesis that the generating set of syzygies be homogeneous is necessary.

We observe that the above proof uses an argument similar to the one used in the proof of Theorem 1.7.4. In fact, as a corollary to the above result, we can recover Theorem 1.7.4.

COROLLARY 3.2.6. *Let $G = \{g_1, \ldots, g_t\}$ be a set of non-zero polynomials in A. Then G is a Gröbner basis if and only if for all $i, j = 1, \ldots, t$, $S(g_i, g_j) \xrightarrow{G}_+ 0$.*

PROOF. Let G be a Gröbner basis. Then, since $S(g_i, g_j)$ is an element of the ideal $\langle g_1, \ldots, g_t\rangle$, we have $S(g_i, g_j) \xrightarrow{G}_+ 0$.

For the converse, we first use Proposition 3.2.3 to see that the set

$$\mathcal{B} = \left\{ \frac{X_{ij}}{\operatorname{lt}(g_i)}\mathbf{e}_i - \frac{X_{ij}}{\operatorname{lt}(g_j)}\mathbf{e}_j \mid i < j, i, j = 1, \ldots, t \right\} \subseteq A^t$$

is a homogeneous generating set of the syzygy module of $\begin{bmatrix} \operatorname{lt}(g_1) & \cdots & \operatorname{lt}(g_t) \end{bmatrix}$. As we have noted after Proposition 3.2.3, each element of $\mathcal{B}$ gives rise to an S-polynomial, which reduces to zero by hypothesis. Therefore, by Theorem 3.2.5, G is a Gröbner basis. □

Exercises

3.2.1. In Theorem 3.2.5 the generating set $\mathcal{B}$ of $\operatorname{Syz}(\operatorname{lt}(g_1), \ldots, \operatorname{lt}(g_t))$ was required to be homogeneous. In this exercise we show that this hypothesis is necessary.

Consider the set $G = \{g_1, g_2\}$, where $g_1 = x + y$, $g_2 = x + 1 \in \mathbb{Q}[x, y]$. We will use lex with $x > y$.

a. Prove that G is not a Gröbner basis for the ideal it generates.

b. Prove that the set $\{(x + 1, -x - 1), (x, -x)\}$ is a generating set for $\operatorname{Syz}(\operatorname{lt}(g_1), \operatorname{lt}(g_2))$.

c. Prove that $(x + 1)g_1 + (-x - 1)g_2 \xrightarrow{G}_+ 0$ and that $xg_1 - xg_2 \xrightarrow{G}_+ 0$.

3.2.2. Give another example that shows that the hypothesis that $\mathcal{B}$ be homogeneous in Theorem 3.2.5 is necessary.

3.2.3. Let $G = \{g_1, \ldots, g_t\}$ be a set of non-zero polynomials in A. Let $\mathcal{B}$ be a homogeneous generating set of $\operatorname{Syz}(\operatorname{lt}(g_1), \ldots, \operatorname{lt}(g_t))$. Prove that G is a Gröbner basis for the ideal $\langle g_1, \ldots, g_t\rangle$ if and only if for all $(h_1, \ldots, h_t) \in \mathcal{B}$, we have

$$h_1g_1 + \cdots + h_tg_t = v_1g_1 + \cdots + v_tg_t,$$

where

$$\operatorname{lp}(h_1g_1 + \cdots + h_tg_t) = \max(\operatorname{lp}(v_1)\operatorname{lp}(g_1), \ldots, \operatorname{lp}(v_t)\operatorname{lp}(g_t)).$$

[Hint: See the proof of Theorem 3.2.5.]

3.2.4. At this point we do not know yet how to compute generators for the syzygy module of $\left[\begin{array}{ccc} f_1 & \cdots & f_s \end{array}\right]$. We will see in Section 3.4 how to do this. However, in certain instances we can easily find some elements of $\operatorname{Syz}(f_1, \ldots, f_s)$. Let I be the ideal of $k[x, y]$ generated by f_1, f_2, f_3, the 2×2 minors of the 2×3 matrix S, where

$$S = \left[\begin{array}{ccc} x & y & x \\ 1 & x & y \end{array}\right].$$

a. Give a method for finding elements of $\operatorname{Syz}(f_1, f_2, f_3)$. [Hint: Think of 3×3 matrices whose determinant must be zero, and obtain 2 syzygies this way.]

b. Generalize this idea to the case of $m \times (m+1)$ matrices with entries in $k[x_1, \ldots, x_n]$.

3.3. Improvements on Buchberger's Algorithm. So far we have not discussed the computational aspects of Buchberger's Algorithm presented in Section 1.7 (Algorithm 1.7.1). A careful look at this issue is outside the scope of this book (see, for example, [**BaSt88, Bu83, GMNRT, Huy, MaMe, Laz91**]). However the computational complexity of Buchberger's Algorithm often makes it difficult to actually compute a Gröbner basis for even small problems and this limits, in practice, the scope of the applications of the theory. So we will now discuss improvements in the algorithm for computing Gröbner bases. We note that the results of this section are rather technical and will only be used occasionally in some of the exercises and examples in the remainder of the book.

Some algebraic results, such as the results presented below, can be used in conjunction with some simple heuristic observations to improve significantly Buchberger's Algorithm (see [**Bu79**]). It can also be shown (but we shall not do it here) that certain algebraic (respectively geometric) properties of the ideal I (respectively of the variety $V(I)$) have a direct influence on the difficulty of the computation of a Gröbner basis for I (see the references above).

Recall that the algorithm has two steps: the computation of S-polynomials and their reduction. A problem that arises is the potentially very large number of S-polynomials that have to be computed. Indeed, as the computation progresses, the number of polynomials in the basis gets larger, and therefore, each time a new polynomial is added to the basis, the number of S-polynomials to compute also increases. Since the algorithm does terminate, the proportion of S-polynomials which reduce to zero eventually increases as we get far in the algorithm. A huge amount of computation might be performed for very little gain, since few new polynomials are added to the basis. In fact, at some point before the algorithm terminates, the desired Gröbner basis is obtained but we do not "know" it.

At that point, the computation of S-polynomials and their reductions are all together useless except for the fact that they verify that we do have a Gröbner basis. One way to improve this situation is to "predict" that some S-polynomials reduce to zero without actually computing these S-polynomials or reducing them. The following results are the basis for two criteria which will allow us to ignore the computation of many S-polynomials.

LEMMA 3.3.1. *Let $f, g \in A = k[x_1, \ldots, x_n]$, both non-zero, and let $d = \gcd(f, g)$. The following statements are equivalent:*

(i) $\mathrm{lp}(\frac{f}{d})$ *and* $\mathrm{lp}(\frac{g}{d})$ *are relatively prime;*

(ii) $S(f,g) \xrightarrow{\{f,g\}}_+ 0$.

In particular, $\{f, g\}$ is a Gröbner basis if and only if $\mathrm{lp}(\frac{f}{d})$ *and* $\mathrm{lp}(\frac{g}{d})$ *are relatively prime.*

PROOF. (i) $\Longrightarrow$ (ii). Assume first that $d = \gcd(f, g) = 1$. We write $f = aX + f'$, $g = bY + g'$, where $\mathrm{lt}(f) = aX$, $\mathrm{lt}(g) = bY$, $a, b \in k$, and X, Y are power products. Then $X = \frac{1}{a}(f - f')$, and $Y = \frac{1}{b}(g - g')$.

CASE 1. $f' = g' = 0$. Then f and g are both terms and $S(f, g) = 0$.

CASE 2. $f' = 0$ and $g' \neq 0$. Then, since $\gcd(\mathrm{lp}(f), \mathrm{lp}(g)) = 1$, we have $S(f,g) = \frac{1}{a}Yf - \frac{1}{b}Xg = \frac{1}{ab}(g - g')f - \frac{1}{ab}fg = -\frac{1}{ab}g'f$. We see that (Exercise 1.5.4) $S(f, g) \xrightarrow{f}_+ 0$.

CASE 3. $f' \neq 0$ and $g' = 0$. This is the same as Case 2.

CASE 4. $f' \neq 0$ and $g' \neq 0$. Then, since $\gcd(\mathrm{lp}(f), \mathrm{lp}(g)) = 1$, we have

$$S(f,g) = \frac{1}{a}Yf - \frac{1}{b}Xg = \frac{1}{ab}(g - g')f - \frac{1}{ab}(f - f')g = \frac{1}{ab}(f'g - g'f).$$

If $\mathrm{lp}(f'g) = \mathrm{lp}(g'f)$, then $\mathrm{lp}(f')\,\mathrm{lp}(g) = \mathrm{lp}(f'g) = \mathrm{lp}(g'f) = \mathrm{lp}(g')\,\mathrm{lp}(f)$, and since $\gcd(\mathrm{lp}(f), \mathrm{lp}(g)) = 1$, we have $\mathrm{lp}(f)$ divides $\mathrm{lp}(f')$ and $\mathrm{lp}(g)$ divides $\mathrm{lp}(g')$. This is a contradiction, since $\mathrm{lp}(f') < \mathrm{lp}(f)$ and $\mathrm{lp}(g') < \mathrm{lp}(g)$. Therefore $\mathrm{lp}(f'g) \neq \mathrm{lp}(g'f)$, and the leading term of $\frac{1}{ab}(f'g - g'f)$ appears in $f'g$ or $g'f$ and hence is a multiple of $\mathrm{lp}(f)$ or $\mathrm{lp}(g)$. If $\mathrm{lp}(f'g) > \mathrm{lp}(g'f)$, then

$$S(f,g) \xrightarrow{g} \frac{1}{ab}((f' - \mathrm{lt}(f'))g - g'f).$$

If $\mathrm{lp}(f'g) < \mathrm{lp}(g'f)$, then

$$S(f,g) \xrightarrow{f} \frac{1}{ab}(f'g - (g' - \mathrm{lt}(g'))f).$$

Using an argument similar to the one above, we see that the leading term of $\frac{1}{ab}((f' - \mathrm{lt}(f'))g - g'f)$ or $\frac{1}{ab}(f'g - (g' - \mathrm{lt}(g'))f)$ is a multiple of $\mathrm{lp}(f)$ or $\mathrm{lp}(g)$. Therefore this reduction process continues using only f or g. At each stage of the reduction the remainder has a leading term which is a multiple of $\mathrm{lp}(f)$ or $\mathrm{lp}(g)$. We see that we can continue this process until we obtain 0, that is, $S(f,g) \xrightarrow{\{f,g\}}_+ 0$.

Now assume that $d = \gcd(f, g) \neq 1$. Then $\gcd(\frac{f}{d}, \frac{g}{d}) = 1$. By assumption we have $\gcd(\mathrm{lp}(\frac{f}{d}), \mathrm{lp}(\frac{g}{d})) = 1$, and hence, by the case above, $\{\frac{f}{d}, \frac{g}{d}\}$ is a Gröbner basis. Thus $\{d\frac{f}{d}, d\frac{g}{d}\}$ is also a Gröbner basis (Exercise 1.6.13). By Theorem 1.7.4, $S(f, g) \xrightarrow{\{f,g\}}_+ 0$.

(ii) $\Longrightarrow$ (i). Let us first assume that $\gcd(f, g) = 1$. We need to show that $\mathrm{lp}(f)$ and $\mathrm{lp}(g)$ are relatively prime. Let $\mathrm{lp}(f) = DX$ and $\mathrm{lp}(g) = DY$, where D, X, and Y are power products in A, $\gcd(X, Y) = 1$. Then

$$S(f, g) = \frac{Y}{\mathrm{lc}(f)} f - \frac{X}{\mathrm{lc}(g)} g.$$

By assumption we have $S(f, g) \xrightarrow{\{f,g\}}_+ 0$, and hence there exist $u, v \in k[x_1, \dots, x_n]$ such that

$$S(f, g) = \frac{Y}{\mathrm{lc}(f)} f - \frac{X}{\mathrm{lc}(g)} g = uf + vg, \tag{3.3.1}$$

where $\mathrm{lp}(uf) \leq \mathrm{lp}(S(f, g))$ and $\mathrm{lp}(vg) \leq \mathrm{lp}(S(f, g))$. From Equation (3.3.1), we obtain

$$\left(\frac{X}{\mathrm{lc}(g)} + v\right) g = \left(\frac{Y}{\mathrm{lc}(f)} - u\right) f.$$

Therefore f divides $(\frac{X}{\mathrm{lc}(g)} + v)$, and g divides $(\frac{Y}{\mathrm{lc}(f)} - u)$, since f and g are relatively prime. Also,

$$\mathrm{lp}(u)DX = \mathrm{lp}(uf) \leq \mathrm{lp}(S(f, g)) < X\,\mathrm{lp}(g) = Y\,\mathrm{lp}(f) = DXY.$$

Thus, $\mathrm{lp}(u) < Y$, and hence $\mathrm{lp}(\frac{Y}{\mathrm{lc}(f)} - u) = Y$. But g divides $(\frac{Y}{\mathrm{lc}(f)} - u)$, so $\mathrm{lp}(g) = DY$ divides $\mathrm{lp}(\frac{Y}{\mathrm{lc}(f)} - u) = Y$, and hence $D = 1$. Therefore $\mathrm{lp}(f)$ and $\mathrm{lp}(g)$ are relatively prime.

Now assume that $d = \gcd(f, g) \neq 1$. Then $\gcd(\frac{f}{d}, \frac{g}{d}) = 1$. It is easy to prove that if $S(f, g) \xrightarrow{\{f,g\}}_+ 0$, then $\frac{1}{d}S(f, g) \xrightarrow{\{\frac{f}{d},\frac{g}{d}\}}_+ 0$ (see Exercise 3.3.3). It is also easy to prove that $\frac{1}{d}S(f, g) = S(\frac{f}{d}, \frac{g}{d})$, since d is monic (see Exercise 3.3.3). Therefore, by the above, we have $\mathrm{lp}(\frac{f}{d})$ and $\mathrm{lp}(\frac{g}{d})$ are relatively prime as desired.

The last statement of the lemma is an immediate consequence of Theorem 1.7.4. □

Lemma 3.3.1 gives a criterion for a priori zero reduction: during Buchberger's Algorithm, whenever f and g are such that $\mathrm{lp}(\frac{f}{d})$ and $\mathrm{lp}(\frac{g}{d})$ are relatively prime, then it is not necessary to compute $S(f, g)$, since $S(f, g)$ will reduce to zero using f and g alone, and hence $S(f, g)$ will not create a new polynomial in the basis. We note that if $\mathrm{lp}(f)$ and $\mathrm{lp}(g)$ are relatively prime, then $d = 1$ and so $S(f, g) \xrightarrow{\{f,g\}}_+ 0$. This is the form in which we will use the criterion below (see crit1).

Now we turn our attention to another criterion that turns out to be remarkably effective in improving the performance of Buchberger's Algorithm.

LEMMA 3.3.2. *Let $X_1, X_2, \ldots, X_s$ be power products in $A = k[x_1, \ldots, x_n]$ and let $c_1, \ldots, c_s \in k - \{0\}$. For $i, j = 1, \ldots, s$, define $X_{ij} = \operatorname{lcm}(X_i, X_j)$, and let*

$$\tau_{ij} = \frac{X_{ij}}{c_i X_i} \boldsymbol{e}_i - \frac{X_{ij}}{c_j X_j} \boldsymbol{e}_j \in \operatorname{Syz}(c_1 X_1, \ldots, c_s X_s) \subseteq A^s,$$

where $\boldsymbol{e}_1, \ldots, \boldsymbol{e}_s$ is the standard basis for A^s. For each $i, j, \ell = 1, \ldots, s$ let $X_{ij\ell} = \operatorname{lcm}(X_i, X_j, X_\ell)$. Then we have

$$\frac{X_{ij\ell}}{X_{ij}} \tau_{ij} + \frac{X_{ij\ell}}{X_{j\ell}} \tau_{j\ell} + \frac{X_{ij\ell}}{X_{\ell i}} \tau_{\ell i} = 0.$$

Moreover, if X_ℓ divides X_{ij}, then τ_{ij} is in the submodule of A^s generated by $\tau_{j\ell}$ and $\tau_{\ell i}$.

PROOF. We have

$$\frac{X_{ij\ell}}{X_{ij}} \tau_{ij} + \frac{X_{ij\ell}}{X_{j\ell}} \tau_{j\ell} + \frac{X_{ij\ell}}{X_{\ell i}} \tau_{\ell i} =$$

$$\frac{X_{ij\ell}}{X_{ij}} \left(\frac{X_{ij}}{c_i X_i} \boldsymbol{e}_i - \frac{X_{ij}}{c_j X_j} \boldsymbol{e}_j\right) + \frac{X_{ij\ell}}{X_{j\ell}} \left(\frac{X_{j\ell}}{c_j X_j} \boldsymbol{e}_j - \frac{X_{j\ell}}{c_\ell X_\ell} \boldsymbol{e}_\ell\right) + \frac{X_{ij\ell}}{X_{\ell i}} \left(\frac{X_{\ell i}}{c_\ell X_\ell} \boldsymbol{e}_\ell - \frac{X_{\ell i}}{c_i X_i} \boldsymbol{e}_i\right)$$

$$= \frac{X_{ij\ell}}{c_i X_i} \boldsymbol{e}_i - \frac{X_{ij\ell}}{c_j X_j} \boldsymbol{e}_j + \frac{X_{ij\ell}}{c_j X_j} \boldsymbol{e}_j - \frac{X_{ij\ell}}{c_\ell X_\ell} \boldsymbol{e}_\ell + \frac{X_{ij\ell}}{c_\ell X_\ell} \boldsymbol{e}_\ell - \frac{X_{ij\ell}}{c_i X_i} \boldsymbol{e}_i = 0.$$

Now if X_ℓ divides X_{ij}, then $X_{ij\ell} = X_{ij}$, and we have

$$\tau_{ij} + \frac{X_{ij}}{X_{j\ell}} \tau_{j\ell} + \frac{X_{ij}}{X_{\ell i}} \tau_{\ell i} = 0.$$

Hence τ_{ij} is in the submodule of A^s generated by $\tau_{j\ell}$ and $\tau_{\ell i}$. □

COROLLARY 3.3.3. *We continue to use the notation of Lemma 3.3.2. Let $\mathcal{B} \subseteq \{\tau_{ij} \mid 1 \leq i < j \leq s\}$ be a generating set for $\operatorname{Syz}(c_1 X_1, \ldots, c_s X_s)$. Suppose we have three distinct indices i, j, ℓ such that $\tau_{i\ell}, \tau_{j\ell}, \tau_{ij} \in \mathcal{B}$, and such that X_ℓ divides $X_{ij} = \operatorname{lcm}(X_i, X_j)$. Then $\mathcal{B} - \{\tau_{ij}\}$ is also a generating set for $\operatorname{Syz}(c_1 X_1, \ldots, c_s X_s)$.*

We will use Corollary 3.3.3 to improve Buchberger's Algorithm in the following way. Let $\{f_1, \ldots, f_s\}$ be a set of generators for an ideal I in $k[x_1, \ldots, x_n]$. Let $c_i X_i = \operatorname{lt}(f_i)$ and use the notation above. We begin by letting $\mathcal{B} = \{\tau_{ij} \mid 1 \leq i < j \leq s\}$. Of course, $\mathcal{B}$ generates $\operatorname{Syz}(\operatorname{lt}(f_1), \ldots, \operatorname{lt}(f_s))$. We apply Corollary 3.3.3 to eliminate as many of the $\tau_{ij} \in \mathcal{B}$ as possible obtaining a possibly smaller set of generators for $\operatorname{Syz}(\operatorname{lt}(f_1), \ldots, \operatorname{lt}(f_s))$. We then compute the S-polynomial, $S(f_i, f_j)$, corresponding to one of the τ_{ij} remaining in $\mathcal{B}$ and reduce it as far as possible; we add the reduction to the set $\{f_1, \ldots, f_s\}$ if it does not reduce to zero, calling it f_{s+1}. We enlarge $\mathcal{B}$ by the set $\{\tau_{i,s+1} \mid 1 \leq i \leq s\}$ to obtain a new set $\mathcal{B}$ which now generates $\operatorname{Syz}(\operatorname{lt}(f_1), \ldots, \operatorname{lt}(f_s), \operatorname{lt}(f_{s+1}))$. We again apply Corollary 3.3.3 to eliminate as many of the $\tau_{ij} \in \mathcal{B}$ as possible obtaining a possibly smaller set of generators for $\operatorname{Syz}(\operatorname{lt}(f_1), \ldots, \operatorname{lt}(f_s), \operatorname{lt}(f_{s+1}))$. We again compute an S-polynomial corresponding to a τ_{ij} remaining in $\mathcal{B}$. We continue this process until all of the S-polynomials corresponding to elements in $\mathcal{B}$ have been computed and reduced to zero, always maintaining $\mathcal{B}$ as a basis of the current syzygy module.

In order to keep track, in the algorithm, of those $\tau_{ij} \in \mathcal{B}$ whose corresponding S-polynomial has been computed and reduced we break up the basis $\mathcal{B}$ into two parts. We also use just the indices. So we will use $\mathcal{NC}$ for the set of all indices $\{i, j\}$ of $\tau_{ij} \in \mathcal{B}$ at any given time for which the S-polynomial has not been computed and use $\mathcal{C}$ for the set of all indices $\{i, j\}$ of $\tau_{ij} \in \mathcal{B}$ at any given time for which the S-polynomial has been computed. We note that at any time in the algorithm after $\mathcal{NC}$ has been initialized, $\{\tau_{ij} \mid \{i, j\} \in \mathcal{NC} \cup \mathcal{C}\}$ is a generating set of the syzygy module of the current set of leading terms (see the proof of Proposition 3.3.4). So we continue the algorithm until $\mathcal{NC} = \emptyset$.

We now give an improved version of Buchberger's Algorithm as Algorithm 3.3.1. We note that the purpose of the first WHILE loop is to initialize $\mathcal{NC}$.

The commands used in the algorithm are defined as follows.

The command $\text{crit1}(i, j)$ returns "TRUE" if and only if $\text{lp}(f_i)$ and $\text{lp}(f_j)$ are relatively prime. If $\text{crit1}(i, j) = \text{TRUE}$, then, by Lemma 3.3.1, we know without computing it, that $S(f_i, f_j)$ reduces to zero. Nevertheless $\{i, j\}$ must be added to $\mathcal{C}$.

The command $\text{crit2}(\mathcal{NC}, \mathcal{C}, s)$ is given as Algorithm 3.3.2.

Algorithm 3.3.2 is an implementation of the ideas in Corollary 3.3.3 and the ensuing discussion. We now make a few technical points about this algorithm. The basic idea is to find triples of indices ν, μ, ρ such that $\{\nu, \mu\}, \{\nu, \rho\}, \{\mu, \rho\}$ are in $\mathcal{NC} \cup \mathcal{C}$ and X_ν divides $\text{lcm}(X_\mu, X_\rho)$. Because of the way we call this procedure in the main algorithm (Algorithm 3.3.1), we need only consider the cases where one of ν, μ, ρ is s. This is because the cases of all triples with ν, μ, ρ all less than s were checked before. Since μ, ρ are interchangeable, it is enough to consider the cases $\rho = s$ and $\nu = s$. These two cases are the two main WHILE loops in Algorithm 3.3.2. Moreover, we note that a pair of the form $\{i, s\}$ cannot lie in $\mathcal{C}$ (this explains why checking membership in $\mathcal{NC} \cup \mathcal{C}$ was often just done by checking membership in $\mathcal{NC}$). Finally, we only check, in the second main WHILE loop, whether $\{i, j\}$ is in $\mathcal{NC}$ since we are only interested in eliminating it from $\mathcal{NC}$.

PROPOSITION 3.3.4. *Given a set of non-zero polynomials $F = \{f_1, \ldots, f_s\}$, the Improved Buchberger's Algorithm (Algorithm 3.3.1) will produce a Gröbner basis for the ideal $I = \langle F \rangle$.*

PROOF. Let $G = \{f_1, \ldots, f_t\}$ $(t \geq s)$ be the output of Algorithm 3.3.1. We first note that the S-polynomials corresponding to every pair in $\mathcal{C}$ reduce to zero. It then suffices to show that $\{\tau_{ij} | \{i, j\} \in \mathcal{C}\}$ is a generating set for $\text{Syz}(\text{lt}(f_1), \ldots, \text{lt}(f_t))$, and for this it suffices to show that at any stage of the algorithm $\{\tau_{ij} | \{i, j\} \in \mathcal{NC} \cup \mathcal{C}\}$ is a generating set for the syzygy module of current leading terms. We see this as follows. At each stage of the algorithm there is one of two possibilities. Either the S-polynomial reduces to zero, and

INPUT: $F = \{f_1, \ldots, f_s\} \subseteq k[x_1, \ldots, x_n]$ with $f_i \neq 0$ $(1 \leq i \leq s)$

OUTPUT: A Gröbner basis G for $\langle f_1, \ldots, f_s \rangle$

INITIALIZATION: $G := F$
$\mathcal{C} := \emptyset$
$\mathcal{NC} := \{\{1, 2\}\}$
$i := 2$

WHILE $i < s$ **DO**

$\mathcal{NC} := \mathcal{NC} \cup \{\{j, i+1\} \mid 1 \leq j \leq i\}$

$\mathcal{NC} :=$ crit2$(\mathcal{NC}, \mathcal{C}, i+1)$

$i := i + 1$

WHILE $\mathcal{NC} \neq \emptyset$ **DO**

Choose $\{i, j\} \in \mathcal{NC}$

$\mathcal{NC} := \mathcal{NC} - \{\{i, j\}\}$

$\mathcal{C} := \mathcal{C} \cup \{\{i, j\}\}$

IF crit1(i, j) = FALSE **THEN**

$S(f_i, f_j) \xrightarrow{G}_+ h$, where h is reduced with respect to G

IF $h \neq 0$ **THEN**

$f_{s+1} := h$

$G := G \cup \{f_{s+1}\}$

$s := s + 1$

$\mathcal{NC} := \mathcal{NC} \cup \{\{i, s\} \mid 1 \leq i \leq s - 1\}$

$\mathcal{NC} :=$ crit2$(\mathcal{NC}, \mathcal{C}, s)$

ALGORITHM 3.3.1. *Improved Buchberger's Algorithm*

$\mathcal{NC} \cup \mathcal{C}$ does not change, or a polynomial is added to G, and the relevant pairs are added to $\mathcal{NC} \cup \mathcal{C}$ and so the set of τ_{ij}'s corresponding to this updated $\mathcal{NC} \cup \mathcal{C}$ is a generating set for the syzygy module of the new set of leading terms. Then we apply crit2 to the new $\mathcal{NC}$ which does not alter this last statement by Corollary 3.3.3.

The algorithm stops for the same reason Algorithm 1.7.1 stopped. □

We note that in Algorithm 3.3.1 we do not give a rule for choosing the pair

INPUT: $\mathcal{NC}, \mathcal{C}, s$ from Algorithm 3.3.1

OUTPUT: $\mathcal{NC}$ with pairs deleted using Corollary 3.3.3

INITIALIZATION: $\ell := 1$

WHILE $\ell < s$ **DO**

  IF $\{\ell, s\} \in \mathcal{NC}$ **THEN**

    $i := 1$

    WHILE $i < s$ **DO**

      IF $\{i, \ell\} \in \mathcal{NC} \cup \mathcal{C}$ **AND** $\{i, s\} \in \mathcal{NC}$ **THEN**

        IF X_ℓ divides $\mathrm{lcm}(X_i, X_s)$ **THEN**

          $\mathcal{NC} := \mathcal{NC} - \{\{i, s\}\}$

      $i := i + 1$

  $\ell := \ell + 1$

$i := 1$

WHILE $i < s$ **DO**

  IF $\{i, s\} \in \mathcal{NC}$ **THEN**

    $j := i + 1$

    WHILE $j < s$ **DO**

      IF $\{j, s\} \in \mathcal{NC}$ **AND** $\{i, j\} \in \mathcal{NC}$ **THEN**

        IF X_s divides $\mathrm{lcm}(X_i, X_j)$ **THEN**

          $\mathcal{NC} := \mathcal{NC} - \{\{i, j\}\}$

      $j := j + 1$

  $i := i + 1$

ALGORITHM 3.3.2. *crit2*($\mathcal{NC}, \mathcal{C}, s$)

$\{i, j\} \in \mathcal{NC}$ for which we compute the corresponding S-polynomial. Some studies have shown that this choice is of vital importance. Often the S-polynomials are computed in such a way that $S(f_i, f_j)$ is computed first if $\mathrm{lcm}(\mathrm{lp}(f_i), \mathrm{lp}(f_j))$ is least (with respect to the current term order) among the $\mathrm{lcm}(\mathrm{lp}(f_\nu), \mathrm{lp}(f_\mu))$. This procedure is called the *normal selection strategy.* Experimental results show that this works very well for degree compatible term orders. It is not so good for the lex term ordering. This can be explained as follows: if at some point the basis contains polynomials $f_1, \ldots, f_\ell$ in which the largest variable (say x_n) does not appear, then the normal selection strategy will first consider only $f_1, \ldots, f_\ell$, disregarding the ones in which x_n appears; in effect, a Gröbner basis for $\langle f_1, \ldots, f_\ell \rangle$ will be computed. Experimental and theoretical results show that this Gröbner basis subcomputation can be worse than the original Gröbner basis computation. There are techniques to get around this problem but they are beyond the scope of this book.

There is another technique that is used to improve the performance of Buchberger's Algorithm. Namely, the polynomials f_i's are inter-reduced at the beginning, and the basis is kept inter-reduced as the algorithm progresses. Even though this may require a lot of computation, it helps to avoid an unmanageable growth in the number of polynomials in the basis, and in the number of divisions necessary throughout the computation. This method is discussed in Buchberger [**Bu85**].

We conclude by noting that the use of crit1 and crit2 help to reduce the number of S-polynomials that have to be computed. In fact, empirical evidence shows that if N is the number of S-polynomials that would be computed without these criteria, the use of these two criteria reduces that number to about $\sqrt{N}$ (see, for example, [**Cz**]).

To illustrate Algorithm 3.3.1 we consider the following example. Since the polynomials generated by the algorithm are scattered throughout the text of the example, we have put boxes around these polynomials for easier reference.

EXAMPLE 3.3.5. Consider polynomials $\boxed{f_1 = x^2y^2 - z^2,}$ $\boxed{f_2 = xy^2z - xyz,}$ and $\boxed{f_3 = xyz^3 - xz^2}$ in $\mathbb{Q}[x, y, z]$. We use the deglex term order with $x < y < z$. We follow Algorithm 3.3.1. However, we will not trace the algorithm crit2 except for one significant example at the end.

We start with $G = \{f_1, f_2, f_3\}$, $\mathcal{C} = \emptyset$, and $\mathcal{NC} = \{\{1,2\},\{1,3\},\{2,3\}\}$. We see that we do not have to consider $\{1,3\}$, that is, we have $\mathcal{NC}$ =crit2($\mathcal{NC}$,$\mathcal{C}$,3) = $\{\{1,2\},\{2,3\}\}$. We choose the pair $\{1,2\}$, changing $\mathcal{NC}$ to $\{\{2,3\}\}$ and $\mathcal{C}$ to $\{\{1,2\}\}$, and compute $S(f_1, f_2) = x^2yz - z^3$ which is reduced with respect to G. Thus we add $\boxed{f_4 = x^2yz - z^3}$ to G. We update $\mathcal{NC}$ to include the pairs with 4 in them and then we use crit2 to compute the new $\mathcal{NC} = \{\{2,3\},\{1,4\},\{3,4\}\}$. We now choose the pair $\{1,4\}$, changing $\mathcal{NC}$ to $\mathcal{NC} = \{\{2,3\},\{3,4\}\}$ and $\mathcal{C}$ to $\mathcal{C} = \{\{1,2\},\{1,4\}\}$, and compute $S(f_1, f_4) = yz^3 - z^3$ which is reduced with respect to G. Thus we add $\boxed{f_5 = yz^3 - z^3}$ to G. After applying crit2 again we obtain $\mathcal{NC} = \{\{2,3\},\{3,4\},\{3,5\}\}$. We choose the pair $\{3,5\}$, so that now $\mathcal{NC} = \{\{2,3\},\{3,4\}\}$ and $\mathcal{C} = \{\{1,2\},\{1,4\},\{3,5\}\}$, and we compute $S(f_3, f_5) = xz^3 - xz^2$ which is reduced with respect to G. Now we add $\boxed{f_6 = xz^3 - xz^2}$ to G. Applying crit2 again we get $\mathcal{NC} = \{\{2,3\},\{3,4\},\{3,6\}\}$. Next we choose the pair $\{3,6\}$, update $\mathcal{NC}$ and $\mathcal{C} = \{\{1,2\},\{1,4\},\{3,5\},\{3,6\}\}$, and we compute $S(f_3, f_6) = xyz^2 - xz^2$ which is reduced with respect to G. Thus we add

$\boxed{f_7 = xyz^2 - xz^2}$ to G. Applying crit2 again we get $\mathcal{NC} = \{\{2,7\},\{3,7\},\{4,7\}\}$. Now we choose the pair $\{4,7\}$ and compute $S(f_4, f_7) = -z^4 + x^2z^2$ which is reduced with respect to G. Thus we add $\boxed{f_8 = -z^4 + x^2z^2}$ to G. After applying crit2 we have $\mathcal{NC} = \{\{2,7\}, \{3,7\}, \{2,8\}, \{4,8\}, \{5,8\},\{6,8\}\}$. Now we choose the pair $\{2,7\}$ and observe that $S(f_2, f_7) = 0$. We choose next the pair $\{3,7\}$ and see that $S(f_3, f_7) \xrightarrow{G}_+ 0$. Then we choose the pair $\{6,8\}$, we update $\mathcal{NC} = \{\{2,8\}, \{4,8\},\{5,8\}\}$ and $\mathcal{C} = \{\{1,2\}, \{1,4\}, \{3,5\}, \{3,6\}, \{4,7\}, \{2,7\}, \{3,7\}, \{6,8\}\}$, and compute $S(f_6, f_8) = x^3z^2 - xz^3$ which is reduced with respect to G. Thus we add $\boxed{f_9 = x^3z^2 - xz^3}$ to G. After applying crit2 we have $\mathcal{NC} = \{\{2,8\}, \{4,8\},\{5,8\}, \{4,9\}, \{6,9\}\}$. The S-polynomials corresponding to the remaining pairs in $\mathcal{NC}$ all reduce to zero. Thus $G = \{f_1, f_2, f_3, f_4, f_5, f_6, f_7, f_8, f_9\}$ is a Gröbner basis for $I = \langle f_1, f_2, f_3 \rangle$.

We now give one example of crit2. We consider the situation right after adding f_6 to G. There we started with $\mathcal{NC} = \{\{2,3\},\{3,4\},\{1,6\},\{2,6\},\{3,6\}, \{4,6\},\{5,6\}\}$ and $\mathcal{C} = \{\{1,2\},\{1,4\},\{3,5\}\}$. For $\ell = 1$ we first note that $\{1,6\}$ is in $\mathcal{NC}$ and we must consider the pairs $\{2,6\}$ and $\{4,6\}$ (we do not consider the pair $\{3,6\}$ because the pair $\{1,3\} \notin \mathcal{NC} \cup \mathcal{C}$, and we do not consider the pair $\{5,6\}$ because $\{1,5\} \notin \mathcal{NC} \cup \mathcal{C}$). For $\{2,6\}$ we see that $\mathrm{lp}(f_1) = x^2y^2$ does not divide $\mathrm{lcm}(\mathrm{lp}(f_2), \mathrm{lp}(f_6)) = xy^2z^3$, and for $\{5,6\}$ we see that $\mathrm{lp}(f_1) = x^2y^2$ does not divide $\mathrm{lcm}(\mathrm{lp}(f_5), \mathrm{lp}(f_6)) = xyz^3$. So we eliminate no pairs from $\mathcal{NC}$. For $\ell = 2$ we note that $\{2,6\} \in \mathcal{NC}$ and we only consider $\{1,6\}$ and $\{3,6\}$. For $\{1,6\}$ we note that $\mathrm{lp}(f_2) = xy^2z$ divides $\mathrm{lcm}(\mathrm{lp}(f_1), \mathrm{lp}(f_6)) = x^2y^2z^3$ and so we eliminate the pair $\{1,6\}$ from $\mathcal{NC}$. We may not eliminate $\{3,6\}$. Now $\mathcal{NC} = \{\{2,3\},\{3,4\},\{2,6\},\{3,6\},\{4,6\},\{5,6\}\}$. For $\ell = 3$ we consider $\{2,6\}$, $\{4,6\}$ and $\{5,6\}$ and all three of them are eliminated giving us $\mathcal{NC} = \{\{2,3\}, \{3,4\}, \{3,6\}\}$. We do not need to consider $\ell = 4,5$, since $\{4,6\},\{5,6\} \notin \mathcal{NC}$. We do not need to consider $\ell = 6$, since there is only one pair with 6 in it left in $\mathcal{NC}$. So we finally arrive at $\mathcal{NC} = \{\{2,3\}, \{3,4\}, \{3,6\}\}$.

Although it might appear in this example that no real saving in computation time has been gained in using crit2, we recall that, in practice, the most time consuming part of Buchberger's Algorithm is the reduction of S-polynomials and we have avoided most of these reductions by using the current algorithm. Indeed, we computed a total of 13 S-polynomials, 6 of which generated elements of the Gröbner basis. Had we not used crit2 we would have had to compute and reduce $\frac{9 \cdot 8}{2} = 36$ S-polynomials. Note that the computations in crit2 are always trivial.

To conclude this section, we mention two other difficulties that arise during the computation of Gröbner bases, namely, the possible rapid growth of the degrees and coefficients of the S-polynomials. Even though the degree and/or size of the coefficients of the original polynomials and the Gröbner basis may be of modest size, the intermediary polynomials generated by the S-polynomial computations and reductions can become quite large. This can dramatically slow down the computation. Doing computations with large coefficients can be very costly because of the large amount of arithmetic that becomes necessary. As an example of this situation, the reduced Gröbner basis for the ideal $\langle 4x^2y^2 + 3x, y^3 + 2xy, 7x^3 + 6y\rangle$ with respect to the lex order with $x > y$ is $\{x, y\}$, while the coefficients during the computation grow as large as 10^8. This can be seen by expressing x as a linear combination of the three original polynomials $f_1 = 4x^2y^2 + 3x$, $f_2 = y^3 + 2xy$, and $f_3 = 7x^3 + 6y$:

$$\begin{aligned}x = &\left(\frac{7}{54}x^2y^5 - \frac{401408}{56428623}y^{10} - \frac{1835008}{56428623}y^9 - \frac{9604}{18809541}y^8 - \frac{43904}{18809541}y^7\right.\\ &\left.-\frac{200704}{18809541}y^6 + \frac{1}{3}\right)f_1 + \left(-\frac{7}{27}x^3y^6 + \frac{1605632}{56428623}x^2y^9 + \frac{7340032}{56428623}x^2y^8\right.\\ &+\frac{38416}{18809541}x^2y^7 + \frac{175616}{18809541}x^2y^6 + \frac{401408}{18809541}xy^7 - \frac{2}{3}xy + \frac{917504}{18809541}y^8\\ &\left.+\frac{4802}{6269847}y^7 + \frac{21952}{6269847}y^6 + \frac{100352}{6269847}y^5 + \frac{1}{3}y^3\right)f_2\\ &-\frac{1}{112857246}y^5(917504y^5 + 14406y^4 + 65856y^3 + 301056y^2 + 6269847)f_3.\end{aligned}$$

There are techniques to try to get around this problem, but again they are beyond the scope of this book.

To illustrate the growth in total degree, consider the ideal $I = \langle x^7 + xy + y, y^5 + yz + z, z^2 + z + 1\rangle \subseteq \mathbb{Q}[x, y, z]$. The generators of this ideal have maximum total degree 7. However the reduced Gröbner basis for I with respect to lex with $z > y > x$ contains a polynomial of degree 70. The problem with polynomials with large total degree is the fact that they can have a very large number of terms. For example, the reduced Gröbner basis for the ideal I above has 3 polynomials with 58, 70, and 35 terms respectively. Again, a large number of terms make any computation very costly.

Exercises

3.3.1. Compute a Gröbner bases for the following ideal using Algorithm 3.3.1 without using a Computer Algebra System.

a. $\langle x^2y - y + x, xy^2 - x\rangle \subseteq \mathbb{Q}[x, y]$ using deglex with $x < y$. Compare with Example 2.1.1.

b. $\langle x^2y + z, xz + y\rangle \subseteq \mathbb{Q}[x, y, z]$ using deglex with $x > y > z$. Compare with Exercise 1.7.3.

c. $\langle x^2y + z, xz + y, y^2z + 1\rangle \subseteq \mathbb{Q}[x, y, z]$ using lex with $x > y > z$.

3.3.2. The following exercise may tax your Computer Algebra System. If so, try to make up your own more modest example that illustrates the same point. Let $I = \langle x^7 + xy + y, y^5 + yz + z, z^2 + z + 1\rangle \subseteq \mathbb{Q}[x, y, z]$.

a. Find the reduced Gröbner basis for I with respect to lex with $x > y > z$. (No computation is needed!)

b. Compute the reduced Gröbner basis for I with respect to lex with $z > y > x$. Compare the two bases.

c. Can you give a reason for the difference between the bases in **a** and **b**?

d. Use **c** to give a method for generating polynomials $f_1, \dots, f_s$ in n variables whose total degree is small, but such that the reduced Gröbner basis for $\langle f_1, \dots, f_s \rangle$ with respect to a certain lex order has polynomials of very high degree. [Hint: In the given example, note that there are 2 solutions for z, and for each such solution, there are 5 solutions for y, etc.]

e. Give examples of polynomials which satisfy **d**.

f. Experiment by changing the degree of the x, y, z terms in the polynomials above, but keeping lex with $z > y > x$. How does it affect the computing time?

g. Experiment by changing the term order and the order on the variables for the examples in this exercise. How does it affect the computation and the computing time?

3.3.3. Let $f, g, d \in k[x_1, \dots, x_n]$ such that d divides both f and g.

a. Show that $S(\frac{f}{d}, \frac{g}{d}) = \frac{\mathrm{lc}(d)}{d} S(f, g)$.

b. Show that if $S(f,g) \xrightarrow{\{f,g\}}_+ 0$, then $\frac{1}{d} S(f,g) \xrightarrow{\{\frac{f}{d}, \frac{g}{d}\}}_+ 0$.

3.4. Computation of the Syzygy Module. In this section we show how to compute $\mathrm{Syz}(f_1, \dots, f_s)$ for $f_1, \dots, f_s \in A = k[x_1, \dots, x_n]$. This is done in two steps. We first compute a Gröbner basis $G = \{g_1, \dots, g_t\}$ for $\langle f_1, \dots, f_s \rangle$ and compute $\mathrm{Syz}(g_1, \dots, g_t)$. For convenience, we will assume that $g_1, \dots, g_t$ are monic. We then show how to obtain $\mathrm{Syz}(f_1, \dots, f_s)$ from $\mathrm{Syz}(g_1, \dots, g_t)$. We will assume that we have a fixed term order on A.

Let $\{g_1, \dots, g_t\}$ be a Gröbner basis, where we assume that the g_i's are monic. For $i \in \{1, \dots, t\}$, we let $\mathrm{lp}(g_i) = X_i$ and for $i \neq j \in \{1, \dots, t\}$, we let $X_{ij} = \mathrm{lcm}(X_i, X_j)$. Then the S-polynomial of g_i and g_j is given by

$$S(g_i, g_j) = \frac{X_{ij}}{X_i} g_i - \frac{X_{ij}}{X_j} g_j.$$

By Theorem 1.6.2, we have

$$S(g_i, g_j) = \sum_{\nu=1}^{t} h_{ij\nu} g_\nu,$$

for some $h_{ij\nu} \in A$, such that

$$\max_{1 \le \nu \le t} (\mathrm{lp}(h_{ij\nu}) \, \mathrm{lp}(g_\nu)) = \mathrm{lp}(S(g_i, g_j)).$$

(The polynomials $h_{ij\nu}$ are obtained using the Division Algorithm.) We now define for $i, j = 1, \dots, t, i \neq j$,

$$\mathbf{s}_{ij} = \frac{X_{ij}}{X_i} \mathbf{e}_i - \frac{X_{ij}}{X_j} \mathbf{e}_j - (h_{ij1}, \dots, h_{ijt}) \in A^t.$$

We note that $\mathbf{s}_{ij} \in \operatorname{Syz}(g_1, \ldots, g_t)$, since

$$\begin{aligned} \begin{bmatrix} g_1 & \cdots & g_t \end{bmatrix} \mathbf{s}_{ij} &= \begin{bmatrix} g_1 & \cdots & g_t \end{bmatrix} \left(\frac{X_{ij}}{X_i}\mathbf{e}_i - \frac{X_{ij}}{X_j}\mathbf{e}_j\right) \\ &\quad - \begin{bmatrix} g_1 & \cdots & g_t \end{bmatrix} (h_{ij1}, \ldots, h_{ijt}) \\ &= S(g_i, g_j) - \sum_{\nu=1}^{t} h_{ij\nu} g_\nu = 0. \end{aligned}$$

THEOREM 3.4.1. *With the notation above, the collection* $\{\mathbf{s}_{ij} \mid 1 \le i < j \le t\}$ *is a generating set for* $\operatorname{Syz}(g_1, \ldots, g_t)$.

PROOF. Suppose to the contrary that there exists $(u_1, \ldots, u_t)$ such that

$$(u_1, \ldots, u_t) \in \operatorname{Syz}(g_1, \ldots, g_t) - \langle \mathbf{s}_{ij} \mid 1 \le i < j \le t \rangle.$$

Then we can choose such a $(u_1, \ldots, u_t)$ with $X = \max_{1 \le i \le t}(\operatorname{lp}(u_i)\operatorname{lp}(g_i))$ least. Let

$$S = \{i \in \{1, \ldots, t\} \mid \operatorname{lp}(u_i)\operatorname{lp}(g_i) = X\}.$$

Now for each $i \in \{1, \ldots, t\}$ we define u_i' as follows:

$$u_i' = \begin{cases} u_i \text{ if } i \notin S \\ u_i - \operatorname{lt}(u_i) \text{ if } i \in S. \end{cases}$$

Also, for $i \in S$, let $\operatorname{lt}(u_i) = c_i X_i'$, where $c_i \in k$ and X_i' is a power product. Since $(u_1, \ldots, u_t) \in \operatorname{Syz}(g_1, \ldots, g_t)$, we see that

$$\sum_{i \in S} c_i X_i' X_i = 0,$$

and so

$$\sum_{i \in S} c_i X_i' \mathbf{e}_i \in \operatorname{Syz}(X_i \mid i \in S).$$

Thus, by Proposition 3.2.3 we have

$$\sum_{i \in S} c_i X_i' \mathbf{e}_i = \sum_{\substack{i<j \\ i,j \in S}} a_{ij} \left(\frac{X_{ij}}{X_i}\mathbf{e}_i - \frac{X_{ij}}{X_j}\mathbf{e}_j\right),$$

for some $a_{ij} \in A$. Since each coordinate of the vector in the left-hand side of the equation above is homogeneous, and since $X_i' X_i = X$, we can choose a_{ij} to be a constant multiple of $\frac{X}{X_{ij}}$. Then we have

$$\begin{aligned}
(u_1, \ldots, u_t) &= \sum_{i \in S} c_i X_i' \boldsymbol{e}_i + (u_1', \ldots, u_t') \\
&= \sum_{\substack{i<j \\ i,j \in S}} a_{ij} \left(\frac{X_{ij}}{X_i} \boldsymbol{e}_i - \frac{X_{ij}}{X_j} \boldsymbol{e}_j\right) + (u_1', \ldots, u_t') \\
&= \sum_{\substack{i<j \\ i,j \in S}} a_{ij} \boldsymbol{s}_{ij} + (u_1', \ldots, u_t') + \sum_{\substack{i<j \\ i,j \in S}} a_{ij} (h_{ij1}, \ldots, h_{ijt}).
\end{aligned}$$

We define $(v_1, \ldots, v_t) = (u_1', \ldots, u_t') + \sum_{\substack{i<j \\ i,j \in S}} a_{ij}(h_{ij1}, \ldots, h_{ijt})$. We note that $(v_1, \ldots, v_t) \in \operatorname{Syz}(g_1, \ldots, g_t) - \langle \boldsymbol{s}_{ij} \mid 1 \leq i < j \leq t \rangle$, since $(u_1, \ldots, u_t), \boldsymbol{s}_{ij} \in \operatorname{Syz}(g_1, \ldots, g_t)$ and $(u_1, \ldots, u_t) \notin \langle \boldsymbol{s}_{ij} \mid 1 \leq i < j \leq t \rangle$. We will obtain the desired contradiction by proving that $\max_{1 \leq \nu \leq t}(\operatorname{lp}(v_\nu) \operatorname{lp}(g_\nu)) < X$. For each $\nu \in \{1, \ldots, t\}$ we have

$$\begin{aligned}
\operatorname{lp}(v_\nu) \operatorname{lp}(g_\nu) &= \operatorname{lp}(u_\nu' + \sum_{\substack{i<j \\ i,j \in S}} a_{ij} h_{ij\nu}) X_\nu \\
&\leq \max(\operatorname{lp}(u_\nu'), \max_{\substack{i<j \\ i,j \in S}}(\operatorname{lp}(a_{ij}) \operatorname{lp}(h_{ij\nu}))) X_\nu.
\end{aligned}$$

But, by definition of u_ν', we have $\operatorname{lp}(u_\nu') X_\nu < X$. Also, as mentioned above, a_{ij} is a constant multiple of $\frac{X}{X_{ij}}$, and hence for all $i, j \in S, i < j$, we have

$$\operatorname{lp}(a_{ij}) \operatorname{lp}(h_{ij\nu}) X_\nu = \frac{X}{X_{ij}} \operatorname{lp}(h_{ij\nu}) X_\nu \leq \frac{X}{X_{ij}} \operatorname{lp}(S(g_i, g_j)) < X.$$

Therefore $\operatorname{lp}(v_\nu) \operatorname{lp}(g_\nu) < X$ for each $\nu \in \{1, \ldots, t\}$ violating the condition that $X = \max_{1 \leq \nu \leq t}(\operatorname{lp}(u_\nu) \operatorname{lp}(g_\nu))$ is least. □

EXAMPLE 3.4.2. We return to Example 3.2.2. Let $g_1 = x^2 - wy$, $g_2 = xy - wz$, and $g_3 = y^2 - xz$. These form a reduced Gröbner basis with respect to the degrevlex ordering with $x > y > z > w$. Using the notation of the above result, we have:

$$X_1 = x^2, X_2 = xy, X_3 = y^2, X_{12} = x^2 y, X_{13} = x^2 y^2, X_{23} = xy^2.$$

Now $S(g_1, g_2) = -wy^2 + xwz = -wg_3$, so $h_{121} = h_{122} = 0$ and $h_{123} = -w$. Therefore

$$\boldsymbol{s}_{12} = \frac{X_{12}}{X_1} \boldsymbol{e}_1 - \frac{X_{12}}{X_2} \boldsymbol{e}_2 - (h_{121}, h_{122}, h_{123}) = (y, -x, w).$$

Also, $S(g_1, g_3) = x^3z - y^3w = xzg_1 - ywg_3$, so that $h_{131} = xz$, $h_{132} = 0$ and $h_{133} = -yw$. Therefore

$$\boldsymbol{s}_{13} = \frac{X_{13}}{X_1}\boldsymbol{e}_1 - \frac{X_{13}}{X_3}\boldsymbol{e}_3 - (h_{131}, h_{132}, h_{133}) = (y^2 - xz, 0, -x^2 + yw).$$

Finally, $S(g_2, g_3) = x^2z - yzw = zg_1$, so that $h_{231} = z$, $h_{232} = 0$ and $h_{233} = 0$. Therefore

$$\boldsymbol{s}_{23} = \frac{X_{23}}{X_2}\boldsymbol{e}_2 - \frac{X_{23}}{X_3}\boldsymbol{e}_3 - (h_{231}, h_{232}, h_{233}) = (-z, y, -x).$$

By Theorem 3.4.1,

$$\mathrm{Syz}(g_1, g_2, g_3) = \langle (y, -x, w), (y^2 - xz, 0, -x^2 + yw), (-z, y, -x) \rangle.$$

We note that $\boldsymbol{s}_{13} = y\boldsymbol{s}_{12} + x\boldsymbol{s}_{23}$, so we have, in fact, that

$$\mathrm{Syz}(g_1, g_2, g_3) = \langle (y, -x, w), (-z, y, -x) \rangle.$$

We now turn our attention to computing $\mathrm{Syz}(f_1, \ldots, f_s)$, for a collection $\{f_1, \ldots, f_s\}$ of non-zero polynomials in A which may not form a Gröbner basis. We first compute a Gröbner basis $\{g_1, \ldots, g_t\}$ for $\langle f_1, \ldots, f_s \rangle$. We again assume that $g_1, \ldots, g_t$ are monic. Set $F = \begin{bmatrix} f_1 & \cdots & f_s \end{bmatrix}$ and $G = \begin{bmatrix} g_1 & \cdots & g_t \end{bmatrix}$. As we saw in Section 2.1, there is a $t \times s$ matrix S and an $s \times t$ matrix T with entries in A such that $F = GS$ and $G = FT$. (Recall that S is obtained using the Division Algorithm, and T is obtained by keeping track of the reductions during Buchberger's Algorithm.) Now using Theorem 3.4.1, we can compute a generating set $\{\boldsymbol{s}_1, \ldots, \boldsymbol{s}_r\}$ for $\mathrm{Syz}(G)$ (the $\boldsymbol{s}_i$'s are column vectors in A^t). Therefore for each $i = 1, \ldots, r$

$$0 = G\boldsymbol{s}_i = (FT)\boldsymbol{s}_i = F(T\boldsymbol{s}_i),$$

and hence

$$\langle T\boldsymbol{s}_i \mid i = 1, \ldots, r \rangle \subseteq \mathrm{Syz}(F).$$

Moreover, if we let I_s be the $s \times s$ identity matrix, we have

$$F(I_s - TS) = F - FTS = F - GS = F - F = \boldsymbol{0},$$

and hence the columns $\boldsymbol{r}_1, \ldots, \boldsymbol{r}_s$ of $I_s - TS$ are also in $\mathrm{Syz}(F)$.

THEOREM 3.4.3. *With the notation above we have*

$$\mathrm{Syz}(f_1, \ldots, f_s) = \langle T\boldsymbol{s}_1, \ldots, T\boldsymbol{s}_r, \boldsymbol{r}_1, \ldots, \boldsymbol{r}_s \rangle \subseteq A^s.$$

PROOF. Let $\boldsymbol{s} = (a_1, \ldots, a_s) \in \mathrm{Syz}(f_1, \ldots, f_s)$. Then $0 = F\boldsymbol{s} = GS\boldsymbol{s}$, and hence $S\boldsymbol{s} \in \mathrm{Syz}(g_1, \ldots, g_t)$. By the definition of $\boldsymbol{s}_1, \ldots, \boldsymbol{s}_r$, we have $S\boldsymbol{s} = \sum_{i=1}^r h_i\boldsymbol{s}_i$ for some $h_i \in A$. Thus we have $TS\boldsymbol{s} = \sum_{i=1}^r h_i(T\boldsymbol{s}_i)$. Finally,

$$\boldsymbol{s} = \boldsymbol{s} - TS\boldsymbol{s} + TS\boldsymbol{s} = (I_s - TS)\boldsymbol{s} + \sum_{i=1}^r h_i(T\boldsymbol{s}_i) = \sum_{i=1}^s a_i\boldsymbol{r}_i + \sum_{i=1}^r h_i(T\boldsymbol{s}_i).$$

Therefore $\mathrm{Syz}(f_1, \ldots, f_s) \subseteq \langle T\boldsymbol{s}_1, \ldots, T\boldsymbol{s}_r, \boldsymbol{r}_1, \ldots, \boldsymbol{r}_s\rangle$. The reverse inclusion has already been noted. $\square$

EXAMPLE 3.4.4. Let $f_1 = x^2y^2 - x^3y$, $f_2 = xy^3 - x^2y^2$, and $f_3 = y^4 - x^3$. Let $F = \begin{bmatrix} f_1 & f_2 & f_3 \end{bmatrix}$. We first compute a Gröbner basis G with respect to the lex term ordering with $x < y$ for $\langle f_1, f_2, f_3\rangle$. We find[2] $G = \begin{bmatrix} g_1 & g_2 & g_3 & g_4 & g_5 \end{bmatrix}$, where $g_1 = y^4 - x^3$, $g_2 = xy^3 - x^3y$, $g_3 = x^2y^2 - x^3y$, $g_4 = x^4y - x^4$, and $g_5 = x^5 - x^4$. Moreover we have

$$\begin{bmatrix} g_1 & g_2 & g_3 & g_4 & g_5 \end{bmatrix} =$$

$$\begin{bmatrix} f_1 & f_2 & f_3 \end{bmatrix} \underbrace{\begin{bmatrix} 0 & 1 & 1 & -(x+y) & -(x+y) \\ 0 & 1 & 0 & -y & -y(y+1) \\ 1 & 0 & 0 & x & -x^2+xy+x \end{bmatrix}}_{T} \quad \text{and}$$

$$\begin{bmatrix} f_1 & f_2 & f_3 \end{bmatrix} = \begin{bmatrix} g_1 & g_2 & g_3 & g_4 & g_5 \end{bmatrix} \underbrace{\begin{bmatrix} 0 & 0 & 1 \\ 0 & 1 & 0 \\ 1 & -1 & 0 \\ 0 & 0 & 0 \\ 0 & 0 & 0 \end{bmatrix}}_{S}.$$

To compute generators for $\mathrm{Syz}(g_1, g_2, g_3, g_4, g_5)$ we need to reduce all $S(g_i, g_j)$'s for $1 \le i < j \le 5$ using g_1, g_2, g_3, g_4, g_5 (Theorem 3.4.1). In view of Corollary 3.3.3 and Exercise 3.4.4, it is enough to reduce $S(g_1, g_2)$, $S(g_2, g_3)$, $S(g_3, g_4)$, and $S(g_4, g_5)$ which give the following respective syzygies:

$$\boldsymbol{s}_1 = (x, -y, -x, -1, 0), \ \boldsymbol{s}_2 = (0, x, -x-y, 0, 0),$$

$$\boldsymbol{s}_3 = (0, 0, x^2, -y, y), \ \boldsymbol{s}_4 = (0, 0, 0, x-1, -y+1).$$

By Theorem 3.4.1, we have

$$\mathrm{Syz}(g_1, g_2, g_3, g_4, g_5) = \langle \boldsymbol{s}_1, \boldsymbol{s}_2, \boldsymbol{s}_3, \boldsymbol{s}_4\rangle.$$

Then

$$T \begin{bmatrix} x \\ -y \\ -x \\ -1 \\ 0 \end{bmatrix} = \begin{bmatrix} 0 \\ 0 \\ 0 \end{bmatrix}, \ T \begin{bmatrix} 0 \\ x \\ -x-y \\ 0 \\ 0 \end{bmatrix} = \begin{bmatrix} -y \\ x \\ 0 \end{bmatrix},$$

[2]We will often use the same letter for a finite set and a row vector corresponding to it, when the context is clear.

$$T\begin{bmatrix} 0 \\ 0 \\ x^2 \\ -y \\ y \end{bmatrix} = \begin{bmatrix} x^2 \\ -y^3 \\ -x^2y+xy^2 \end{bmatrix}, T\begin{bmatrix} 0 \\ 0 \\ 0 \\ x-1 \\ -y+1 \end{bmatrix} = \begin{bmatrix} -x^2+y^2 \\ -xy+y^3 \\ x^2y-xy^2 \end{bmatrix}.$$

We note that $(-x^2+y^2, -xy+y^3, x^2y-xy^2)$ is in the submodule of A^3 generated by $(-y, x, 0)$ and $(x^2, -y^3, -x^2y+xy^2)$.

Finally, it is easy to verify that

$$I_3 - TS = \begin{bmatrix} 0 & 0 & 0 \\ 0 & 0 & 0 \\ 0 & 0 & 0 \end{bmatrix}.$$

Therefore

$$\mathrm{Syz}(f_1, f_2, f_3) = \langle(-y, x, 0), (x^2, -y^3, -x^2y+xy^2)\rangle.$$

Using the notation of Theorem 3.4.3, we note that in Example 3.4.4 the rows of $I_s - TS$ did not give any syzygies that were not already in the submodule $\langle T\mathbf{s}_1, \dots, T\mathbf{s}_r\rangle$ of A^s. This is not always the case (see Exercise 3.4.2). However it is easy to see (Exercise 3.4.3) that if the polynomials $f_1, \dots, f_s$ appear in the list $g_1, \dots, g_t$, then $\mathrm{Syz}(f_1, \dots, f_s) = \langle T\mathbf{s}_1, \dots, T\mathbf{s}_r\rangle$. If the Gröbner basis for $\langle f_1, \dots, f_s\rangle$ is computed using Buchberger's Algorithm (Algorithm 1.7.1) or by the improved Buchberger's Algorithm (Algorithm 3.3.1), then the polynomials $f_1, \dots, f_s$ do appear in the list $g_1, \dots, g_t$. However, if a reduced Gröbner basis is computed, as is done in all Computer Algebra Systems, then the polynomials $f_1, \dots, f_s$ may not appear in the list $g_1, \dots, g_t$.

Exercises

3.4.1. Compute generators for $\mathrm{Syz}(f_1, \dots, f_s)$ in the following examples:

a. $f_1 = x^2y+z$, $f_2 = xz+y \in \mathbb{Q}[x,y,z]$.

b. $f_1 = x^2y-y+x$, $f_2 = xy^2-x \in \mathbb{Q}[x,y]$.

c. $f_1 = x^2y+z$, $f_2 = xz+y$, $f_3 = y^2z+1 \in \mathbb{Q}[x,y,z]$.

3.4.2. In Theorem 3.4.3 we had to include the columns of the matrix $I_s - TS$ in the set of generators for $\mathrm{Syz}(f_1, \dots, f_s)$. In this exercise, we show that these vectors are necessary. Consider the polynomials $f_1 = xy+1$, $f_2 = xz+1$, and $f_3 = yz+1 \in \mathbb{Q}[x,y,z]$.

a. Verify that the reduced Gröbner basis for $I = \langle f_1, f_2, f_3\rangle$ with respect to lex with $x > y > z$ is $G = \{g_1, g_2, g_3\}$, where $g_1 = y-z$, $g_2 = x-z$, and $g_3 = z^2+1$. Also, show that

$$\begin{bmatrix} g_1 & g_2 & g_3 \end{bmatrix} = \begin{bmatrix} f_1 & f_2 & f_3 \end{bmatrix} \underbrace{\begin{bmatrix} -z & -z & z^2 \\ y & 0 & -yz \\ 0 & x & 1 \end{bmatrix}}_{T}.$$

b. Compute the matrix S such that $\left[\begin{array}{lll} f_1 & f_2 & f_3 \end{array}\right] = \left[\begin{array}{lll} g_1 & g_2 & g_3 \end{array}\right] S$.

c. Compute the 3 generators, say $\boldsymbol{s}_1, \boldsymbol{s}_2, \boldsymbol{s}_3$, for $\mathrm{Syz}(g_1, g_2, g_3)$.

d. Compute $I_3 - TS$. This matrix has 2 non-zero columns, say $\boldsymbol{r}_1$, $\boldsymbol{r}_2$.

e. Verify that $\langle T\boldsymbol{s}_1, T\boldsymbol{s}_2, T\boldsymbol{s}_3\rangle \neq \langle T\boldsymbol{s}_1, T\boldsymbol{s}_2, T\boldsymbol{s}_3, \boldsymbol{r}_1, \boldsymbol{r}_2\rangle$.

3.4.3. Assume that $\{f_1, \dots, f_s\} \subseteq \{g_1, \dots, g_t\}$ and $\{g_1, \dots, g_t\}$ is a Gröbner basis for $I = \langle f_1, \dots, f_s\rangle$. Prove that, with the notation of Theorem 3.4.3, $\mathrm{Syz}(f_1, \dots, f_s) = \langle T\boldsymbol{s}_1, \dots, T\boldsymbol{s}_r\rangle$, that is, the columns of the matrix $I_s - TS$ are not necessary.

3.4.4. Generalize Theorem 3.4.1 as follows: Let $\{g_1, \dots, g_t\}$ be a Gröbner basis and let τ_{ij} be as in Section 3.3. Let $\mathcal{B} \subseteq \{\tau_{ij} \mid 1 \leq i < j \leq t\}$ be a generating set for $\mathrm{Syz}(\mathrm{lt}(g_1), \dots, \mathrm{lt}(g_t))$. Prove that $\{\boldsymbol{s}_{ij} \mid \tau_{ij} \in \mathcal{B}\}$ is a generating set for $\mathrm{Syz}(g_1, \dots, g_t)$.

3.4.5. Apply Exercise 3.4.4 and Corollary 3.3.3 to the computations in Exercise 3.4.1.

3.4.6. Let $f_1, \dots, f_s, g \in k[x_1, \dots, x_n]$ and consider the linear equation

$$h_1 f_1 + h_2 f_2 + \cdots + h_s f_s = g,$$

with unknowns $h_1, \dots, h_s \in k[x_1, \dots, x_n]$. Let $\boldsymbol{S}$ be the set of solutions; i.e.

$$\boldsymbol{S} = \{(h_1, \dots, h_s) \in A^s \mid h_1 f_1 + h_2 f_2 + \cdots + h_s f_s = g\}.$$

a. Prove that $\boldsymbol{S}$ is not empty if and only if $g \in \langle f_1, \dots, f_s\rangle$.

b. Prove that if $\boldsymbol{S} \neq \emptyset$ then $\boldsymbol{S} = \boldsymbol{h} + \mathrm{Syz}(f_1, \dots, f_s) = \{\boldsymbol{h} + \boldsymbol{s} \mid \boldsymbol{s} \in \mathrm{Syz}(f_1, \dots, f_s)\}$, where $\boldsymbol{h}$ is a particular solution. Give a method for computing $\boldsymbol{h}$.

c. Use the above to find the solution set for the equation

$$h_1(x^2y^2 - x^3y) + h_2(xy^3 - x^2y^2) + h_3(y^4 - x^3) = y^7 - y^6.$$

3.5. Gröbner Bases for Modules. As before, let $A = k[x_1, \dots, x_n]$ for a field k. We have seen in the previous sections of this chapter that certain submodules of A^m are important. In this section, we continue to study such submodules, but now from the point of view adopted for ideals in the earlier parts of this book. Namely, we will generalize the theory of Gröbner bases to submodules of A^m. As a result we will be able to compute with submodules of A^m in a way similar to the way we computed with ideals previously.

The idea is to mimic the constructions we used in Chapter 1 as much as possible. Let us recall the ingredients for the methods we used before for ideals. First, in Section 1.4 we defined the concept of a power product and then defined a term order on these power products, that is, a total order with special properties with respect to divisibility (we will need to define a concept of divisibility). Using these ideas, in Section 1.5 we defined the concept of reduction which in turn led to the Division Algorithm. Then in Section 1.6 we defined the concept of a Gröbner basis, giving the equivalent conditions for a Gröbner basis in Theorem 1.6.2 (and further in Theorem 1.9.1). The next issue, discussed in Section 1.7, was how to compute Gröbner bases, and for that we developed S-polynomials and Buchberger's Algorithm. We will follow this development very closely. Indeed many of the proofs in this and the next two sections are very similar to the corresponding ones for ideals and will be left to the exercises.

We again need the standard basis

$$\boldsymbol{e}_1 = (1, 0, \dots, 0), \boldsymbol{e}_2 = (0, 1, 0, \dots, 0), \dots, \boldsymbol{e}_m = (0, 0, \dots, 0, 1)$$

of A^m. Then by a *monomial*[3] in A^m we mean a vector of the type $X\boldsymbol{e}_i$ $(1 \le i \le m)$, where X is a power product in A. That is, a monomial is a column vector with all coordinates equal to zero except for one which is a power product of A. Monomials in A^m will replace the notion of power products in the ring A. So, for example, $(0, x_1^2x_3, 0)$ and $(0, 0, x_2)$ are monomials in A^3, but $(0, x_1 + x_2, 0)$ and $(0, x_2, x_1)$ are not. If $\boldsymbol{X} = X\boldsymbol{e}_i$ and $\boldsymbol{Y} = Y\boldsymbol{e}_j$ are monomials in A^m, we say that $\boldsymbol{X}$ *divides* $\boldsymbol{Y}$ provided that $i = j$ and X divides Y. Thus in A^3 we see that $(0, x_1^2x_3, 0)$ divides $(0, x_1^2x_3^3, 0)$, but does not divide $(0, x_1x_3, 0)$ or $(x_1^2x_3^3, 0, 0)$. We note that in case $\boldsymbol{X}$ divides $\boldsymbol{Y}$ there is a power product Z in the ring A such that $\boldsymbol{Y} = Z\boldsymbol{X}$. In this case we define[4]

$$\frac{\boldsymbol{Y}}{\boldsymbol{X}} = \frac{Y}{X} = Z.$$

So, for example

$$\frac{(0, x_1^2x_3^2, 0)}{(0, x_1^2x_3, 0)} = \frac{x_1^2x_3^2}{x_1^2x_3} = x_3.$$

Similarly, by a *term*, we mean a vector of the type $c\boldsymbol{X}$, where $c \in k - \{0\}$ and $\boldsymbol{X}$ is a monomial. Thus $(0, 5x_1^2x_3^4, 0, 0) = 5\boldsymbol{X}$, where $\boldsymbol{X} = (0, x_1^2x_3^4, 0, 0) = x_1^2x_3^4\boldsymbol{e}_2$, is a term of A^4 but not a monomial. Also, if $\boldsymbol{X} = cX\boldsymbol{e}_i$ and $\boldsymbol{Y} = dY\boldsymbol{e}_j$ are terms of A^m, we say that $\boldsymbol{X}$ *divides* $\boldsymbol{Y}$ provided that $i = j$ and X divides Y. We write

$$\frac{\boldsymbol{Y}}{\boldsymbol{X}} = \frac{dY}{cX}.$$

So, for example,

$$\frac{(0, 5x_1^2x_3^2, 0)}{(0, 2x_1^2x_3, 0)} = \frac{5x_1^2x_3^2}{2x_1^2x_3} = \frac{5}{2}x_3.$$

We now can define a term order on the monomials of A^m.

DEFINITION 3.5.1. *By a* term order *on the monomials of A^m we mean a total order, $<$, on these monomials satisfying the following two conditions:*

(i) $\boldsymbol{X} < Z\boldsymbol{X}$, *for every monomial $\boldsymbol{X}$ of A^m and power product $Z \neq 1$ of A;*

---

[3]In the case that $m = 1$ we have now called a monomial what we referred to before as a power product. From now on in the book we will use monomial and power product interchangeably for such elements in the ring A.

[4]Be careful about what we are doing here. We are "dividing" two *vectors* in A^m to obtain *an element in* A; but we are only doing this in the very special case where each of the *vectors* has only one non-zero coordinate in the same spot which is a power product and one power product divides the other. The "quotient" is defined to be the quotient of those two power products.

(ii) *If* $\boldsymbol{X} < \boldsymbol{Y}$, *then* $Z\boldsymbol{X} < Z\boldsymbol{Y}$ *for all monomials* $\boldsymbol{X}, \boldsymbol{Y} \in A^m$ *and every power product* $Z \in A$.

Looking at Definition 1.4.1 we see that Conditions (i) and (ii) there correspond to Conditions (i) and (ii) in Definition 3.5.1. Condition (i), specialized to the case of $m = 1$, simply says that $Z > 1$, since the monomial $\boldsymbol{X}$ can be cancelled. This is Condition (i) of Definition 1.4.1. The corresponding second conditions are exactly parallel.

If we are given a term order on A there are two natural ways of obtaining a term order on A^m. These are given in the following two definitions.

DEFINITION 3.5.2. *For monomials* $\boldsymbol{X} = X\boldsymbol{e}_i$ *and* $\boldsymbol{Y} = Y\boldsymbol{e}_j$ *of* A^m, *we say that*

$$\boldsymbol{X} < \boldsymbol{Y} \Longleftrightarrow \begin{cases} X < Y \\ \text{or} \\ X = Y \text{ and } i < j. \end{cases}$$

We call this order TOP *for "term over position", since it places more importance on the term order on* A *than on the position in the vector.*

So, for example, in the case of two variables and $m = 2$, using deglex on the power products of A with $x < y$, we see that

$$(x, 0) < (0, x) < (y, 0) < (xy, 0).$$

DEFINITION 3.5.3. *For monomials* $\boldsymbol{X} = X\boldsymbol{e}_i$ *and* $\boldsymbol{Y} = Y\boldsymbol{e}_j$ *of* A^m, *we say that*

$$\boldsymbol{X} < \boldsymbol{Y} \Longleftrightarrow \begin{cases} i < j \\ \text{or} \\ i = j \text{ and } X < Y. \end{cases}$$

We call this order POT *for "position over term", since it places more importance on the position in the vector than on the term order on* A.

So in this case we have, again for the case of two variables and $m = 2$, using deglex on the power products of A with $x < y$,

$$(x, 0) < (y, 0) < (xy, 0) < (0, x).$$

It is easily verified that these two orders satisfy the two conditions of Definition 3.5.1 (Exercise 3.5.1). Of course, each of these two orders could just as well have been defined with a different ordering on the subscripts $\{1, \ldots, m\}$. In order to indicate which order we are using we will write, for example, $\boldsymbol{e}_1 < \boldsymbol{e}_2 < \cdots < \boldsymbol{e}_m$. There are many other examples of orders and we will use an order different from either one of the above in the next section.

We note that we are using the symbol "<" in two different ways, both for a term order on the power products of A and for a term order on the monomials of A^m. The meaning will always be clear from the context.

In analogy to Theorem 1.4.6 we have the following

LEMMA 3.5.4. *Every term order on the monomials of A^m is a well-ordering.*

PROOF. The proof of this lemma is exactly the same as the proof of Theorem 1.4.6 except that the Hilbert Basis Theorem (Theorem 1.1.1) used there must be replaced by Theorem 3.1.1 (Exercise 3.5.2). □

We now adopt some notation. We first fix a term order $<$ on the monomials of A^m. Then for all $\boldsymbol{f} \in A^m$, with $\boldsymbol{f} \neq \boldsymbol{0}$, we may write

$$\boldsymbol{f} = a_1\boldsymbol{X}_1 + a_2\boldsymbol{X}_2 + \cdots + a_r\boldsymbol{X}_r,$$

where, for $1 \leq i \leq r$, $0 \neq a_i \in k$ and $\boldsymbol{X}_i$ is a monomial in A^m satisfying $\boldsymbol{X}_1 > \boldsymbol{X}_2 > \cdots > \boldsymbol{X}_r$. We define

- $\mathrm{lm}(\boldsymbol{f}) = \boldsymbol{X}_1$, the *leading monomial* of $\boldsymbol{f}$;
- $\mathrm{lc}(\boldsymbol{f}) = a_1$, the *leading coefficient* of $\boldsymbol{f}$;
- $\mathrm{lt}(\boldsymbol{f}) = a_1\boldsymbol{X}_1$, the *leading term* of $\boldsymbol{f}$.

We define $\mathrm{lt}(\boldsymbol{0}) = \boldsymbol{0}, \mathrm{lm}(\boldsymbol{0}) = \boldsymbol{0}$, and $\mathrm{lc}(\boldsymbol{0}) = 0$.

Note that, consistent with using monomials in A^m instead of the power products in A, we now use leading "monomials" instead of leading "power products", and use the symbol "lm" instead of "lp".

EXAMPLE 3.5.5. Let $\boldsymbol{f} = (2x^3y - y^3 + 5x, 3xy^2 + 4x + y^2) \in A^2$, with the lex ordering on $A = \mathbb{Q}[x, y]$ with $x < y$. Then in the TOP ordering with $\boldsymbol{e}_1 < \boldsymbol{e}_2$ of Definition 3.5.2 above, we have

$$\boldsymbol{f} = -y^3\boldsymbol{e}_1 + 3xy^2\boldsymbol{e}_2 + y^2\boldsymbol{e}_2 + 2x^3y\boldsymbol{e}_1 + 4x\boldsymbol{e}_2 + 5x\boldsymbol{e}_1,$$

and so $\mathrm{lm}(\boldsymbol{f}) = y^3\boldsymbol{e}_1$, $\mathrm{lc}(\boldsymbol{f}) = -1$, and $\mathrm{lt}(\boldsymbol{f}) = -y^3\boldsymbol{e}_1$. On the other hand, in the POT ordering with $\boldsymbol{e}_1 < \boldsymbol{e}_2$ of Definition 3.5.3 above, we have

$$\boldsymbol{f} = 3xy^2\boldsymbol{e}_2 + y^2\boldsymbol{e}_2 + 4x\boldsymbol{e}_2 - y^3\boldsymbol{e}_1 + 2x^3y\boldsymbol{e}_1 + 5x\boldsymbol{e}_1,$$

and so $\mathrm{lm}(\boldsymbol{f}) = xy^2\boldsymbol{e}_2$, $\mathrm{lc}(\boldsymbol{f}) = 3$ and $\mathrm{lt}(\boldsymbol{f}) = 3xy^2\boldsymbol{e}_2$.

We note that in the case of TOP and POT above, lm, lc and lt are multiplicative, in the following sense: $\mathrm{lm}(f\boldsymbol{g}) = \mathrm{lp}(f)\,\mathrm{lm}(\boldsymbol{g})$, $\mathrm{lc}(f\boldsymbol{g}) = \mathrm{lc}(f)\,\mathrm{lc}(\boldsymbol{g})$, and $\mathrm{lt}(f\boldsymbol{g}) = \mathrm{lt}(f)\,\mathrm{lt}(\boldsymbol{g})$, for all $f \in A$ and $\boldsymbol{g} \in A^m$ (Exercise 3.5.6).

We now move on to the second ingredient in our construction of Gröbner bases for modules, namely, reduction and the Division Algorithm. It should be emphasized that now that we have defined monomials (in place of power products), divisibility and quotients of monomials, and term orders, the definitions can be lifted word for word from Section 1.5. The basic idea behind the algorithm is the same as for polynomials: when dividing $\boldsymbol{f}$ by $\boldsymbol{f}_1, \ldots, \boldsymbol{f}_s$, we want to cancel monomials of $\boldsymbol{f}$ using the leading terms of the $\boldsymbol{f}_i$'s, and continue this process until it cannot be done anymore.

DEFINITION 3.5.6. *Given $\boldsymbol{f}, \boldsymbol{g}, \boldsymbol{h}$ in A^m, $\boldsymbol{g} \neq \boldsymbol{0}$, we say that $\boldsymbol{f}$* reduces *to $\boldsymbol{h}$ modulo $\boldsymbol{g}$ in one step, written*

$$\boldsymbol{f} \xrightarrow{\boldsymbol{g}} \boldsymbol{h},$$

if and only if $\operatorname{lt}(\boldsymbol{g})$ *divides a term* $\boldsymbol{X}$ *that appears in* $\boldsymbol{f}$ *and* $\boldsymbol{h} = \boldsymbol{f} - \frac{\boldsymbol{X}}{\operatorname{lt}(\boldsymbol{g})}\boldsymbol{g}$.

We can think of $\boldsymbol{h}$ in the definition as the remainder of a one step division of $\boldsymbol{f}$ by $\boldsymbol{g}$ similar to the one seen in Section 1.5. Observe that the terms introduced by subtracting $\frac{\boldsymbol{X}}{\operatorname{lt}(\boldsymbol{g})}\boldsymbol{g}$ from $\boldsymbol{f}$ are smaller than the term $\boldsymbol{X}$ subtracted out of $\boldsymbol{f}$. We can continue this process and subtract off all possible terms in $\boldsymbol{f}$ divisible by $\operatorname{lt}(\boldsymbol{g})$.

EXAMPLE 3.5.7. Let $\boldsymbol{f} = (-y^3+2x^3y, 3xy^2+y^2+4x)$ and $\boldsymbol{g} = (x+1, y^2+x)$ be in A^2. We use the lex ordering on $A = \mathbb{Q}[x,y]$, with $x < y$, and TOP with $\boldsymbol{e}_1 < \boldsymbol{e}_2$ in A^2. Then, $\operatorname{lt}(\boldsymbol{g}) = (0, y^2) = y^2\boldsymbol{e}_2$, and

$$\begin{aligned} \boldsymbol{f} &\xrightarrow{\boldsymbol{g}} (-y^3+2x^3y-3x^2-3x, y^2-3x^2+4x) \\ &\xrightarrow{\boldsymbol{g}} (-y^3+2x^3y-3x^2-4x-1, -3x^2+3x). \end{aligned}$$

DEFINITION 3.5.8. *Let* $\boldsymbol{f}, \boldsymbol{h}$, *and* $\boldsymbol{f}_1, \dots, \boldsymbol{f}_s$ *be vectors in* A^m, *with* $\boldsymbol{f}_1, \dots, \boldsymbol{f}_s$ *non-zero and let* $F = \{\boldsymbol{f}_1, \dots, \boldsymbol{f}_s\}$. *We say that* $\boldsymbol{f}$ reduces *to* $\boldsymbol{h}$ *modulo* F, *denoted*

$$\boldsymbol{f} \xrightarrow{F}_+ \boldsymbol{h},$$

if and only if there exists a sequence of indices $i_1, i_2, \dots, i_t \in \{1, \dots, s\}$ *and vectors* $\boldsymbol{h}_1, \dots, \boldsymbol{h}_{t-1} \in A^m$ *such that*

$$\boldsymbol{f} \xrightarrow{\boldsymbol{f}_{i_1}} \boldsymbol{h}_1 \xrightarrow{\boldsymbol{f}_{i_2}} \boldsymbol{h}_2 \xrightarrow{\boldsymbol{f}_{i_3}} \cdots \xrightarrow{\boldsymbol{f}_{i_{t-1}}} \boldsymbol{h}_{t-1} \xrightarrow{\boldsymbol{f}_{i_t}} \boldsymbol{h}.$$

EXAMPLE 3.5.9. Let $\boldsymbol{f}_1 = (xy-y, x^2), \boldsymbol{f}_2 = (x, y^2-x) \in A^2$. We use the lex ordering on $A = \mathbb{Q}[x,y]$, with $x < y$, and TOP with $\boldsymbol{e}_1 < \boldsymbol{e}_2$ in A^2. Let $F = \{\boldsymbol{f}_1, \boldsymbol{f}_2\}$ and $\boldsymbol{f} = (y^2+2x^2y, y^2)$. Then

$$\boldsymbol{f} \xrightarrow{F}_+ (y^2+2y-x, -2x^3-2x^2+x),$$

since

$$\boldsymbol{f} = (y^2+2x^2y, y^2) \xrightarrow{\boldsymbol{f}_2} (y^2+2x^2y-x, x) \xrightarrow{\boldsymbol{f}_1}$$

$$(y^2+2xy-x, -2x^3+x) \xrightarrow{\boldsymbol{f}_1} (y^2+2y-x, -2x^3-2x^2+x).$$

Notice that this last vector $\boldsymbol{h} = (y^2+2y-x, -2x^3-2x^2+x)$ cannot be reduced further by $\boldsymbol{f}_1$ or $\boldsymbol{f}_2$. This is because $\operatorname{lm}(\boldsymbol{f}_1) = (xy, 0)$ and no power product in the first coordinate of $\boldsymbol{h}$ is divisible by xy and $\operatorname{lm}(\boldsymbol{f}_2) = (0, y^2)$ and no power product in the second coordinate of $\boldsymbol{h}$ is divisible by y^2.

DEFINITION 3.5.10. *A vector* $\boldsymbol{r}$ *in* A^m *is called* reduced *with respect to a set* $F = \{\boldsymbol{f}_1, \dots, \boldsymbol{f}_s\}$ *of non-zero vectors in* A^m *if* $\boldsymbol{r} = \boldsymbol{0}$ *or no monomial that appears in* $\boldsymbol{r}$ *is divisible by any one of the* $\operatorname{lm}(\boldsymbol{f}_i), i = 1, \dots, s$. *If* $\boldsymbol{f} \xrightarrow{F}_+ \boldsymbol{r}$ *and* $\boldsymbol{r}$ *is reduced with respect to* F, *then we call* $\boldsymbol{r}$ *a* remainder *for* $\boldsymbol{f}$ *with respect to* F.

The reduction process allows us to give a Division Algorithm that mimics the Division Algorithm for polynomials. Given $\boldsymbol{f}, \boldsymbol{f}_1, \dots, \boldsymbol{f}_s \in A^m$, with $\boldsymbol{f}_1, \dots, \boldsymbol{f}_s \neq \boldsymbol{0}$, this algorithm returns quotients $a_1, \dots, a_s \in A = k[x_1, \dots, x_n]$, and a remainder $\boldsymbol{r} \in A^m$, which is reduced with respect to F, such that

$$\boldsymbol{f} = a_1\boldsymbol{f}_1 + \cdots + a_s\boldsymbol{f}_s + \boldsymbol{r}.$$

This algorithm is given as Algorithm 3.5.1.

INPUT: $\boldsymbol{f}, \boldsymbol{f}_1, \dots, \boldsymbol{f}_s \in A^m$ with $\boldsymbol{f}_i \neq \boldsymbol{0}$ $(1 \leq i \leq s)$

OUTPUT: $a_1, \dots, a_s \in A, \boldsymbol{r} \in A^m$ with $\boldsymbol{f} = a_1\boldsymbol{f}_1 + \cdots + a_s\boldsymbol{f}_s + \boldsymbol{r}$

and $\boldsymbol{r}$ is reduced with respect to $\{\boldsymbol{f}_1, \dots, \boldsymbol{f}_s\}$ and

$\max(\operatorname{lm}(a_1\boldsymbol{f}_1), \dots, \operatorname{lm}(a_s\boldsymbol{f}_s), \operatorname{lm}(\boldsymbol{r})) = \operatorname{lm}(\boldsymbol{f})$

INITIALIZATION: $a_1 := 0, a_2 := 0, \dots, a_s := 0, \boldsymbol{r} := \boldsymbol{0}, \boldsymbol{g} := \boldsymbol{f}$

WHILE $\boldsymbol{g} \neq \boldsymbol{0}$ **DO**

IF there exists i such that $\operatorname{lm}(\boldsymbol{f}_i)$ divides $\operatorname{lm}(\boldsymbol{g})$ **THEN**

Choose i least such that $\operatorname{lm}(\boldsymbol{f}_i)$ divides $\operatorname{lm}(\boldsymbol{g})$

$a_i := a_i + \dfrac{\operatorname{lt}(\boldsymbol{g})}{\operatorname{lt}(\boldsymbol{f}_i)}$

$\boldsymbol{g} := \boldsymbol{g} - \dfrac{\operatorname{lt}(\boldsymbol{g})}{\operatorname{lt}(\boldsymbol{f}_i)}\boldsymbol{f}_i$

ELSE

$\boldsymbol{r} := \boldsymbol{r} + \operatorname{lt}(\boldsymbol{g})$

$\boldsymbol{g} := \boldsymbol{g} - \operatorname{lt}(\boldsymbol{g})$

ALGORITHM 3.5.1. *Division Algorithm in* A^m

Note that the step $\boldsymbol{r} := \boldsymbol{r} + \operatorname{lt}(\boldsymbol{g})$ in the ELSE part of the algorithm is used to put in the remainder the terms that are not divisible by any $\operatorname{lt}(\boldsymbol{f}_i)$ and the step $\boldsymbol{g} := \boldsymbol{g} - \operatorname{lt}(\boldsymbol{g})$ is used to continue in the algorithm to attempt to divide into the next lower term of $\boldsymbol{g}$.

EXAMPLE 3.5.11. We will redo Example 3.5.9 going step by step through Algorithm 3.5.1.

INITIALIZATION: $a_1 := 0, a_2 := 0, \boldsymbol{r} := \boldsymbol{0}, \boldsymbol{g} := (y^2 + 2x^2y, y^2)$

First pass through the WHILE loop:

$(xy, 0) = \operatorname{lm}(\boldsymbol{f}_1)$ does not divides $\operatorname{lm}(\boldsymbol{g}) = (0, y^2)$

$(0, y^2) = \operatorname{lm}(\boldsymbol{f}_2)$ divides $\operatorname{lm}(\boldsymbol{g}) = (0, y^2)$

$$\begin{aligned}
a_2 &:= a_2 + \tfrac{\mathrm{lt}(\boldsymbol{g})}{\mathrm{lt}(\boldsymbol{f}_2)} = 1\\
\boldsymbol{g} &:= \boldsymbol{g} - \tfrac{\mathrm{lt}(\boldsymbol{g})}{\mathrm{lt}(\boldsymbol{f}_2)}\boldsymbol{f}_2 = (y^2+2x^2y, y^2) - \tfrac{(0,y^2)}{(0,y^2)}(x, y^2-x)\\
&= (y^2+2x^2y-x, x)
\end{aligned}$$

Second pass through the WHILE loop:

Neither $\mathrm{lm}(\boldsymbol{f}_1)$ nor $\mathrm{lm}(\boldsymbol{f}_2)$ divides $\mathrm{lm}(\boldsymbol{g}) = (y^2, 0)$

$$\begin{aligned}
\boldsymbol{r} &:= \boldsymbol{r} + \mathrm{lt}(\boldsymbol{g}) = (y^2, 0)\\
\boldsymbol{g} &:= \boldsymbol{g} - \mathrm{lt}(\boldsymbol{g}) = (2x^2y - x, x)
\end{aligned}$$

Third pass through the WHILE loop:

$(xy, 0) = \mathrm{lm}(\boldsymbol{f}_1)$ divides $\mathrm{lm}(\boldsymbol{g}) = (x^2y, 0)$

$$\begin{aligned}
a_1 &:= a_1 + \tfrac{\mathrm{lt}(\boldsymbol{g})}{\mathrm{lt}(\boldsymbol{f}_1)} = 2x\\
\boldsymbol{g} &:= (2x^2y - x, x) - \tfrac{(2x^2y,0)}{(xy,0)}(xy-y, x^2)\\
&= (2xy - x, -2x^3 + x)
\end{aligned}$$

Fourth pass through the WHILE loop:

$(xy, 0) = \mathrm{lm}(\boldsymbol{f}_1)$ divides $\mathrm{lm}(\boldsymbol{g}) = (xy, 0)$

$$\begin{aligned}
a_1 &:= a_1 + \tfrac{\mathrm{lt}(\boldsymbol{g})}{\mathrm{lt}(\boldsymbol{f}_1)} = 2x + 2\\
\boldsymbol{g} &:= (2xy - x, -2x^3 + x) - \tfrac{(2xy,0)}{(xy,0)}(xy-y, x^2)\\
&= (2y - x, -2x^3 - 2x^2 + x)
\end{aligned}$$

Fifth pass through the WHILE loop:

Neither $\mathrm{lm}(\boldsymbol{f}_1)$ nor $\mathrm{lm}(\boldsymbol{f}_2)$ divide $\mathrm{lm}(\boldsymbol{g}) = (y, 0)$

$$\begin{aligned}
\boldsymbol{r} &:= \boldsymbol{r} + \mathrm{lt}(\boldsymbol{g}) = (y^2 + 2y, 0)\\
\boldsymbol{g} &:= \boldsymbol{g} - \mathrm{lt}(\boldsymbol{g}) = (-x, -2x^3 - 2x^2 + x).
\end{aligned}$$

The remaining four passes through the WHILE loop, one for each of the four remaining terms in $\boldsymbol{g}$, will be similar to the last one, since neither $\mathrm{lm}(\boldsymbol{f}_1)$ nor $\mathrm{lm}(\boldsymbol{f}_2)$ divides any of the remaining terms of $\boldsymbol{g}$. So we finally get, as we did in Example 3.5.9, that

$$\boldsymbol{f} \xrightarrow{F}_+ (y^2 + 2y - x, -2x^3 - 2x^2 + x) = \boldsymbol{r}$$

and, moreover

$$\boldsymbol{f} = (2x+2)\boldsymbol{f}_1 + \boldsymbol{f}_2 + \boldsymbol{r}.$$

THEOREM 3.5.12. *Given a set $F = \{\boldsymbol{f}_1, \ldots, \boldsymbol{f}_s\}$ of non-zero vectors and $\boldsymbol{f}$ in A^m, the Division Algorithm (Algorithm 3.5.1) will produce polynomials $a_1, \ldots, a_s, \in A$ and a vector $\boldsymbol{r} \in A^m$ such that*

$$\boldsymbol{f} = a_1\boldsymbol{f}_1 + \cdots + a_s\boldsymbol{f}_s + \boldsymbol{r},$$

with $\boldsymbol{r}$ reduced with respect to F, and

$$\mathrm{lm}(\boldsymbol{f}) = \max(\mathrm{lm}(a_1\boldsymbol{f}_1), \ldots, \mathrm{lm}(a_s\boldsymbol{f}_s), \mathrm{lm}(\boldsymbol{r})).$$

PROOF. The proof is exactly the same as the proof of Theorem 1.5.9 except that we use Lemma 3.5.4 instead of Theorem 1.4.6. □

We now have what we need to define Gröbner bases in modules. So let M be a submodule of A^m.

DEFINITION 3.5.13. *A set of non-zero vectors* $G = \{\boldsymbol{g}_1, \dots, \boldsymbol{g}_t\}$ *contained in the submodule* M *is called a* Gröbner basis *for* M *if and only if for all* $\boldsymbol{f} \in M$, *there exists* $i \in \{1, \dots, t\}$ *such that* $\mathrm{lm}(\boldsymbol{g}_i)$ *divides* $\mathrm{lm}(\boldsymbol{f})$. *We say that the set* G *is a* Gröbner basis *provided* G *is a Gröbner basis for the submodule,* $\langle G\rangle$, *it generates.*

We now give the characterizations of a Gröbner basis analogous to those in Theorems 1.6.2 and Theorem 1.6.7. The proof is basically the same as the one for the ideal case and is left to the exercises (Exercise 3.5.9). We first define, for a subset W of A^m, the leading term module of W to be the submodule of A^m,

$$\mathrm{Lt}(W) = \langle \mathrm{lt}(\boldsymbol{w}) \mid \boldsymbol{w} \in W\rangle \subseteq A^m.$$

THEOREM 3.5.14. *The following statements are equivalent for a submodule* $M \subseteq A^m$ *and* $G = \{\boldsymbol{g}_1, \dots, \boldsymbol{g}_t\} \subseteq M$ *with* $\boldsymbol{g}_i \neq \boldsymbol{0}$ $(1 \le i \le t)$.

(i) *For all* $\boldsymbol{f} \in M$, *there exists* i *such that* $\mathrm{lm}(\boldsymbol{g}_i)$ *divides* $\mathrm{lm}(\boldsymbol{f})$ *(that is,* G *is a Gröbner basis for* M*).*

(ii) $\boldsymbol{f} \in M$ *if and only if* $\boldsymbol{f} \xrightarrow{G}_+ \boldsymbol{0}$.

(iii) *For all* $\boldsymbol{f} \in M$, *there exists* $h_1, \dots, h_t \in A$ *such that* $\boldsymbol{f} = h_1\boldsymbol{g}_1 + \cdots + h_t\boldsymbol{g}_t$ *and* $\mathrm{lm}(\boldsymbol{f}) = \max_{1\le i\le t}(\mathrm{lm}(h_i\boldsymbol{g}_i))$.

(iv) $\mathrm{Lt}(G) = \mathrm{Lt}(M)$.

(v) *For all* $\boldsymbol{f} \in A^m$, *if* $\boldsymbol{f} \xrightarrow{G}_+ \boldsymbol{r}_1$, $\boldsymbol{f} \xrightarrow{G}_+ \boldsymbol{r}_2$, *and* $\boldsymbol{r}_1, \boldsymbol{r}_2$ *are reduced with respect to* G, *then* $\boldsymbol{r}_1 = \boldsymbol{r}_2$.

For completeness sake, we note the Corollaries of this result which correspond to the Corollaries of Theorem 1.6.2 (Exercises 3.5.10 and 3.5.11).

COROLLARY 3.5.15. *If* $G = \{\boldsymbol{g}_1, \dots, \boldsymbol{g}_t\}$ *is a Gröbner basis for the submodule* M *of* A^m, *then* $M = \langle \boldsymbol{g}_1, \dots, \boldsymbol{g}_t\rangle$.

COROLLARY 3.5.16. *Every non-zero submodule* M *of* A^m *has a Gröbner basis.*

We will now introduce the analogue of S-polynomials. We continue to emphasize that we need only copy what was done in the polynomial case, except that it is not yet clear what the least common multiple of two monomials should be. So let $\boldsymbol{X} = X\boldsymbol{e}_i$ and $\boldsymbol{Y} = Y\boldsymbol{e}_j$ be two monomials in A^m. Then by the *least common multiple* of $\boldsymbol{X}$ and $\boldsymbol{Y}$ (denoted $\mathrm{lcm}(\boldsymbol{X}, \boldsymbol{Y})$), we mean

- $\boldsymbol{0}$, if $i \neq j$;
- $L\boldsymbol{e}_i$ where $L = \mathrm{lcm}(X, Y)$, if $i = j$.

For example, $\mathrm{lcm}((x^2yz, 0), (xy^3, 0)) = (x^2y^3z, 0)$ and $\mathrm{lcm}((x^2y, 0), (0, xy^3)) = (0, 0)$.

DEFINITION 3.5.17. *Let* $\mathbf{0} \neq \boldsymbol{f}, \boldsymbol{g} \in A^m$. *Let* $\boldsymbol{L} = \operatorname{lcm}(\operatorname{lm}(\boldsymbol{f}), \operatorname{lm}(\boldsymbol{g}))$. *The vector*

$$S(\boldsymbol{f}, \boldsymbol{g}) = \frac{\boldsymbol{L}}{\operatorname{lt}(\boldsymbol{f})}\boldsymbol{f} - \frac{\boldsymbol{L}}{\operatorname{lt}(\boldsymbol{g})}\boldsymbol{g}$$

is called the S-polynomial[5] *of* $\boldsymbol{f}$ *and* $\boldsymbol{g}$.

The motivation for this definition is the same as the one used for the definition of S-polynomials in Definition 1.7.1. Namely, we are interested in linear combinations of $\boldsymbol{f}$ and $\boldsymbol{g}$ in which the leading monomials cancel out. This cannot happen if $\operatorname{lm}(\boldsymbol{f})$ and $\operatorname{lm}(\boldsymbol{g})$ have their respective non-zero entries in different coordinates and so, in this case, we define their least common multiple to be the zero vector making their S-polynomial the zero vector. On the other hand, if $\operatorname{lm}(\boldsymbol{f})$ and $\operatorname{lm}(\boldsymbol{g})$ have their respective non-zero entries in the same coordinate, then $S(\boldsymbol{f}, \boldsymbol{g})$ is set up to cancel out these leading monomials in the most efficient way.

EXAMPLE 3.5.18. We consider $A = \mathbb{Q}[x, y]$ with the deglex ordering with $x < y$ and A^2 with the TOP ordering with $\boldsymbol{e}_1 < \boldsymbol{e}_2$. Then,

$$S((x^2+1, 5xy^3+x), (x^2y, 3x^3y+y)) =$$

$$\frac{(0, x^3y^3)}{(0, 5xy^3)}(x^2+1, 5xy^3+x) - \frac{(0, x^3y^3)}{(0, 3x^3y)}(x^2y, 3x^3y+y) =$$

$$\frac{x^2}{5}(x^2+1, 5xy^3+x) - \frac{y^2}{3}(x^2y, 3x^3y+y) = (\frac{1}{5}x^4 + \frac{1}{5}x^2 - \frac{1}{3}x^2y^3, \frac{1}{5}x^3 - \frac{1}{3}y^3).$$

Note that the leading monomial of each of the summands is $(0, x^3y^3)$ while the leading monomial of the S-polynomial is $(x^2y^3, 0) < (0, x^3y^3)$.

THEOREM 3.5.19. *Let* $G = \{\boldsymbol{g}_1, \ldots, \boldsymbol{g}_t\}$ *be a set of non-zero vectors in* A^m. *Then* G *is a Gröbner basis for the submodule* $M = \langle \boldsymbol{g}_1, \ldots, \boldsymbol{g}_t \rangle$ *of* A^m *if and only if for all* $i \neq j$,

$$S(\boldsymbol{g}_i, \boldsymbol{g}_j) \xrightarrow{G}_+ \mathbf{0}.$$

We will leave the proof of this Theorem to the exercises (Exercise 3.5.13); it follows the proof of Theorem 1.7.4 exactly.

This last Theorem allows us to give the analog of Buchberger's Algorithm, Algorithm 1.7.1, for computing Gröbner bases. It is given as Algorithm 3.5.2. Although this algorithm is exactly the same as Algorithm 1.7.1, we will restate it here for the convenience of the reader. The proof of the correctness of the algorithm is left to the exercises (Exercise 3.5.14).

[5] We use the term "S-polynomial" even though the result is clearly a vector with polynomial coordinates.

INPUT: $F = \{\boldsymbol{f}_1, \dots, \boldsymbol{f}_s\} \subseteq A^m$ with $\boldsymbol{f}_i \neq \boldsymbol{0}$ $(1 \leq i \leq s)$

OUTPUT: $G = \{\boldsymbol{g}_1, \dots, \boldsymbol{g}_t\}$, a Gröbner basis for $\langle \boldsymbol{f}_1, \dots, \boldsymbol{f}_s \rangle$

INITIALIZATION: $G := F$, $\mathcal{G} := \{\{\boldsymbol{f}_i, \boldsymbol{f}_j\} \mid \boldsymbol{f}_i \neq \boldsymbol{f}_j \in G\}$

WHILE $\mathcal{G} \neq \emptyset$ **DO**

  Choose any $\{\boldsymbol{f}, \boldsymbol{g}\} \in \mathcal{G}$

  $\mathcal{G} := \mathcal{G} - \{\{\boldsymbol{f}, \boldsymbol{g}\}\}$

  $S(\boldsymbol{f}, \boldsymbol{g}) \xrightarrow{G}_+ \boldsymbol{h}$, where $\boldsymbol{h}$ is reduced with respect to G

  IF $\boldsymbol{h} \neq \boldsymbol{0}$ **THEN**

    $\mathcal{G} := \mathcal{G} \cup \{\{\boldsymbol{u}, \boldsymbol{h}\} \mid \text{for all } \boldsymbol{u} \in G\}$

    $G := G \cup \{\boldsymbol{h}\}$

ALGORITHM 3.5.2. *Buchberger's Algorithm for Modules*

EXAMPLE 3.5.20. Consider the following vectors of $(\mathbb{Q}[x,y])^3$:

$$\boldsymbol{f}_1 = (0, y, x),\ \boldsymbol{f}_2 = (0, x, xy - x),\ \boldsymbol{f}_3 = (x, y^2, 0),\ \boldsymbol{f}_4 = (y, 0, x).$$

We use the deglex term order on $\mathbb{Q}[x,y]$ with $x > y$ and the TOP order on $(\mathbb{Q}[x,y])^3$ with $\boldsymbol{e}_1 > \boldsymbol{e}_2 > \boldsymbol{e}_3$. We compute a Gröbner basis for $M = \langle \boldsymbol{f}_1, \boldsymbol{f}_2, \boldsymbol{f}_3, \boldsymbol{f}_4 \rangle$ using Algorithm 3.5.2. Let $G = \{\boldsymbol{f}_1, \boldsymbol{f}_2, \boldsymbol{f}_3, \boldsymbol{f}_4\}$. The only S-polynomials with non-zero $\boldsymbol{L}$ in Definition 3.5.17 are the ones corresponding to $\mathcal{G} = \{\{\boldsymbol{f}_1, \boldsymbol{f}_2\}, \{\boldsymbol{f}_1, \boldsymbol{f}_4\}, \{\boldsymbol{f}_2, \boldsymbol{f}_4\}\}$. We compute

$$S(\boldsymbol{f}_1, \boldsymbol{f}_2) = y\boldsymbol{f}_1 - \boldsymbol{f}_2 = (0, y^2 - x, x) \xrightarrow{G}_+ (-x, -x - y, 0).$$

This vector is reduced with respect to G, so we set $\boldsymbol{f}_5 = (-x, -x - y, 0)$, and we add it to G. Next,

$$S(\boldsymbol{f}_1, \boldsymbol{f}_4) = \boldsymbol{f}_1 - \boldsymbol{f}_4 = (-y, y, 0).$$

Again, this vector is reduced with respect to G, so we set $\boldsymbol{f}_6 = (-y, y, 0)$, and we add it to G. One readily sees that $S(\boldsymbol{f}_2, \boldsymbol{f}_4) \xrightarrow{G}_+ \boldsymbol{0}$. The new vectors $\boldsymbol{f}_5$ and $\boldsymbol{f}_6$ generate only one S-polynomial that needs to be considered,

$$S(\boldsymbol{f}_5, \boldsymbol{f}_6) = y\boldsymbol{f}_5 - x\boldsymbol{f}_6 = (0, -2xy - y^2, 0) \xrightarrow{G}_+ (0, -2xy - x - y, 0).$$

This vector is reduced with respect to G, so we set $\boldsymbol{f}_7 = (0, -2xy - x - y, 0)$, and we add it to G. This new vector generates only one S-polynomial that needs to be considered,

$$S(\boldsymbol{f}_3, \boldsymbol{f}_7) = 2x\boldsymbol{f}_3 + y\boldsymbol{f}_7 = (2x^2, -xy - y^2, 0) \xrightarrow{G}_+ (0, -2x^2 + \frac{1}{2}x + \frac{1}{2}y, 0).$$

We set $\boldsymbol{f}_8 = (0, -2x^2 + \frac{1}{2}x + \frac{1}{2}y, 0)$ and add it to G. This new vector generates two S-polynomials we need to consider,

$$S(\boldsymbol{f}_3, \boldsymbol{f}_8) = 2x^2\boldsymbol{f}_3 + y^2\boldsymbol{f}_8 = (2x^3, \frac{1}{2}xy^2 + \frac{1}{2}y^3, 0) \xrightarrow{G}_+ \mathbf{0}$$

and

$$S(\boldsymbol{f}_7, \boldsymbol{f}_8) = x\boldsymbol{f}_7 - y\boldsymbol{f}_8 = (0, -x^2 - \frac{3}{2}xy - \frac{1}{2}y^2, 0) \xrightarrow{G}_+ \mathbf{0}.$$

Therefore $G = \{\boldsymbol{f}_1, \boldsymbol{f}_2, \boldsymbol{f}_3, \boldsymbol{f}_4, \boldsymbol{f}_5, \boldsymbol{f}_6, \boldsymbol{f}_7, \boldsymbol{f}_8\}$ is a Gröbner basis for M.

Finally, we conclude this section by noting that the results in Chapter 1 concerning reduced Gröbner bases hold in this context as well. We will not prove the results here as, again, they exactly parallel the ones in Chapter 1, but we will state the main Definition and Theorem here for completeness. The proof will be left for the exercises (Exercise 3.5.17).

DEFINITION 3.5.21. *A Gröbner basis $G = \{\boldsymbol{g}_1, \dots, \boldsymbol{g}_t\} \subseteq A^m$ is a* reduced *Gröbner basis if, for all i, $\boldsymbol{g}_i$ is reduced with respect to $G - \{\boldsymbol{g}_i\}$ and $\text{lc}(g_i) = 1$ for all $i = 1, \dots, t$. Thus for all i, no non-zero term in $\boldsymbol{g}_i$ is divisible by any $\text{lm}(\boldsymbol{g}_j)$ for any $j \neq i$.*

THEOREM 3.5.22. *Fix a term order. Then every non-zero submodule M of A^m has a unique reduced Gröbner basis with respect to this term order. This Gröbner basis is effectively computable once M has been given as generated by a finite set of vectors in A^m.*

EXAMPLE 3.5.23. We go back to Example 3.5.20. In that example we had that $G = \{\boldsymbol{f}_1,\ \boldsymbol{f}_2,\ \boldsymbol{f}_3,\ \boldsymbol{f}_4,\ \boldsymbol{f}_5,\ \boldsymbol{f}_6,\ \boldsymbol{f}_7,\ \boldsymbol{f}_8\}$ is a Gröbner basis for M. We first observe that since $\text{lm}(\boldsymbol{f}_1)$ divides $\text{lm}(\boldsymbol{f}_2)$ and $\text{lm}(\boldsymbol{f}_4)$ we may eliminate $\boldsymbol{f}_2$ and $\boldsymbol{f}_4$ from G and still have a Gröbner basis. Moreover, $\boldsymbol{f}_3 \longrightarrow_+ (0, y^2 - x - y, 0)$. Therefore the reduced Gröbner basis for M is

$$\Big\{(0, y, x),\ (0, y^2 - x - y, 0),\ (0, x^2 - \frac{1}{4}x - \frac{1}{4}y, 0),$$
$$(0, xy + \frac{1}{2}x + \frac{1}{2}y, 0),\ (y, -y, 0),\ (x, x + y, 0)\Big\}.$$

Exercises

3.5.1. Prove that the POT and TOP orders of Definitions 3.5.3 and 3.5.2 are term orders on A^m.

3.5.2. Complete the proof of Lemma 3.5.4.

3.5.3. Prove the analog of Proposition 1.4.5: Let $<$ be a term order on A^m. For $\boldsymbol{X}, \boldsymbol{Y}$ monomials in A^m, if $\boldsymbol{X}$ divides $\boldsymbol{Y}$, then $\boldsymbol{X} \leq \boldsymbol{Y}$.

3.5.4. Prove the analog of Exercise 1.4.6: Let $<$ be a total order on the monomials of A^m satisfying Condition (ii) of Definition 3.5.1, and assume that $<$ is also a well-ordering. Prove that $\boldsymbol{X} < Z\boldsymbol{X}$ for every monomial $\boldsymbol{X}$ of A^m and power product $Z \neq 1$ of A.

3.5.5. Write the following vectors as the sum of terms in decreasing order according to the indicated term orders:

a. $\boldsymbol{f} = (x^2y - xy^2, x^3+1, y^3-1)$, with the deglex term order on $A = \mathbb{Q}[x, y]$ with $x > y$, and the TOP ordering on A^3 with $\boldsymbol{e}_1 > \boldsymbol{e}_2 > \boldsymbol{e}_3$. Then change the order to deglex with $y > x$ and the POT ordering with $\boldsymbol{e}_1 < \boldsymbol{e}_2 < \boldsymbol{e}_3$. Finally, change the order to lex with $y > x$ and the TOP ordering with $\boldsymbol{e}_1 < \boldsymbol{e}_2 < \boldsymbol{e}_3$.

b. $\boldsymbol{f} = (x^2 + xy, y^2 + yz, z^2 + xz)$, with the degrevlex term order on $A = \mathbb{Q}[x, y, z]$ with $x > y > z$, and the TOP ordering on A^3 with $\boldsymbol{e}_1 < \boldsymbol{e}_2 < \boldsymbol{e}_3$. Then change the order to degrevlex with $z > y > x$ and the POT ordering with $\boldsymbol{e}_1 > \boldsymbol{e}_2 > \boldsymbol{e}_3$. Finally, change the order to lex with $z > y > x$ and the TOP ordering with $\boldsymbol{e}_1 < \boldsymbol{e}_2 < \boldsymbol{e}_3$.

3.5.6. Prove that when using a POT or TOP ordering, we have that lt, lm, and lc are multiplicative. Namely, prove that for all $f \in A$ and $\boldsymbol{g} \in A^m$ we have $\mathrm{lt}(f\boldsymbol{g}) = \mathrm{lt}(f)\,\mathrm{lt}(\boldsymbol{g})$, $\mathrm{lm}(f\boldsymbol{g}) = \mathrm{lp}(f)\,\mathrm{lm}(\boldsymbol{g})$, and $\mathrm{lc}(f\boldsymbol{g}) = \mathrm{lc}(f)\,\mathrm{lc}(\boldsymbol{g})$.

3.5.7. Complete the proof of Theorem 3.5.12.

3.5.8. As in Example 3.5.11, follow Algorithm 3.5.1 to find the quotients and remainders of the following divisions:

a. Divide $\boldsymbol{f} = (x^2y+y, xy+y^2, xy^2+x^2)$, by $F = \{\boldsymbol{f}_1, \boldsymbol{f}_2, \boldsymbol{f}_3, \boldsymbol{f}_4\}$, where $\boldsymbol{f}_1 = (x^2, xy, y^2)$, $\boldsymbol{f}_2 = (y, 0, x)$, $\boldsymbol{f}_3 = (0, x, y)$, and $\boldsymbol{f}_4 = (y, 1, 0)$. Use lex with $x > y$ on $A = \mathbb{Q}[x, y]$ with the TOP ordering on A^3 with $\boldsymbol{e}_1 > \boldsymbol{e}_2 > \boldsymbol{e}_3$.

b. Divide $\boldsymbol{f} = (x^2y+y, xy+y^2, xy^2+x^2)$, by $F = \{\boldsymbol{f}_1, \boldsymbol{f}_2, \boldsymbol{f}_3, \boldsymbol{f}_4\}$, where $\boldsymbol{f}_1 = (xy, 0, x)$, $\boldsymbol{f}_2 = (y, x, 0)$, $\boldsymbol{f}_3 = (x + y, 0, 0)$, and $\boldsymbol{f}_4 = (x, y, 0)$. Use deglex with $x > y$ on $A = \mathbb{Q}[x, y]$ with the POT ordering on A^3 with $\boldsymbol{e}_1 > \boldsymbol{e}_2 > \boldsymbol{e}_3$.

3.5.9. Prove Theorem 3.5.14.

3.5.10. Prove Corollary 3.5.15.

3.5.11. Prove Corollary 3.5.16.

3.5.12. Prove the analog of Exercise 1.6.13: Let $\{\boldsymbol{g}_1, \ldots, \boldsymbol{g}_t\}$ be non-zero vectors in A^m and $0 \neq h \in A$. Prove that $\{\boldsymbol{g}_1, \ldots, \boldsymbol{g}_t\}$ is a Gröbner basis if and only if $\{h\boldsymbol{g}_1, \ldots, h\boldsymbol{g}_t\}$ is a Gröbner basis.

3.5.13. Prove Theorem 3.5.19.

3.5.14. Prove that Algorithm 3.5.2 produces a Gröbner basis for $\langle \boldsymbol{f}_1, \ldots, \boldsymbol{f}_s \rangle$.

3.5.15. Compute the Gröbner bases for the following modules with respect to the indicated term orders. You should do the computations without a Computer Algebra System.

a. $M = \langle \boldsymbol{f}_1, \boldsymbol{f}_2, \boldsymbol{f}_3, \boldsymbol{f}_4 \rangle \subseteq A^3$, where $\boldsymbol{f}_1 = (x^2, xy, y^2)$, $\boldsymbol{f}_2 = (y, 0, x)$, $\boldsymbol{f}_3 = (0, x, y)$, and $\boldsymbol{f}_4 = (y, 1, 0)$. Use lex on $A = \mathbb{Q}[x, y]$ with $x > y$ and the TOP ordering on A^3 with $\boldsymbol{e}_1 > \boldsymbol{e}_2 > \boldsymbol{e}_3$.

b. $M = \langle \boldsymbol{f}_1, \boldsymbol{f}_2, \boldsymbol{f}_3, \boldsymbol{f}_4 \rangle \subseteq A^3$, where $\boldsymbol{f}_1 = (x - y, x, x)$, $\boldsymbol{f}_2 = (yx, y, y)$, $\boldsymbol{f}_3 = (y, x, x)$, and $\boldsymbol{f}_4 = (y, x, 0)$. Use deglex on $A = \mathbb{Q}[x, y]$ with $y > x$ and the TOP ordering on A^3 with $\boldsymbol{e}_1 > \boldsymbol{e}_2 > \boldsymbol{e}_3$.

3.5.16. Give an example that shows that the analog of Lemma 3.3.1 is false in the module case; that is, crit1 of Section 3.3 cannot be used in Algorithm 3.5.2.

3.5.17. Prove Theorem 3.5.22.

3.5.18. Find the reduced Gröbner bases for the examples in Exercise 3.5.15.

3.5.19. In this exercise, we will take a different view of the module A^m which allows us to implement Gröbner basis theory in Computer Algebra Systems that don't have a built-in module facility. Let $e_1, e_2, \ldots, e_m$ be new variables and consider the polynomial ring $k[x_1, \ldots, x_n, e_1, \ldots, e_m]$. We identify A^m with the A-submodule of $k[x_1, \ldots, x_n, e_1, \ldots, e_m]$ generated by $e_1, e_2, \ldots, e_m$ simply by sending $(f_1, \ldots, f_s)$ to $f_1e_1 + \cdots + f_me_m \in k[x_1, \ldots, x_n, e_1, \ldots, e_m]$. Fix a term order on $k[x_1, \ldots, x_n]$. Consider any order on the variables $e_1, e_2, \ldots, e_m$ such that $e_1 < e_2 < \cdots < e_m$. Prove the following:

a. Consider an elimination order between the x and e variables. Then, in the above correspondence, if the x variables are larger than the e variables we have the TOP ordering in A^m and if the e variables are larger than the x variables we have the POT ordering in A^m.

b. Note that division and quotient of monomials in A^m just mean the usual division and quotient in $k[x_1, \ldots, x_n, e_1, \ldots, e_m]$. Then note that Definitions 3.5.6, 3.5.8, and 3.5.10 become the usual ones in $k[x_1, \ldots, x_n, e_1, \ldots, e_m]$. Show that the Division Algorithm for modules (Algorithm 3.5.1) corresponds directly to the polynomial Division Algorithm (Algorithm 1.5.1).

c. Show that the usual Buchberger's Algorithm (Algorithm 1.7.1) performed in $k[x_1, \ldots, x_n, e_1, \ldots, e_m]$ can be used to compute Gröbner bases in A^m with the following modification: In the definition of the least common multiple, we must set $\operatorname{lcm}(Xe_i, Ye_j) = 0$ for power products X and Y in the x variables when $i \neq j$.

d. Redo the computations in Exercise 3.5.15 using the above method.

3.6. Elementary Applications of Gröbner Bases for Modules. Let $M = \langle \boldsymbol{f}_1, \ldots, \boldsymbol{f}_s \rangle$ be a submodule of A^m. As in the case of ideals in A (see Section 2.1), we show in the present section how to perform effectively the following tasks:

(i) Given $\boldsymbol{f} \in A^m$, determine whether $\boldsymbol{f}$ is in M (this is the module membership problem), and if so, find $h_1, \ldots, h_s \in A$ such that $\boldsymbol{f} = h_1\boldsymbol{f}_1 + \cdots + h_s\boldsymbol{f}_s$;

(ii) Given M', another submodule of A^m, determine whether M' is contained in M, and if $M' \subseteq M$, whether $M' = M$;

(iii) Find coset representatives for the elements of A^m/M;

(iv) Find a basis for the k-vector space A^m/M.

Moreover, we show how the theory of elimination we introduced in Section 2.3 is carried over to the module setting.

Let $F = \{\boldsymbol{f}_1, \dots, \boldsymbol{f}_s\}$ be a set of non-zero vectors in A^m and let $M = \langle \boldsymbol{f}_1, \dots, \boldsymbol{f}_s \rangle$. Let $G = \{\boldsymbol{g}_1, \dots, \boldsymbol{g}_t\}$ be a Gröbner basis for M with respect to some term order.

We start with Task (i). Let $\boldsymbol{f} \in A^m$. We have already noted in Theorem 3.5.14 that

$$\boldsymbol{f} \in M \Longleftrightarrow \boldsymbol{f} \xrightarrow{G}_+ \mathbf{0}.$$

So we can determine algorithmically whether $\boldsymbol{f} \in M$ or not. Moreover, if $\boldsymbol{f} \in M$, we apply the Division Algorithm for modules presented in Section 3.5 (Algorithm 3.5.1) to get

$$\boldsymbol{f} = h_1\boldsymbol{g}_1 + \cdots + h_t\boldsymbol{g}_t. \tag{3.6.1}$$

As in the ideal case, we can find an $s \times t$ matrix T with polynomial entries such that, if we view F and G as matrices whose columns are the $\boldsymbol{f}_i$'s and the $\boldsymbol{g}_j$'s respectively, we have $G = FT$. (T is obtained, as in the ideal case, by keeping track of the reductions during Buchberger's Algorithm for modules.) Therefore Equation (3.6.1) can be transformed to express the vector $\boldsymbol{f}$ as a linear combination of the vectors $\boldsymbol{f}_1, \dots, \boldsymbol{f}_s$.

EXAMPLE 3.6.1. Let $A = \mathbb{Q}[x, y]$. We consider the submodule M of A^3 generated by $F = \{\boldsymbol{f}_1, \boldsymbol{f}_2, \boldsymbol{f}_3, \boldsymbol{f}_4\}$, where

$$\boldsymbol{f}_1 = (xy, y, x),\ \boldsymbol{f}_2 = (x^2 + x, y + x^2, y),$$
$$\boldsymbol{f}_3 = (-y, x, y),\ \boldsymbol{f}_4 = (x^2, x, y).$$

We use the lex term ordering on A with $x < y$ and TOP with $\boldsymbol{e}_1 > \boldsymbol{e}_2 > \boldsymbol{e}_3$ on A^3. To indicate the leading term of a vector we will underline it; e.g.

$$\boldsymbol{f}_1 = (\underline{xy}, y, x),\ \boldsymbol{f}_2 = (x^2 + x, \underline{y} + x^2, y),$$
$$\boldsymbol{f}_3 = (\underline{-y}, x, y),\ \boldsymbol{f}_4 = (x^2, x, \underline{y}).$$

The reduced Gröbner basis for M is $G = \{\boldsymbol{g}_1, \boldsymbol{g}_2, \boldsymbol{g}_3, \boldsymbol{g}_4, \boldsymbol{g}_5, \boldsymbol{g}_6\}$, where

$$\boldsymbol{g}_1 = (\underline{x^3} + x, x^2 - x, -x),\ \boldsymbol{g}_2 = (x, \underline{y} + x^2 - x, 0),$$
$$\boldsymbol{g}_3 = (\underline{y} + x^2, 0, 0),\ \boldsymbol{g}_4 = (x^2, x, \underline{y}),$$
$$\boldsymbol{g}_5 = (x^2, \underline{x^3}, -x^3),\ \boldsymbol{g}_6 = (x^2 - 2x, -x^2 + 2x, \underline{x^5} - x^4 - 3x^3 + x^2 + 2x).$$

Consider the vector $\boldsymbol{f} = (-2x, x - 1, \underline{xy} + x)$. To determine whether $\boldsymbol{f}$ is in M, we perform the Division Algorithm presented in Section 3.5 (Algorithm 3.5.1)[6]:

$$\begin{aligned} \boldsymbol{f} &\xrightarrow{x, \boldsymbol{g}_4} (\underline{-x^3} - 2x, -x^2 + x - 1, x) \\ &\xrightarrow{-1, \boldsymbol{g}_1} (\underline{-x}, -1, 0). \end{aligned}$$

Since $x\boldsymbol{e}_1$, and $\boldsymbol{e}_2$ cannot be divided by any $\operatorname{lm}(\boldsymbol{g}_i)$, the vector $(-x, -1, 0)$ is reduced with respect to G, and hence $\boldsymbol{f}$ is not in M.

---

[6]We remind the reader that when we write $\boldsymbol{f} \xrightarrow{X, \boldsymbol{g}} \boldsymbol{h}$ we mean that $\boldsymbol{h} = \boldsymbol{f} - X\boldsymbol{g}$.

Now consider the vector $\boldsymbol{g} = (\underline{yx^5} - yx + x, yx^4 + y + 2x^2 - x, x^5 + yx)$. Again we perform the Division Algorithm:

$$\begin{aligned}
\boldsymbol{g} &\xrightarrow{x^5,\boldsymbol{g}_3} (-yx - x^7 + x, \underline{yx^4} + y + 2x^2 - x, yx + x^5)\\
&\xrightarrow{x^4,\boldsymbol{g}_2} (\underline{-yx} - x^7 - x^5 + x, y - x^6 + x^5 + 2x^2 - x, xy + x^5)\\
&\xrightarrow{-x,\boldsymbol{g}_3} (-x^7 - x^5 + x^3 + x, y - x^6 + x^5 + 2x^2 - x, \underline{xy} + x^5)\\
&\xrightarrow{x,\boldsymbol{g}_4} (-x^7 - x^5 + x, \underline{y} - x^6 + x^5 + x^2 - x, x^5)\\
&\xrightarrow{1,\boldsymbol{g}_2} (\underline{-x^7} - x^5, -x^6 + x^5, x^5)\\
&\xrightarrow{-x^4,\boldsymbol{g}_1} (0,0,0).
\end{aligned}$$

Therefore $\boldsymbol{g}$ is in M. Moreover, we get

$$\begin{aligned}
\boldsymbol{g} &= x^5\boldsymbol{g}_3 + x^4\boldsymbol{g}_2 - x\boldsymbol{g}_3 + x\boldsymbol{g}_4 + \boldsymbol{g}_2 - x^4\boldsymbol{g}_1\\
&= -x^4\boldsymbol{g}_1 + (x^4+1)\boldsymbol{g}_2 + (x^5 - x)\boldsymbol{g}_3 + x\boldsymbol{g}_4.
\end{aligned}$$

We now want to express $\boldsymbol{g}$ as a linear combination of the original $\boldsymbol{f}_i$'s. So we consider the matrix T that transforms G into F. We have

(3.6.2) $$\begin{bmatrix} \boldsymbol{g}_1 & \boldsymbol{g}_2 & \boldsymbol{g}_3 & \boldsymbol{g}_4 & \boldsymbol{g}_5 & \boldsymbol{g}_6 \end{bmatrix}$$

$$= \begin{bmatrix} \boldsymbol{f}_1 & \boldsymbol{f}_2 & \boldsymbol{f}_3 & \boldsymbol{f}_4 \end{bmatrix} \underbrace{\begin{bmatrix} -1 & 0 & 0 & 0 & -x^2 - y & h_1\\ 1 & 1 & 0 & 0 & x + y & h_2\\ -x & 0 & -1 & 0 & x(1-y) & h_3\\ x-1 & -1 & 1 & 1 & xy - y - x & h_4 \end{bmatrix}}_{T},$$

where

(3.6.3) $$\begin{aligned}
h_1 &= y^2 + 2x^2y - xy - 2y + x^4 - x^3 - 3x^2 + x + 2\\
h_2 &= -y^2 - x^2y + 2y + 2x^2 + x - 2\\
h_3 &= xy^2 + x^3y - x^2y - 3xy - 2x^3 + x^2 + 4x\\
h_4 &= -xy^2 + y^2 - x^3y + 2x^2y + 2xy - 2y - 2x^2 - 3x + 2.
\end{aligned}$$

Therefore we have

$$\begin{aligned}
\boldsymbol{g} &= -x^4\boldsymbol{g}_1 + (x^4+1)\boldsymbol{g}_2 + (x^5 - x)\boldsymbol{g}_3 + x\boldsymbol{g}_4\\
&= -x^4(-\boldsymbol{f}_1 + \boldsymbol{f}_2 - x\boldsymbol{f}_3 + (x-1)\boldsymbol{f}_4) + (x^4+1)(\boldsymbol{f}_2 - \boldsymbol{f}_4)\\
&\quad + (x^5 - x)(-\boldsymbol{f}_3 + \boldsymbol{f}_4) + x\boldsymbol{f}_4\\
&= x^4\boldsymbol{f}_1 + \boldsymbol{f}_2 + x\boldsymbol{f}_3 - \boldsymbol{f}_4.
\end{aligned}$$

Now we turn our attention to Task (ii). Let M' be another submodule of A^m, say $M' = \langle \boldsymbol{f}'_1, \ldots, \boldsymbol{f}'_\ell \rangle$. Then $M' \subseteq M$ if and only if $\boldsymbol{f}'_i \in M$ for $i = 1, \ldots, \ell$. This can be verified algorithmically using the method described above.. Moreover, if $M' \subseteq M$, then $M' = M$ if and only if $M \subseteq M'$, and this, again, can be verified algorithmically using the above method. Alternatively, we can compute reduced

Gröbner bases for M and M' and use the fact that reduced Gröbner bases for submodules of A^m are unique (Theorem 3.5.22) to determine whether $M = M'$.

EXAMPLE 3.6.2. We go back to Example 3.6.1. Let M' be the submodule of A^3 generated by $\{\boldsymbol{f}_2, \boldsymbol{f}_3, \boldsymbol{f}_4, \boldsymbol{g}\}$. M' is a submodule of M, since we showed in Example 3.6.1 that $\boldsymbol{g} \in M$. However, using the same order we did in Example 3.6.1, the reduced Gröbner basis for M' is given by $\{\boldsymbol{g}_1', \boldsymbol{g}_2', \boldsymbol{g}_3', \boldsymbol{g}_4', \boldsymbol{g}_5', \boldsymbol{g}_6'\}$, where

$$\boldsymbol{g}_1' = (x, y + x^2 - x, 0), \ \boldsymbol{g}_2' = (y + x^2, 0, 0)$$

$$\boldsymbol{g}_3' = (x^6, x^7, -x^7), \ \boldsymbol{g}_4' = (x^7 + x^5, x^6 - x^5, -x^5),$$

$$\boldsymbol{g}_5' = (x^2, x, y), \ \boldsymbol{g}_6' = (x^6 - 2x^5, -x^6 + 2x^5, x^9 - x^8 - 3x^7 + x^6 + 2x^5).$$

Since the reduced Gröbner basis for M' is not equal to the one for M, the modules M and M' are not equal.

Now if we consider the submodule M'' of A^3 generated by $\{\boldsymbol{f}_1, \boldsymbol{f}_2, \boldsymbol{f}_3, \boldsymbol{g}\}$, then again M'' is a submodule of M. Moreover, we can compute that the reduced Gröbner basis for M'' is the same as the reduced Gröbner basis for M, and therefore $M'' = M$.

We next turn our attention to Task (iii), that is, we find coset representatives for the quotient module A^m/M. Let G be a Gröbner basis for M and let $\boldsymbol{f} \in A^m$. We know from Theorems 3.5.12 and 3.5.14 that there exists a unique vector $\boldsymbol{r} \in A^m$, reduced with respect to G, such that

$$\boldsymbol{f} \xrightarrow{G}_+ \boldsymbol{r}.$$

As in the ideal case, we call this vector $\boldsymbol{r}$ the *normal form* of $\boldsymbol{f}$ with respect to G, and we denote it by $N_G(\boldsymbol{f})$.

PROPOSITION 3.6.3. *Let $\boldsymbol{f}$ and $\boldsymbol{g}$ be vectors in A^m. Then*

$$\boldsymbol{f} + M = \boldsymbol{g} + M \text{ in } A^m/M \iff N_G(\boldsymbol{f}) = N_G(\boldsymbol{g}).$$

Therefore $\{N_G(\boldsymbol{f}) \mid \boldsymbol{f} \in A^m\}$ is a set of coset representatives for the quotient module A^m/M. Moreover the map

$$\begin{array}{rccc} N_G\colon & A^m & \longrightarrow & A^m \\ & \boldsymbol{f} & \longmapsto & N_G(\boldsymbol{f}) \end{array}$$

is k-linear.

The proof of this result is similar to the one for the ideal case (Proposition 2.1.4) and is left to the reader as an exercise (Exercise 3.6.4).

As in the ideal case (Proposition 2.1.6), we have the following Proposition whose proof we also leave to the exercises (Exercise 3.6.5). It solves Task (iv).

PROPOSITION 3.6.4. *A basis for the k-vector space A^m/M consists of all the cosets of monomials $\boldsymbol{X} \in A^m$ such that no $\operatorname{lm}(\boldsymbol{g}_i)$ divides $\boldsymbol{X}$.*

EXAMPLE 3.6.5. We again go back to Example 3.6.1. The leading terms of the reduced Gröbner basis are:

$$x^3\mathbf{e}_1, y\mathbf{e}_2, y\mathbf{e}_1, y\mathbf{e}_3, x^3\mathbf{e}_2, x^5\mathbf{e}_3.$$

Therefore a basis for the $\mathbb{Q}$-vector space A^3/M is

$$\{\mathbf{e}_1, x\mathbf{e}_1, x^2\mathbf{e}_1, \mathbf{e}_2, x\mathbf{e}_2, x^2\mathbf{e}_2, \mathbf{e}_3, x\mathbf{e}_3, x^2\mathbf{e}_3, x^3\mathbf{e}_3, x^4\mathbf{e}_3\}.$$

To conclude this section, we consider the theory of elimination presented in Section 2.3, but now in the module setting. Again the proofs are very similar to the ones for the ideal case and, except for the proof of Theorem 3.6.6, are left to the reader.

Let $y_1, \ldots, y_\ell$ be new variables, and consider a non-zero module

$$M \subseteq (A[y_1, \ldots, y_\ell])^m = (k[x_1, \ldots, x_n, y_1, \ldots, y_\ell])^m.$$

As in the ideal case (Section 2.3), we wish to "eliminate" some of the variables. For example, we wish to compute generators (and a Gröbner basis) for the module $M \cap A^m$, that is, we wish to eliminate the variables $y_1, \ldots, y_\ell$. First, we choose an elimination order on the power products of $A[y_1, \ldots, y_\ell]$ with the y variables larger than the x variables (see Section 2.3). The next result is the analog of Theorem 2.3.4 in the module context (in fact Theorem 2.3.4 is the special case $m = 1$ in Theorem 3.6.6).

THEOREM 3.6.6. *With the notation as above, let G be a Gröbner basis for M with respect to the TOP monomial ordering on $(A[y_1, \ldots, y_\ell])^m$. Then $G \cap A^m$ is a Gröbner basis for $M \cap A^m$.*

PROOF. Clearly $\langle G \cap A^m\rangle \subseteq M \cap A^m$. So let $\mathbf{0} \neq \boldsymbol{f} \in M \cap A^m$. Then there is a $\boldsymbol{g} \in G$ such that $\operatorname{lm}(\boldsymbol{g})$ divides $\operatorname{lm}(\boldsymbol{f})$. Since the coordinates of $\boldsymbol{f}$ involve only the x variables, we see that $\operatorname{lm}(\boldsymbol{g})$ can only involve the x variables as well. Then since we are using an elimination order with the y variables larger than the x variables we see that the polynomial in the coordinate of $\boldsymbol{g}$ giving rise to $\operatorname{lm}(\boldsymbol{g})$ can contain only x variables. Finally, since the order is TOP on A^m we see that the polynomials in all of the coordinates of $\boldsymbol{g}$ must contain only x variables. □

We will give an example in the exercises where the above result is false if the TOP ordering is replaced by the POT ordering on A^m (see Exercise 3.6.8).

EXAMPLE 3.6.7. In Example 3.6.1 we saw that the reduced Gröbner basis for M has three vectors in x alone, namely $\boldsymbol{g}_1, \boldsymbol{g}_5$, and $\boldsymbol{g}_6$. Therefore by Theorem 3.6.6, $M \cap (\mathbb{Q}[x])^3$ is generated by $\{\boldsymbol{g}_1, \boldsymbol{g}_5, \boldsymbol{g}_6\}$.

We can use this result to compute intersection of submodules of A^m and ideal quotients of two submodules of A^m. First, as in the ideal case (Proposition 2.3.5) we have

PROPOSITION 3.6.8. *Let $M = \langle \boldsymbol{f}_1, \dots, \boldsymbol{f}_s\rangle$ and $N = \langle \boldsymbol{g}_1, \dots, \boldsymbol{g}_t\rangle$ be submodules of A^m and let w be a new variable. Consider the module*

$$L = \langle w\boldsymbol{f}_1, \dots, w\boldsymbol{f}_s, (1-w)\boldsymbol{g}_1, \dots, (1-w)\boldsymbol{g}_t\rangle \subseteq (A[w])^m.$$

Then $M \cap N = L \cap A^m$.

We note that in Proposition 3.6.8, neither $\{\boldsymbol{f}_1, \dots, \boldsymbol{f}_s\}$ nor $\{\boldsymbol{g}_1, \dots, \boldsymbol{g}_t\}$ need be a Gröbner basis.

As a consequence of this result we obtain a method for computing generators for the module $M \cap N \subseteq A^m$: we first compute a Gröbner basis G for L with respect to the TOP ordering on monomials of $(A[w])^m$, using an elimination order on the power products in $A[w]$ with w larger than the x variables; a Gröbner basis for $M \cap N$ is then given by $G \cap A^m$.

EXAMPLE 3.6.9. Let M be the submodule of A^3 of Example 3.6.1, and let N be the submodule of A^3 generated by the vector $\boldsymbol{g}_1 = (y, x, xy)$. The reduced Gröbner basis for $\langle w\boldsymbol{f}_1, w\boldsymbol{f}_2, w\boldsymbol{f}_3, w\boldsymbol{f}_4, (1-w)\boldsymbol{g}_1\rangle \subseteq (A[w])^3$ with respect to the TOP term ordering with $\boldsymbol{e}_1 > \boldsymbol{e}_2 > \boldsymbol{e}_3$ using the lex ordering in $A[w]$ with $w > y > x$ has 8 vectors, two of which are in A^3:

$$\begin{aligned}
\boldsymbol{h}_1 &= (9y^2 - 7yx^6 + 2yx^5 + 25yx^4 + 7yx^3 - 9yx^2 - 9yx,\\
&\quad\; 9yx - 7x^7 + 2x^6 + 25x^5 + 7x^4 - 9x^3 - 9x^2,\\
&\quad\; 9y^2x - 7yx^7 + 2yx^6 + 25yx^5 + 7yx^4 - 9yx^3 - 9yx^2)\\
\boldsymbol{h}_2 &= (yx^7 - yx^6 - 3yx^5 + yx^4 + 2yx^3, x^8 - x^7 - 3x^6 + x^5 + 2x^4,\\
&\quad\; yx^8 - yx^7 - 3yx^6 + yx^5 + 2yx^4).
\end{aligned}$$

Therefore $M \cap N = \langle \boldsymbol{h}_1, \boldsymbol{h}_2\rangle \subseteq A^3$.

DEFINITION 3.6.10. *Let M and N be two submodules of A^m. The* ideal quotient *$N\colon M$ is defined to be*

$$N\colon M = \{f \in A \mid fM \subseteq N\} \subseteq A.$$

Note that $N\colon M$ is an ideal in A.

As in the ideal case (Lemmas 2.3.10 and 2.3.11) we have

LEMMA 3.6.11. *Let $M = \langle \boldsymbol{f}_1, \dots, \boldsymbol{f}_s\rangle \subseteq A^m$ and let N be any other submodule of A^m. Then*

$$N\colon M = \bigcap_{i=1}^{s} N\colon \langle \boldsymbol{f}_i\rangle.$$

Since we have a method for computing intersection of ideals (Proposition 2.3.5 or equivalently Proposition 3.6.8 with $m = 1$), we only need to show how to compute $N\colon \langle \boldsymbol{f}\rangle$ for a single vector $\boldsymbol{f} \in A^m$.

LEMMA 3.6.12. *Let N be a submodule of A^m and let $\boldsymbol{f}$ be a vector in A^m. Then*

$$N : \langle \boldsymbol{f} \rangle = \{a \in A \mid \boldsymbol{g} = a\boldsymbol{f} \in N \cap \langle \boldsymbol{f} \rangle\}.$$

Thus we may compute $N : \langle \boldsymbol{f} \rangle$ by first computing a set of generators for $N \cap \langle \boldsymbol{f} \rangle$ and dividing these generators by $\boldsymbol{f}$ using the Division Algorithm. The quotients obtained are then a set of generators for $N : \langle \boldsymbol{f} \rangle$.

EXAMPLE 3.6.13. In Example 3.6.9 we saw that

$$M \cap \langle \boldsymbol{g}_1 \rangle = \langle \boldsymbol{h}_1, \boldsymbol{h}_2 \rangle.$$

Note that

$$\begin{aligned} \boldsymbol{h}_1 &= (9y - 7x^6 + 2x^5 + 25x^4 + 7x^3 - 9x^2 - 9x)\boldsymbol{g}_1 \\ \boldsymbol{h}_2 &= (x^7 - x^6 - 3x^5 + x^4 + 2x^3)\boldsymbol{g}_1. \end{aligned}$$

Therfore, by Lemma 3.6.12,

$$M : \langle \boldsymbol{g}_1 \rangle = \langle 9y - 7x^6 + 2x^5 + 25x^4 + 7x^3 - 9x^2 - 9x, x^7 - x^6 - 3x^5 + x^4 + 2x^3 \rangle \subseteq A.$$

Now let $\boldsymbol{g}_2 = (y + x^2, y, x^2) \in A^3$. We wish to compute $M : \langle \boldsymbol{g}_1, \boldsymbol{g}_2 \rangle$. By Lemma 3.6.11, we first need to compute $M : \langle \boldsymbol{g}_2 \rangle$. We proceed as in Example 3.6.9 using the same term order, and we find

$$M \cap \langle \boldsymbol{g}_2 \rangle = \langle \boldsymbol{h}_3, \boldsymbol{h}_4 \rangle,$$

where

$$\begin{aligned} \boldsymbol{h}_3 &= (x^7 - x^6 - 3x^5 + x^4 + 2x^3)\boldsymbol{g}_2 \\ \boldsymbol{h}_4 &= (9y + 2x^6 - 7x^5 - 2x^4 + 16x^3 + 9x^2 - 9x)\boldsymbol{g}_2. \end{aligned}$$

Therefore, by Lemma 3.6.12,

$$M : \langle \boldsymbol{g}_2 \rangle = \langle x^7 - x^6 - 3x^5 + x^4 + 2x^3, 9y + 2x^6 - 7x^5 - 2x^4 + 16x^3 + 9x^2 - 9x \rangle \subseteq A.$$

Now, by Lemma 3.6.11,

$$M : \langle \boldsymbol{g}_1, \boldsymbol{g}_2 \rangle = (M : \langle \boldsymbol{g}_1 \rangle) \cap (M : \langle \boldsymbol{g}_2 \rangle).$$

To compute this intersection, we find the Gröbner basis for the ideal

$$\langle w(9y - 7x^6 + 2x^5 + 25x^4 + 7x^3 - 9x^2 - 9x), w(x^7 - x^6 - 3x^5 + x^4 + 2x^3),$$
$$(1-w)(x^7 - x^6 - 3x^5 + x^4 + 2x^3), (1-w)(9y + 2x^6 - 7x^5 - 2x^4 + 16x^3 + 9x^2 - 9x) \rangle$$

of $A[w]$ with respect to the lex term ordering with $w > y > x$. The Gröbner basis has five polynomials, three of which do not contain the variable w,

$$\begin{aligned} u_1 &= x^7 - x^6 - 3x^5 + x^4 + 2x^3 \\ u_2 &= 9yx - 5x^6 + 4x^5 + 14x^4 - 5x^3 - 9x^2 \\ u_3 &= 3y^2 - x^6 + 2x^5 + x^4 - 2x^3 - 3x^2. \end{aligned}$$

Therefore $M : \langle \boldsymbol{g}_1, \boldsymbol{g}_2 \rangle = \langle u_1, u_2, u_3 \rangle \subseteq A$.

Some of the computations we have performed in this section can also be done more efficiently using the syzygy module of the matrix $\left[\begin{array}{ccc} \boldsymbol{f}_1 & \cdots & \boldsymbol{f}_s \end{array} \right]$, where $\boldsymbol{f}_i \in A^m$. In the next section, we introduce these syzygies, and then, in Section 3.8, we give these applications.

Exercises

3.6.1. Consider the following 3 vectors in $(\mathbb{Q}[x,y])^2$:

$$\boldsymbol{f}_1 = (xy - x, y), \ \boldsymbol{f}_2 = (x, y), \ \boldsymbol{f}_3 = (y, xy^2).$$

You should do the following without the use of a Computer Algebra System.

a. Let $\boldsymbol{f} = (-x, -x^3y^2 + 2xy - y^2 + y)$. Show that $\boldsymbol{f} \in \langle \boldsymbol{f}_1, \boldsymbol{f}_2, \boldsymbol{f}_3 \rangle$. Moreover, express $\boldsymbol{f}$ as a linear combination of $\boldsymbol{f}_1, \boldsymbol{f}_2$, and $\boldsymbol{f}_3$.

b. Let $\boldsymbol{g} = (-x, x^3y^2 + 2xy - y^2 + y)$. Show that $\boldsymbol{g} \notin \langle \boldsymbol{f}_1, \boldsymbol{f}_2, \boldsymbol{f}_3 \rangle$.

c. Clearly we have $\langle \boldsymbol{f}_2, \boldsymbol{g} \rangle \subseteq \langle \boldsymbol{f}_1, \boldsymbol{f}_2, \boldsymbol{g} \rangle$. Prove that this inclusion is strict.

3.6.2. Consider the following vectors in $(\mathbb{Q}[x,y])^3$:

$$\boldsymbol{f}_1 = (0, y, x), \ \boldsymbol{f}_2 = (0, x, xy - x), \ \boldsymbol{f}_3 = (y, x, 0), \ \boldsymbol{f}_4 = (y^2, y, 0).$$

You should do the following without the use of a Computer Algebra System.

a. Compute a Gröbner basis for $M = \langle \boldsymbol{f}_1, \boldsymbol{f}_2, \boldsymbol{f}_3, \boldsymbol{f}_4 \rangle$ with respect to the TOP ordering on $(\mathbb{Q}[x,y])^3$ with $\boldsymbol{e}_1 > \boldsymbol{e}_2 > \boldsymbol{e}_3$, using deglex on $\mathbb{Q}[x,y]$ with $x > y$. Use this to compute the matrix T which gives the Gröbner basis vectors in terms of the original vectors.

b. Use **a** to show that the vector $(x^2y - y^2 + xy^2, xy^2 - y^2 + x^2 + 2xy - x - y, x^2y + xy^2 - 3xy + x)$ is in M and express it as a linear combination of $\boldsymbol{f}_1, \boldsymbol{f}_2, \boldsymbol{f}_3, \boldsymbol{f}_4$.

3.6.3. Consider the following two submodules of $(\mathbb{Q}[x,y,z])^3$:

$$M = \langle (x^2 - y, y, xz - y), (xz + x, yx + y, yz + z), (x, 0, x) \rangle \text{ and}$$

$$M' = \langle (-y, y, zx - x^2 - y), (y^2 + y, yx^2 - y^2 + 2xy - y, yx^2 - x^3 + y^2 + y), (x, 0, x), (0, xy + y, -xz - x + yz + z) \rangle.$$

Determine whether any of the following holds:

$$M \subseteq M', \ M' \subseteq M, \ M = M'.$$

3.6.4. Prove Proposition 3.6.3.

3.6.5. Prove Proposition 3.6.4.

3.6.6. Consider the module M of Example 3.6.1. Determine whether $\boldsymbol{f} + M = \boldsymbol{g} + M$ for the following examples:

a. $\boldsymbol{f} = (2x, y^2x + y^2 + yx + 2y - 3x, -y + x)$, $\boldsymbol{g} = (-y^2 + x - y, y^3 + 2y^2, y^2 - x)$.

b. $\boldsymbol{f} = (y^3 + y^2 + x - y, y^2 + y, y + x)$, $\boldsymbol{g} = (x, y^3 + 2y^2 - yx - x, 0)$.

3.6.7. Find a $\mathbb{Q}$-basis for the vector space A^3/M for each module M in Exercise 3.5.15.

3.6.8. In Theorem 3.6.6 we required that the order be TOP. In this exercise we show that this is necessary. Consider the vectors $\boldsymbol{f}_1 = (y, xy)$, $\boldsymbol{f}_2 = (0, x+1) \in (\mathbb{Q}[x,y])^2$. We use the POT ordering on $(\mathbb{Q}[x,y])^2$ with $\boldsymbol{e}_1 > \boldsymbol{e}_2$ and lex on $\mathbb{Q}[x,y]$ with $x > y$.

a. Prove that $G = \{\boldsymbol{f}_1, \boldsymbol{f}_2\}$ is a Gröbner basis and note $G \cap (\mathbb{Q}[y])^2 = \emptyset$.

b. Prove that $M \cap (\mathbb{Q}[y])^2$ contains a non-zero vector.

3.6.9. Consider the submodule M of $(\mathbb{Q}[x,y,z])^3$:

$$M = \langle (x, xz, z^2), (x, x^2, x+y), (y, 0, x), (x, 0, z)\rangle.$$

Compute generators for the following modules:

a. $M \cap (\mathbb{Q}[x,y])^3$.

b. $M \cap (\mathbb{Q}[y,z])^3$.

c. $M \cap (\mathbb{Q}[x,z])^3$.

3.6.10. Prove Proposition 3.6.8.

3.6.11. Compute generators for the intersection of the following two submodules of $(\mathbb{Q}[x,y])^3$:

$$M = \langle (x, x^2, x+y), (y, 0, x)\rangle \text{ and } N = \langle (x^2, xy, y^2), (x-y, x, x)\rangle.$$

3.6.12. State and prove the analog of Exercise 2.3.8 for the computation of the intersection of more than two submodules of A^m. Use this to compute generators for the intersection of the following three submodules of $(\mathbb{Q}[x,y])^3$: $M_1 = \langle (x, x, -y),\ (x, y, -x)\rangle$, $M_2 = \langle (x, y, y),\ (x, x, x)\rangle$, $M_3 = \langle (y, y, y), (y, x, x)\rangle$.

3.6.13. Prove Lemma 3.6.11.

3.6.14. Prove Lemma 3.6.12.

3.6.15. Compute generators for $M : N$, where $M = \langle (0, y, x), (0, x, xy-x), (y, x, 0), (y^2, y, 0)\rangle$ and $N = \langle (y, x, x),\ (x, y, y)\rangle \subseteq (\mathbb{Q}[x,y])^3$.

3.6.16. Consider module homomorphisms $\phi\colon A^s \longrightarrow A^\ell$ and $\gamma\colon A^m \longrightarrow A^\ell$, such that $\operatorname{im}(\gamma) \subseteq \operatorname{im}(\phi)$. A theorem of Module Theory states that there exists a "lifting", that is, a homomorphism $\psi\colon A^m \longrightarrow A^s$ such that $\phi \circ \psi = \gamma$; i.e. the following diagram commutes:

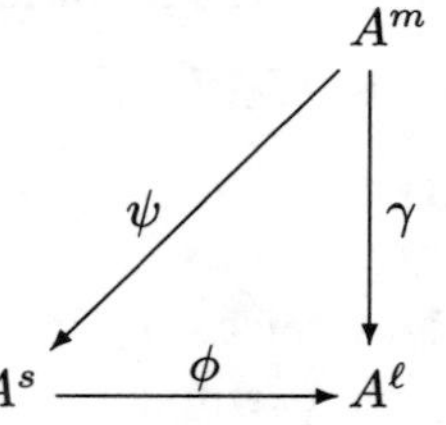

a. Show how to compute ψ. That is, show how to compute $\psi(\boldsymbol{e}_i)$, $i = 1, \ldots, m$. [Hint: Let $\{\boldsymbol{g}_1, \ldots, \boldsymbol{g}_t\}$ be a Gröbner basis for $\operatorname{im}(\phi)$ and apply the method used to solve Task (i) at the beginning of the section.]

b. Compute ψ in the following example. Let $A = \mathbb{Q}[x,y]$. We define $\gamma\colon A \longrightarrow A^3$ by $\gamma(1) = (x^2y^2+x^2, y^3+xy+y^2, xy^2-1)$, and we define $\phi\colon A^3 \longrightarrow A^3$ by $\phi(\mathbf{e}_1) = (x^2+xy, y^2, xy-1)$, $\phi(\mathbf{e}_2) = (x^2y-x, y^2+x, xy-x)$, and $\phi(\mathbf{e}_3) = (y^2+x+y, x^2, x-y)$. You should first verify that $\mathrm{im}(\gamma) \subseteq \mathrm{im}(\phi)$.

3.7. Syzygies for Modules. We now turn our attention to computing the syzygy module of a matrix $[\ \boldsymbol{f}_1 \ \cdots \ \boldsymbol{f}_s\ ]$ of column vectors in A^m. This computation is very similar to the computation of the syzygy module of the $1 \times s$ matrix $[\ f_1 \ \cdots \ f_s\ ]$ of polynomials in A (see Section 3.4).

Recall that in Section 3.1 we defined the map $\phi\colon A^s \longrightarrow A^m$ by $\phi(h_1, \ldots, h_s) = \sum_{i=1}^{s} h_i \boldsymbol{f}_i$. As in the case where $m = 1$ (see Section 3.2), the kernel of this map is called the syzygy module of $[\ \boldsymbol{f}_1 \ \cdots \ \boldsymbol{f}_s\ ]$ and is a submodule of A^s. More formally we have

DEFINITION 3.7.1. *Let $\boldsymbol{f}_1, \ldots, \boldsymbol{f}_s \in A^m$. A* syzygy *of the $m \times s$ matrix $F = [\ \boldsymbol{f}_1 \ \cdots \ \boldsymbol{f}_s\ ]$ is a vector $(h_1, \ldots, h_s) \in A^s$ such that*

$$\sum_{i=1}^{s} h_i \boldsymbol{f}_i = \mathbf{0}.$$

The set of all such syzygies is called the syzygy module *of F and is denoted by* $\mathrm{Syz}(\boldsymbol{f}_1, \ldots, \boldsymbol{f}_s)$ *or by* $\mathrm{Syz}(F)$.

In other words $\mathrm{Syz}(F) = \mathrm{Syz}(\boldsymbol{f}_1, \ldots, \boldsymbol{f}_s)$ can be viewed as the set of all polynomial solutions $\boldsymbol{h} \in A^s$ of the system of homogeneous linear equations $F\boldsymbol{h} = \mathbf{0}$ with polynomial coefficients. That is, if $\boldsymbol{f}_1 = (f_{11}, \ldots, f_{m1}), \ldots, \boldsymbol{f}_s = (f_{1s}, \ldots, f_{ms})$, then $\mathrm{Syz}(\boldsymbol{f}_1, \ldots, \boldsymbol{f}_s)$ is the set of all simultaneous polynomial solutions $\chi_1, \ldots, \chi_s$ of the system

$$\left\{\begin{array}{lcl} f_{11}\chi_1 + \cdots + f_{1s}\chi_s & = & 0 \\ f_{21}\chi_1 + \cdots + f_{2s}\chi_s & = & 0 \\ & \vdots & \\ f_{m1}\chi_1 + \cdots + f_{ms}\chi_s & = & 0. \end{array}\right.$$

As in the case of the syzygy module of a $1 \times s$ matrix $[\ f_1 \ \cdots \ f_s\ ]$ of polynomials in A, the computation of $\mathrm{Syz}(\boldsymbol{f}_1, \ldots, \boldsymbol{f}_s)$ is done in two steps. We first compute a Gröbner basis $\{\boldsymbol{g}_1, \ldots, \boldsymbol{g}_t\}$ for $\langle \boldsymbol{f}_1, \ldots, \boldsymbol{f}_s \rangle \subseteq A^m$ and compute $\mathrm{Syz}(\boldsymbol{g}_1, \ldots, \boldsymbol{g}_t) \subseteq A^t$. We then obtain $\mathrm{Syz}(\boldsymbol{f}_1, \ldots, \boldsymbol{f}_s) \subseteq A^s$ from $\mathrm{Syz}(\boldsymbol{g}_1, \ldots, \boldsymbol{g}_t)$.

So let us first start with $G = [\ \boldsymbol{g}_1 \ \cdots \ \boldsymbol{g}_t\ ]$. As we did in the polynomial case, we assume that $\mathrm{lc}(\boldsymbol{g}_i) = 1$. We follow closely the construction we used for the ideal case (see Section 3.4). Let $i \neq j \in \{1, \ldots, t\}$. Let $\mathrm{lm}(\boldsymbol{g}_i) = \boldsymbol{X}_i$ and

$\boldsymbol{X}_{ij} = \operatorname{lcm}(\boldsymbol{X}_i, \boldsymbol{X}_j)$. Then the S-polynomial of $\boldsymbol{g}_i$ and $\boldsymbol{g}_j$ is given by

$$S(\boldsymbol{g}_i, \boldsymbol{g}_j) = \frac{\boldsymbol{X}_{ij}}{\boldsymbol{X}_i}\boldsymbol{g}_i - \frac{\boldsymbol{X}_{ij}}{\boldsymbol{X}_j}\boldsymbol{g}_j.$$

By Theorems 3.5.12 and 3.5.19, we have

$$S(\boldsymbol{g}_i, \boldsymbol{g}_j) = \sum_{\nu=1}^{t} h_{ij\nu}\boldsymbol{g}_\nu,$$

for some $h_{ij\nu} \in A$, such that

$$\max_{1\leq\nu\leq t}(\operatorname{lp}(h_{ij\nu})\operatorname{lm}(\boldsymbol{g}_\nu)) = \operatorname{lm}(S(\boldsymbol{g}_i, \boldsymbol{g}_j)). \tag{3.7.1}$$

For $i, j \in \{1, \dots, t\}$, we define

$$\boldsymbol{s}_{ij} = \frac{\boldsymbol{X}_{ij}}{\boldsymbol{X}_i}\boldsymbol{e}_i - \frac{\boldsymbol{X}_{ij}}{\boldsymbol{X}_j}\boldsymbol{e}_j - (h_{ij1}, \dots, h_{ijt}) \in A^t.$$

We easily see that $\boldsymbol{s}_{ij} \in \operatorname{Syz}(\boldsymbol{g}_1, \dots, \boldsymbol{g}_t)$.

We first state the analog of Proposition 3.2.3.

PROPOSITION 3.7.2. $\operatorname{Syz}(\boldsymbol{X}_1, \dots, \boldsymbol{X}_s)$ *is generated by*

$$\left\{\frac{\boldsymbol{X}_{ij}}{\boldsymbol{X}_i}\boldsymbol{e}_i - \frac{\boldsymbol{X}_{ij}}{\boldsymbol{X}_j}\boldsymbol{e}_j \in A^s \mid i, j \in \{1, \dots, s\}\right\}.$$

We now give the analog of Theorem 3.4.1. The proof is identical except that instead of Proposition 3.2.3 we use Proposition 3.7.2. We leave the proofs of both the Proposition and the Theorem as exercises (Exercises 3.7.4 and 3.7.5).

THEOREM 3.7.3. *With the notation above, the collection* $\{\boldsymbol{s}_{ij} \mid 1 \leq i < j \leq t\}$ *is a generating set for* $\operatorname{Syz}(G) = \operatorname{Syz}(\boldsymbol{g}_1, \dots, \boldsymbol{g}_t)$.

EXAMPLE 3.7.4. We go back to Example 3.5.23. Recall that the set $\{\boldsymbol{g}_1, \boldsymbol{g}_2, \boldsymbol{g}_3, \boldsymbol{g}_4, \boldsymbol{g}_5, \boldsymbol{g}_6\}$ is a Gröbner basis with respect to the deglex term order on $\mathbb{Q}[x, y]$ with $x > y$ and the TOP order on $(\mathbb{Q}[x, y])^3$ with $\boldsymbol{e}_1 > \boldsymbol{e}_2 > \boldsymbol{e}_3$, where

$$\boldsymbol{g}_1 = (0, y, x),\ \boldsymbol{g}_2 = (0, y^2 - x - y, 0),$$

$$\boldsymbol{g}_3 = (x, x + y, 0),\ \boldsymbol{g}_4 = (y, -y, 0),$$

$$\boldsymbol{g}_5 = (0, xy + \frac{1}{2}x + \frac{1}{2}y, 0),\ \boldsymbol{g}_6 = (0, x^2 - \frac{1}{4}x - \frac{1}{4}y, 0).$$

To obtain the generators for $\operatorname{Syz}(\boldsymbol{g}_1, \boldsymbol{g}_2, \boldsymbol{g}_3, \boldsymbol{g}_4, \boldsymbol{g}_5, \boldsymbol{g}_6)$, we follow Theorem 3.7.3 and so we compute and reduce all S-polynomials. The S-polynomials we need to compute are $S(\boldsymbol{g}_2, \boldsymbol{g}_5), S(\boldsymbol{g}_2, \boldsymbol{g}_6), S(\boldsymbol{g}_5, \boldsymbol{g}_6)$ and $S(\boldsymbol{g}_3, \boldsymbol{g}_4)$. For example, since

$S(\boldsymbol{g}_3, \boldsymbol{g}_4) = y\boldsymbol{g}_3 - x\boldsymbol{g}_4 = \boldsymbol{g}_2 + 2\boldsymbol{g}_5$, we have $\boldsymbol{s}_{34} = (0, -1, y, -x, -2, 0)$. The other syzygies are computed in a similar way (Exercise 3.7.2) to obtain

$$\begin{aligned} \boldsymbol{s}_{25} &= (0, x + \frac{1}{2}, 0, 0, -y + \frac{3}{2}, 1) \\ \boldsymbol{s}_{56} &= (0, -\frac{1}{4}, 0, 0, x - \frac{3}{4}, -y - \frac{1}{2}). \end{aligned}$$

Note that we have not included $\boldsymbol{s}_{26}$ because it is in $\langle \boldsymbol{s}_{34}, \boldsymbol{s}_{25}, \boldsymbol{s}_{56} \rangle$ (See Exercise 3.7.7).

EXAMPLE 3.7.5. We consider the submodule M of $A^3 = (\mathbb{Q}[x, y])^3$ of Example 3.6.1. We saw that $\{\boldsymbol{g}_1, \boldsymbol{g}_2, \boldsymbol{g}_3, \boldsymbol{g}_4, \boldsymbol{g}_5, \boldsymbol{g}_6\}$ forms a Gröbner basis with respect to the TOP term ordering in A^3 with $\boldsymbol{e}_1 > \boldsymbol{e}_2 > \boldsymbol{e}_3$ and with lex in A with $y > x$, where

$$\boldsymbol{g}_1 = (\underline{x^3} + x, x^2 - x, -x),\ \boldsymbol{g}_2 = (x, \underline{y} + x^2 - x, 0),$$
$$\boldsymbol{g}_3 = (\underline{y} + x^2, 0, 0),\ \boldsymbol{g}_4 = (x^2, x, \underline{y}),$$
$$\boldsymbol{g}_5 = (x^2, \underline{x^3}, -x^3),\ \boldsymbol{g}_6 = (x^2 - 2x, -x^2 + 2x, \underline{x^5} - x^4 - 3x^3 + x^2 + 2x).$$

The syzygy module $\mathrm{Syz}(\boldsymbol{g}_1, \boldsymbol{g}_2, \boldsymbol{g}_3, \boldsymbol{g}_4, \boldsymbol{g}_5, \boldsymbol{g}_6)$ can be computed using Theorem 3.7.3. For example

$$\begin{aligned} S(\boldsymbol{g}_1, \boldsymbol{g}_3) &= y\boldsymbol{g}_1 - x^3\boldsymbol{g}_3 \\ &= (yx - x^5, \underline{yx^2} - yx, -yx) \\ &\xrightarrow{x^2, \boldsymbol{g}_2} (\underline{yx} - x^5 - x^3, -yx - x^4 + x^3, -yx) \\ &\xrightarrow{x, \boldsymbol{g}_3} (-x^5 - 2x^3, \underline{-yx} - x^4 + x^3, -yx) \\ &\xrightarrow{-x, \boldsymbol{g}_2} (-x^5 - 2x^3 + x^2, -x^4 + 2x^3 - x^2, \underline{-yx}) \\ &\xrightarrow{-x, \boldsymbol{g}_4} (\underline{-x^5} - x^3 + x^2, -x^4 + 2x^3, 0) \\ &\xrightarrow{-x^2, \boldsymbol{g}_1} (x^2, \underline{x^3}, -x^3) \\ &\xrightarrow{1, \boldsymbol{g}_5} (0, 0, 0). \end{aligned}$$

So $S(\boldsymbol{g}_1, \boldsymbol{g}_3) = y\boldsymbol{g}_1 - x^3\boldsymbol{g}_3 = x^2\boldsymbol{g}_2 + x\boldsymbol{g}_3 - x\boldsymbol{g}_2 - x\boldsymbol{g}_4 - x^2\boldsymbol{g}_1 + \boldsymbol{g}_5$, and therefore

$$\boldsymbol{s}_{13} = (y + x^2, -x^2 + x, -x^3 - x, x, -1, 0).$$

The other generators of the syzygy module are obtained similarly (Exercise 3.7.3):

$$\begin{aligned} \boldsymbol{s}_{25} &= (-x^2 + x + 2, -x^3, -x^2, x^3, y + x^2 - x - 2, 1) \\ \boldsymbol{s}_{46} &= (x^4 - x^3 - 4x^2 + 2x + 4, x^2 - 2x, -x^2 + 2x, \\ &\quad -x^5 + x^4 + 3x^3 - x^2 - 2x, x^2 - x - 2, y + 2). \end{aligned}$$

These are the only generators needed, since the remaining S-polynomials are all zero.

We now consider the computation of $\mathrm{Syz}(\boldsymbol{f}_1, \ldots, \boldsymbol{f}_s)$, for a non-zero matrix $F = \left[\begin{array}{ccc} \boldsymbol{f}_1 & \cdots & \boldsymbol{f}_s \end{array}\right]$ of column vectors in A^m. We first compute a Gröbner basis $\{\boldsymbol{g}_1, \ldots, \boldsymbol{g}_t\}$ for $\langle F \rangle$ and set $G = \left[\begin{array}{ccc} \boldsymbol{g}_1 & \cdots & \boldsymbol{g}_t \end{array}\right]$. As in the ideal case, there is a $t \times s$ matrix S and an $s \times t$ matrix T with entries in A such that $F = GS$ and $G = FT$ (S is obtained using the Division Algorithm and T is obtained by keeping track of the reductions during Buchberger's Algorithm for modules.) As in the ideal case we compute generators of $\mathrm{Syz}(\boldsymbol{g}_1, \ldots, \boldsymbol{g}_t)$, say $\boldsymbol{s}_1, \ldots, \boldsymbol{s}_r$, and we let $\boldsymbol{r}_1, \ldots, \boldsymbol{r}_s$ be the columns of the matrix $I_s - TS$, where I_s is the $s \times s$ identity matrix. The proof of the following result is similar to the one for the ideal case (Theorem 3.4.3) and we leave it to the reader (Exercise 3.7.9).

THEOREM 3.7.6. *With the notation above we have*

$$\mathrm{Syz}(\boldsymbol{f}_1, \ldots, \boldsymbol{f}_s) = \langle T\boldsymbol{s}_1, \ldots, T\boldsymbol{s}_r, \boldsymbol{r}_1, \ldots, \boldsymbol{r}_s \rangle \subseteq A^s.$$

EXAMPLE 3.7.7. We go back to Example 3.7.5. Recall from Example 3.6.1 that the original vectors are

$$\boldsymbol{f}_1 = (xy, y, x), \ \boldsymbol{f}_2 = (x^2 + x, y + x^2, y),$$

$$\boldsymbol{f}_3 = (-y, x, y), \ \boldsymbol{f}_4 = (x^2, x, y).$$

The Gröbner basis G for $\langle \boldsymbol{f}_1, \boldsymbol{f}_2, \boldsymbol{f}_3, \boldsymbol{f}_4 \rangle$ given in Example 3.7.5 consists of the six vectors $\boldsymbol{g}_i$, $i = 1, \ldots, 6$. Also, in that example we saw that a basis for $\mathrm{Syz}(\boldsymbol{g}_1, \boldsymbol{g}_2, \boldsymbol{g}_3, \boldsymbol{g}_4, \boldsymbol{g}_5, \boldsymbol{g}_6)$ is $\boldsymbol{s}_{13}$, $\boldsymbol{s}_{25}$, and $\boldsymbol{s}_{46}$. Recall also that the matrix T which gives G in terms of F is given in Example 3.6.1 in Equations (3.6.2) and (3.6.3). To compute $\mathrm{Syz}(\boldsymbol{f}_1, \boldsymbol{f}_2, \boldsymbol{f}_3, \boldsymbol{f}_4)$ we use Theorem 3.7.6 and compute

$$\begin{aligned}
T\boldsymbol{s}_{13} &= (0,0,0,0) \\
T\boldsymbol{s}_{25} &= (0,0,0,0) \\
T\boldsymbol{s}_{46} &= (y^3 + 2y^2x^2 - y^2x + yx^4 - yx^3, -y^3 - y^2x^2 + yx^2 + x^4, \\
& \quad\ y^3x + y^2x^3 - y^2x^2 - y^2x - yx^3 - x^5 + x^4 + x^3, \\
& \quad\ -y^3x + y^3 - y^2x^3 + 2y^2x^2 - yx^3 - x^4 - x^3).
\end{aligned}$$

Now we need the matrix S that expresses F in terms of G. We have

$$\left[\begin{array}{cccc} \boldsymbol{f}_1 & \boldsymbol{f}_2 & \boldsymbol{f}_3 & \boldsymbol{f}_4 \end{array}\right] = \left[\begin{array}{cccccc} \boldsymbol{g}_1 & \boldsymbol{g}_2 & \boldsymbol{g}_3 & \boldsymbol{g}_4 & \boldsymbol{g}_5 & \boldsymbol{g}_6 \end{array}\right] \underbrace{\left[\begin{array}{cccc} -1 & 0 & 0 & 0 \\ 1 & 1 & 0 & 0 \\ x & 0 & -1 & 0 \\ 0 & 1 & 1 & 1 \\ 0 & 0 & 0 & 0 \\ 0 & 0 & 0 & 0 \end{array}\right]}_{S}.$$

Then

$$I_4 - TS = \begin{bmatrix} 0 & 0 & 0 & 0 \\ 0 & 0 & 0 & 0 \\ 0 & 0 & 0 & 0 \\ 0 & 0 & 0 & 0 \end{bmatrix}.$$

Therefore

$$\mathrm{Syz}(\boldsymbol{f}_1, \boldsymbol{f}_2, \boldsymbol{f}_3, \boldsymbol{f}_4) = \langle T\boldsymbol{s}_{46} \rangle \subseteq A^4.$$

In the last example, the rows of the matrix $I_s - TS$ did not contribute to the syzygy module $\mathrm{Syz}(\boldsymbol{f}_1, \boldsymbol{f}_2, \boldsymbol{f}_3, \boldsymbol{f}_4)$. It is not true in general that the $T\boldsymbol{s}_i$ in Theorem 3.7.6 give a complete set of generators for $\mathrm{Syz}(F)$ as the next example shows.

EXAMPLE 3.7.8. Let $A = \mathbb{Q}[x, y]$ and let $\boldsymbol{f}_1 = (y + 2x^2 + x, y)$, $\boldsymbol{f}_2 = (-y + x, y)$, and $\boldsymbol{f}_3 = (x^2 + x, y)$ be vectors in A^2. Then the reduced Gröbner basis G for $\langle \boldsymbol{f}_1, \boldsymbol{f}_2, \boldsymbol{f}_3 \rangle$ with respect to the TOP term ordering on A^2 with $\boldsymbol{e}_1 > \boldsymbol{e}_2$ and the lex ordering on A with $y > x$ is $\{\boldsymbol{g}_1, \boldsymbol{g}_2\}$, where

$$\begin{aligned} \boldsymbol{g}_1 &= (y + x^2, 0) \\ \boldsymbol{g}_2 &= (x^2 + x, y). \end{aligned}$$

Since $\mathrm{lm}(\boldsymbol{g}_1) = y\boldsymbol{e}_1$ and $\mathrm{lm}(\boldsymbol{g}_2) = y\boldsymbol{e}_2$, we see that $\mathrm{Syz}(\boldsymbol{g}_1, \boldsymbol{g}_2) = \langle (0, 0) \rangle$. Also, we have

$$\begin{bmatrix} \boldsymbol{g}_1 & \boldsymbol{g}_2 \end{bmatrix} = \begin{bmatrix} \boldsymbol{f}_1 & \boldsymbol{f}_2 & \boldsymbol{f}_3 \end{bmatrix} \underbrace{\begin{bmatrix} 0 & 0 \\ -1 & 0 \\ 1 & 1 \end{bmatrix}}_{T},$$

and

$$\begin{bmatrix} \boldsymbol{f}_1 & \boldsymbol{f}_2 & \boldsymbol{f}_3 \end{bmatrix} = \begin{bmatrix} \boldsymbol{g}_1 & \boldsymbol{g}_2 \end{bmatrix} \underbrace{\begin{bmatrix} 1 & -1 & 0 \\ 1 & 1 & 1 \end{bmatrix}}_{S}.$$

We have

$$I_3 - TS = \begin{bmatrix} 1 & 0 & 0 \\ 1 & 0 & 0 \\ 2 & 0 & 0 \end{bmatrix}.$$

Therefore

$$\mathrm{Syz}(\boldsymbol{f}_1, \boldsymbol{f}_2, \boldsymbol{f}_3) = \langle (1, 1, -2) \rangle \subseteq A^3.$$

We conclude this section by showing that the generators for $\mathrm{Syz}(\boldsymbol{g}_1, \ldots, \boldsymbol{g}_t)$ computed in Theorem 3.7.3 (or in the polynomial case Theorem 3.4.1) form a Gröbner basis for $\mathrm{Syz}(\boldsymbol{g}_1, \ldots, \boldsymbol{g}_t)$ with respect to a certain order which we define next (see Schreyer [**Schre**]). This result is technical and will only be used in Section 3.10.

LEMMA 3.7.9. *Let $\boldsymbol{g}_1, \ldots, \boldsymbol{g}_t$ be non-zero vectors in A^m and let $<$ be a term order in A^m. We define an order $<$ on the monomials of A^t as follows:*

$$X\boldsymbol{e}_i < Y\boldsymbol{e}_j \iff \begin{cases} \operatorname{lm}(X\boldsymbol{g}_i) < \operatorname{lm}(Y\boldsymbol{g}_j) \text{ or} \\ \operatorname{lm}(X\boldsymbol{g}_i) = \operatorname{lm}(Y\boldsymbol{g}_j) \text{ and } j < i. \end{cases}$$

Then $<$ is a term ordering on A^t.

The reader should note that when $\operatorname{lm}(X\boldsymbol{g}_i) = \operatorname{lm}(Y\boldsymbol{g}_j)$ in the lemma, we have $X\boldsymbol{e}_i < Y\boldsymbol{e}_j$ when $j < i$, that is, the j and i are reversed. Note that the hypotheses of the lemma do not require that $\{\boldsymbol{g}_1, \ldots, \boldsymbol{g}_t\}$ be a Gröbner basis.

DEFINITION 3.7.10. *The term order defined in Lemma* 3.7.9 *is called the* order on A^t induced by $[\ \boldsymbol{g}_1 \ \cdots \ \boldsymbol{g}_t\ ]$ *(and of course, implicitly, by the term order $<$ on A^m).*

PROOF OF LEMMA 3.7.9. We first show that $<$ is a total order. Let X, Y be power products in A. First, if $i \neq j \in \{1, \ldots, t\}$, then either $\operatorname{lm}(X\boldsymbol{g}_i) = \operatorname{lm}(Y\boldsymbol{g}_j)$ and one of $i < j$ or $j < i$ holds, or one of $\operatorname{lm}(X\boldsymbol{g}_i) < \operatorname{lm}(Y\boldsymbol{g}_j)$ or $\operatorname{lm}(Y\boldsymbol{g}_j) < \operatorname{lm}(X\boldsymbol{g}_i)$ holds. In any case, one of $X\boldsymbol{e}_i < Y\boldsymbol{e}_j$ or $Y\boldsymbol{e}_j < X\boldsymbol{e}_i$ holds. If $i = j \in \{1, \ldots, t\}$, and $X \neq Y$, then one of $\operatorname{lm}(X\boldsymbol{g}_i) < \operatorname{lm}(Y\boldsymbol{g}_i)$ or $\operatorname{lm}(Y\boldsymbol{g}_i) < \operatorname{lm}(X\boldsymbol{g}_i)$ holds, for otherwise, if $\operatorname{lm}(X\boldsymbol{g}_i) = \operatorname{lm}(Y\boldsymbol{g}_i)$, then

$$X \operatorname{lm}(\boldsymbol{g}_i) = \operatorname{lm}(X\boldsymbol{g}_i) = \operatorname{lm}(Y\boldsymbol{g}_i) = Y \operatorname{lm}(\boldsymbol{g}_i),$$

and hence $X = Y$, since $\boldsymbol{g}_i \neq \boldsymbol{0}$. Therefore, we have one of $X\boldsymbol{e}_i < Y\boldsymbol{e}_i$ or $Y\boldsymbol{e}_i < X\boldsymbol{e}_i$.

We now verify that $<$ is a term order as defined in Definition 3.5.1. Let X, Z be power products in A such that $Z \neq 1$. Let $i \in \{1, \ldots, t\}$. Then $\operatorname{lm}(X\boldsymbol{g}_i) < Z \operatorname{lm}(X\boldsymbol{g}_i) = \operatorname{lm}(ZX\boldsymbol{g}_i)$, and hence $X\boldsymbol{e}_i < ZX\boldsymbol{e}_j$. Finally, let X, Y, Z be power products in A, and let $i, j \in \{1, \ldots, t\}$. Assume that $X\boldsymbol{e}_i < Y\boldsymbol{e}_j$. If $\operatorname{lm}(X\boldsymbol{g}_i) < \operatorname{lm}(Y\boldsymbol{g}_j)$, then

$$\operatorname{lm}(ZX\boldsymbol{g}_i) = Z \operatorname{lm}(X\boldsymbol{g}_i) < Z \operatorname{lm}(Y\boldsymbol{g}_j) = \operatorname{lm}(ZY\boldsymbol{g}_j),$$

and hence $ZX\boldsymbol{e}_i < ZY\boldsymbol{e}_j$. If $\operatorname{lm}(X\boldsymbol{g}_i) = \operatorname{lm}(Y\boldsymbol{g}_j)$ and $j < i$, then

$$\operatorname{lm}(ZX\boldsymbol{g}_i) = Z \operatorname{lm}(X\boldsymbol{g}_i) = Z \operatorname{lm}(Y\boldsymbol{g}_j) = \operatorname{lm}(ZY\boldsymbol{g}_j)$$

and $j < i$, so $ZX\boldsymbol{e}_i < ZY\boldsymbol{e}_j$. □

EXAMPLE 3.7.11. We first consider the case where $m = 1$ and polynomials $g_1 = x^2y^2 - x^3y$, $g_2 = xy^3 - x^2y^2$, and $g_3 = y^4 - x^3$. We use the lex term ordering in $\mathbb{Q}[x, y]$ with $x < y$. In the notation of Lemma 3.7.9, we have

$$x\boldsymbol{e}_2 < y\boldsymbol{e}_1 < x\boldsymbol{e}_3,$$

since

$$\operatorname{lp}(xg_2) = \operatorname{lp}(yg_1) = x^2y^3 < xy^4 = \operatorname{lp}(xg_3).$$

Note that $<$ is neither TOP nor POT as defined in Section 3.5.

EXAMPLE 3.7.12. Let us use the six vectors $\boldsymbol{g}_i$, $i = 1, \ldots, 6$ of Example 3.7.5. These vectors determine a term ordering on A^6 as described in Lemma 3.7.9. In order to distinguish the basis vectors $\mathbf{e}_i$ in A^3 and A^6, we will use $\mathbf{e}_{3i}$ and $\mathbf{e}_{6i}$ for the ith basis vector in A^3 and A^6 respectively. In A^3 we are using lex with $y > x$ and TOP with $\mathbf{e}_{33} < \mathbf{e}_{32} < \mathbf{e}_{31}$. Using the order induced by $\left[\begin{array}{ccc} \boldsymbol{g}_1 & \cdots & \boldsymbol{g}_6 \end{array}\right]$ we have

$$x\mathbf{e}_{66} < \mathbf{e}_{62} < x\mathbf{e}_{64} < y\mathbf{e}_{61},$$

since $\operatorname{lm}(x\boldsymbol{g}_6) = x^6\mathbf{e}_{33}$, $\operatorname{lm}(\boldsymbol{g}_2) = y\mathbf{e}_{32}$, $\operatorname{lm}(x\boldsymbol{g}_4) = xy\mathbf{e}_{33}$, and $\operatorname{lm}(y\boldsymbol{g}_1) = yx^3\mathbf{e}_{31}$, and

$$x^6\mathbf{e}_{33} < y\mathbf{e}_{32} < xy\mathbf{e}_{33} < yx^3\mathbf{e}_{31}.$$

THEOREM 3.7.13. *Let $G = \{\boldsymbol{g}_1, \ldots, \boldsymbol{g}_t\}$ be a Gröbner basis. With the notation of Theorem 3.7.3, the collection $\{\boldsymbol{s}_{ij} \mid 1 \le i < j \le t\}$ is a Gröbner basis for* $\operatorname{Syz}(\boldsymbol{g}_1, \ldots, \boldsymbol{g}_t)$ *with respect to the term order $<$ on monomials of A^t induced by* $\left[\begin{array}{ccc} \boldsymbol{g}_1 & \cdots & \boldsymbol{g}_t \end{array}\right]$. *Moreover,* $\operatorname{lm}(\boldsymbol{s}_{ij}) = \dfrac{\boldsymbol{X}_{ij}}{\boldsymbol{X}_i}\mathbf{e}_i$ *for each $1 \le i < j \le t$.*

PROOF. We first prove that for $1 \le i < j \le t$, we have $\operatorname{lm}(\boldsymbol{s}_{ij}) = \dfrac{\boldsymbol{X}_{ij}}{\boldsymbol{X}_i}\mathbf{e}_i$. Note that we have $\operatorname{lm}\left(\dfrac{\boldsymbol{X}_{ij}}{\boldsymbol{X}_i}\boldsymbol{g}_i\right) = \operatorname{lm}\left(\dfrac{\boldsymbol{X}_{ij}}{\boldsymbol{X}_j}\boldsymbol{g}_j\right) = \boldsymbol{X}_{ij}$. Therefore

$$\frac{\boldsymbol{X}_{ij}}{\boldsymbol{X}_i}\mathbf{e}_j < \frac{\boldsymbol{X}_{ij}}{\boldsymbol{X}_i}\mathbf{e}_i,$$

since $i < j$. Now let $X\mathbf{e}_\ell$ be a monomial that appears in $(h_{ij1}, \ldots, h_{ijt})$. Then $\operatorname{lm}(X\boldsymbol{g}_\ell) \le \operatorname{lm}(S(\boldsymbol{g}_i, \boldsymbol{g}_j))$, by Equation (3.7.1). But $\operatorname{lm}(S(\boldsymbol{g}_i, \boldsymbol{g}_j)) < \operatorname{lm}\left(\dfrac{\boldsymbol{X}_{ij}}{\boldsymbol{X}_i}\boldsymbol{g}_i\right)$, therefore $X\mathbf{e}_\ell < \dfrac{\boldsymbol{X}_{ij}}{\boldsymbol{X}_i}\mathbf{e}_i$.

We now show that $\{\boldsymbol{s}_{ij} \mid 1 \le i < j \le t\}$ is a Gröbner basis for $\operatorname{Syz}(\boldsymbol{g}_1, \ldots, \boldsymbol{g}_t)$ with respect to $<$. Let $\boldsymbol{s} \in \operatorname{Syz}(\boldsymbol{g}_1, \ldots, \boldsymbol{g}_t)$. By Defintion 3.5.13 we need to show that there exist i, j such that $1 \le i < j \le t$ and $\operatorname{lm}(\boldsymbol{s}_{ij})$ divides $\operatorname{lm}(\boldsymbol{s})$. First we write $\boldsymbol{s} = \sum_{\ell=1}^t a_\ell \mathbf{e}_\ell$, where $a_\ell \in A$. Let $Y_\ell = \operatorname{lp}(a_\ell)$ and $c_\ell = \operatorname{lc}(a_\ell)$. Note that we have $\operatorname{lm}(\boldsymbol{s}) = Y_i\mathbf{e}_i$ for some $i \in \{1, \ldots, t\}$. For this i define

$$S = \{\ell \in \{1, \ldots, t\} \mid \operatorname{lm}(Y_\ell\boldsymbol{g}_\ell) = \operatorname{lm}(Y_i\boldsymbol{g}_i)\}.$$

We observe that if $\ell \in S$, then $\ell \ge i$, by the definition of $<$. Define a new vector

$$\boldsymbol{s}' = \sum_{\ell \in S} c_\ell Y_\ell \mathbf{e}_\ell.$$

Since $\boldsymbol{s}$ is a syzygy of $\left[\begin{array}{ccc} \boldsymbol{g}_1 & \cdots & \boldsymbol{g}_t \end{array}\right]$, we have

$$\sum_{\ell \in S} c_\ell Y_\ell \operatorname{lt}(\boldsymbol{g}_\ell) = \mathbf{0}.$$

Therefore $\boldsymbol{s}'$ is a syzygy of $\left[\ \operatorname{lt}(\boldsymbol{g}_1) \ \cdots \ \operatorname{lt}(\boldsymbol{g}_t)\ \right]$. Noting that the indices of the non-zero coordinates of $\boldsymbol{s}'$ are in S, we have, by Proposition 3.7.2, that $\boldsymbol{s}'$ is in the submodule of A^t generated by $\mathcal{B} = \left\{ \frac{\boldsymbol{X}_{\ell m}}{\boldsymbol{X}_\ell}\boldsymbol{e}_\ell - \frac{\boldsymbol{X}_{\ell m}}{\boldsymbol{X}_m}\boldsymbol{e}_m \in A^t \mid \ell, m \in S, \ell < m \right\}$. Thus we have

$$\boldsymbol{s}' = \sum_{\substack{\ell,m\in S\\ \ell<m}} a_{\ell m}\left(\frac{\boldsymbol{X}_{\ell m}}{\boldsymbol{X}_\ell}\boldsymbol{e}_\ell - \frac{\boldsymbol{X}_{\ell m}}{\boldsymbol{X}_m}\boldsymbol{e}_m\right),$$

for some $a_{\ell m} \in A$. Now $\operatorname{lm}(\boldsymbol{s}') = \operatorname{lm}(\boldsymbol{s}) = Y_i\boldsymbol{e}_i$ and thus, since $j > i$ for all $j \neq i \in S$, we see that

$$c_iY_i\boldsymbol{e}_i = \operatorname{lt}(\boldsymbol{s}') = \sum \operatorname{lt}(a_{ij})\frac{X_{ij}}{X_i}\boldsymbol{e}_i,$$

where the sum is over all $j \in S$, $j \neq i$ such that $Y_i = \operatorname{lp}(a_{ij})\frac{X_{ij}}{X_i}$. It follows immediately that for some $j \in S$, $\frac{X_{ij}}{X_i}\boldsymbol{e}_i = \operatorname{lm}(\boldsymbol{s}_{ij})$ divides $\operatorname{lm}(\boldsymbol{s}') = \operatorname{lm}(\boldsymbol{s})$ as desired. □

We might hope that a result similar to Theorem 3.7.13 holds for finding a Gröbner basis for $\operatorname{Syz}(\boldsymbol{f}_1, \ldots, \boldsymbol{f}_s)$, where the corresponding order on monomials of A^s would be the one induced by $\left[\ \boldsymbol{f}_1 \ \cdots \ \boldsymbol{f}_s\ \right]$ where $\{\boldsymbol{f}_1, \ldots, \boldsymbol{f}_s\}$ is not necessarily a Gröbner basis. We saw in Theorem 3.7.6 how to compute generators for $\operatorname{Syz}(\boldsymbol{f}_1, \ldots, \boldsymbol{f}_s)$. These generators do not form, in general, a Gröbner basis with respect to the order induced by $\left[\ \boldsymbol{f}_1 \ \cdots \ \boldsymbol{f}_s\ \right]$ (see Exercise 3.7.15). So, to obtain a Gröbner basis for $\operatorname{Syz}(\boldsymbol{f}_1, \ldots, \boldsymbol{f}_s)$, we would use the algorithm presented in Section 3.5 (Algorithm 3.5.2) starting with the generators of $\operatorname{Syz}(\boldsymbol{f}_1, \ldots, \boldsymbol{f}_s)$ given in Theorem 3.7.6.

Exercises

3.7.1. Let $f, f_1, \ldots, f_s \in k[x_1, \ldots, x_n]$. Show that $f \in \langle f_1, \ldots, f_s\rangle$ if and only if in the Gröbner basis, G, of $\operatorname{Syz}(f, f_1, \ldots, f_s)$ with respect to the POT ordering with the first coordinate largest, there is a vector $(u, u_1, \ldots, u_s) \in G$ with $u \neq 0$ and $u \in k$. In this case we obtain $f = -\frac{1}{u}\sum_{i=1}^{s} u_if_i$.

3.7.2. Complete the computations in Example 3.7.4.

3.7.3. Complete the computations in Example 3.7.5.

3.7.4. Prove Proposition 3.7.2.

3.7.5. Prove Theorem 3.7.3.

3.7.6. State and prove the analog of Theorem 3.2.5 in the module case; that is, give a definition of Gröbner bases for modules in terms of the syzygy module of the leading terms.

3.7.7. State and prove the analog of Corollary 3.3.3, and use it to describe how crit2 can be implemented in Algorithm 3.5.2. Note that we have already seen in Exercise 3.5.16 that crit1 cannot be used in the module case.

3.7.8. Use Exercise 3.7.7 to state and prove the analog of Exercise 3.4.4 in the module case (generalizing Theorem 3.7.3).

3.7.9. Prove Theorem 3.7.6.

3.7.10. Compute generators for $\operatorname{Syz}(\boldsymbol{f}_1, \boldsymbol{f}_2, \boldsymbol{f}_3, \boldsymbol{f}_4)$ in the following examples. You should do this without the use of a Computer Algebra System.

a. $\boldsymbol{f}_1 = (xy + y, x)$, $\boldsymbol{f}_2 = (-y + x, y)$, $\boldsymbol{f}_3 = (x, y + x)$, $\boldsymbol{f}_4 = (-x, y) \in (\mathbb{Q}[x, y])^2$.

b. $\boldsymbol{f}_1 = (xy - x, y)$, $\boldsymbol{f}_2 = (x^2 - y, x)$, $\boldsymbol{f}_3 = (x^3 - x, y + x^2)$, $\boldsymbol{f}_4 = (-x + y^2, y) \in (\mathbb{Q}[x, y])^2$.

3.7.11. State and prove the analog of Exercise 3.4.3 in the module case.

3.7.12. Consider the following analog of Exercise 3.4.6. Let $\boldsymbol{f}_1, \ldots, \boldsymbol{f}_s, \boldsymbol{g} \in A^m$. We consider the linear equation

$$h_1 \boldsymbol{f}_1 + h_2 \boldsymbol{f}_2 + \cdots + h_s \boldsymbol{f}_s = \boldsymbol{g},$$

with unknowns $h_1, \ldots, h_s \in A$. Let $\boldsymbol{S} \subseteq A^s$ be the set of all solutions $(h_1, \ldots, h_s)$.

a. Prove that $\boldsymbol{S}$ is not empty if and only if $\boldsymbol{g} \in \langle \boldsymbol{f}_1, \ldots, \boldsymbol{f}_s \rangle$.

b. Prove that if $\boldsymbol{S} \neq \emptyset$ then $\boldsymbol{S} = \boldsymbol{h} + \operatorname{Syz}(\boldsymbol{f}_1, \ldots, \boldsymbol{f}_s) = \{\boldsymbol{h} + \boldsymbol{s} \mid \boldsymbol{s} \in \operatorname{Syz}(\boldsymbol{f}_1, \ldots, \boldsymbol{f}_s)\}$, where $\boldsymbol{h}$ is a particular solution. Give a method for computing $\boldsymbol{h}$.

c. Use the above to find the solution set for the equation

$$h_1(x+y, y, x) + h_2(x, y, x) + h_3(-x, -x+y, x) + h_4(x, x, y) = (y, 0, x^3).$$

3.7.13. Consider $\boldsymbol{g}_1 = (x^2, y)$, $\boldsymbol{g}_2 = (xy + y, y^3)$, and $\boldsymbol{g}_3 = (x^2, -x + y^3) \in (\mathbb{Q}[x, y])^2$. We use the lex order on $\mathbb{Q}[x, y]$ with $x > y$ and the POT ordering on $(\mathbb{Q}[x, y])^2$ with $\boldsymbol{e}_1 > \boldsymbol{e}_2$. Consider the vector $\boldsymbol{f} = (xy^2 + y^3, x^2y + xy^3, y^3 + x^2y) \in (\mathbb{Q}[x, y])^3$. Write $\boldsymbol{f}$ as the sum of terms in descending order according to the order on $(\mathbb{Q}[x, y])^3$ induced by $\begin{bmatrix} \boldsymbol{g}_1 & \boldsymbol{g}_2 & \boldsymbol{g}_3 \end{bmatrix}$. Repeat the exercise using deglex with $x > y$ on $\mathbb{Q}[x, y]$.

3.7.14. Verify that the following vectors form a Gröbner basis with respect to the indicated term order, compute generators for $\operatorname{Syz}(\boldsymbol{g}_1, \ldots, \boldsymbol{g}_t)$, and verify Theorem 3.7.13.

a. $\boldsymbol{g}_1 = (0, x^2)$, $\boldsymbol{g}_2 = (y, x)$, $\boldsymbol{g}_3 = (2x, x)$, $\boldsymbol{g}_4 = (0, 2y + x)$. Use the deglex order on $\mathbb{Q}[x, y]$ with $y > x$ and the POT ordering on $(\mathbb{Q}[x, y])^2$ with $\boldsymbol{e}_1 > \boldsymbol{e}_2$.

b. $\boldsymbol{g}_1 = (0, x^2)$, $\boldsymbol{g}_2 = (y^2 - y, x)$, $\boldsymbol{g}_3 = (0, xy)$, $\boldsymbol{g}_4 = (x - y, x - y)$, $\boldsymbol{g}_5 = (0, y^2)$. Use the deglex order on $\mathbb{Q}[x, y]$ with $x > y$ and the POT ordering on $(\mathbb{Q}[x, y])^2$ with $\boldsymbol{e}_1 > \boldsymbol{e}_2$.

3.7.15. We mentioned at the end of the section that a result similar to Theorem 3.7.13 does not hold for the generators of $\operatorname{Syz}(\boldsymbol{f}_1, \ldots, \boldsymbol{f}_s)$ obtained in Theorem 3.7.6. Namely, the generators for $\operatorname{Syz}(\boldsymbol{f}_1, \ldots, \boldsymbol{f}_s)$ obtained in Theorem 3.7.6 do not, in general, form a Gröbner basis for $\operatorname{Syz}(\boldsymbol{f}_1, \ldots, \boldsymbol{f}_s)$ with respect to the order induced by $\begin{bmatrix} \boldsymbol{f}_1, \ldots, \boldsymbol{f}_s \end{bmatrix}$. Consider the vectors

$$\boldsymbol{f}_1 = (x + y, y, x), \quad \boldsymbol{f}_2 = (x - y, x, y), \quad \boldsymbol{f}_3 = (x + y, x, y),$$

$$\boldsymbol{f}_4 = (-x+y, y, x), \ \boldsymbol{f}_5 = (x, x, x) \in (\mathbb{Q}[x,y])^3.$$

a. Verify that the reduced Gröbner basis for $\langle \boldsymbol{f}_1, \boldsymbol{f}_2, \boldsymbol{f}_3, \boldsymbol{f}_4, \boldsymbol{f}_5 \rangle$ with respect to the TOP ordering with $\boldsymbol{e}_1 > \boldsymbol{e}_2 > \boldsymbol{e}_3$ and lex on $\mathbb{Q}[x,y]$ with $y > x$ is given by the vectors

$$\boldsymbol{g}_1 = (0, y, x), \ \boldsymbol{g}_2 = (0, 0, y-x), \ \boldsymbol{g}_3 = (0, x, x),$$

$$\boldsymbol{g}_4 = (y, 0, 0), \ \boldsymbol{g}_5 = (x, 0, 0).$$

b. Verify that $\mathrm{Syz}(\boldsymbol{g}_1, \boldsymbol{g}_2, \boldsymbol{g}_3, \boldsymbol{g}_4, \boldsymbol{g}_5) = \langle (-x, -x, y, 0, 0), (0, 0, 0, -x, y) \rangle$.

c. Verify that Theorem 3.7.6 gives

$$\mathrm{Syz}(\boldsymbol{f}_1, \boldsymbol{f}_2, \boldsymbol{f}_3, \boldsymbol{f}_4, \boldsymbol{f}_5) = \langle (-x, -x, -x, -x, 2y+2x), (y, x, -x, -y, 0) \rangle.$$

d. Verify that the two vectors given in **c** do not form a Gröbner basis with respect to the order induced by $\left[\ \boldsymbol{f}_1 \ \ \boldsymbol{f}_2 \ \ \boldsymbol{f}_3 \ \ \boldsymbol{f}_4 \ \ \boldsymbol{f}_5 \ \right]$.

3.7.16. Let $M = \langle \boldsymbol{f}_1, \ldots, \boldsymbol{f}_s \rangle$ be a submodule of A^m. Assume that we have a generating set for $\mathrm{Syz}(\boldsymbol{f}_1, \ldots, \boldsymbol{f}_s)$, say $\mathrm{Syz}(\boldsymbol{f}_1, \ldots, \boldsymbol{f}_s) = \langle \boldsymbol{s}_1, \ldots, \boldsymbol{s}_\ell \rangle \subseteq A^s$. Now consider vectors $\boldsymbol{g}_i = \sum_{j=1}^{s} a_{ij} \boldsymbol{f}_j$, for some $a_{ij} \in A$, for $i = 1, \ldots, t$.

a. Use Theorem 3.5.22 to give a method to decide if $M = \langle \boldsymbol{g}_1, \ldots, \boldsymbol{g}_t \rangle$. If so, give a method to find $b_{ij} \in A$ such that $\boldsymbol{f}_i = \sum_{j=1}^{t} b_{ij} \boldsymbol{g}_j$. Use the proof of Theorem 3.7.6 to find a generating set for $\mathrm{Syz}(\boldsymbol{g}_1, \ldots, \boldsymbol{g}_t)$. (Note that the proof of Theorem 3.7.6 has nothing to do with the theory of Gröbner bases.)

b. Apply the methods given in **a** to the following example. Consider $\boldsymbol{f}_1 = (xy, y, x)$, $\boldsymbol{f}_2 = (x, y+x, y)$, $\boldsymbol{f}_3 = (-y, x, y)$, and $\boldsymbol{f}_4 = (x, x, y) \in (\mathbb{Q}[x,y])^3$. It is easy to verify that $\mathrm{Syz}(\boldsymbol{f}_1, \boldsymbol{f}_2, \boldsymbol{f}_3, \boldsymbol{f}_4) = \langle (y^3 + y^2 x, -y^3 - y^2 x + yx^2 + x^3, y^3 x - yx^2, -y^3 x + y^3 - yx^2 - x^3) \rangle$. Now consider the polynomials $\boldsymbol{g}_1 = (0, y, 0) = \boldsymbol{f}_2 - \boldsymbol{f}_4$, $\boldsymbol{g}_2 = (-y-x, 0, 0) = \boldsymbol{f}_3 - \boldsymbol{f}_4$, $\boldsymbol{g}_3 = (x, x, y) = \boldsymbol{f}_4$, $\boldsymbol{g}_4 = (-x^2, 0, x) = \boldsymbol{f}_1 - \boldsymbol{f}_2 + x\boldsymbol{f}_3 + (-x+1)\boldsymbol{f}_4$, $\boldsymbol{g}_5 = (0, x^2, -x^2+x) = (-x-y+1)\boldsymbol{f}_1 + (x+y-1)\boldsymbol{f}_2 + (-yx+x)\boldsymbol{f}_3 + (yx-y-x+1)\boldsymbol{f}_4$, and $\boldsymbol{g}_6 = (-x^2, -x^2, x^3) = (x^2 - y^2 + y)\boldsymbol{f}_1 + (-2x^2 + y^2 - y)\boldsymbol{f}_2 + (yx^2 - xy^2 + xy)\boldsymbol{f}_3 + (-x^2 y + xy^2 + 2x^2 - y^2 - x + y)\boldsymbol{f}_4$. You should first verify that

$$M = \langle \boldsymbol{f}_1, \boldsymbol{f}_2, \boldsymbol{f}_3, \boldsymbol{f}_4 \rangle = \langle \boldsymbol{g}_1, \boldsymbol{g}_2, \boldsymbol{g}_3, \boldsymbol{g}_4, \boldsymbol{g}_5, \boldsymbol{g}_6 \rangle.$$

3.7.17. (Möller [**Mö90**]) In this exercise, we give a more efficient way to solve the problem raised in Exercise 3.7.16. We have the same hypotheses as in Exercise 3.7.16: we are given a generating set $\{\boldsymbol{f}_1, \ldots, \boldsymbol{f}_s\}$ for a submodule M of A^m (not necessarily a Gröbner basis), a generating set $\{\boldsymbol{s}_1, \ldots, \boldsymbol{s}_\ell\} \subseteq A^s$ for $\mathrm{Syz}(\boldsymbol{f}_1, \ldots, \boldsymbol{f}_s)$, and a collection of vectors $\{\boldsymbol{g}_1, \ldots, \boldsymbol{g}_t\}$ such that $\boldsymbol{g}_i = \sum_{j=1}^{s} a_{ij} \boldsymbol{f}_j$, for some $a_{ij} \in A$, for $i = 1, \ldots, t$. As in Exercise 3.7.16, we wish to determine whether $M = \langle \boldsymbol{g}_1, \ldots, \boldsymbol{g}_t \rangle$ and if so to find generators for $\mathrm{Syz}(\boldsymbol{g}_1, \ldots, \boldsymbol{g}_t)$. Let $\boldsymbol{e}_{si}, i = 1, \ldots, s$ and

$\boldsymbol{e}_{tj}, j = 1, \ldots, t$ be the standard basis for A^s and A^t respectively. Finally, let $\boldsymbol{a}_i = (a_{i1}, \ldots, a_{is}) \in A^s$, for $i = 1, \ldots, t$.

a. Let $W = \{(\boldsymbol{u}, \boldsymbol{v}) \in A^{s+t} \mid \boldsymbol{u} = (u_1, \ldots, u_s) \in A^s, \boldsymbol{v} = (v_1, \ldots, v_t) \in A^t, \sum_{i=1}^s u_i \boldsymbol{f}_i = \sum_{j=1}^t v_j \boldsymbol{g}_j\}$. Prove that

$$\{(\boldsymbol{a}_1, \boldsymbol{e}_{t1}), \ldots, (\boldsymbol{a}_t, \boldsymbol{e}_{tt}), (\boldsymbol{s}_1, \boldsymbol{0}), \ldots, (\boldsymbol{s}_\ell, \boldsymbol{0})\}$$

generates W.

b. Prove that $M = \langle \boldsymbol{g}_1, \ldots, \boldsymbol{g}_t \rangle$ if and only if there are $\boldsymbol{b}_1, \ldots, \boldsymbol{b}_s \in A^t$ such that $(\boldsymbol{e}_{si}, \boldsymbol{b}_i) \in W$, for $i = 1, \ldots, s$.

c. Fix an order $<_s$ and an order $<_t$ in A^s and A^t respectively. Consider a new order $<$ on A^{s+t} defined as follows: for $\boldsymbol{u}_1, \boldsymbol{u}_2 \in A^s$ and $\boldsymbol{v}_1, \boldsymbol{v}_2 \in A^t$,

$$(\boldsymbol{u}_1, \boldsymbol{v}_1) < (\boldsymbol{u}_2, \boldsymbol{v}_2) \Longleftrightarrow \begin{cases} \boldsymbol{u}_1 <_s \boldsymbol{u}_2 \text{ or} \\ \boldsymbol{u}_1 = \boldsymbol{u}_2 \text{ and } \boldsymbol{v}_1 <_t \boldsymbol{v}_2. \end{cases}$$

Prove that $M = \langle \boldsymbol{g}_1, \ldots, \boldsymbol{g}_t \rangle$ if and only the reduced Gröbner basis for W with respect to $<$ is the union of the two sets $\{(\boldsymbol{e}_{si}, \boldsymbol{b}_i) \mid i = 1, \ldots, s\}$ and $\{(\boldsymbol{0}, \boldsymbol{t}_i) \mid i = 1, \ldots, \ell\}$. Prove that the set $\{(\boldsymbol{0}, \boldsymbol{t}_i) \mid i = 1, \ldots, \ell\}$ is a generating set for $\operatorname{Syz}(\boldsymbol{g}_1, \ldots, \boldsymbol{g}_t)$.

d. Redo the computation in Exercise 3.7.16 **b** using this method.

One can think of the vectors $\boldsymbol{a}_i$ as the rows of a matrix A which defines a linear transformation $T\colon M \longrightarrow M$. This exercise answers the question of whether T is onto and whether there is a linear transformation $T'\colon M \longrightarrow M$ defined by a matrix B such that $(T' \circ T)(M) = M$ (B is an "inverse" of A in the sense that $BA - I_s \subseteq \operatorname{Syz}(\boldsymbol{f}_1, \ldots, \boldsymbol{f}_s)$, where I_s is the $s \times s$ identity matrix). The method presented here is a generalization of a linear algebra method to compute the inverse of the matrix of a linear transformation. In linear algebra over fields, to compute the inverse of an $s \times s$ matrix A, one reduces the matrix $\left[\begin{array}{c|c} A & I_s \end{array}\right]$ to the row reduced echelon form $\left[\begin{array}{c|c} I_s & B \end{array}\right]$, where I_s is the $s \times s$ identity matrix. The matrix B is then the inverse of A. In the case of a module M, we use the Gröbner Basis Algorithm to transform the matrix

$$\left[\begin{array}{c|c} A & I_t \\ \hline \operatorname{Syz}(\boldsymbol{f}_1, \ldots, \boldsymbol{f}_s) & 0 \end{array}\right]$$

into the matrix

$$\left[\begin{array}{c|c} I_s & B \\ \hline 0 & \operatorname{Syz}(\boldsymbol{g}_1, \ldots, \boldsymbol{g}_t) \end{array}\right].$$

3.8. Applications of Syzygies. The purpose of this section is to reconsider and solve more efficiently various problems, such as computing the intersection of submodules of A^m (where, again, $A = k[x_1, \ldots, x_n]$ for a field k). These problems were solved previously using elimination. As we have noted before, the lex term ordering (or more generally, any elimination ordering) is very inefficient from a computational point of view. Indeed, much effort in the theory of Gröbner bases has been expended in order to avoid the use of elimination in computations involving Gröbner bases. Here, we will show how the computation of intersections, ideal quotients and kernels of homomorphisms can all be done using syzygies, where the latter may be computed using any term order whatsoever. Even in A (when $m = 1$) these computations using syzygies turn out to be more efficient than using elimination orders in A.

We begin by considering the simplest case, the intersection of two ideals in A. This case, nevertheless, contains all of the essential ingredients of the most general situation. Let I and J be ideals in A. Assume that $I = \langle f_1, \ldots, f_s \rangle$ and $J = \langle g_1, \ldots, g_t \rangle$ (we will not assume that either $\{f_1, \ldots, f_s\}$ or $\{g_1, \ldots, g_t\}$ is a Gröbner basis). Then a polynomial h is in $I \cap J$ if and only if

$$h = a_1 f_1 + a_2 f_2 + \cdots + a_s f_s \text{ and } h = b_1 g_1 + b_2 g_2 + \cdots + b_t g_t,$$

for some polynomials $a_1, \ldots, a_s, b_1, \ldots, b_t \in A$. This is the same as the *two* conditions that $(-h, a_1, \ldots, a_s)$ is a syzygy of the matrix $\begin{bmatrix} 1 & f_1 & \cdots & f_s \end{bmatrix}$ and $(-h, b_1, \ldots, b_t)$ is a syzygy of the matrix $\begin{bmatrix} 1 & g_1 & \cdots & g_t \end{bmatrix}$. It is then easy to put these two conditions into a single condition for syzygies of vectors in A^2. Let

$$\boldsymbol{i} = \begin{bmatrix} 1 \\ 1 \end{bmatrix}, \boldsymbol{f}_1 = \begin{bmatrix} f_1 \\ 0 \end{bmatrix}, \ldots, \boldsymbol{f}_s = \begin{bmatrix} f_s \\ 0 \end{bmatrix}, \boldsymbol{g}_1 = \begin{bmatrix} 0 \\ g_1 \end{bmatrix}, \ldots, \boldsymbol{g}_t = \begin{bmatrix} 0 \\ g_t \end{bmatrix}.$$

Then it is easily seen that h is in $I \cap J$ if and only if there are polynomials $a_1, \ldots, a_s, b_1, \ldots, b_t \in A$ such that

$$(-h, a_1, \ldots, a_s, b_1, \ldots, b_t) \tag{3.8.1}$$

is a syzygy of

$$\begin{bmatrix} \boldsymbol{i} & \boldsymbol{f}_1 & \cdots & \boldsymbol{f}_s & \boldsymbol{g}_1 & \cdots & \boldsymbol{g}_t \end{bmatrix} = \begin{bmatrix} 1 & f_1 & \ldots & f_s & 0 & \ldots & 0 \\ 1 & 0 & \ldots & 0 & g_1 & \ldots & g_t \end{bmatrix}.$$

Thus we have essentially shown

PROPOSITION 3.8.1. *Using the notation above,*

$$I \cap J = \big\{h \mid \text{ there exist polynomials } a_1, \ldots, a_s, b_1, \ldots, b_t \text{ such that }$$

$$(h, a_1, \ldots, a_s, b_1, \ldots, b_t) \text{ is a syzygy of } \begin{bmatrix} \boldsymbol{i} & \boldsymbol{f}_1 & \cdots & \boldsymbol{f}_s & \boldsymbol{g}_1 & \cdots & \boldsymbol{g}_t \end{bmatrix}\big\}.$$

Moreover, if $\boldsymbol{h}_1, \ldots, \boldsymbol{h}_r$ *is a generating set for* $\operatorname{Syz}(\boldsymbol{i}, \boldsymbol{f}_1, \ldots, \boldsymbol{f}_s, \boldsymbol{g}_1, \ldots, \boldsymbol{g}_t)$ *and the first coordinate of* $\boldsymbol{h}_i$ *is* h_i *for* $1 \leq i \leq r$, *then* $\{h_1, \ldots, h_r\}$ *is a generating set for* $I \cap J$.

PROOF. For the first statement we simply note that $h \in I \cap J$ if and only if $-h \in I \cap J$, and this accounts for the difference between the statement of the proposition and Expression (3.8.1). The second statement follows from the first statement, since if a vector is a linear combination of other vectors, then the first coordinate of the vector is a linear combination of the first coordinates of the other vectors. $\square$

EXAMPLE 3.8.2. We will consider the following ideals in $\mathbb{Q}[x, y]$:

$$I = \langle xy - x - y - 1, x^2 + 1\rangle \text{ and } J = \langle -x^2 + xy, x^2y + y, y^3 + x\rangle.$$

We wish to compute $I \cap J$. We use the deglex ordering with $x > y$ in A. So we let

$$\boldsymbol{i} = \begin{bmatrix} 1 \\ 1 \end{bmatrix}, \boldsymbol{f}_1 = \begin{bmatrix} xy - x - y - 1 \\ 0 \end{bmatrix}, \boldsymbol{f}_2 = \begin{bmatrix} x^2 + 1 \\ 0 \end{bmatrix},$$

$$\boldsymbol{g}_1 = \begin{bmatrix} 0 \\ -x^2 + xy \end{bmatrix}, \boldsymbol{g}_2 = \begin{bmatrix} 0 \\ x^2y + y \end{bmatrix}, \boldsymbol{g}_3 = \begin{bmatrix} 0 \\ y^3 + x \end{bmatrix}.$$

Then we can compute that the module of syzygies with respect to TOP and $\boldsymbol{e}_1 > \boldsymbol{e}_2$ is generated by $(x^2y + y, 0, -y, 0, -1, 0)$, $(xy^3 + xy^2 + 3y^3 - x^2 + xy + y^2 + x + 3y, 2xy^2 + y^2 + 2y + 1, -2y^3 + 2y^2 + 1, 2xy^2 - y^2 - 3y - 1, 2xy - 2y^2 - y - 3, -1)$, $(x^3 + 2y^3 + 2x^2 - 2y^2 + x + 2y, xy^2 - 2xy + y^2 + x - 1, -y^3 + 3y^2 - x - 3y - 1, xy^2 - 2xy + x - y + 2, xy - y^2 - 2x + 2y - 2, -1)$, and $(-4xy^2 - 10y^3 - 4x - 10y, x^2y^2 - x^2y - 5xy^2 + 3xy - 4y^2 - 2x - 4y - 2, -xy^3 + 2xy^2 + 6y^3 - xy - 8y^2 + 4y - 2, x^2y^2 - x^2y - 6xy^2 + xy + 2y^2 + 9y - 1, x^2y - xy^2 - x^2 - 5xy + 6y^2 + x + 10, -x + 4)$. So the first coordinates of these vectors form a generating set for $I \cap J$, that is,

$$I \cap J = \langle x^2y + y, xy^3 + xy^2 + 3y^3 - x^2 + xy + y^2 + x + 3y,$$

$$x^3 + 2y^3 + 2x^2 - 2y^2 + x + 2y, -4xy^2 - 10y^3 - 4x - 10y\rangle.$$

These polynomials do not form a Gröbner basis for $I \cap J$ with respect to the given term order. Computing a Gröbner basis for this ideal from the given generators, we get the much simpler generating set

$$I \cap J = \langle xy^2 + x, y^3 + y, x^2 - y^2\rangle.$$

We note that in Proposition 3.8.1, as we just saw, we only obtained a generating set for $I \cap J$. Of course, we could obtain a Gröbner basis for $I \cap J$ by applying Buchberger's Algorithm to this set of generators. Alternatively we have

THEOREM 3.8.3. *We use the same notation as above. Assume we have a fixed term order $<$ on A and consider the corresponding POT order on A^{1+s+t} with $\boldsymbol{e}_1$ largest. Let $\{\boldsymbol{h}_1, \ldots, \boldsymbol{h}_r\}$ be a Gröbner basis for $\operatorname{Syz}(\boldsymbol{i}, \boldsymbol{f}_1, \ldots, \boldsymbol{f}_s, \boldsymbol{g}_1, \ldots, \boldsymbol{g}_t)$. Assume that the first coordinate of $\boldsymbol{h}_i$ is h_i for $1 \le i \le r$. Then $\{h_1, \ldots, h_r\}$ is a Gröbner basis for $I \cap J$ with respect to $<$.*

PROOF. We need to show that $\langle \operatorname{lt}(h_1), \ldots, \operatorname{lt}(h_r)\rangle = \operatorname{Lt}(I \cap J)$. One containment is clear, so let $h \in I \cap J$, where we may assume that $h \neq 0$. From Proposition 3.8.1 we can choose

$$\boldsymbol{h} = (h, a_1, \ldots, a_s, b_1, \ldots, b_t) \in \operatorname{Syz}(\boldsymbol{i}, \boldsymbol{f}_1, \ldots, \boldsymbol{f}_s, \boldsymbol{g}_1, \ldots, \boldsymbol{g}_t).$$

Since $\boldsymbol{h}_1, \ldots, \boldsymbol{h}_r$ is a Gröbner basis for this module, we can find polynomials $c_1, \cdots, c_r$ such that $\boldsymbol{h} = \sum_{i=1}^r c_i \boldsymbol{h}_i$ and $\operatorname{lm}(\boldsymbol{h}) = \max_{1 \le i \le r}(\operatorname{lp}(c_i)\operatorname{lm}(\boldsymbol{h}_i))$. Then, since the order is POT, $h \neq 0$, and the first coordinate is largest, we see that $(\operatorname{lt}(h), 0, \ldots, 0) = \operatorname{lt}(\boldsymbol{h}) = \sum \operatorname{lt}(c_i)\operatorname{lt}(\boldsymbol{h}_i)$, where the sum is over all i such that $\operatorname{lm}(\boldsymbol{h}) = \operatorname{lp}(c_i)\operatorname{lm}(\boldsymbol{h}_i)$. It is then clear that we have $\operatorname{lt}(h) = \sum \operatorname{lt}(c_i)\operatorname{lt}(h_i)$ where i ranges over the same i's as before. This gives the desired result. □

We note that in Theorem 3.8.3, as should be evident from the proof of the theorem, we made essential use of the POT ordering. In particular, the TOP ordering would not work, as an example in Exercise 3.8.3 shows.

EXAMPLE 3.8.4. We redo the computation of the previous example. For simplicity we use the Gröbner bases for the ideals I and J with respect to deglex with $x > y$. We compute that $I = \langle x + y, y^2 + 1\rangle$ and $J = \langle y^3 + y, x - y\rangle$. So using

$$\boldsymbol{i} = \begin{bmatrix} 1 \\ 1 \end{bmatrix}, \boldsymbol{f}_1 = \begin{bmatrix} x+y \\ 0 \end{bmatrix}, \boldsymbol{f}_2 = \begin{bmatrix} y^2+1 \\ 0 \end{bmatrix}, \boldsymbol{g}_1 = \begin{bmatrix} 0 \\ y^3+y \end{bmatrix}, \boldsymbol{g}_2 = \begin{bmatrix} 0 \\ x-y \end{bmatrix},$$

we compute a Gröbner basis for $\operatorname{Syz}(\boldsymbol{i}, \boldsymbol{f}_1, \boldsymbol{f}_2, \boldsymbol{g}_1, \boldsymbol{g}_2)$ with respect to POT with $\mathbf{e}_1 > \mathbf{e}_2 > \mathbf{e}_3 > \mathbf{e}_4 > \mathbf{e}_5$, and obtain $(0, 0, 0, x-y, -y^3-y)$, $(0, y^2+1, -x-y, 0, 0)$, $(y^3+y, 0, -y, -1, 0)$, $(xy^2+x, 0, -x, -1, -y^2-1)$, $(x^2-y^2, -x+y, 0, 0, -x-y)$. Reading off the first coordinates of these vectors, we obtain the same Gröbner basis we did in the previous example.

At this point it is convenient to compactify the notation. Let H_1 be an $s_1 \times t_1$ matrix and H_2 be an $s_2 \times t_2$ matrix. We define the $(s_1 + s_2) \times (t_1 + t_2)$ matrix $H_1 \oplus H_2$ (called the *direct sum* of H_1 and H_2) to be the matrix whose upper left-hand corner matrix is the $s_1 \times t_1$ matrix H_1, whose lower right-hand corner matrix is the $s_2 \times t_2$ matrix H_2 and the rest of whose entries consist entirely of zeros. Thus, for example,

$$\begin{bmatrix} f_1 & f_2 \\ f_3 & f_4 \end{bmatrix} \oplus \begin{bmatrix} g_1 \\ g_2 \\ g_3 \end{bmatrix} = \begin{bmatrix} f_1 & f_2 & 0 \\ f_3 & f_4 & 0 \\ 0 & 0 & g_1 \\ 0 & 0 & g_2 \\ 0 & 0 & g_3 \end{bmatrix}.$$

Similarly, let $H_1, \ldots, H_r$ be $s_i \times t_i$ matrices for $1 \le i \le r$. We define $H_1 \oplus H_2 \oplus \cdots \oplus H_r$ to be the $(s_1 + \cdots + s_r) \times (t_1 + \cdots + t_r)$ matrix with the matrices $H_1, \ldots, H_r$ down the diagonal and zeros elsewhere.

Further, let $H_1, \ldots, H_r$ be matrices where, this time, each H_i is an $s \times t_i$ matrix for $1 \le i \le r$. We define $\begin{bmatrix} H_1 | H_2 | \cdots | H_r \end{bmatrix}$ as the $s \times (t_1 + \cdots + t_r)$

matrix whose first t_1 columns are the columns of H_1, whose next t_2 columns are the columns of H_2, etc.

For example, using the notation above in the discussion of $I \cap J$, let F be the $1 \times s$ matrix $\left[\begin{array}{ccc} f_1 & \cdots & f_s \end{array}\right]$ and let G be the $1 \times t$ matrix $\left[\begin{array}{ccc} g_1 & \cdots & g_t \end{array}\right]$. If we let

$$H = \left[\begin{array}{c} \boldsymbol{i}|F \oplus G \end{array}\right] = \left[\begin{array}{ccccccc} 1 & f_1 & \cdots & f_s & 0 & \cdots & 0 \\ 1 & 0 & \cdots & 0 & g_1 & \cdots & g_t \end{array}\right],$$

then the set of first coordinates of $\mathrm{Syz}(H)$ is $I \cap J$.

As a second illustration of this notation we consider the intersection of more than two ideals. So let $I_1, I_2, \ldots, I_r$ be ideals in A, $I_j = \langle f_{j1}, f_{j2}, \ldots, f_{jt_j}\rangle$, and define the $1 \times t_j$ matrix $F_j = \left[\begin{array}{cccc} f_{j1} & f_{j2} & \cdots & f_{jt_j} \end{array}\right]$, for $1 \leq j \leq r$. Let $\boldsymbol{i}$ be the $r \times 1$ matrix (column vector) all of whose entries are 1. Set $H = \left[\begin{array}{c} \boldsymbol{i}|F_1 \oplus F_2 \oplus \cdots \oplus F_r \end{array}\right]$. Then, in exactly the same way as above, we see that the set of first coordinates of a generating set for $\mathrm{Syz}(H)$ is a generating set for $I_1 \cap I_2 \cap \ldots \cap I_r$, and if the POT order is used on $A^{1+t_1+t_2+\cdots+t_r}$ to compute a Gröbner basis for $\mathrm{Syz}(H)$, then the set of first coordinates is, in fact, a Gröbner basis for $I_1 \cap I_2 \cap \cdots \cap I_r$.

EXAMPLE 3.8.5. Let $A = \mathbb{Q}[x, y, z]$ and consider the term order deglex with $x > y > z$. Let $I_1 = \langle x-y, y-z, z-x\rangle, I_2 = \langle x-1, y\rangle$ and $I_3 = \langle y+1, x+1, z-1\rangle$. To compute $I_1 \cap I_2 \cap I_3$ we set

$$H = \left[\begin{array}{ccccccccc} 1 & x-y & y-z & z-x & 0 & 0 & 0 & 0 & 0 \\ 1 & 0 & 0 & 0 & x-1 & y & 0 & 0 & 0 \\ 1 & 0 & 0 & 0 & 0 & 0 & y+1 & x+1 & z-1 \end{array}\right].$$

We then compute, using the TOP ordering on A^3 with $\boldsymbol{e}_1 > \boldsymbol{e}_2 > \boldsymbol{e}_3$, that $\mathrm{Syz}(H) = \langle (0,1,1,1,0,0,0,0,0), (xy-y^2,\ 0,\ y,\ y,\ -y,\ y-1,\ y,\ -y,\ 0), (x^2 - y^2 - x + y,\ 0,\ y+z-1,\ x+z-1,\ -x,\ y-1,\ y+z-3,\ -x-z+3,\ x-y), (y^2z - yz^2 - y^2 + yz,\ 0,\ -yz+y,\ 0,\ 0,\ -yz+z^2+y-z,\ -yz+y+2z-2,\ 0,\ yz-y-2), (xz+y^2-2yz+y-z,\ 0,\ -y+z-1,\ z,\ -z,\ -y+2z-1,\ -y+z+1,\ -z,\ y+1), (0,\ 0,\ x-z,\ y-z,\ -y,\ x-1,\ x-z+2,\ -y+z-2,\ -x+y)\rangle$, from which we read off that

$$I_1 \cap I_2 \cap I_3 = \langle xy-y^2, x^2-y^2-x+y, y^2z-yz^2-y^2+yz, xz+y^2-2yz+y-z\rangle.$$

We note that this is not a Gröbner basis for $I_1 \cap I_2 \cap I_3$.

We next consider the case of the intersection of two submodules M, N of A^m. Assume that $M = \langle \boldsymbol{f}_1, \ldots, \boldsymbol{f}_s\rangle$ and $N = \langle \boldsymbol{g}_1, \ldots, \boldsymbol{g}_t\rangle$. Then a vector $\boldsymbol{h} \in A^m$ is in $M \cap N$ if and only if

$$\boldsymbol{h} = a_1\boldsymbol{f}_1 + a_2\boldsymbol{f}_2 + \cdots + a_s\boldsymbol{f}_s \text{ and } \boldsymbol{h} = b_1\boldsymbol{g}_1 + b_2\boldsymbol{g}_2 + \cdots + b_t\boldsymbol{g}_t,$$

for some polynomials $a_1, \ldots, a_s, b_1, \ldots, b_t \in A$. This is the same as the two conditions that $(-\boldsymbol{h}, a_1, \ldots, a_s)$ is a syzygy of the matrix $\left[\begin{array}{c} I_m|\boldsymbol{f}_1, \ldots, \boldsymbol{f}_s \end{array}\right]$ and $(-\boldsymbol{h}, b_1, \ldots, b_t)$ is a syzygy of the matrix $\left[\begin{array}{c} I_m|\boldsymbol{g}_1, \ldots, \boldsymbol{g}_t \end{array}\right]$ (here, I_m denotes the $m \times m$ identity matrix, and $(-\boldsymbol{h}, a_1, \ldots, a_s)$ is the vector in A^{m+s} whose first

m coordinates are those of $-\boldsymbol{h}$). It is then easy to put these two conditions into a single condition for syzygies of vectors in A^{m+s+t}. Let

$$J = \begin{bmatrix} I_m \\ I_m \end{bmatrix}$$

(so that[7] $J = {}^t\left[\ I_m | I_m\ \right]$), and let F be the $m \times s$ matrix $F = \left[\ \boldsymbol{f}_1 \ \cdots \ \boldsymbol{f}_s\ \right]$ and G be the $m \times t$ matrix $G = \left[\ \boldsymbol{g}_1 \ \cdots \ \boldsymbol{g}_t\ \right]$. Set $H = \left[\ J | F \oplus G\ \right]$. Then, as in Proposition 3.8.1, we see that the set of vectors $\boldsymbol{h}$ which are the first m coordinates of vectors in $\operatorname{Syz}(H)$ is $M \cap N$. Moreover, the set of vectors which consist of the first m coordinates of each of the vectors of a set of generators of $\operatorname{Syz}(H)$ is a generating set for $M \cap N$.

EXAMPLE 3.8.6. We again let $A = \mathbb{Q}[x, y, z]$ and consider the term order deglex with $x > y > z$. Let $M = \langle (x-y, z), (x, y) \rangle$, and $N = \langle (x+1, y), (x-1, z) \rangle$. So we set

$$H = \begin{bmatrix} 1 & 0 & x-y & x & 0 & 0 \\ 0 & 1 & z & y & 0 & 0 \\ 1 & 0 & 0 & 0 & x+1 & x-1 \\ 0 & 1 & 0 & 0 & y & z \end{bmatrix}.$$

Then, using the TOP ordering on A^4 with $\boldsymbol{e}_1 > \boldsymbol{e}_2 > \boldsymbol{e}_3 > \boldsymbol{e}_4$, we obtain $\operatorname{Syz}(H) = \langle (xy - 2x + y,\ y^2 - y - z,\ 1,\ -y+1,\ -y+1,\ 1), (x^2z - 2x^2 + xz + 4x - 2y,\ xyz - xy - xz - y^2 + yz + 2y + 2z,\ x-2,\ -xz + x + y - z - 2,\ -xz + x + y - 2,\ x - y - 2) \rangle$, from which we read off that

$$M \cap N = \langle (xy - 2x + y, y^2 - y - z),$$
$$(x^2z - 2x^2 + xz + 4x - 2y, xyz - xy - xz - y^2 + yz + 2y + 2z) \rangle.$$

In general, if we have r submodules M_i generated by the columns of the $m \times t_i$ matrices F_i $(1 \le i \le r)$ and

$$H = \left[\ {}^t\underbrace{\left[\ I_m | I_m | \cdots | I_m\ \right]}_{r \text{ copies}} | F_1 \oplus F_2 \oplus \cdots \oplus F_r\ \right],$$

where again I_m denotes the $m \times m$ identity matrix, then the set of vectors $\boldsymbol{h}$ which are the first m coordinates of vectors in $\operatorname{Syz}(H)$ is the intersection of the M_i's. Moreover, as in the case of the intersection of two modules, the set of vectors which consists of the first m coordinates of each of the vectors of a set of generators for $\operatorname{Syz}(H)$ is a generating set for the intersection of the r modules M_i.

We now turn to the ideal quotient. This can be computed as a special case of the computation of intersection of ideals as we did in Section 2.3. On the other hand it is easy to recognize it directly as a syzygy computation. So let $I = \langle f_1, \ldots, f_s \rangle$ and $J = \langle g_1, \ldots, g_t \rangle$ be ideals in A. Recall that $I \colon J = \{h \in$

[7]For a matrix S, we denote its transpose by ${}^tS$.

$A \mid hJ \subseteq I\} = \{h \in A \mid hg_i, i = 1, \dots, t,$ is a linear combination of $f_1, \dots, f_s\}$. Let $\boldsymbol{g}$ be the column vector $\boldsymbol{g} = (g_1, \dots, g_t)$ and let F be the row vector $F = \left[\begin{array}{ccc} f_1 & \cdots & f_s \end{array}\right]$. In this case, we set

$$H = \left[\begin{array}{c} \boldsymbol{g} | \underbrace{F \oplus F \oplus \cdots \oplus F}_{t \text{ copies}} \end{array}\right].$$

Then $I \colon J$ is the set of all first coordinates of $\operatorname{Syz}(H)$.

EXAMPLE 3.8.7. We let $A = \mathbb{Q}[x, y]$ and consider the term order lex with $x > y$. Let $I = \langle x(x+y)^2, y \rangle$ and $J = \langle x^2, x+y \rangle$. Now consider

$$H = \left[\begin{array}{ccccc} x^2 & x(x+y)^2 & y & 0 & 0 \\ x+y & 0 & 0 & x(x+y)^2 & y \end{array}\right].$$

Then using the TOP ordering on A^2 with $\boldsymbol{e}_1 > \boldsymbol{e}_2$, we compute $\operatorname{Syz}(H) = \langle (-y,$ $0,\ x^2,\ 0,\ x+y),\ (-xy - 2y^2,\ y,\ -xy^2,\ 0,\ x^2 + 3xy + 2y^2),\ (x^2 + xy - y^2,\ -x + y,$ $-xy^2,\ -1,\ xy + y^2) \rangle$, from which we read off that

$$I \colon J = \langle -y, -xy - 2y^2, x^2 + xy - y^2 \rangle.$$

Computing a Gröbner basis for this ideal (or by simply looking at it), we obtain $I \colon J = \langle y, x^2 \rangle$, in agreement with the computation of the same ideal in Example 2.3.12.

We will leave the computation of the ideal quotient of two submodules of A^m (Definition 3.6.10) to the exercises (Exercise 3.8.5).

We now consider modules which are given by a presentation as defined in Section 3.1. Let $\boldsymbol{f}_1, \dots, \boldsymbol{f}_s$ be in A^m, let $N = \langle \boldsymbol{f}_1, \dots, \boldsymbol{f}_s \rangle$ and consider $M = A^m/N$. We wish to show how to compute the intersection of two (or more) submodules of M, find the ideal quotient of two submodules of M and find the annihilator of M.

We begin with the latter as it is a special case of what we have already done. We define the *annihilator* of M to be the ideal

$$\operatorname{ann}(M) = \{h \in A \mid hM = \{0\}\}.$$

Since $M = A^m/N$ we see that $\operatorname{ann}(M) = N \colon A^m$. Although we have relegated the computation of the ideal quotient of two modules to the exercises, in this special case it is easy to write down the correct matrix whose syzygies allow us to compute $\operatorname{ann}(M)$, and we will do so now. We simply observe that A^m is generated by the usual standard basis $\boldsymbol{e}_1, \dots, \boldsymbol{e}_m$ and so $h \in \operatorname{ann}(M)$ if and only if $h\boldsymbol{e}_i \in N$ for all $i = 1, \dots, m$. Thus letting $F = \left[\begin{array}{ccc} \boldsymbol{f}_1 & \cdots & \boldsymbol{f}_s \end{array}\right]$ (an $m \times s$ matrix), and

$$H = \left[\begin{array}{c} {}^t\left[\begin{array}{c} {}^t\boldsymbol{e}_1 | \cdots | {}^t\boldsymbol{e}_m \end{array}\right] | \underbrace{F \oplus F \oplus \cdots \oplus F}_{m \text{ copies}} \end{array}\right],$$

we see that $\operatorname{ann}(M)$ is the set of all first coordinates of $\operatorname{Syz}(H)$. (We note that the matrix ${}^t\left[\ {}^t\mathbf{e}_1 | \cdots | {}^t\mathbf{e}_m\ \right]$ is the $m^2 \times 1$ matrix of the vectors $\mathbf{e}_i$ stacked up on top of each other.)

EXAMPLE 3.8.8. We consider $N = \langle (x^2 + y, xz - y), (xy - yz, z - x) \rangle \subset A^2$ and let $M = A^2/N$. So in this case

$$H = \begin{bmatrix} 1 & x^2+y & xy-yz & 0 & 0 \\ 0 & xz-y & z-x & 0 & 0 \\ 0 & 0 & 0 & x^2+y & xy-yz \\ 1 & 0 & 0 & xz-y & z-x \end{bmatrix}.$$

Then using TOP with $\mathbf{e}_1 > \mathbf{e}_2 > \mathbf{e}_3 > \mathbf{e}_4$ and degrevlex with $x > y > z$, we see that $\operatorname{Syz}(H) = \langle (x^2yz - xyz^2 + x^3 - xy^2 - x^2z + y^2z + xy - yz, -x + z, -xz + y, -xy + yz, x^2 + y) \rangle$, from which we conclude that

$$\operatorname{ann}(M) = \langle x^2yz - xyz^2 + x^3 - xy^2 - x^2z + y^2z + xy - yz \rangle.$$

We now consider two submodules M_1, M_2 of $M = A^m/N$ and we wish to compute their intersection. There are submodules K_1, K_2 of A^m such that $M_1 = K_1/N$ and $M_2 = K_2/N$. We note that, $K_1 = \{h \in A^m \mid h + N \in M_1\}$, or alternatively if $M_1 = \langle \mathbf{g}_1 + N, \ldots, \mathbf{g}_t + N \rangle$ then

$$K_1 = \langle \mathbf{g}_1, \ldots, \mathbf{g}_t \rangle + N = \langle \mathbf{g}_1, \ldots, \mathbf{g}_t, \mathbf{f}_1, \ldots, \mathbf{f}_s \rangle.$$

We wish to compute $M_1 \cap M_2$. Clearly $M_1 \cap M_2 = (K_1 \cap K_2)/N$. Since we know how to compute $K_1 \cap K_2$ we can compute $M_1 \cap M_2$. Alternatively, continuing to use the notation above, and setting $M_2 = \langle \mathbf{h}_1 + N, \ldots, \mathbf{h}_r + N \rangle$, we see that a coset $\mathbf{p} + N \in M_1 \cap M_2$ if and only if there are $a_1, \ldots, a_t \in A$ and $b_1, \ldots, b_r \in A$ such that

$$\mathbf{p} + N = \sum_{i=1}^{t} a_i(\mathbf{g}_i + N) \text{ and } \mathbf{p} + N = \sum_{i=1}^{r} b_i(\mathbf{h}_i + N).$$

This last statement is equivalent to

$$\mathbf{p} = \sum_{i=1}^{t} a_i\mathbf{g}_i + \sum_{j=1}^{s} a'_j\mathbf{f}_j \text{ and } \mathbf{p} = \sum_{i=1}^{r} b_i\mathbf{h}_i + \sum_{j=1}^{s} b'_j\mathbf{f}_j,$$

for some $a_1, \ldots, a_t, a'_1, \ldots, a'_s, b_1, \ldots, b_r, b'_1, \ldots b'_s \in A$. Thus, setting

$$F = \left[\ \mathbf{f}_1 \ \cdots \ \mathbf{f}_s\ \right], G = \left[\ \mathbf{g}_1 \ \cdots \ \mathbf{g}_t\ \right], H = \left[\ \mathbf{h}_1 \ \cdots \ \mathbf{h}_r\ \right]$$

and I_m the $m \times m$ identity matrix, we see that the set of all such $\mathbf{p}$'s is just the set of first m coordinates of $\operatorname{Syz}(S)$, where

$$S = \left[\ {}^t\left[\ I_m | I_m\ \right] \middle| \left[\ G|F\ \right] \oplus \left[\ H|F\ \right]\ \right].$$

The same sort idea will allow one to construct $M_1 \colon M_2$ (Exercise 3.8.8).

We now go on to consider the following question. Let $N \subset M$ be submodules of A^m. We would like to determine a presentation of M/N (see Definition 3.1.4).

So assume that $M = \langle \boldsymbol{f}_1, \ldots, \boldsymbol{f}_s \rangle$ and $N = \langle \boldsymbol{g}_1, \ldots, \boldsymbol{g}_t \rangle$, where we do not assume that either generating set is a Gröbner basis. We define an A-module homomorphism

$$\begin{array}{rccc} \phi: & A^s & \longrightarrow & M/N \\ & \boldsymbol{e}_i & \longmapsto & \boldsymbol{f}_i + N \end{array}$$

(for $1 \le i \le s$). It is clear that ϕ maps A^s onto M/N since $M = \langle \boldsymbol{f}_1, \ldots, \boldsymbol{f}_s \rangle$. Let $K = \ker(\phi)$. Then the desired presentation is $A^s/K \cong M/N$. So we need to compute an explicit set of generators for K. We note that $\boldsymbol{h} = (h_1, \ldots, h_s)$ is in K if and only if $h_1 \boldsymbol{f}_1 + \cdots + h_s \boldsymbol{f}_s$ is in N, which, in turn, is true if and only if there are polynomials $a_1, \ldots, a_t \in A$ such that

$$h_1 \boldsymbol{f}_1 + \cdots + h_s \boldsymbol{f}_s = a_1 \boldsymbol{g}_1 + \cdots + a_t \boldsymbol{g}_t.$$

Let

$$H = \left[\begin{array}{cccccc} \boldsymbol{f}_1 & \cdots & \boldsymbol{f}_s & \boldsymbol{g}_1 & \cdots & \boldsymbol{g}_t \end{array} \right].$$

We have proved

THEOREM 3.8.9. *With the notation above, let* $\boldsymbol{p}_1, \ldots, \boldsymbol{p}_r \in A^{s+t}$ *be a generating set for* $\mathrm{Syz}(H)$, *and let* $\boldsymbol{h}_i \in k[x_1, \ldots, x_n]^s$ *denote the vector whose coordinates are the first s coordinates of* $\boldsymbol{p}_i$, $1 \le i \le r$. *Then*

$$K = \langle \boldsymbol{h}_1, \ldots, \boldsymbol{h}_r \rangle.$$

EXAMPLE 3.8.10. Let $A = \mathbb{Q}[x,y]$, $M = \langle (xy, y), (-y, x), (x^2, x) \rangle \subset A^2$ and $N = \langle (-x^3, y), (0, y + x^2) \rangle$. Then, since

$$\begin{array}{rcl} (-x^3, y) & = & (xy, y) + x(-y, x) - x(x^2, x) \text{ and} \\ (0, y + x^2) & = & (xy, y) + x(-y, x), \end{array}$$

we see that $N \subset M$. So to determine $K \subset A^3$ such that $M/N \cong A^3/K$, we set

$$H = \left[\begin{array}{ccccc} xy & -y & x^2 & -x^3 & 0 \\ y & x & x & y & y + x^2 \end{array} \right]$$

and compute

$$\mathrm{Syz}(H) = \langle (0, 0, x, 1, -1), (1, x, 0, 0, -1), (-x, 0, y, 0, 0) \rangle$$

(this last computation was done using TOP with $\boldsymbol{e}_1 > \boldsymbol{e}_2$ and lex with $y > x$). Thus

$$K = \langle (0, 0, x), (1, x, 0), (-x, 0, y) \rangle.$$

Therefore $M/N \cong A^3/K$.

We close this section by generalizing this last result to the case of an affine algebra. In this case we let $A = k[x_1, \ldots, x_n]/I$, for an ideal $I = \langle d_1, \ldots, d_\ell \rangle \subset A$. We again consider $N \subset M$ to be submodules of A^m, but now, of course, the coordinates of the elements of A^m are cosets of I. So, for a vector $\boldsymbol{b} = (b_1, \ldots, b_m) \in k[x_1, \ldots, x_n]^m$, we set $\overline{\boldsymbol{b}} = (b_1 + I, \ldots, b_m + I) \in A^m$. So we may assume that $N = \langle \overline{\boldsymbol{g}}_1, \ldots, \overline{\boldsymbol{g}}_t \rangle$ for $\boldsymbol{g}_1, \ldots, \boldsymbol{g}_t \in k[x_1, \ldots, x_n]^m$ and $M = \langle \overline{\boldsymbol{f}}_1, \ldots, \overline{\boldsymbol{f}}_s \rangle$ for $\boldsymbol{f}_1, \ldots, \boldsymbol{f}_s \in k[x_1, \ldots, x_n]^m$. We define an A-module homomorphism

$$\begin{array}{rccc} \phi: & A^s & \longrightarrow & M/N \\ & \overline{\boldsymbol{e}}_i & \longmapsto & \overline{\boldsymbol{f}}_i + N \end{array}$$

(for $1 \leq i \leq s$). It is clear that ϕ maps A^s onto M/N since $M = \langle \overline{\boldsymbol{f}}_1, \ldots, \overline{\boldsymbol{f}}_s \rangle$. Let $K = \ker(\phi)$. Then the desired presentation is $A^s/K \cong M/N$. So we need to compute an explicit set of generators of K. We note that $\overline{\boldsymbol{h}} = (h_1 + I, \ldots, h_s + I)$ is in K if and only if $(h_1 + I)\overline{\boldsymbol{f}}_1 + \cdots + (h_s + I)\overline{\boldsymbol{f}}_s$ is in N, which is true if and only if there are polynomials $a_1, \ldots, a_t \in k[x_1, \ldots, x_n]$ such that

$$(h_1 + I)\overline{\boldsymbol{f}}_1 + \cdots + (h_s + I)\overline{\boldsymbol{f}}_s = (a_1 + I)\overline{\boldsymbol{g}}_1 + \cdots + (a_t + I)\overline{\boldsymbol{g}}_t.$$

This last statement is readily seen to be equivalent to the statement that every coordinate of

$$h_1\boldsymbol{f}_1 + \cdots + h_s\boldsymbol{f}_s - (a_1\boldsymbol{g}_1 + \cdots + a_t\boldsymbol{g}_t)$$

is an element of I. Thus in this case, we let $F = \left[\begin{array}{ccc} \boldsymbol{f}_1 & \cdots & \boldsymbol{f}_s \end{array} \right]$, $G = \left[\begin{array}{ccc} \boldsymbol{g}_1 & \cdots & \boldsymbol{g}_t \end{array} \right]$, and $D = \left[\begin{array}{ccc} d_1 & \cdots & d_\ell \end{array} \right]$ and set

$$H = \left[\; F|G|\underbrace{D \oplus D \oplus \cdots \oplus D}_{m \text{ copies}} \; \right].$$

We have proved

THEOREM 3.8.11. *With the notation above, let*

$$\boldsymbol{p}_1, \ldots, \boldsymbol{p}_r \in k[x_1, \ldots, x_n]^{s+t+m\ell}$$

be a generating set of $\operatorname{Syz}(H)$, *and let* $\boldsymbol{h}_i \in k[x_1, \ldots, x_n]^s$ *denote the vector whose coordinates are the first* s *coordinates of* $\boldsymbol{p}_i$ $(1 \leq i \leq r)$. *Then*

$$K = \langle \overline{\boldsymbol{h}}_1, \ldots, \overline{\boldsymbol{h}}_r \rangle.$$

We note that all the computations that we have done in this section could be formulated using Theorem 3.8.11 (see Exercise 3.8.10).

EXAMPLE 3.8.12. We will redo Example 3.8.10, except that we will now assume that the ring is $A = \mathbb{Q}[x, y]/I$ where $I = \langle x^3 + y^2, x^2 \rangle$. So now $M = \langle (\overline{xy}, \overline{y}), (-\overline{y}, \overline{x}), (\overline{x}^2, \overline{x}) \rangle \subset A^2$ and $N = \langle (-\overline{x}^3, \overline{y}), (\overline{0}, \overline{y} + \overline{x}^2) \rangle$. We have that $N \subset M$ as before. So to determine $K \subset A^3$ such that $M/N \cong A^3/K$, we set

$$H = \left[\begin{array}{ccccccccc} xy & -y & x^2 & -x^3 & 0 & x^3 + y^2 & x^2 & 0 & 0 \\ y & x & x & y & y + x^2 & 0 & 0 & x^3 + y^2 & x^2 \end{array} \right]$$

and compute Syz(H) to be generated by (0, 0, x, 1, -1, 0, 0, 0, 0), (1, x, 0, 0, -1, 0, 0, 0, 0), (0, 0, 0, 1, -1, 0, x, 0, 1), ($-x$, 0, 0, 0, x, 0, y, 0, $-x$), (0, y, 0, 1, $-x-1$, 1, 0, 0, $x+1$), ($-x$, 0, y, 0, 0, 0, 0, 0, 0), and (0, 0, 0, 0, $y-x^2$, 0, 0, -1, x^2+x) (this last computation using TOP with $e_1 > e_2$ and lex with $y > x$). Thus

$$K = \langle(\overline{0}, \overline{0}, \overline{x}), (\overline{1}, \overline{x}, \overline{0}), (-\overline{x}, \overline{0}, \overline{0}), (\overline{0}, \overline{y}, \overline{0}), (-\overline{x}, \overline{0}, \overline{y},)\rangle.$$

Exercises

3.8.1. Redo the computation of the intersection of the ideals given in Example 2.3.6, Exercises 2.3.6 and 2.3.7 using the technique of syzygies presented in this section.

3.8.2. Redo the computation of the intersection of the modules given in Example 3.6.9, Exercises 3.6.11 and 3.6.12 using the technique of syzygies presented in this section.

3.8.3. In Theorem 3.8.3 we required that the order be POT. In this exercise, we show that the TOP ordering would not give rise to a Gröbner basis for the intersection of two ideals. Consider the ideals $I = \langle x^2, yx+x, xy^2+y\rangle$ and $J = \langle y^2, x-xy, x^2-y\rangle$ of $\mathbb{Q}[x, y]$. Compute generators for $I \cap J$ by computing generators for the syzygy module of

$$H = \begin{bmatrix} 1 & x^2 & xy+x & xy^2+y & 0 & 0 & 0 \\ 1 & 0 & 0 & 0 & y^2 & x-xy & x^2-y \end{bmatrix}$$

using the TOP ordering on $(\mathbb{Q}[x, y])^2$ with $e_1 > e_2$ and deglex on $\mathbb{Q}[x, y]$ with $y > x$. Show that these generators do not form a Gröbner basis for $I \cap J$ with respect to deglex with $y > x$. One might think that the problem comes from the fact that the generators for I and J above did not form a Gröbner basis with respect to deglex with $y > x$ to start with. This is not the case as is easily seen by repeating the exercise with the ideals $I = \langle x+y, x^2+1\rangle$ and $J = \langle y^2, x^2-y\rangle$ using the same orders as above.

3.8.4. Redo the computation of the ideal quotient given in Example 2.3.12 and Exercise 2.3.11 using the technique of syzygies presented in this section.

3.8.5. Give the analog of the matrix H used to compute generators for $I\colon J$, where I and J are ideals, in the module case. That is, find H such that $M\colon N$ can be obtained from Syz(H), where M and N are modules. Redo the computation of the ideal quotient of the modules given in Example 3.6.13 and Exercise 3.6.15 using this technique.

3.8.6. Consider the submodule N of $(\mathbb{Q}[x, y])^3$,

$$N = \langle(xy-y, x, y+x), (xy-x, x, xy+x), (x+y, y, xy+y)\rangle.$$

Compute generators for $\operatorname{ann}((\mathbb{Q}[x, y])^3/N)$.

3.8.7. Consider the following two submodules of $(\mathbb{Q}[x, y])^3$,

$$K_1 = \langle(xy, x, y), (-y, x, x), (x^2, x, y)\rangle \text{ and}$$

$$K_2 = \langle (0, y^2+y, y^2+y), (0, 0, x^2-2xy-x+y^2+y), (0, x-y, x-y), (y, y, y)\rangle.$$

a. Prove that $N = \langle (y^2+y, 0, x^2-2xy+y^2-x+y), (0, xy+x, y^2-xy+x^2+y)\rangle$ is a submodule of K_1 and K_2.

b. Compute generators for the module $(K_1/N) \cap (K_2/N) \subseteq (\mathbb{Q}[x,y])^3/N$.

3.8.8. Give the analog of the matrix H used to compute generators for $(K_1/N) \cap (K_2/N)$ which can be used to compute generators for $(K_1/N)\colon (K_2/N)$. Compute generators for the ideal quotient $(K_1/N)\colon (K_2/N)$, where K_1, K_2 and N are the modules of Exercise 3.8.7.

3.8.9. Find the presentation of K_2/N, where K_2 and N are as in Exercise 3.8.7.

3.8.10. Follow the construction of Theorem 3.8.11 to generalize all the computations performed in this section to the case where A is an affine ring. Namely, for $A = k[x_1, \ldots, x_n]/I$, where $I = \langle d_1, \ldots, d_\ell\rangle$, give the analog of the matrix H that is used to compute generators for the intersection of two ideals in A, the intersection of two submodules of A^m, the ideal quotient of two submodules of A^m, the annihilator of $M = A^m/N$, the intersection of submodules of A^m/N, and the ideal quotient of two submodules of A^m/N.

3.8.11. Let $I = \langle f_1, \ldots, f_s\rangle \subseteq k[x_1, \ldots, x_n]$ be an ideal. Show that $f + I$ has an inverse in $k[x_1, \ldots, x_n]/I$ if and only if there is an element in the reduced Gröbner basis for $\mathrm{Syz}(1, f_1, \ldots, f_s, f)$, with respect to any POT order with the first coordinate largest, of the form $(1, h_1, \ldots, h_s, h)$. Show that, in this case, $h + I$ is the inverse of $f + I$. Redo the computations in Example 2.1.10 and Exercise 2.1.6 using this technique.

3.8.12. All the computations done in this section can be done using only Theorem 3.8.9. That is, all the computations done in this section can be viewed as the computations of kernels of certain A-module homomorphisms, where $A = k[x_1, \ldots, x_n]$. We illustrate this in the present exercise.
Consider the following diagram of free modules,

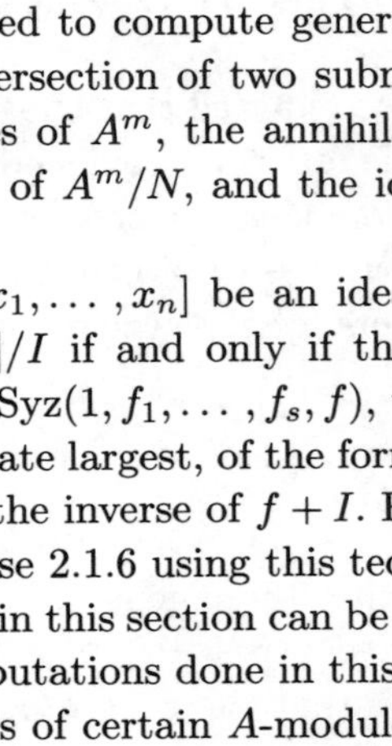

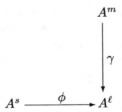

We define the *pullback* of this diagram to be the submodule PB of $A^s \oplus A^m$ defined as follows

$$PB = \{(\boldsymbol{g}, \boldsymbol{h}) \in A^s \oplus A^m \mid \phi(\boldsymbol{g}) = \gamma(\boldsymbol{h})\}.$$

We get a commutative diagram

$$\begin{array}{ccc} PB & \xrightarrow{\pi_m} & A^m \\ {\scriptstyle \pi_s}\downarrow & & \downarrow{\scriptstyle \gamma} \\ A^s & \xrightarrow{\phi} & A^\ell \end{array}$$

where π_m and π_s are the compositions of the inclusion $PB \longrightarrow A^s \oplus A^m$ with the projection onto A^m and A^s respectively.

a. Show how to compute generators for PB.

b. Show how to compute generators for the intersection of two submodules of a free module using the pullback.

c. Show how to compute generators for the annihilator of an element of A^s/N using the pullback, where N is a submodule of A^s. Then show how to compute generators for $\operatorname{ann}(A^s/N)$.

d. Show how to compute generators for $M: N$, where M and N are two submodules of A^m.

3.9. Computation of Hom. As before we let $A = k[x_1, \ldots, x_n]$. Also, we consider two A-modules M and N. We are interested in the study of the set of all A-module homomorphisms between M and N. We define

$$\operatorname{Hom}(M, N) = \{\phi \colon M \longrightarrow N \mid \phi \text{ is an } A\text{-module homomorphism }\}.$$

We define the usual addition of two elements ϕ and ψ in $\operatorname{Hom}(M, N)$:

$$(\phi + \psi)(\boldsymbol{m}) = \phi(\boldsymbol{m}) + \psi(\boldsymbol{m}) \text{ for all } \boldsymbol{m} \in M.$$

It is easily verified that $\operatorname{Hom}(M, N)$ is an abelian group under this addition. We can also define multiplication of elements in $\operatorname{Hom}(M, N)$ by elements of A: for $a \in A$ and $\phi \in \operatorname{Hom}(M, N)$, we define $a\phi$ by

$$(a\phi)(\boldsymbol{m}) = a(\phi(\boldsymbol{m})) = \phi(a\boldsymbol{m}) \text{ for all } \boldsymbol{m} \in M.$$

Again, it is easily verified that, with this multiplication, $\operatorname{Hom}(M, N)$ is an A-module.

Given two finitely generated A-modules M and N, we wish to "describe" $\operatorname{Hom}(M, N)$. As mentioned earlier, there are two ways one can "describe" an A-module V (see the discussion following Lemma 3.1.3): by computing a presentation of V, or, if V is a submodule of some quotient A^m/U, by giving generators (elements of A^m/U) for V. We will first give generators for $\operatorname{Hom}(M, N)$ as a submodule of some quotient A^m/U. We will then obtain a presentation for $\operatorname{Hom}(M, N)$ using Theorem 3.8.9.

Since we will need it later on, we first consider the easy case of computing $\operatorname{Hom}(A^s, A^t)$. Let $\boldsymbol{e}_i, i = 1, \ldots, s$ be the standard basis for A^s. An element ϕ of $\operatorname{Hom}(A^s, A^t)$ is uniquely determined by the images $\phi(\boldsymbol{e}_i) \in A^t, i = 1, \ldots, s$.

Then, as in linear algebra, ϕ is given by matrix multiplication by the matrix whose columns are the $\phi(\mathbf{e}_i)$'s. We will denote this matrix by ϕ also. Note that ϕ is a $t \times s$ matrix. We further identify the matrix ϕ in $\text{Hom}(A^s, A^t)$ with the column vector ϕ^* in A^{st} formed by concatenating the columns of ϕ in order. This identification gives an explicit isomorphism $\text{Hom}(A^s, A^t) \cong A^{st}$, since it is readily verified that it gives a one to one and onto A-module homomorphism

$$\text{Hom}(A^s, A^t) \longrightarrow A^{st}.$$

EXAMPLE 3.9.1. As an example of the above identification, we identify the matrix $\begin{bmatrix} f_1 & f_3 & f_5 \\ f_2 & f_4 & f_6 \end{bmatrix}$ with the column vector[8] $(f_1, f_2, f_3, f_4, f_5, f_6)$. Or, as another example, suppose we have

$$\phi\colon \mathbb{Q}[x]^2 \longrightarrow \mathbb{Q}[x]^3$$

given by $\phi(f_1, f_2) = ((2x+1)f_1 - x^2 f_2, x f_1 + f_2, (x^2 - x) f_2)$. Then it is readily verified that ϕ is a $\mathbb{Q}[x]$-module homomorphism. Moreover, $\phi(1, 0) = (2x+1, x, 0)$ and $\phi(0, 1) = (-x^2, 1, x^2 - x)$. Hence the matrix associated with ϕ is

$$\begin{bmatrix} 2x+1 & -x^2 \\ x & 1 \\ 0 & x^2 - x \end{bmatrix}$$

and so the vector in $\mathbb{Q}[x]^6$ associated with ϕ is $(2x+1, x, 0, -x^2, 1, x^2 - x)$.

Now, in order to describe $\text{Hom}(M, N)$ explicitly, we need to assume that we are given M and N explicitly. That is, we assume that we are given presentations of M and N, say

$$M \cong A^s/L \qquad \text{and} \qquad N \cong A^t/K.$$

Then the idea is to compute $\text{Hom}(A^s/L, A^t/K)$ by adapting the ideas in the computation of $\text{Hom}(A^s, A^t)$ above.

To do this, suppose that we have a presentation of the A-module M, $M \cong A^s/L$. Then we have a sequence of A-module homomorphisms

$$0 \longrightarrow L \xrightarrow{\,i\,} A^s \xrightarrow{\,\pi\,} M \longrightarrow 0, \tag{3.9.1}$$

where the map $i\colon L \longrightarrow A^s$ is the inclusion map, and $\pi\colon A^s \longrightarrow M$ is the map which sends the standard basis of A^s onto the generating set of M corresponding to the standard basis in the isomorphism $M \cong A^s/L$. Sequence (3.9.1) is called a *short exact sequence*. In general a sequence of A-modules and A-module homomorphisms

$$\cdots \longrightarrow N_{i+1} \xrightarrow{\alpha_{i+1}} N_i \xrightarrow{\alpha_i} N_{i-1} \longrightarrow \cdots$$

[8] Recall our space saving notation for column vectors!

is called *exact* if $\text{im}(\alpha_{i+1}) = \ker(\alpha_i)$ for each i. It is easy to check that Sequence (3.9.1) is exact, which, in this case, means that i is one to one, $\ker(\pi) = \text{im}(i)$, and π is onto.

Now we find a presentation of L, say

$$L \cong A^{s_1}/L_1,$$

and thus we have another short exact sequence

$$0 \longrightarrow L_1 \xrightarrow{i_1} A^{s_1} \xrightarrow{\pi_1} L \longrightarrow 0.$$

This leads to an exact sequence

$$0 \longrightarrow L_1 \xrightarrow{i_1} A^{s_1} \xrightarrow{\Gamma} A^s \xrightarrow{\pi} M \longrightarrow 0,$$

where $\Gamma = i \circ \pi_1$ (the exactness is easily checked). We will only be interested in the exact sequence

$$A^{s_1} \xrightarrow{\Gamma} A^s \xrightarrow{\pi} M \longrightarrow 0.$$

We use this last sequence, and a similar one for N to compute $\text{Hom}(M, N)$. In order to do this we review the elementary homological properties of Hom. What we need is summarized in Lemma 3.9.2. We refer the reader to [**Hun**].

We assume we are given A-modules M_1, M_2, and N and an A-module homomorphism ϕ: $M_1 \longrightarrow M_2$. We define two maps

$$\begin{array}{ccc} \text{Hom}(M_2, N) & \xrightarrow{\circ\phi} & \text{Hom}(M_1, N) \\ \psi & \longmapsto & \psi \circ \phi \end{array} \quad \text{and} \quad \begin{array}{ccc} \text{Hom}(N, M_1) & \xrightarrow{\phi_\circ} & \text{Hom}(N, M_2) \\ \rho & \longmapsto & \phi \circ \rho. \end{array}$$

It is easy to verify that ${}_\circ\phi$ and $\phi_\circ$ are A-module homomorphisms. For example, if $M_1 = A^{s_1}$, $M_2 = A^{s_2}$, and $N = A^t$, and the maps are given by matrices, then ${}_\circ\phi$ and $\phi_\circ$ are given by matrix multiplication, e.g. $\psi \xrightarrow{\circ\phi} \psi\phi$.

LEMMA 3.9.2. *Assume that we have an exact sequence of A-modules and A-module homomorphisms*

$$M_1 \xrightarrow{\phi} M_2 \xrightarrow{\psi} M_3 \longrightarrow 0. \tag{3.9.2}$$

Let P be another A-module. Then the two sequences

$$0 \longrightarrow \text{Hom}(M_3, P) \xrightarrow{\circ\psi} \text{Hom}(M_2, P) \xrightarrow{\circ\phi} \text{Hom}(M_1, P) \tag{3.9.3}$$

and

$$\text{Hom}(P, M_1) \xrightarrow{\phi_\circ} \text{Hom}(P, M_2) \xrightarrow{\psi_\circ} \text{Hom}(P, M_3) \longrightarrow 0 \tag{3.9.4}$$

are exact. In particular,

$$\text{Hom}(M_3, P) \cong \ker({}_\circ\phi) \tag{3.9.5}$$

and

$$\text{Hom}(P, M_3) \cong \text{Hom}(P, M_2)/\ker(\psi_\circ) = \text{Hom}(P, M_2)/\text{im}(\phi_\circ). \tag{3.9.6}$$

We now return to the computation of $\operatorname{Hom}(M, N)$ for two A-modules M and N. From now on, to ease notation, we will identify M and N with their respective presentations, that is, we will assume that $M = A^s/L$ and $N = A^t/K$. (However, if M and N are not given initially by presentations, we will have to take into account the isomorphisms between M (resp. N) and A^s/L (resp. A^t/K).) So, as noted above, we have the two exact sequences

$$A^{s_1} \xrightarrow{\Gamma} A^s \xrightarrow{\pi} M \longrightarrow 0 \tag{3.9.7}$$

$$A^{t_1} \xrightarrow{\Delta} A^t \xrightarrow{\pi'} N \longrightarrow 0. \tag{3.9.8}$$

As before, the map Γ (resp. Δ) is given by a matrix which we will also denote Γ (resp. Δ). The columns of Γ (resp. Δ) are easily seen to be the vectors which generate L (resp. K), since the image of Γ (resp. Δ) is L (resp. K). We have that Γ is an $s \times s_1$ matrix and Δ is a $t \times t_1$ matrix.

Now in Sequence (3.9.3) we let $P = N$ and replace Sequence (3.9.2) with Sequence (3.9.7) to obtain the exact sequence

$$0 \longrightarrow \operatorname{Hom}(M, N) \xrightarrow{\alpha} \operatorname{Hom}(A^s, N) \xrightarrow{\gamma} \operatorname{Hom}(A^{s_1}, N), \tag{3.9.9}$$

where $\alpha = _\circ\pi$ and $\gamma = _\circ\Gamma$. Thus from Equation (3.9.5) we see that

$$\operatorname{Hom}(M, N) \cong \ker(\gamma). \tag{3.9.10}$$

Thus to get a presentation of $\operatorname{Hom}(M, N)$ it suffices to obtain a presentation of $\ker(\gamma)$. For this we first compute presentations of $\operatorname{Hom}(A^s, N)$ and $\operatorname{Hom}(A^{s_1}, N)$, then compute the map γ^* corresponding to γ between the two presentations, and then we can compute $\ker(\gamma^*)$ using Theorem 3.8.9.

Now in Sequence (3.9.4) we let $P = A^s$ and $P = A^{s_1}$ respectively, replacing Sequence (3.9.2) with Sequence (3.9.8) to obtain the exact sequences

$$\operatorname{Hom}(A^s, A^{t_1}) \xrightarrow{\delta} \operatorname{Hom}(A^s, A^t) \xrightarrow{\mu} \operatorname{Hom}(A^s, N) \longrightarrow 0, \tag{3.9.11}$$

where $\delta = \Delta_\circ$ and $\mu = \pi'_\circ$ for $P = A^s$ and

$$\operatorname{Hom}(A^{s_1}, A^{t_1}) \xrightarrow{\delta'} \operatorname{Hom}(A^{s_1}, A^t) \xrightarrow{\mu'} \operatorname{Hom}(A^{s_1}, N) \longrightarrow 0, \tag{3.9.12}$$

where $\delta' = \Delta_\circ$ and $\mu' = \pi'_\circ$ for $P = A^{s_1}$. These give, using (3.9.6)

$$\begin{array}{rcll} \operatorname{Hom}(A^s, N) & \cong & \operatorname{Hom}(A^s, A^t)/\operatorname{im}(\delta) & \text{and} \\ \operatorname{Hom}(A^{s_1}, N) & \cong & \operatorname{Hom}(A^{s_1}, A^t)/\operatorname{im}(\delta'). & \end{array} \tag{3.9.13}$$

Since, $\operatorname{Hom}(A^s, A^t) \cong A^{st}$, it now suffices to describe $\operatorname{im}(\delta)$ as a submodule of A^{st} in order to obtain a presentation of $\operatorname{Hom}(A^s, N)$. Of course, a similar statement holds for $\mathrm{Hom}(A^{s_1}, N)$. The map $\delta\colon \mathrm{Hom}(A^s, A^{t_1}) \longrightarrow \mathrm{Hom}(A^s, A^t)$ corresponds to a map δ^* which is given by an $st \times st_1$ matrix S. That is

$$\begin{array}{ccc} A^{st_1} & \xrightarrow{\delta^*} & A^{st} \\ \phi^* & \longmapsto & S\phi^*. \end{array}$$

In order to facilitate the statements of the following results, we will adopt the notation for making new matrices out of old ones presented right after Example 3.8.2.

LEMMA 3.9.3. *The matrix S above is the matrix given by* $\underbrace{\Delta \oplus \cdots \oplus \Delta}_{s \text{ copies}}$.

PROOF. A basis for $\mathrm{Hom}(A^s, A^{t_1})$ is the set of $t_1 \times s$ matrices ψ_{ij} whose entries are all zero except for the ij entry which is equal to one. These matrices correspond to the column vectors $\psi^*_{ij} \in A^{st_1}$ obtained, as before, by concatenating the columns of the matrix ψ_{ij}. The vectors ψ^*_{ij} form the standard basis of $A^{st_1} \cong \mathrm{Hom}(A^s, A^{t_1})$. The columns of the matrix S are then the images of the vectors ψ^*_{ij} under δ^*. That is, the columns of S are the vectors obtained from the matrices $\delta(\psi_{ij}) \in \mathrm{Hom}(A^s, A^t)$ by concatenating the columns of $\delta(\psi_{ij})$. We note that by concatenating the columns the way we do, we have, in effect ordered the ψ^*_{ij}'s (or equivalently the ψ_{ij}'s) as follows:

$$\psi^*_{11}, \psi^*_{21}, \ldots, \psi^*_{t_1 1}, \psi^*_{12}, \ldots, \psi^*_{t_1 2}, \ldots, \psi^*_{t_1 s}.$$

Now for each i and j, $\delta(\psi_{ij}) = \Delta \circ \psi_{ij}$. Because of the identification of the homomorphisms Δ and ψ_{ij} with the corresponding matrices Δ and ψ_{ij} respectively, we have $\delta(\psi_{ij}) = \Delta\psi_{ij}$, where the right-hand side expression is a matrix multiplication. Now the $t \times s$ matrix $\Delta\psi_{ij}$ is the matrix whose j-th column is the i-th column of Δ with all other entries equal to zero. Therefore, after concatenation of columns, the matrices $\Delta\psi_{11}, \Delta\psi_{21}, \ldots, \Delta\psi_{t_1 1}$ correspond to the first t_1 columns of $\underbrace{\Delta \oplus \cdots \oplus \Delta}_{s \text{ copies}}$. Similarly the matrices $\Delta\psi_{12}, \Delta\psi_{22}, \ldots, \Delta\psi_{t_1 2}$ correspond to columns $t_1 + 1$ through $2t_1$ of $\underbrace{\Delta \oplus \cdots \oplus \Delta}_{s \text{ copies}}$. This analysis can be continued and we see that $S = \underbrace{\Delta \oplus \cdots \oplus \Delta}_{s \text{ copies}}$. $\square$

Now the image of δ is the column space of S; that is, the submodule of A^{st} generated by the columns of S (in general, we denote the module generated by the columns of a matrix T by $\langle T \rangle$). Thus we have

$$\mathrm{Hom}(A^s, N) \cong A^{st}/\langle \underbrace{\Delta \oplus \cdots \oplus \Delta}_{s \text{ copies}} \rangle. \tag{3.9.14}$$

Similarly

$$\operatorname{Hom}(A^{s_1}, N) \cong A^{s_1 t}/\langle\underbrace{\Delta \oplus \cdots \oplus \Delta}_{s_1 \text{ copies}}\rangle. \tag{3.9.15}$$

Therefore we have a presentation of $\operatorname{Hom}(A^s, N)$ and of $\operatorname{Hom}(A^{s_1}, N)$, as desired.

We now return to the map

$$\begin{array}{ccc} \operatorname{Hom}(A^s, N) & \xrightarrow{\gamma} & \operatorname{Hom}(A^{s_1}, N) \\ \phi & \longmapsto & \phi \circ \Gamma. \end{array}$$

It corresponds to a map γ^*

$$A^{st}/\langle\underbrace{\Delta \oplus \cdots \oplus \Delta}_{s \text{ copies}}\rangle \xrightarrow{\gamma^*} A^{s_1 t}/\langle\underbrace{\Delta \oplus \cdots \oplus \Delta}_{s_1 \text{ copies}}\rangle.$$

LEMMA 3.9.4. *There is an $s_1 t \times st$ matrix T which defines γ^* in the sense that $\gamma^*(\phi^* + \langle\underbrace{\Delta \oplus \cdots \oplus \Delta}_{s \text{ copies}}\rangle) = T\phi^* + \langle\underbrace{\Delta \oplus \cdots \oplus \Delta}_{s_1 \text{ copies}}\rangle$. Moreover, T is the transpose of the tensor product $\Gamma \otimes I_t$. The matrix $\Gamma \otimes I_t$ is defined to be the $st \times s_1 t$ matrix obtained by replacing each entry γ_{ij} of Γ by the square matrix $\gamma_{ij} I_t$, where I_t is the $t \times t$ identity matrix.*

PROOF. We first define the map

$$\begin{array}{ccc} \operatorname{Hom}(A^s, A^t) & \xrightarrow{\overline{\gamma}} & \operatorname{Hom}(A^{s_1}, A^t) \\ \phi & \longmapsto & \phi \circ \Gamma. \end{array}$$

Using the isomorphisms $A^{st} \cong \operatorname{Hom}(A^s, A^t)$ and $A^{s_1 t} \cong \operatorname{Hom}(A^{s_1}, A^t)$, we see that $\overline{\gamma}$ corresponds to a map $\overline{\gamma}^* : A^{st} \longrightarrow A^{s_1 t}$. Let ψ_{ij} and ψ_{ij}^* be defined as in Lemma 3.9.3 with t_1 replaced by t. Then $\overline{\gamma}^*$ is given by a matrix T whose columns are the vectors $\overline{\gamma}^*(\psi_{ij}^*) \in A^{s_1 t}$. These vectors are obtained by concatenating the columns of $\overline{\gamma}(\psi_{ij})$.

Now from Equation (3.9.13) we have $\operatorname{Hom}(A^s, N) \cong \operatorname{Hom}(A^s, A^t)/\operatorname{im}(\delta)$ and $\operatorname{Hom}(A^{s_1}, N) \cong \operatorname{Hom}(A^{s_1}, A^t)/\operatorname{im}(\delta')$. We next show that the map

$$\gamma\colon \operatorname{Hom}(A^s, N) \longrightarrow \operatorname{Hom}(A^{s_1}, N)$$

is induced by the map $\overline{\gamma}$; that is, $\gamma(\phi + \operatorname{im}(\delta)) = \overline{\gamma}(\phi) + \operatorname{im}(\delta')$. To do this, it suffices to show that $\overline{\gamma}(\operatorname{im}(\delta)) \subseteq \operatorname{im}(\delta')$. So let $\psi \in \operatorname{Hom}(A^s, A^{t_1})$; then

$$\overline{\gamma}(\delta(\psi)) = \overline{\gamma}(\Delta \circ \psi) = (\Delta \circ \psi) \circ \Gamma = \delta'(\psi \circ \Gamma) \in \operatorname{im}(\delta').$$

Thus γ^* is induced by $\overline{\gamma}^*$ and

$$\gamma^*(\phi^* + \langle\underbrace{\Delta \oplus \cdots \oplus \Delta}_{s \text{ copies}}\rangle) = \overline{\gamma}^*(\phi^*) + \langle\underbrace{\Delta \oplus \cdots \oplus \Delta}_{s_1 \text{ copies}}\rangle = T\phi^* + \langle\underbrace{\Delta \oplus \cdots \oplus \Delta}_{s_1 \text{ copies}}\rangle.$$

Hence γ^* is given by the matrix T. We now wish to describe T.

As mentioned above, the columns of T are the vectors $\overline{\gamma}^*(\psi_{ij}^*) \in A^{s_1 t}$ obtained from the matrices $\overline{\gamma}(\psi_{ij}) = \psi_{ij} \circ \Gamma$ by concatenating columns. Again, because of the identification of the maps ψ_{ij} and Γ with the matrices ψ_{ij} and Γ respectively, we have $\overline{\gamma}(\psi_{ij}) = \psi_{ij}\Gamma$, where the right-hand side expression is a matrix multiplication. For each i and j, $\psi_{ij}\Gamma$ is the $t \times s_1$ matrix whose i-th row is the j-th row of Γ. So if we order the ψ_{ij}'s as before, the matrices $\psi_{11}\Gamma, \psi_{21}\Gamma, \dots, \psi_{t1}\Gamma$ have their respective first, second, etc. row equal to the first row of Γ. The corresponding vectors in $A^{s_1 t}$ obtained by concatenation of columns are then the columns of the transpose of the matrix

$$\left[\ \gamma_{11}I_t|\gamma_{12}I_t|\cdots|\gamma_{1s}I_t\ \right],$$

where $\left[\ \gamma_{11}\quad \gamma_{12}\quad \cdots\quad \gamma_{1s}\ \right]$ is the first row of Γ and I_t is the $t \times t$ identity matrix. Similarly, the matrices $\psi_{12}\Gamma, \psi_{22}\Gamma, \dots, \psi_{t2}\Gamma$ correspond to vectors which form the columns of

$$\left[\ \gamma_{21}I_t|\gamma_{22}I_t|\cdots|\gamma_{2s}I_t\ \right],$$

where $\left[\ \gamma_{21}\quad \gamma_{22}\quad \cdots\quad \gamma_{2s}\ \right]$ is the second row of Γ. We can continue this analysis and we see that T is obtained as indicated in the statement of the lemma. □

Recall that $\mathrm{Hom}(M, N)$ is isomorphic to $\ker(\gamma)$ and hence to $\ker(\gamma^*)$. So we now compute the kernel of γ^*. We first compute the kernel of the homomorphism

$$A^{st} \longrightarrow A^{s_1 t}/\langle\underbrace{\Delta \oplus \cdots \oplus \Delta}_{s_1 \text{ copies}}\rangle$$

given by

$$\phi^* \longmapsto T\phi^* + \langle\Delta \oplus \cdots \oplus \Delta\rangle.$$

Let U be the matrix whose columns generate this kernel. Thus the columns of U are given by the first st coordinates of the generators of the syzygy module of the columns of T and those of $\Delta \oplus \cdots \oplus \Delta$ (see Theorem 3.8.9).

Therefore

$$\mathrm{Hom}(M, N) \cong \langle U\rangle/\langle\underbrace{\Delta \oplus \cdots \oplus \Delta}_{s \text{ copies}}\rangle. \tag{3.9.16}$$

A presentation of $\mathrm{Hom}(M, N)$ can then be computed again using the method given in Theorem 3.8.9.

To summarize we now state

THEOREM 3.9.5. *Let M and N be A-modules given by explicit presentations $M \cong A^s/L$ and $N \cong A^t/K$. We compute* $\mathrm{Hom}(M, N)$ *as follows.*

(i) *Use the generators of L and K as columns to define the matrices Γ and Δ respectively;*
(ii) *Let $T = {}^t(\Gamma \otimes I_t)$;*
(iii) *Compute the matrix U defined by the kernel of the composite map*

$$A^{st} \xrightarrow{T} A^{s_1 t} \longrightarrow A^{s_1 t}/\langle\underbrace{\Delta \oplus \cdots \oplus \Delta}_{s_1 \text{ copies}}\rangle,$$

using Theorem 3.8.9. *This gives Equation* (3.9.16)*;*

(iv) *Compute a presentation of* Hom(M, N) *using Theorem* 3.8.9.

To illustrate the Theorem, we now give an example.

EXAMPLE 3.9.6. We let $A = \mathbb{Q}[x, y, z]$ with the lex order with $x > y > z$, using TOP with $\boldsymbol{e}_1 > \boldsymbol{e}_2 > \cdots$ on all the modules considered. Let $M = \langle \boldsymbol{f}_1, \boldsymbol{f}_2, \boldsymbol{f}_3, \boldsymbol{f}_4 \rangle \subseteq A^3$, where

$$\boldsymbol{f}_1 = \begin{bmatrix} xy \\ xz \\ yz \end{bmatrix}, \quad \boldsymbol{f}_2 = \begin{bmatrix} y \\ x \\ y \end{bmatrix}, \quad \boldsymbol{f}_3 = \begin{bmatrix} 0 \\ x^3 - x^2 z \\ x^2 y - xyz \end{bmatrix}, \quad \boldsymbol{f}_4 = \begin{bmatrix} yz \\ x^2 \\ xy \end{bmatrix},$$

and let $N = \langle \boldsymbol{g}_1, \boldsymbol{g}_2, \boldsymbol{g}_3 \rangle \subseteq A^2$, where

$$\boldsymbol{g}_1 = \begin{bmatrix} x^2 \\ y^2 \end{bmatrix}, \quad \boldsymbol{g}_2 = \begin{bmatrix} x^2 \\ yz \end{bmatrix}, \quad \boldsymbol{g}_3 = \begin{bmatrix} x^2 z \\ xy^2 + yz^2 \end{bmatrix}.$$

We first need to compute presentations of M and N. To do this we use Theorem 3.8.9. Let $L = \mathrm{Syz}(\boldsymbol{f}_1, \boldsymbol{f}_2, \boldsymbol{f}_3, \boldsymbol{f}_4)$ and compute L to be

$$L = \langle (-1, x+z, 0, -1), (-z, z^2, -1, x-z) \rangle \subseteq A^4.$$

We also let $K = \mathrm{Syz}(\boldsymbol{g}_1, \boldsymbol{g}_2, \boldsymbol{g}_3)$ and compute K to be

$$K = \langle (yx, -yx + yz - z^2, -y + z) \rangle \subseteq A^3.$$

So we have the presentations

$$M \cong A^4/L \qquad \text{and} \qquad N \cong A^3/K.$$

We let

$$\Gamma = \begin{bmatrix} -1 & -z \\ x+z & z^2 \\ 0 & -1 \\ -1 & x-z \end{bmatrix} \qquad \text{and} \qquad \Delta = \begin{bmatrix} yx \\ -yx + yz - z^2 \\ -y + z \end{bmatrix}.$$

Thus the matrix T is

$$\begin{bmatrix} -1 & 0 & 0 & x+z & 0 & 0 & 0 & 0 & 0 & -1 & 0 & 0 \\ 0 & -1 & 0 & 0 & x+z & 0 & 0 & 0 & 0 & 0 & -1 & 0 \\ 0 & 0 & -1 & 0 & 0 & x+z & 0 & 0 & 0 & 0 & 0 & -1 \\ -z & 0 & 0 & z^2 & 0 & 0 & -1 & 0 & 0 & x-z & 0 & 0 \\ 0 & -z & 0 & 0 & z^2 & 0 & 0 & -1 & 0 & 0 & x-z & 0 \\ 0 & 0 & -z & 0 & 0 & z^2 & 0 & 0 & -1 & 0 & 0 & x-z \end{bmatrix}.$$

To compute the kernel of the map $A^{12} \longrightarrow A^6/\langle \Delta \oplus \Delta \rangle$ given by T composed with the projection $A^6 \longrightarrow A^6/\langle \Delta \oplus \Delta \rangle$, we compute the syzygy module of the columns of T and of

$$\Delta \oplus \Delta = \begin{bmatrix} yx & 0 \\ -yx + yz - z^2 & 0 \\ -y + z & 0 \\ 0 & yx \\ 0 & -yx + yz - z^2 \\ 0 & -y + z \end{bmatrix}.$$

The first 12 entries of the generators of that syzygy module are the columns of the matrix U, where

$$U = \begin{bmatrix} -1 & 0 & 0 & x & 0 & 0 & 0 & y \\ 0 & -1 & 0 & 0 & x & 0 & yz - z^2 & -y \\ 0 & 0 & -1 & 0 & 0 & x & -y + z & 0 \\ 0 & 0 & 0 & 1 & 0 & 0 & -y & 0 \\ 0 & 0 & 0 & 0 & 1 & 0 & y & 0 \\ 0 & 0 & 0 & 0 & 0 & 1 & 0 & 0 \\ x & 0 & 0 & 0 & 0 & 0 & 0 & 0 \\ 0 & x & 0 & 0 & 0 & 0 & 0 & yz - z^2 \\ 0 & 0 & x & 0 & 0 & 0 & 0 & -y + z \\ 1 & 0 & 0 & z & 0 & 0 & -yz & -y \\ 0 & 1 & 0 & 0 & z & 0 & yz & y \\ 0 & 0 & 1 & 0 & 0 & z & 0 & 0 \end{bmatrix}.$$

From Equation (3.9.16) we have

$$\mathrm{Hom}(M, N) \cong \mathrm{Hom}(A^4/L, A^3/K) \cong \langle U \rangle / \langle \Delta \oplus \Delta \oplus \Delta \oplus \Delta \rangle.$$

Finally to compute a presentation of $\langle U \rangle / \langle \Delta \oplus \Delta \oplus \Delta \oplus \Delta \rangle$ we compute the syzygy module of the eight columns of U and the four columns of $\Delta \oplus \Delta \oplus \Delta \oplus \Delta =$

$$\begin{bmatrix} yx & 0 & 0 & 0 \\ -yx + yz - z^2 & 0 & 0 & 0 \\ -y + z & 0 & 0 & 0 \\ 0 & yx & 0 & 0 \\ 0 & -yx + yz - z^2 & 0 & 0 \\ 0 & -y + z & 0 & 0 \\ 0 & 0 & yx & 0 \\ 0 & 0 & -yx + yz - z^2 & 0 \\ 0 & 0 & -y + z & 0 \\ 0 & 0 & 0 & yx \\ 0 & 0 & 0 & -yx + yz - z^2 \\ 0 & 0 & 0 & -y + z \end{bmatrix}.$$

There are 4 generators for this syzygy module and we need the first 8 coordinates of each of them, which give

$$\boldsymbol{t}_1 = \begin{bmatrix} 0 \\ 0 \\ 0 \\ y \\ -y \\ 0 \\ 1 \\ 0 \end{bmatrix}, \quad \boldsymbol{t}_2 = \begin{bmatrix} y \\ -y \\ 0 \\ 0 \\ 0 \\ 0 \\ 0 \\ 1 \end{bmatrix}, \quad \boldsymbol{t}_3 = \begin{bmatrix} 0 \\ 0 \\ 0 \\ 0 \\ yz - z^2 \\ -y+z \\ -x \\ 0 \end{bmatrix}, \quad \boldsymbol{t}_4 = \begin{bmatrix} 0 \\ yz - z^2 \\ -y+z \\ 0 \\ 0 \\ 0 \\ 0 \\ -x \end{bmatrix}.$$

Therefore

$$\operatorname{Hom}(M, N) \cong \operatorname{Hom}(A^4/L, A^3/K) \cong A^8/\langle \boldsymbol{t}_1, \boldsymbol{t}_2, \boldsymbol{t}_3, \boldsymbol{t}_4 \rangle.$$

We now show how to generate A-module homomorphisms from A^4/L to A^3/K, and hence from M to N. For $\boldsymbol{a} = (a_1, \ldots, a_8) \in A^8$, the product $U\boldsymbol{a}$ is a vector in A^{12}. Using this vector we construct a 3×4 matrix whose columns are the 4 consecutive 3-tuples of the vector $U\boldsymbol{a}$. This matrix defines a homomorphism from A^4/L to A^3/K. The fact that this homomorphism is well-defined follows from the construction of U.

Let $\boldsymbol{e}_i, i = 1, \ldots, 8$ be the standard basis of A^8. Then $\operatorname{Hom}(A^4/L, A^3/K)$ is generated by the matrices obtained from $U\boldsymbol{e}_i$, $i = 1, \ldots, 8$, that is, by the matrices obtained from the columns of U.

Therefore $\operatorname{Hom}(A^4/L, A^3/K)$ is generated by the cosets of the eight matrices

$$\phi_1 = \begin{bmatrix} -1 & 0 & x & 1 \\ 0 & 0 & 0 & 0 \\ 0 & 0 & 0 & 0 \end{bmatrix}, \quad \phi_2 = \begin{bmatrix} 0 & 0 & 0 & 0 \\ -1 & 0 & x & 1 \\ 0 & 0 & 0 & 0 \end{bmatrix}, \quad \phi_3 = \begin{bmatrix} 0 & 0 & 0 & 0 \\ 0 & 0 & 0 & 0 \\ -1 & 0 & x & 1 \end{bmatrix}$$

$$\phi_4 = \begin{bmatrix} x & 1 & 0 & z \\ 0 & 0 & 0 & 0 \\ 0 & 0 & 0 & 0 \end{bmatrix}, \quad \phi_5 = \begin{bmatrix} 0 & 0 & 0 & 0 \\ x & 1 & 0 & z \\ 0 & 0 & 0 & 0 \end{bmatrix}, \quad \phi_6 = \begin{bmatrix} 0 & 0 & 0 & 0 \\ 0 & 0 & 0 & 0 \\ x & 1 & 0 & z \end{bmatrix}$$

$$\phi_7 = \begin{bmatrix} 0 & -y & 0 & -yz \\ yz - z^2 & y & 0 & yz \\ -y+z & 0 & 0 & 0 \end{bmatrix}, \quad \phi_8 = \begin{bmatrix} y & 0 & 0 & -y \\ -y & 0 & yz - z^2 & y \\ 0 & 0 & -y+z & 0 \end{bmatrix}.$$

It can easily be verified that, for $i = 1, \ldots, 8$, we have $\phi_i L \subseteq K$.

Thus any $\phi \in \operatorname{Hom}(A^4/L, A^3/L)$ can be expressed in terms of the ϕ_i's

$$\phi = \sum_{i=1}^{8} a_i \phi_i,$$

where $a_i \in A$. We recall that we identify the homomorphism ϕ with the matrix that defines it.

To obtain a description of $\operatorname{Hom}(M, N)$ from $\operatorname{Hom}(A^4/L, A^3/K)$ we make use of the isomorphisms

$$\begin{array}{ccccccc} M & \xrightarrow{\cong} & A^4/L & \quad & A^3/K & \xrightarrow{\cong} & N \\ \boldsymbol{f}_1 & \longmapsto & \boldsymbol{e}_1 + L & & \boldsymbol{e}_1' + K & \longmapsto & \boldsymbol{g}_1 \\ \boldsymbol{f}_2 & \longmapsto & \boldsymbol{e}_2 + L & & \boldsymbol{e}_2' + K & \longmapsto & \boldsymbol{g}_2 \\ \boldsymbol{f}_3 & \longmapsto & \boldsymbol{e}_3 + L & & \boldsymbol{e}_3' + K & \longmapsto & \boldsymbol{g}_3 \\ \boldsymbol{f}_4 & \longmapsto & \boldsymbol{e}_4 + L & & & & \end{array}$$

where the vectors $\boldsymbol{e}_1$, $\boldsymbol{e}_2$, $\boldsymbol{e}_3$, and $\boldsymbol{e}_4$ are the standard basis vectors for A^4, and the vectors $\boldsymbol{e}_1'$, $\boldsymbol{e}_2'$, and $\boldsymbol{e}_3'$ are the standard basis vectors for A^3. Let ϕ be any homomorphism in $\operatorname{Hom}(A^4/L, A^3/K)$, say $\phi = \sum_{i=1}^{8} a_i\phi_i$. Then we define a homomorphism ϕ' in $\operatorname{Hom}(M, N)$ as follows. First, it is enough to determine $\phi'(\boldsymbol{f}_i)$ for $i = 1, 2, 3, 4$. But $\phi'(\boldsymbol{f}_i)$ is just the image of $\phi(\boldsymbol{e}_i + L)$ under the isomorphism $A^3/K \cong N$. In particular, $\phi'(\boldsymbol{f}_1)$ is the image under the isomorphism $A^3/K \cong N$ of

$$\begin{aligned} \phi(\boldsymbol{e}_1 + L) &= \sum_{i=1}^{8} a_i\phi_i(\boldsymbol{e}_1 + L) \\ &= a_1\begin{bmatrix} -1 \\ 0 \\ 0 \end{bmatrix} + a_2\begin{bmatrix} 0 \\ -1 \\ 0 \end{bmatrix} + a_3\begin{bmatrix} 0 \\ 0 \\ -1 \end{bmatrix} + a_4\begin{bmatrix} x \\ 0 \\ 0 \end{bmatrix} \\ &\quad + a_5\begin{bmatrix} 0 \\ x \\ 0 \end{bmatrix} + a_6\begin{bmatrix} 0 \\ 0 \\ x \end{bmatrix} + a_7\begin{bmatrix} 0 \\ yz - z^2 \\ -y + z \end{bmatrix} + a_8\begin{bmatrix} y \\ -y \\ 0 \end{bmatrix} + K \\ &= \begin{bmatrix} -a_1 + xa_4 + ya_8 \\ -a_2 + xa_5 + (yz - z^2)a_7 - ya_8 \\ -a_3 + xa_6 + (-y + z)a_7 \end{bmatrix} + K \\ &= (-a_1 + xa_4 + ya_8)(\boldsymbol{e}_1' + K) \\ &\quad + (-a_2 + xa_5 + (yz - z^2)a_7 - ya_8)(\boldsymbol{e}_2' + K) \\ &\quad + (-a_3 + xa_6 + (-y + z)a_7)(\boldsymbol{e}_3' + K). \end{aligned}$$

That is

$$\begin{aligned} \phi'(\boldsymbol{f}_1) &= (-a_1 + xa_4 + ya_8)\boldsymbol{g}_1 + (-a_2 + xa_5 + (yz - z^2)a_7 - ya_8)\boldsymbol{g}_2 \\ &\quad + (-a_3 + xa_6 + (-y + z)a_7)\boldsymbol{g}_3. \end{aligned}$$

The images of $\boldsymbol{f}_2$, $\boldsymbol{f}_3$, and $\boldsymbol{f}_4$ can be computed in a similar way.

In particular $\operatorname{Hom}(M, N)$ is generated by the homomorphisms ϕ_i' corresponding to the homomorphisms ϕ_i which generate $\mathrm{Hom}(A^4/L, A^3/K)$:

$$\begin{array}{llll} \phi_1': & M & \longrightarrow & N \\ & \boldsymbol{f}_1 & \longmapsto & -\boldsymbol{g}_1 \\ & \boldsymbol{f}_2 & \longmapsto & \boldsymbol{0} \\ & \boldsymbol{f}_3 & \longmapsto & x\boldsymbol{g}_1 \\ & \boldsymbol{f}_4 & \longmapsto & \boldsymbol{g}_1 \end{array} \quad \begin{array}{llll} \phi_2': & M & \longrightarrow & N \\ & \boldsymbol{f}_1 & \longmapsto & -\boldsymbol{g}_2 \\ & \boldsymbol{f}_2 & \longmapsto & \boldsymbol{0} \\ & \boldsymbol{f}_3 & \longmapsto & x\boldsymbol{g}_2 \\ & \boldsymbol{f}_4 & \longmapsto & \boldsymbol{g}_2 \end{array} \quad \begin{array}{llll} \phi_3': & M & \longrightarrow & N \\ & \boldsymbol{f}_1 & \longmapsto & -\boldsymbol{g}_3 \\ & \boldsymbol{f}_2 & \longmapsto & \boldsymbol{0} \\ & \boldsymbol{f}_3 & \longmapsto & x\boldsymbol{g}_3 \\ & \boldsymbol{f}_4 & \longmapsto & \boldsymbol{g}_3 \end{array}$$

$$\begin{array}{llll} \phi_4': & M & \longrightarrow & N \\ & \boldsymbol{f}_1 & \longmapsto & x\boldsymbol{g}_1 \\ & \boldsymbol{f}_2 & \longmapsto & \boldsymbol{g}_1 \\ & \boldsymbol{f}_3 & \longmapsto & \boldsymbol{0} \\ & \boldsymbol{f}_4 & \longmapsto & z\boldsymbol{g}_1 \end{array} \quad \begin{array}{llll} \phi_5': & M & \longrightarrow & N \\ & \boldsymbol{f}_1 & \longmapsto & x\boldsymbol{g}_2 \\ & \boldsymbol{f}_2 & \longmapsto & \boldsymbol{g}_2 \\ & \boldsymbol{f}_3 & \longmapsto & \boldsymbol{0} \\ & \boldsymbol{f}_4 & \longmapsto & z\boldsymbol{g}_2 \end{array} \quad \begin{array}{llll} \phi_6': & M & \longrightarrow & N \\ & \boldsymbol{f}_1 & \longmapsto & x\boldsymbol{g}_3 \\ & \boldsymbol{f}_2 & \longmapsto & \boldsymbol{g}_3 \\ & \boldsymbol{f}_3 & \longmapsto & \boldsymbol{0} \\ & \boldsymbol{f}_4 & \longmapsto & z\boldsymbol{g}_3 \end{array}$$

$$\begin{array}{llll} \phi_7': & M & \longrightarrow & N \\ & \boldsymbol{f}_1 & \longmapsto & (yz-z^2)\boldsymbol{g}_2 \\ & & & +(-y+z)\boldsymbol{g}_3 \\ & \boldsymbol{f}_2 & \longmapsto & -y\boldsymbol{g}_1 + y\boldsymbol{g}_2 \\ & \boldsymbol{f}_3 & \longmapsto & \boldsymbol{0} \\ & \boldsymbol{f}_4 & \longmapsto & -yz\boldsymbol{g}_1 + yz\boldsymbol{g}_2 \end{array} \qquad \begin{array}{llll} \phi_8': & M & \longrightarrow & N \\ & \boldsymbol{f}_1 & \longmapsto & y\boldsymbol{g}_1 - y\boldsymbol{g}_2 \\ & \boldsymbol{f}_2 & \longmapsto & \boldsymbol{0} \\ & \boldsymbol{f}_3 & \longmapsto & (yz-z^2)\boldsymbol{g}_2 \\ & & & +(-y+z)\boldsymbol{g}_3 \\ & \boldsymbol{f}_4 & \longmapsto & -y\boldsymbol{g}_1 + y\boldsymbol{g}_2. \end{array}$$

Exercises

3.9.1. Let $A = \mathbb{Q}[x, y, z]$, $M = A^3/L$, and $N = A^4/K$, where

$$L = \langle (x, x+y, z), (x, x^2, x-z), (-xz, x-z, y-z) \rangle \text{ and}$$

$$K = \langle (x+y, x^2, xy+yz, y-z), (xy, x-z, y-z, xz^2), (xy, y-z, x-z, yz^2) \rangle.$$

Compute generators for, and a presentation of $\mathrm{Hom}(M, N)$.

3.9.2. Consider the following modules

$$M = \langle (0, y, x), (y, 0, y), (y, y, x), (x, -y, xy), (0, x, x) \rangle \subseteq (\mathbb{Q}[x, y])^3 \text{ and}$$

$$N = \langle (xy, y, y, y), (x^2, x, x, x) \rangle \subseteq (\mathbb{Q}[x, y])^4.$$

Compute generators for, and a presentation of $\mathrm{Hom}(M, N)$.

3.10. Free Resolutions. In this section we first show how to compute an explicit free resolution of a finitely generated A-module, where $A = k[x_1, \dots, x_n]$. We will then use this free resolution to prove a theorem of commutative algebra concerning A. Finally, the free resolution together with the results from the last section will be used to give an outline of the method for the computation of the Ext functor.

Let M be a finitely generated A-module. We saw in Section 3.1 that M has a presentation, that is, $M \cong A^{s_0}/M_0$ for some s_0 and some submodule M_0 of A^{s_0}. We also have the following short exact sequence

$$0 \longrightarrow M_0 \xrightarrow{i_0} A^{s_0} \xrightarrow{\pi_0} M \longrightarrow 0 \tag{3.10.1}$$

to indicate that A^{s_0}/M_0 is a presentation of M. The map $i_0\colon M_0 \longrightarrow A^{s_0}$ is the inclusion map, and $\pi_0\colon A^{s_0} \longrightarrow M$ is the map which sends the standard basis of A^{s_0} onto the generating set of M corresponding to the standard basis in the isomorphism $M \cong A^{s_0}/M_0$.

Now we find a presentation of M_0, say

$$M_0 \cong A^{s_1}/M_1,$$

and thus we have another short exact sequence

$$0 \longrightarrow M_1 \xrightarrow{i_1} A^{s_1} \xrightarrow{\pi_1} M_0 \longrightarrow 0.$$

This leads to an exact sequence

$$0 \longrightarrow M_1 \xrightarrow{i_1} A^{s_1} \xrightarrow{\phi_1} A^{s_0} \xrightarrow{\phi_0} M \longrightarrow 0,$$

where $\phi_0 = \pi_0$, and $\phi_1 = i_0 \circ \pi_1$ (the exactness is easily checked).

We continue this process recursively: at the jth step, we find a presentation A^{s_j}/M_j of M_{j-1}. We then obtain a sequence of module homomorphisms as in Figure 3.3.

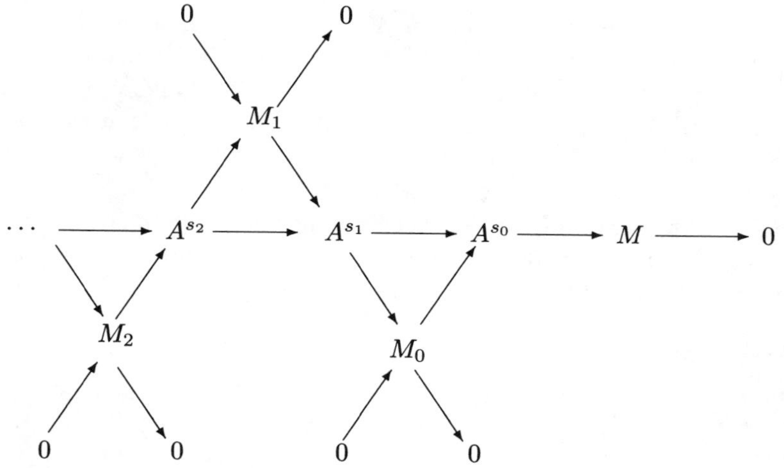

FIGURE 3.3. Presentations of M, M_0, and M_1

Thus we have obtained an exact sequence

$$\cdots \longrightarrow A^{s_t} \longrightarrow A^{s_{t-1}} \longrightarrow \cdots \longrightarrow A^{s_1} \longrightarrow A^{s_0} \longrightarrow M \longrightarrow 0. \tag{3.10.2}$$

DEFINITION 3.10.1. *The exact sequence obtained in* (3.10.2) *is called a* free resolution *of the module* M.

In module theory, free modules are the analog of vector spaces. A free resolution of a module M is used to "measure" how far away from a free module M is and to give some useful numerical invariants for the module M and the ring A.

If in Sequence (3.10.2) it turns out that there exists an ℓ such that $A^{s_j} = 0$ for all $j \geq \ell$, we say that the free resolution has *finite length* and that its length is $\leq \ell$. We will show below that finitely generated A-modules have free resolutions of finite length. In fact, for each A-module M, there exists a finite resolution of length at most n, the number of variables in A. We say that A has *global dimension* less than or equal to n. We will now use the theory we have developed so far to prove this result (Theorem 3.10.4). First we have a technical lemma.

Let $\{\boldsymbol{g}_1, \dots, \boldsymbol{g}_t\} \subseteq A^m$ be a Gröbner basis for $M = \langle \boldsymbol{f}_1, \dots, \boldsymbol{f}_s \rangle$ with respect to some term order on the monomials of A^m. Let $\{\boldsymbol{s}_{ij} \mid 1 \leq i < j \leq t\} \subseteq A^t$ be as in Theorem 3.7.3. Recall that it is a Gröbner basis for Syz$(\boldsymbol{g}_1, \dots, \boldsymbol{g}_t)$ with respect to the term order $<$ on monomials of A^t induced by $\left[\begin{array}{ccc} \boldsymbol{g}_1 & \cdots & \boldsymbol{g}_t \end{array}\right]$, by Theorem 3.7.13. We note that we are using two different term orders: one in A^m which is used to compute the Gröbner basis $\{\boldsymbol{g}_1, \dots, \boldsymbol{g}_t\}$ for M, and the other is defined in A^t with respect to which the set $\{\boldsymbol{s}_{ij} \mid 1 \leq i < j \leq t\}$ is a Gröbner basis for Syz$(\boldsymbol{g}_1, \dots, \boldsymbol{g}_t)$. We will use the same notation for both orders, that is, we will use $<$ and we will write $\mathrm{lm}(\boldsymbol{g}_i)$ and $\mathrm{lm}(\boldsymbol{s}_{ij})$. The context should indicate which module we are working in and which order we are using. In fact, we need to be careful which order we put on A^t. Notice that the order on A^t changes if we reorder the vectors $\boldsymbol{g}_1, \dots, \boldsymbol{g}_t$. Also, since we will be using the standard bases for both A^t and A^m, we will denote by $\boldsymbol{e}_1, \dots, \boldsymbol{e}_t$ the standard basis for A^t, and by $\boldsymbol{e}'_1, \dots, \boldsymbol{e}'_m$ the standard basis for A^m. We will assume that they are ordered as follows: the $\boldsymbol{g}_i$'s are arranged in such a way that if $\nu < \mu$ and $\mathrm{lm}(\boldsymbol{g}_\nu) = X_\nu \boldsymbol{e}'_j$, $\mathrm{lm}(\boldsymbol{g}_\mu) = X_\mu \boldsymbol{e}'_j$ for some power products $X_\nu, X_\mu \in A$ and some $j \in \{1, \dots, m\}$ (that is, $\mathrm{lm}(\boldsymbol{g}_\nu)$ and $\mathrm{lm}(\boldsymbol{g}_\mu)$ have the same non-zero coordinate) then $X_\nu > X_\mu$ with respect to the lex ordering with $x_1 > x_2 > \cdots > x_n$ (regardless of the order that was used to compute the Gröbner basis for M).

LEMMA 3.10.2. *Let $i \in \{1, \dots, n-1\}$. If the variables $x_1, \dots, x_i$ do not appear in $\mathrm{lm}(\boldsymbol{g}_\nu)$, for some $\nu \in \{1, \dots, t\}$, then the variables $x_1, \dots, x_i, x_{i+1}$ do not appear in $\mathrm{lm}(\boldsymbol{s}_{\nu\mu})$, for every μ such that $\nu < \mu \leq t$. So, if $x_1, \dots, x_i$ do not appear in any $\mathrm{lm}(\boldsymbol{g}_\nu)$ for $\nu = 1, \dots, t$, then $x_1, \dots, x_i, x_{i+1}$ do not appear in any $\mathrm{lm}(\boldsymbol{s}_{\nu\mu})$ for $1 \leq \nu < \mu \leq t$. In particular, x_1 does not appear in any $\mathrm{lm}(\boldsymbol{s}_{j\mu})$, for $1 \leq j < \mu \leq t$.*

PROOF. Let $\nu, \mu \in \{1, \dots, t\}$, $\nu < \mu$. If $\mathrm{lm}(\boldsymbol{g}_\nu)$ and $\mathrm{lm}(\boldsymbol{g}_\mu)$ do not involve the same coordinate, then $\boldsymbol{X}_{\nu\mu} = \boldsymbol{0}$, and hence $\boldsymbol{s}_{\nu\mu} = \boldsymbol{0}$, so no variables appear in $\boldsymbol{s}_{\nu\mu}$ at all. Otherwise, let $\boldsymbol{X}_\nu = \mathrm{lm}(\boldsymbol{g}_\nu) = X_\nu \boldsymbol{e}'_j$ and $\boldsymbol{X}_\mu = \mathrm{lm}(\boldsymbol{g}_\mu) = X_\mu \boldsymbol{e}'_j$ for some $j \in \{1, \dots, m\}$. Then, by Theorem 3.7.13,

$$\mathrm{lm}(\boldsymbol{s}_{\nu\mu}) = \frac{X_{\nu\mu}}{X_\nu} \boldsymbol{e}_\nu,$$

where $X_{\nu\mu}$ is, as usual, $\operatorname{lcm}(X_\mu, X_\nu)$. First, let us assume that $x_1, \ldots, x_i$ do not appear in X_ν. Since $X_\nu > X_\mu$ with respect to the lex term ordering with $x_1 > x_2 > \cdots > x_n$, then $x_1, \ldots, x_i$ do not appear in X_μ and the power of x_{i+1} in X_ν is at least as large as the power of x_{i+1} in X_μ. Therefore the power of x_{i+1} in $X_{\nu\mu}$ is the same as the power of x_{i+1} in X_ν, and hence the variables $x_1, \ldots, x_i, x_{i+1}$ do not appear in $\dfrac{X_{\nu\mu}}{X_\nu}$ and so do not appear in $\operatorname{lm}(\boldsymbol{s}_{\nu\mu})$. Thus if $x_1, \ldots, x_i$ do not appear in any $\operatorname{lm}(\boldsymbol{g}_\nu)$, $\nu = 1, \ldots, t$, then $x_1, \ldots, x_i, x_{i+1}$ do not appear in any $\operatorname{lm}(\boldsymbol{s}_{\nu\mu})$, for $1 \le \nu < \mu \le t$. The last statement is proved in exactly the same way. □

EXAMPLE 3.10.3. As in Example 3.7.5 we let M be the submodule of $A^3 = (\mathbb{Q}[x,y])^3$ of Example 3.6.1. We saw that the vectors $\boldsymbol{g}_1, \boldsymbol{g}_2, \boldsymbol{g}_3, \boldsymbol{g}_4, \boldsymbol{g}_5, \boldsymbol{g}_6$ form a Gröbner basis with respect to the TOP term ordering in A^3 with $\boldsymbol{e}_1 > \boldsymbol{e}_2 > \boldsymbol{e}_3$ and with lex in A and $y > x$, where

$$\boldsymbol{g}_1 = (x^3 + x, x^2 - x, -x),\ \boldsymbol{g}_2 = (x, y + x^2 - x, 0),$$

$$\boldsymbol{g}_3 = (y + x^2, 0, 0),\ \boldsymbol{g}_4 = (x^2, x, y),$$

$$\boldsymbol{g}_5 = (x^2, x^3, -x^3),\ \boldsymbol{g}_6 = (x^2 - 2x, -x^2 + 2x, x^5 - x^4 - 3x^3 + x^2 + 2x).$$

We re-order the $\boldsymbol{g}$'s as required in Lemma 3.10.2 with $y > x$, and we get

$$\boldsymbol{g}_1 = (y + x^2, 0, 0),\ \boldsymbol{g}_2 = (x^3 + x, x^2 - x, -x),$$

$$\boldsymbol{g}_3 = (x, y + x^2 - x, 0),\ \boldsymbol{g}_4 = (x^2, x^3, -x^3),$$

$$\boldsymbol{g}_5 = (x^2, x, y),\ \boldsymbol{g}_6 = (x^2 - 2x, -x^2 + 2x, x^5 - x^4 - 3x^3 + x^2 + 2x).$$

The syzygies are now

$$\begin{aligned}
\boldsymbol{s}_{12} &= (x^3 + x, -y - x^2, x^2 - x, 1, -x, 0)\\
\boldsymbol{s}_{34} &= (x^2, x^2 - x - 2, x^3, -y - x^2 + x + 2, -x^3, -1)\\
\boldsymbol{s}_{56} &= (x^2 - 2x, -x^4 + x^3 + 4x^2 - 2x - 4, -x^2 + 2x,\\
&\qquad -x^2 + x + 2, x^5 - x^4 - 3x^3 + x^2 + 2x, -y - 2).
\end{aligned}$$

Using the order induced by the $\boldsymbol{g}$'s we have

$$\begin{aligned}
\operatorname{lm}(\boldsymbol{s}_{12}) &= x^3\boldsymbol{e}_1\\
\operatorname{lm}(\boldsymbol{s}_{34}) &= x^3\boldsymbol{e}_3\\
\operatorname{lm}(\boldsymbol{s}_{56}) &= x^5\boldsymbol{e}_5.
\end{aligned}$$

THEOREM 3.10.4. *Every finitely generated A-module has a free resolution of length less than or equal to n, where n is the number of variables in A.*

PROOF. Let M be any finitely generated A-module. Then M has a presentation $M \cong A^{s_0}/M_0$, where $M_0 = \langle \boldsymbol{g}_1, \ldots, \boldsymbol{g}_t \rangle \subseteq A^{s_0}$. If $M_0 = \{\boldsymbol{0}\}$, then we are done; otherwise we may assume that $G = \{\boldsymbol{g}_1, \ldots, \boldsymbol{g}_t\}$ forms a Gröbner basis for M_0 with respect to some term order in A^{s_0}. Assume also that the $\boldsymbol{g}_i$'s are arranged as in Lemma 3.10.2. Let $x_1, \ldots, x_i$ be the variables that do not appear in any of the $\operatorname{lm}(\boldsymbol{g}_\nu)$'s ($i = 0$ is possible). We will prove that M has a free resolution of length less than or equal to $n - i$.

CASE 1. $n = i$. Then none of the variables $x_1, \ldots, x_n$ appear in any of the $\operatorname{lm}(\boldsymbol{g}_j)$'s. We need to show that $M \cong A^{s_0}/M_0$ is isomorphic to a free module. Since the leading monomials of the $\boldsymbol{g}_i$'s are of the form $a\boldsymbol{e}_j$ for some $a \in k$, and $j \in \{1, \ldots, s_0\}$, we see that the module $\operatorname{Lt}(\boldsymbol{g}_1, \ldots, \boldsymbol{g}_t)$ is the free submodule of A^{s_0} generated by the $\boldsymbol{e}_j$'s that appear in some $\operatorname{lm}(\boldsymbol{g}_i)$. Let M' be the free submodule of A^{s_0} generated by the other $\boldsymbol{e}_j$'s. We will show that $M \cong M'$. Consider the map

$$\begin{array}{rccc} \pi\colon & M' & \longrightarrow & A^{s_0}/M_0 \cong M \\ & \boldsymbol{f} & \longmapsto & \boldsymbol{f} + M_0. \end{array}$$

It is easy to see that π is an A-module homomorphism. Also, if $\boldsymbol{f} \in M'$ and $\boldsymbol{f} \in M_0$, then, by Theorem 3.5.14, $\operatorname{lm}(\boldsymbol{f})$ is divisible by some $\operatorname{lm}(\boldsymbol{g}_i)$ for some $i \in \{1, \ldots, t\}$. But this is impossible unless $\boldsymbol{f} = 0$, since M' is generated by those $\boldsymbol{e}_j$'s which do not appear as leading monomials of any $\boldsymbol{g}_i$. Therefore π is one to one. Finally, by Proposition 3.6.3, we see that for all $\boldsymbol{f} \in A^{s_0}$, $\boldsymbol{f} + M_0 = N_G(\boldsymbol{f}) + M_0$, where $N_G(\boldsymbol{f})$ is the remainder of $\boldsymbol{f}$ under division by G. Moreover $N_G(\boldsymbol{f}) \in M'$, since the monomials of $N_G(\boldsymbol{f})$ are exactly those monomials which are not divisible by any $\boldsymbol{e}_j$ which appear in some $\operatorname{lm}(\boldsymbol{g}_i)$. Therefore π is onto, and hence π is an isomorphism.

CASE 2. $n - i > 0$. We construct a free resolution

$$A^{s_j} \xrightarrow{\phi_j} A^{s_{j-1}} \xrightarrow{\phi_{j-1}} \cdots \xrightarrow{\phi_2} A^{s_1} \xrightarrow{\phi_1} A^{s_0} \xrightarrow{\phi_0} M \cong A^{s_0}/M_0 \longrightarrow 0$$

recursively as follows. At the jth step, we choose a monomial order on the free module A^{s_j} and find a Gröbner basis G for $\ker(\phi_j)$. We arrange the elements of G according to Lemma 3.10.2. We then choose $A^{s_{j+1}}$ to be a free module whose basis maps onto G and we let ϕ_{j+1} be the projection map. Note that $\ker(\phi_0) = M_0 = \langle \boldsymbol{g}_1, \ldots, \boldsymbol{g}_t \rangle$ and $\ker(\phi_1) = \operatorname{Syz}(\boldsymbol{g}_1, \ldots, \boldsymbol{g}_t)$. If $x_1, \ldots, x_i$ do not appear in the leading monomials of the $\boldsymbol{g}_\nu$'s, then, by Lemma 3.10.2, $x_1, \ldots, x_i, x_{i+1}$ do not appear in the leading monomials of the elements of the Gröbner basis for $\ker(\phi_1)$. If x_1 appears in $\operatorname{lm}(\boldsymbol{g}_j)$ for some j, then, by Lemma 3.10.2, x_1 does not appear in the leading monomials of the elements of the Gröbner basis for $\ker(\phi_1)$. So if we apply Lemma 3.10.2 recursively, we see that no variables appear in the leading monomials of the elements of the Gröbner basis for $\ker(\phi_{n-i})$. By the case $n - i = 0$, we see that

$$A^{s_{n-i}}/\ker(\phi_{n-i}) \cong \phi_{n-i}(A^{s_{n-i}})$$

is free. Therefore if we replace $A^{s_{n-i}}$ by $A^{s_{n-i}}/\ker(\phi_{n-i})$, we get the desired resolution. □

EXAMPLE 3.10.5. We continue Example 3.10.3. Since the leading term of $\boldsymbol{s}_{12}$, $\boldsymbol{s}_{34}$, and $\boldsymbol{s}_{56}$ involve different basis vectors, we see that $\mathrm{Syz}(\boldsymbol{s}_{12}, \boldsymbol{s}_{34}, \boldsymbol{s}_{56}) = (0,0,0)$. Therefore $A^3 \cong \langle \boldsymbol{s}_{12}, \boldsymbol{s}_{34}, \boldsymbol{s}_{56}\rangle$ and so we have the following free resolution for $M = \langle \boldsymbol{g}_1, \boldsymbol{g}_2, \boldsymbol{g}_3, \boldsymbol{g}_4, \boldsymbol{g}_5, \boldsymbol{g}_6\rangle = \langle \boldsymbol{f}_1, \boldsymbol{f}_2, \boldsymbol{f}_3, \boldsymbol{f}_4\rangle \subseteq A^3$:

$$0 \longrightarrow A^3 \xrightarrow{\phi_1} A^6 \xrightarrow{\phi_0} M \longrightarrow 0,$$

where

$$\begin{array}{llcl} \phi_0: & A^6 & \longrightarrow & M \\ & (h_1, h_2, h_3, h_4, h_5, h_6) & \longmapsto & \sum_{i=1}^6 h_i \boldsymbol{g}_i. \end{array}$$

The kernel of this map is $\mathrm{Syz}(\boldsymbol{g}_1, \boldsymbol{g}_2, \boldsymbol{g}_3, \boldsymbol{g}_4, \boldsymbol{g}_5, \boldsymbol{g}_6)$. Also,

$$\begin{array}{llcl} \phi_1: & A^3 & \longrightarrow & A^6 \\ & (\ell_1, \ell_2, \ell_3) & \longmapsto & \ell_1 \boldsymbol{s}_{12} + \ell_2 \boldsymbol{s}_{34} + \ell_3 \boldsymbol{s}_{56}. \end{array}$$

As we saw above, the kernel of ϕ_1 is $(0,0,0)$.

To conclude this section, we use the techniques developed so far to outline the computation of $\mathrm{Ext}^n(M, N)$. We will assume that the reader is familiar with this concept and we will only give an indication on how to go about the computation.

We begin with a free resolution for an A-module M

$$\cdots \xrightarrow{\Gamma_{i+2}} A^{s_{i+1}} \xrightarrow{\Gamma_{i+1}} A^{s_i} \xrightarrow{\Gamma_i} A^{s_{i-1}} \longrightarrow \cdots \xrightarrow{\Gamma_2} A^{s_1} \xrightarrow{\Gamma_1} A^{s_0} \longrightarrow M \longrightarrow 0,$$

which we compute as above. For the A-module N we form the usual complex which at the ith position looks like

$$\cdots \longrightarrow \mathrm{Hom}(A^{s_{i+1}}, N) \xrightarrow{\gamma_{i+1}} \mathrm{Hom}(A^{s_i}, N) \xrightarrow{\gamma_i} \mathrm{Hom}(A^{s_{i-1}}, N) \longrightarrow \cdots.$$

As in Lemma 3.9.2 and Lemma 3.9.3 we can compute presentations of these Hom modules and using Lemma 3.9.4 we can compute the maps between them yielding another complex which at the ith spot looks like

$$\cdots \longrightarrow A^{u_{i+1}}/L_{i+1} \xrightarrow{T_{i+1}} A^{u_i}/L_i \xrightarrow{T_i} A^{u_{i-1}}/L_{i-1} \longrightarrow \cdots.$$

We can use Theorem 3.8.9 to compute $\ker(T_i)$. Also, $\mathrm{im}(T_{i+1})$ is obtained using the columns of the matrix that determines T_{i+1}. Thus we can compute

$$\mathrm{Ext}^i(M, N) \cong \ker(T_i)/\mathrm{im}(T_{i+1}),$$

again using Theorem 3.8.9.

Exercises

3.10.1. Compute a free resolution for the module $M = \langle (x,y,z), (y,x,z), (y,z,x), (x,z,y), (y, x-z, z), (y, z, x-z)\rangle \subseteq (\mathbb{Q}[x,y,z])^3$.

Chapter 4. Gröbner Bases over Rings

In the previous three chapters we considered the theory of Gröbner bases in the ring $A = k[x_1, \ldots, x_n]$, where k is a field. We are now going to generalize the theory to the case where $A = R[x_1, \ldots, x_n]$ for a Noetherian commutative ring R. Sometimes we will need to be more specific and require R to be an integral domain, a unique factorization domain (UFD), or a principal ideal domain (PID). We will give many of the same type of applications we gave in the previous chapters in this more general context. Moreover, the theory of Gröbner bases over rings will allow us to use inductive techniques on the number of variables; for example, for a field k, $k[x, y]$ can be viewed as a polynomial ring in one variable y over the ring $k[x]$ (i.e. $k[x, y] = (k[x])[y]$). We give an example of this technique in Section 4.4 where it is used to test whether ideals are prime.

The theory of Gröbner bases over rings has complications that did not appear in the theory over fields. Indeed, many of the results will not hold in this generality. Moreover, many of the basic techniques will become more complicated because we now have to deal with ideals of coefficients in the ring R.

In Section 4.1 we give the basic definitions and lay the foundations for the theory of Gröbner bases over rings. An algorithm for constructing Gröbner bases will be presented in Section 4.2. We will have to assume certain computability conditions on R in order for this algorithm to be effective. We use a method presented by Möller [**Mö88**] to compute the appropriate syzygies needed for this algorithm. We then give examples of computing Gröbner bases over the rings $\mathbb{Z}$, $\mathbb{Z}_{20}$ and $\mathbb{Z}[\sqrt{-5}]$. In Section 4.3 we give the usual applications including elimination, computing syzygy modules, and a result of Zacharias [**Za**] which gives a method for computing a complete set of coset representatives of A modulo an ideal. We then go on, in Section 4.4, to discuss questions related to rings of quotients and use this material to give an algorithm to determine whether an ideal in A is a prime ideal. Next, in Section 4.5 we specialize the ring R to be a PID and show that in this case we may again use the notion of S-polynomials to compute Gröbner bases. We conclude that section by giving a structure theorem of Lazard [**Laz85**] for Gröbner bases in polynomial rings in one variable over a PID. In the last section, we use Lazard's result to compute the primary decomposition of ideals in such rings.

4.1. Basic Definitions. In this section we develop the theory of Gröbner bases for polynomials with coefficients in a Noetherian ring R. As we did for modules in Section 3.5, we will mimic the constructions of Chapter 1 as much as we can. We will assume that we have a term order $<$ on the power products in the variables $x_1, \ldots, x_n$. With respect to this term order, we have the usual concepts of leading power product, leading term, and leading coefficient of a polynomial in $A = R[x_1, \ldots, x_n]$. Next, we need the concept of reduction, see Section 1.5 and 3.5. In those sections we required the notion of divisibility of leading terms. Actually, there we were not concerned about whether we were dividing leading terms or dividing leading power products since one was a non-zero element of the field k times the other and this had no effect on the divisibility. When the coefficients are not elements of a field, this becomes a very important issue. It turns out that in order to have a reasonable theory of reduction and Gröbner bases using the same ideas of reduction as in the field case, we need R to be a PID. We will explore this in Section 4.5. Since we want to define reduction and Gröbner bases in the context of rings more general than PID's, we must modify our previous concept of reduction. The correct way to do this is to work with syzygies in the ring R.

After we define this new concept of reduction, we will be able to pattern our results again on what we did in Chapter 1. There is one major exception and that is in the definition of Gröbner basis itself. In the case of ideals of polynomials with coefficients in a field, and also in the case of modules, the definition again involved the concept of dividing one leading term by another. So we need to change our very definition of Gröbner basis. It turns out that many, but not all, of the equivalent conditions for a Gröbner basis over a field given in Theorem 1.9.1 will work for us in our new situation.

So to reiterate the setup, we assume that we are given a Noetherian commutative ring R and we let $A = R[x_1, \ldots, x_n]$. We then have from the Hilbert Basis Theorem 1.1.3, that A is also a Noetherian ring. We assume that we have a term order $<$ on the power products, $\mathbb{T}^n$, in the variables $x_1, \ldots, x_n$. Then from Theorem 1.4.6, we know that $<$ is a well-ordering on $\mathbb{T}^n$ (the point here is that this is a property of the power products $\mathbb{T}^n$, not of the polynomial ring $k[x_1, \ldots, x_n]$, for a field k, even though the proof of Theorem 1.4.6 used the Hilbert Basis Theorem in $k[x_1, \ldots, x_n]$). For $f \in A$, $f \neq 0$, we write $f = a_1X_1 + \cdots + a_sX_s$, where $a_1, \ldots, a_s \in R$, $a_i \neq 0$ and $X_1, \ldots, X_s$ are power products in $x_1, \ldots, x_n$ with $X_1 > X_2 > \cdots > X_s$. We define as before, $\operatorname{lp}(f) = X_1$, $\operatorname{lc}(f) = a_1$ and $\operatorname{lt}(f) = a_1X_1$ (called the leading power product, leading coefficient and leading term of f, respectively).

We now turn our attention to the concept of reduction. Recall, that in the case where $R = k$ is a field, given three polynomials f, g, h in $k[x_1, \ldots, x_n]$, with $g \neq 0$, we say that $f \xrightarrow{g} h$, if and only if $\operatorname{lp}(g)$ divides a term X that appears in f and $h = f - \frac{X}{\operatorname{lt}(g)}g$. The case where R is not a field differs in two ways. The first difference is that in the case of rings, it is convenient to only reduce the leading term of f. It is readily seen that all the results on Gröbner bases in Chapter 1, excluding the ones making explicit use of reduced polynomials, are valid with this restricted concept of reduction. In the case of rings, however, the results involving reduced polynomials are no longer valid and so reducing terms that are not leading terms is unnecessary (see Exercise 4.1.6).

With this in mind, we could rephrase our definition of reduction over a field k by saying that $f \xrightarrow{g} h$, provided that $\operatorname{lp}(g)$ divides $\operatorname{lp}(f)$ and $h = f - \frac{\operatorname{lt}(f)}{\operatorname{lt}(g)}g$ (note that $\operatorname{lt}(f) = \operatorname{lt}(\frac{\operatorname{lt}(f)}{\operatorname{lt}(g)}g)$ and so the leading term of f has been canceled). We see that this notion requires that we divide by $\operatorname{lt}(g) = \operatorname{lc}(g)\operatorname{lp}(g)$. The problem with this is that over a ring R we may not be able to divide by the ring element $\operatorname{lc}(g)$. We could build this into our definition just as we must require that $\operatorname{lp}(g)$ divide $\operatorname{lp}(f)$, but this turns out to be too restrictive when we are attempting to divide by more than one polynomial, as, of course, the theory requires. The key idea in resolving this difficulty is to use a linear combination of the the leading terms of the divisors, whose leading power products divide $\operatorname{lp}(f)$, to eliminate $\operatorname{lt}(f)$.

So we now assume that we are considering polynomials f and $f_1, \dots, f_s$ in $A = R[x_1, \dots, x_n]$ with $f_1, \dots, f_s \neq 0$, and we want to divide f by $f_1, \dots, f_s$. That is, we want to cancel the leading term of f using the leading terms of $f_1, \dots, f_s$. We should be allowed to use any f_i which has the property that $\operatorname{lp}(f_i)$ divides $\operatorname{lp}(f)$ and so what we desire is that $\operatorname{lt}(f)$ be a linear combination of these $\operatorname{lt}(f_i)$. We thus arrive at

DEFINITION 4.1.1. *Given two polynomials f and h and a set of non-zero polynomials $F = \{f_1, \dots, f_s\}$ in A, we say that f* reduces *to h modulo F in one step, denoted*

$$f \xrightarrow{F} h,$$

if and only if

$$h = f - (c_1X_1f_1 + \cdots + c_sX_sf_s)$$

for $c_1, \dots, c_s \in R$ and power products $X_1, \dots, X_s$ where $\operatorname{lp}(f) = X_i\operatorname{lp}(f_i)$ for all i such that $c_i \neq 0$ and $\operatorname{lt}(f) = c_1X_1\operatorname{lt}(f_1) + \cdots + c_sX_s\operatorname{lt}(f_s)$.

EXAMPLE 4.1.2. Let $R = \mathbb{Z}$ and let $f = xy - 1, f_1 = 7x + 3, f_2 = 11x^3 - 2y^2 + 1$ and $f_3 = 3y - 5$. We will use the deglex ordering with $x < y$. So with $F = \{f_1, f_2, f_3\}$, we see that $f \xrightarrow{F} h$, where $h = -3y - 10x - 1$ since $h = f - (yf_1 - 2xf_3)$ and $xy = \operatorname{lt}(f) = y\operatorname{lt}(f_1) - 2x\operatorname{lt}(f_3)$. (Here $c_2 = 0$, as it must, since $\operatorname{lp}(f) = xy$ is not divisible by $\operatorname{lp}(f_2) = x^3$.) So we have done what we said we wanted to do, namely, we have canceled out the leading term of f using the polynomials in the set F. Also, f could not have been reduced using only one of the polynomials f_1, f_2, f_3.

We draw attention to the condition $\text{lp}(f) = X_i \,\text{lp}(f_i)$ for all i such that $c_i \neq 0$. Its purpose is to ensure that each $c_i X_i \,\text{lt}(f_i)$ in the difference $h = f - (c_1 X_1 f_1 + \cdots + c_s X_s f_s)$ with $c_i \neq 0$ is actually used to help cancel $\text{lt}(f)$. Because of the possibility of zero divisors in the ring R, we must be careful about this. For example, if we only required $\text{lp}(f) = \text{lp}(c_i X_i f_i)$ for all i such that $c_i \neq 0$, we could end up with a term $c\,\text{lp}(f)$ remaining in h. To see this consider the case where $R = \mathbb{Z}_{10}$ (recall that $\mathbb{Z}_n$ denotes the integers modulo n) with deglex and $x > y$ and let $f = 3y$, $f_1 = 5x^2 + y$, and $f_2 = y$. Then $\text{lt}(f) = 2\,\text{lt}(f_1) + 3\,\text{lt}(f_2)$ and $\text{lp}(f) = \text{lp}(2f_1) = \text{lp}(3f_2) = y$ whereas $h = f - (2f_1 + 3f_2) = -2y$. Having been careful with the definition we have the following crucial lemma whose easy proof we leave to the exercises (Exercise 4.1.13).

LEMMA 4.1.3. *With the notation of Definition* 4.1.1 *we have* $\text{lp}(h) < \text{lp}(f)$.

Let $f \in A$ and let $F = \{f_1, \ldots, f_s\}$ be a set of non-zero polynomials in A. We now examine how we would determine whether f is reducible modulo F. We first find the set $J = \{j \mid \text{lp}(f_j) \text{ divides } \text{lp}(f), 1 \leq j \leq s\}$, which is readily done. We are restricted to such J by the requirement that $\text{lp}(f) = X_i \,\text{lp}(f_i)$ in Definition 4.1.1.Then we must solve the equation

$$\text{lc}(f) = \sum_{j \in J} b_j \,\text{lc}(f_j) \tag{4.1.1}$$

for b_j's in R. This equation can be solved if and only if $\text{lc}(f) \in \langle \text{lc}(f_j) \mid j \in J \rangle_R$, where for a subset $C \subseteq R$ we denote by $\langle C \rangle_R$ the ideal in the ring R generated by the elements of C. Once we have the b_j's then we can reduce f:

$$f \xrightarrow{F} f - \sum_{j \in J} b_j \frac{\text{lp}(f)}{\text{lp}(f_j)} f_j.$$

EXAMPLE 4.1.4. We put Example 4.1.2 in this context. We have $\text{lp}(f) = xy$ and so $J = \{1, 3\}$. Thus we need to solve $\text{lc}(f) = 1 = b_1 \,\text{lc}(f_1) + b_3 \,\text{lc}(f_3) = 7b_1 + 3b_3$. We choose the solution $b_1 = 1, b_3 = -2$ and thus we reduce f as $h = f - (b_1 y f_1 + b_3 x f_3) = f - (y f_1 - 2x f_3) = -3y - 10x - 1$.

It is thus clear that we must be able to solve linear equations in the ring R in order to be able to reduce in A. This condition is one of two conditions R must satisfy in order to compute the objects of interest to us in this chapter. We list these two conditions in the following

DEFINITION 4.1.5. *We will say that* linear equations are solvable in R *provided that*

(i) *Given* $a, a_1, \ldots, a_m \in R$, *there is an algorithm to determine whether* $a \in \langle a_1, \ldots, a_m \rangle_R$ *and if it is, to compute* $b_1, \ldots, b_m \in R$ *such that* $a = a_1 b_1 + \cdots + a_m b_m$;

(ii) *Given* $a_1, \ldots, a_m \in R$, *there is an algorithm that computes a set of generators for the* R*-module*

$$\mathrm{Syz}_R(a_1, \ldots, a_m) = \{(b_1, \ldots, b_m) \in R^m \mid a_1b_1 + \cdots + a_mb_m = 0\}.$$

Examples of such rings include $\mathbb{Z}$, $\mathbb{Z}_m$, $\mathbb{Q}[x_1, \ldots, x_n]$, $\mathbb{Z}[i]$ where $i^2 = -1$, and $\mathbb{Z}[\sqrt{-5}]$.

The first condition in Definition 4.1.5 is the one discussed above which we found necessary in order to make the reduction process computable. The second condition is needed to guarantee that the algorithm (Algorithm 4.2.1) to be presented in the next section for computing Gröbner bases in A is actually implementable. We will always assume that linear equations are solvable in R when an algorithm is presented. However, as has been the case throughout this book, we will otherwise be informal about our assumptions on the ring. In particular, without the assumption that linear equations are solvable in R, what we present in this chapter is valid, and should be viewed as mathematical existence statements.

As in the case of ideals and modules where the coefficients lie in a field, we need to iterate our reduction process.

DEFINITION 4.1.6. *Let* f, h, *and* $f_1, \ldots, f_s$ *be polynomials in* A, *with* $f_1, \ldots, f_s \neq 0$, *and let* $F = \{f_1, \ldots, f_s\}$. *We say that* f reduces *to* h *modulo* F, *denoted*

$$f \xrightarrow{F}_+ h,$$

if and only if there exist polynomials $h_1, \ldots, h_{t-1} \in A$ *such that*

$$f \xrightarrow{F} h_1 \xrightarrow{F} h_2 \xrightarrow{F} \cdots \xrightarrow{F} h_{t-1} \xrightarrow{F} h.$$

We note that if $f \xrightarrow{F}_+ h$, then $f - h \in \langle f_1, \ldots, f_s \rangle$.

EXAMPLE 4.1.7. We continue Example 4.1.2, where $R = \mathbb{Z}$ and $f = xy - 1, f_1 = 7x + 3, f_2 = 11x^3 - 2y^2 + 1, f_3 = 3y - 5$ and $F = \{f_1, f_2, f_3\}$. We see that $f \xrightarrow{F}_+ -10x - 6$, since $f \xrightarrow{F} -3y - 10x - 1 \xrightarrow{F} -10x - 6$. The first reduction is the one noted in the previous example and the second is obtained by simply adding f_3 to $-3y - 10x - 1$. We note that this reduction could not have been done in one step.

DEFINITION 4.1.8. *A polynomial* r *is called* minimal[1] *with respect to a set of non-zero polynomials* $F = \{f_1, \ldots, f_s\}$ *if* r *cannot be reduced.*

[1]In the case of ideals and modules where the coefficients were in a field we required that every term in r could not be reduced. Here we are only requiring that $\mathrm{lt}(r)$ cannot be reduced modulo F. In the literature this latter concept is sometimes refered to as "reduced", but we adopt the word "minimal" so that we are consistent with the similar concepts in the rest of this book.

We recall the notation that for a subset $W \subseteq A$, the leading term ideal of W is denoted

$$\operatorname{Lt}(W) = \langle \{\operatorname{lt}(w) \mid w \in W\} \rangle.$$

We have the following easy

LEMMA 4.1.9. *A polynomial $r \in A$, with $r \neq 0$, is minimal with respect to a set of non-zero polynomials $F = \{f_1, \dots, f_s\} \subseteq A$ if and only if $\operatorname{lt}(r) \notin \operatorname{Lt}(F)$.*

PROOF. If r is not minimal, then r can be reduced and so from Definition 4.1.1 we have that $\operatorname{lt}(r) = c_1 X_1 \operatorname{lt}(f_1) + \cdots + c_s X_s \operatorname{lt}(f_s)$ for $c_i \in R$ and power products X_i. It immediately follows that $\operatorname{lt}(r) \in \operatorname{Lt}(F)$. Conversely, if $\operatorname{lt}(r) \in \operatorname{Lt}(F)$, then $\operatorname{lt}(r) = h_1 \operatorname{lt}(f_1) + \cdots + h_s \operatorname{lt}(f_s)$ for some polynomials $h_i \in A$. If we expand this equation out into its individual terms, we see that the only power product that can occur with a non-zero coefficient is $\operatorname{lp}(r)$; that is, we may assume that each h_i is a term $c_i X_i$. It is then clear that $r - (c_1 X_1 f_1 + \cdots + c_s X_s f_s)$ is a reduction of r. □

We note that in Example 4.1.7, the polynomial $r = -10x - 6$ is minimal with respect to $F = \{f_1, f_2, f_3\}$ since only f_1 has the property that $\operatorname{lp}(f_i)$ divides $\operatorname{lp}(r) = x$ and $\operatorname{lc}(r) = -10 \notin \langle \operatorname{lc}(f_1) \rangle_{\mathbb{Z}} = \langle 7 \rangle_{\mathbb{Z}}$.

We have

THEOREM 4.1.10. *Let $f, f_1, \dots, f_s \in A$ with $f_1, \dots, f_s \neq 0$, and set $F = \{f_1, \dots, f_s\}$. Then there is an $r \in A$, minimal with respect to F, such that $f \xrightarrow{F}_+ r$. Moreover, there are $h_1, \dots, h_s \in A$ such that*

$$f = h_1 f_1 + \cdots + h_s f_s + r$$

with

$$\operatorname{lp}(f) = \max\left(\left(\max_{1 \leq i \leq s} \operatorname{lp}(h_i) \operatorname{lp}(f_i)\right), \operatorname{lp}(r)\right).$$

If linear equations are solvable in R, then $h_1, \dots, h_s, r$ are computable.

PROOF. Either f is minimal with respect to F or $f \xrightarrow{F} r_1$. Similarly, either r_1 is minimal with respect to F or $r_1 \xrightarrow{F} r_2$. Continuing in this way we get

$$f \xrightarrow{F} r_1 \xrightarrow{F} r_2 \xrightarrow{F} \cdots,$$

where we have, by Lemma 4.1.3, $\operatorname{lp}(f) > \operatorname{lp}(r_1) > \operatorname{lp}(r_2) > \cdots$. This process must end since the order on the power products is a well-ordering, and so we obtain the desired polynomial r. We now have

$$f \xrightarrow{F} r_1 \xrightarrow{F} r_2 \xrightarrow{F} \cdots \xrightarrow{F} r_{\ell-1} \xrightarrow{F} r.$$

By the definition of reduction,

$$f - r_1 = c_{11} X_{11} f_1 + \cdots + c_{1s} X_{1s} f_s$$

for some $c_{11}, \ldots, c_{1s} \in R$ and some power products $X_{11}, \ldots, X_{1s}$, where $\operatorname{lt}(f) = c_{11}X_{11}\operatorname{lt}(f_1) + \cdots + c_{1s}X_{1s}\operatorname{lt}(f_s)$ and $\operatorname{lp}(f) = X_{1i}\operatorname{lp}(f_i)$, for all i such that $c_{1i} \neq 0$. This gives a representation of the desired type for $f - r_1$. Similarly, $r_1 - r_2 = c_{21}X_{21}f_1 + \cdots + c_{2s}X_{2s}f_s$ for some $c_{21}, \ldots, c_{2s} \in R$ and some power products $X_{21}, \ldots, X_{2s}$, where $\operatorname{lt}(r_1) = c_{21}X_{21}\operatorname{lt}(f_1) + \cdots + c_{2s}X_{2s}\operatorname{lt}(f_s)$ and $\operatorname{lp}(r_1) = X_{2i}\operatorname{lp}(f_i)$, for all i such that $c_{2i} \neq 0$. Since $\operatorname{lp}(f) > \operatorname{lp}(r_1) > \operatorname{lp}(r_2)$, we get a representation of the desired type for $f - r_2$, namely

$$f - r_2 = (c_{11}X_{11} + c_{21}X_{21})f_1 + \cdots + (c_{1s}X_{1s} + c_{2s}X_{2s})f_s.$$

Continuing in this way we eventually obtain a representation for $f - r$ of the desired type. The last statement is clear. □

The method for computing r is given as Algorithm 4.1.1. We note that in obtaining $\operatorname{lt}(r) = c_1X_1 + \cdots + c_sX_s$ in Algorithm 4.1.1, we are assuming that $c_i = 0$ for all i such that $\operatorname{lp}(f_i)$ does not divide $\operatorname{lp}(r)$. Assuming that linear equations are solvable in R we can determine whether the c_i's exist and compute them if they do.

INPUT: $f, f_1, \ldots, f_s \in A$ with $f_i \neq 0$ $(1 \leq i \leq s)$

OUTPUT: $h_1, \ldots, h_s, r$, where $f = h_1f_1 + \cdots + h_sf_s + r$ with

$\operatorname{lp}(f) = \max(\max_{1 \leq i \leq s}(\operatorname{lp}(h_i)\operatorname{lp}(f_i)), \operatorname{lp}(r))$

and r is minimal with respect to $\{f_1, \ldots, f_s\}$

INITIALIZATION: $h_1 := 0, \ldots, h_s := 0, r := f$

WHILE there is an i such that $\operatorname{lp}(f_i)$ divides $\operatorname{lp}(r)$ and there

are $c_1, \ldots, c_s \in R$ and power products $X_1, \ldots, X_s$

such that $\operatorname{lt}(r) = c_1X_1\operatorname{lt}(f_1) + \cdots + c_sX_s\operatorname{lt}(f_s)$ and

$\operatorname{lp}(r) = X_i\operatorname{lp}(f_i)$ for all i such that $c_i \neq 0$ **DO**

$r := r - (c_1X_1f_1 + \cdots + c_sX_sf_s)$

FOR $i := 1$ to s **DO**

$h_i := h_i + c_iX_i$

ALGORITHM 4.1.1. *Division Algorithm over Rings*

EXAMPLE 4.1.11. We reconsider Examples 4.1.2 and 4.1.7 using Algorithm 4.1.1. The first pass through the WHILE loop was done in Example 4.1.2 and gave us $r = f - (yf_1 + (-2x)f_3) = -3y - 10x - 1$, $h_1 = y$, $h_2 = 0$, and $h_3 = -2x$. The second pass through the WHILE was done in Example 4.1.7 and gave us

$r = (-3y - 10x - 1) - (-f_3) = -10x - 6$, $h_1 = y$, $h_2 = 0$, and $h_3 = -2x - 1$. The WHILE loop stops, since only $\operatorname{lp}(f_1)$ divides $\operatorname{lp}(r)$ but there is no c_1 such that $-10x = \operatorname{lt}(r) = c_1 \operatorname{lt}(f_1) = c_1(7x)$. Thus, $f = yf_1 + (-2x - 1)f_3 + (-10x - 6)$.

We may now give the first characteristic properties of Gröbner bases over $A = R[x_1, \ldots, x_n]$, for a ring R.

THEOREM 4.1.12. *Let I be an ideal of A and let $G = \{g_1, \ldots, g_t\}$ be a set of non-zero polynomials in I. Then the following are equivalent.*

(i) $\operatorname{Lt}(G) = \operatorname{Lt}(I)$.

(ii) *For any polynomial $f \in A$ we have*

$$f \in I \text{ if and only if } f \xrightarrow{G}_+ 0.$$

(iii) *For all $f \in I$, $f = h_1 g_1 + \cdots + h_t g_t$ for some polynomials $h_1, \ldots, h_t \in A$ such that $\operatorname{lp}(f) = \max_{1 \le i \le t}(\operatorname{lp}(h_i)\operatorname{lp}(g_i))$.*

PROOF. (i)$\Longrightarrow$ (ii). We know that if $f \xrightarrow{G}_+ 0$, then $f \in I$. Conversely assume that $f \in I$. Then we know from Theorem 4.1.10, that $f \xrightarrow{G}_+ r$ with r minimal. If $r \neq 0$ then, by Lemma 4.1.9, $\operatorname{lt}(r) \notin \operatorname{Lt}(G)$. But $f \in I$ and $f - r \in I$ imply that $r \in I$ and so $\operatorname{lt}(r) \in \operatorname{Lt}(I) = \operatorname{Lt}(G)$, which is a contradiction.

(ii)$\Longrightarrow$ (iii). This is the special case of $r = 0$ of Theorem 4.1.10.

(iii)$\Longrightarrow$(i). For $f \in I$ we need to show that $\operatorname{lt}(f) \in \operatorname{Lt}(G)$. We have that $f = h_1 g_1 + \cdots + h_t g_t$ such that $\operatorname{lp}(f) = \max_{1 \le i \le t}(\operatorname{lp}(h_i)\operatorname{lp}(g_i))$. It is then easily seen that $\operatorname{lt}(f) = \sum \operatorname{lt}(h_i)\operatorname{lt}(f_i)$ where the sum is over all i satisfying $\operatorname{lp}(f) = \operatorname{lp}(h_i)\operatorname{lp}(g_i)$. Thus $\operatorname{lt}(f) \in \operatorname{Lt}(G)$, as desired. □

It is important to notice the form of Statement (iii) in the Theorem. We use $\operatorname{lp}(f) = \max_{1 \le i \le t}(\operatorname{lp}(h_i)\operatorname{lp}(g_i))$ instead of $\operatorname{lp}(f) = \max_{1 \le i \le t} \operatorname{lp}(h_i g_i)$, since over a ring R with zero divisors, we need not have $\operatorname{lp}(h_i)\operatorname{lp}(g_i) = \operatorname{lp}(h_i g_i)$.

DEFINITION 4.1.13. *A set G of non-zero polynomials contained in an ideal I is called a* Gröbner basis *for I provided that G satisfies any one of the three equivalent conditions of Theorem* 4.1.12. *A set G of non-zero polynomials contained in A is called a* Gröbner basis *provided that G is a Gröbner basis for $\langle G \rangle$.*

EXAMPLE 4.1.14. Let $R = \mathbb{Z}$ and $A = \mathbb{Z}[x, y]$ with the deglex ordering with $x < y$. Let $f_1 = 4x + 1$, $f_2 = 6y + 1$ and $I = \langle f_1, f_2 \rangle$. Then $3yf_1 - 2xf_2 = 3y - 2x \in I$ while $\operatorname{lt}(3y - 2x) = 3y \notin \langle \operatorname{lt}(f_1), \operatorname{lt}(f_2) \rangle = \langle 4x, 6y \rangle$, and thus $\{f_1, f_2\}$ is not a Gröbner basis for I. On the other hand, as another example, let $g_1 = 2x + 1, g_2 = 3y + 1$ and set $I' = \langle g_1, g_2 \rangle$. Then by simply looking at all linear combinations of g_1 and g_2, it is easily seen that $\operatorname{Lt}(I') = \langle 2x, 3y, xy \rangle = \langle 2x, 3y \rangle = \langle \operatorname{lt}(g_1), \operatorname{lt}(g_2) \rangle$, and so $\{g_1, g_2\}$ is a Gröbner basis for I'.

COROLLARY 4.1.15. *If G is a Gröbner basis for the ideal I in A, then $I = \langle G \rangle$.*

PROOF. This is immediate from part (iii) of Theorem 4.1.12. □

We note that the remainder r obtained in Theorem 4.1.10 is not necessarily unique, even in the case where F is a Gröbner basis (see Exercise 4.1.6). We have, however, in Theorem 4.1.12 that for G a Gröbner basis and $f \in \langle G \rangle$, the only possible remainder for f is 0 with respect to G. That is,

COROLLARY 4.1.16. *If G is a Gröbner basis and $f \in \langle G \rangle$ and $f \xrightarrow{G}_+ r$, where r is minimal, then $r = 0$.*

PROOF. Since $f \in \langle G \rangle$ we have that $r \in \langle G \rangle$ and so, since G is a Gröbner basis, we see that if $r \neq 0$ then r cannot be minimal by Lemma 4.1.9. □

We further note that the Noetherian property of the ring R, and hence of the ring $R[x_1, \ldots, x_n]$, yields the following Corollary, whose proof we leave to the exercises (Exercise 4.1.5).

COROLLARY 4.1.17. *Let $I \subseteq R[x_1, \ldots, x_n]$ be a non-zero ideal. Then I has a finite Gröbner basis.*

We will develop a method for computing Gröbner bases over rings in the next section. However, in the special case where $R = k[y]$, for a field k and a single variable y, we see from the next theorem that we can compute a Gröbner basis over R using the theory presented in Chapter 1. This result will not be used until Section 4.5.

THEOREM 4.1.18. *If $G = \{g_1, \ldots, g_t\}$ is a Gröbner basis in $k[y, x_1, \ldots, x_n]$ with respect to an elimination order with the x variables larger than y, then G is a Gröbner basis in $(k[y])[x_1, \ldots, x_n]$.*

PROOF. Let $f \in I = \langle g_1, \ldots, g_t \rangle$. We will denote by $\mathrm{Lt}(I), \mathrm{lt}(f), \mathrm{lp}(f)$, and $\mathrm{lc}(f)$ the leading term ideal of I, the leading term, leading power product, and leading coefficient of f with respect to the elimination order $<$, and by $\mathrm{Lt}_x(I), \mathrm{lt}_x(f), \mathrm{lp}_x(f)$, and $\mathrm{lc}_x(f)$ the leading term ideal of I, the leading term, leading power product, and leading coefficient of f in $(k[y])[x_1, \ldots, x_n]$ (i.e., here the order is the one on the x variables alone which we will denote by $<_x$). Note that $\mathrm{lc}_x(f) \in k[y]$. We will denote the leading term of a polynomial $a \in k[y]$ by $\mathrm{lt}_y(a)$.

We need to show that $\mathrm{Lt}_x(I) = \langle \mathrm{lt}_x(g_1), \ldots, \mathrm{lt}_x(g_t) \rangle$. One inclusion is trivial. For the reverse inclusion, let $f \in I$. We write $g_i = a_i X_i +$ lower x terms, where X_i is a power product in the x variables alone and $a_i \in k[y]$; so $\mathrm{lt}_x(g_i) = a_i X_i$. Since G is a Gröbner basis in $k[y, x_1, \ldots, x_n]$ we can apply the Division Algorithm (see Theorem 1.5.9 and its proof) to write

$$f = \alpha_1 y^{\nu_1} T_1 g_{i_1} + \alpha_2 y^{\nu_2} T_2 g_{i_2} + \cdots + \alpha_N y^{\nu_N} T_N g_{i_N}, \tag{4.1.2}$$

where T_j is a power product in the x variables alone, $\alpha_j \in k$, for $1 \le j \le N$ and, since $\mathrm{lp}(\alpha_j y^{\nu_j} T_j g_{i_j}) = y^{\nu_j} T_j \, \mathrm{lp}_y(a_{i_j}) X_{i_j}$,

(4.1.3)
$$y^{\nu_1} T_1 \, \mathrm{lp}_y(a_{i_1}) X_{i_1} > y^{\nu_2} T_2 \, \mathrm{lp}_y(a_{i_2}) X_{i_2} > \cdots > y^{\nu_N} T_N \, \mathrm{lp}_y(a_{i_N}) X_{i_N}.$$

Since we are using an elimination order with y smallest, we must have

$$T_1 X_{i_1} \ge T_2 X_{i_2} \ge \cdots \ge T_N X_{i_N}.$$

Choose j_0 least such that $T_{j_0} X_{i_{j_0}} > T_{j_0+1} X_{i_{j_0+1}}$. Then $T_1 X_{i_1} = T_j X_{i_j}$ for $1 \le j \le j_0$, and $T_1 X_{i_1} > X$ for all other x power products, X, appearing in the right side of Equation (4.1.2). Thus

$$\mathrm{lt}_x(f) = \Big(\sum_{j=1}^{j_0} \alpha_j y^{\nu_j} a_{i_j}\Big) T_1 X_{i_1}, \tag{4.1.4}$$

provided that $h = \sum_{j=1}^{j_0} \alpha_j y^{\nu_j} a_{i_j} \ne 0$. But from Equation (4.1.3), looking at the first j_0 terms and canceling $T_j X_{i_j} = T_1 X_{i_1}$ we get

$$y^{\nu_1} \, \mathrm{lp}_y(a_{i_1}) > y^{\nu_2} \, \mathrm{lp}_y(a_{i_2}) > \cdots > y^{\nu_{j_0}} \, \mathrm{lp}_y(a_{i_{j_0}}),$$

and so $\mathrm{lp}_y(h) = y^{\nu_1} \, \mathrm{lp}_y(a_{i_1}) \ne 0$. Finally, from Equation (4.1.4) we see

$$\mathrm{lt}_x(f) = \sum_{j=1}^{j_0} \alpha_j T_j \, \mathrm{lp}_y(a_{i_j}) X_{i_j} = \sum_{j=1}^{j_0} \alpha_j T_j \, \mathrm{lt}_x(g_{i_j}),$$

as desired. □

The converse of Theorem 4.1.18 is not true as the following example shows.

EXAMPLE 4.1.19. Consider the ideal $I = \langle y(y+1)x, y^2x \rangle \subseteq R[x]$, where $R = k[y]$. Then $\{y(y+1)x, y^2x\}$ is a Gröbner basis in $R[x]$, since, in that ring, the polynomials $y(y+1)x$ and y^2x are both terms. However, $\{y(y+1)x, y^2x\}$ is not a Gröbner basis in $k[x,y]$ with respect to the lex order with $x > y$ (or any other order for that matter), since the polynomial yx is in I but its leading term (itself) is divisible by neither $\mathrm{lt}(y(y+1)x) = y^2x$ nor $\mathrm{lt}(y^2x) = y^2x$.

Exercises

4.1.1. In $\mathbb{Z}[x,y]$, let $f_1 = 6x^2 + y + 1$, $f_2 = 10xy - y - x$, $f_3 = 15y^2 + x$, and $F = \{f_1, f_2, f_3\}$. Consider deglex with $x > y$.

a. Show that $2x^3y \xrightarrow{F} -x^3 - x^2y - 2xy^2 - 2xy$.

b. Show that $x^3y^2 \xrightarrow{F} x^4 + x^3y + x^2y^2 - xy^3 - xy^2$.

c. Show that $x^3y^2 + 5x^4 + x^3y \xrightarrow{F}_+ -xy^3 - x^2y - 2xy^2 - y^3 - x^2 - 2xy - y^2$.

In each case note that the remainder polynomial is minimal with respect to F. There are many ways one can reduce the above polynomials; the reductions above are just one possibility.

4.1.2. In $\mathbb{Z}[x,y,z]$, let $f_1 = x^2yz+yz+1$, $f_2 = 5yz^2-xy+z$, $f_3 = 7yz-z-7y-2$, and $F = \{f_1, f_2, f_3\}$. Consider deglex with $x < y < z$.

a. Use the Division Algorithm (Algorithm 4.1.1) on $f = -3xy + 2xy^2 - 2yz + 2z^2 + 3x^2y^2z^2$ to write $f = h_1f_1 + h_2f_2 + h_3f_3 + r$ where r is minimal with respect to F and

$$\operatorname{lp}(f) = \max(\operatorname{lp}(h_1)\operatorname{lp}(f_1), \operatorname{lp}(h_2)\operatorname{lp}(f_2), \operatorname{lp}(h_3)\operatorname{lp}(f_3), \operatorname{lp}(r)).$$

(There are two obvious ways to begin this exercise; try them both.)

b. Show that the set F is not a Gröbner basis.

4.1.3. In $\mathbb{Z}_{10}[x,y]$, let $f_1 = 3x^2y+3x$, $f_2 = 7xy^2+y$, and $F = \{f_1, f_2\}$. Consider lex with $x > y$.

a. Use the Division Algorithm (Algorithm 4.1.1) on $f = x^3y^3 + 5x^2y^2 + x^2y + 1$ to write $f = h_1f_1 + h_2f_2 + r$, where r is minimal with respect to F and $\operatorname{lp}(f) = \max(\operatorname{lp}(h_1)\operatorname{lp}(f_1), \operatorname{lp}(h_2)\operatorname{lp}(f_2), \operatorname{lp}(r))$. [One answer is $r = -x + 1$.]

b. Show that the set F is not a Gröbner basis.

4.1.4. Show that the subset $F = \{3x^2y - x, 2z^2 - x\} \subseteq \mathbb{Z}[x,y,z]$ is a Gröbner basis with respect to deglex where $x > y > z$ and is not a Gröbner basis with respect to lex where $x > y > z$.

4.1.5. Prove Corollary 4.1.17.

4.1.6. If you attempted to mimic the definition of reduced Gröbner basis of Chapter 1, what would be the significance of the equality of ideals in $\mathbb{Z}[x,y]$, $\langle 2x^2, 3y^2 + x^2\rangle = \langle 2x^2, 3y^2 + 3x^2\rangle$? Alternatively think about this example for the idea of uniqueness of reduction. [Hint: Either set is a Gröbner basis with respect to deglex with $y > x$. Also, try reducing $6x^2y^2 - x^4$.]

4.1.7. Show that for any polynomials $f, g \in R[x_1, \ldots, x_n]$, for any finite set of non-zero polynomials F in $R[x_1, \ldots, x_n]$, and for any power product $X \in R[x_1, \ldots, x_n]$, we have

a. If $f \in F$, then $fg \xrightarrow{F}_+ 0$, provided R is an integral domain.

b. If $f \xrightarrow{F}_+ g$, then $Xf \xrightarrow{F}_+ Xg$.

4.1.8. Let $\{g_1, \ldots, g_t\} \subseteq R[x_1, \ldots, x_n]$ and $0 \neq h \in R[x_1, \ldots, x_n]$, where R is an integral domain. Prove that $\{g_1, \ldots, g_t\}$ is a Gröbner basis if and only if $\{hg_1, \ldots, hg_t\}$ is a Gröbner basis.

4.1.9. Let I be a non-zero ideal in $R[x_1, \ldots, x_n]$ and let G be a Gröbner basis for I. We say that G is a *minimal* Gröbner basis provided that for all $g \in G$, g is minimal with respect to the set $G - \{g\}$. Prove that every Gröbner basis for I contains a minimal Gröbner basis for I.

4.1.10. Let I be a non-zero ideal in $R[x_1, \ldots, x_n]$ and let G be a Gröbner basis for I.

a. Prove that $G \cap R$ generates $I \cap R$.

b. Call a generating set F for an ideal $J \subseteq R$ a *minimal generating* set provided that for all $r \in F$ we have $\langle F - \{r\}\rangle \neq J$. Show that if G

is a minimal Gröbner basis for I (see Exercise 4.1.9), then $G \cap R$ is a minimal generating set for $I \cap R$.

4.1.11. Let I be an ideal in $R[x_1, \dots, x_n]$ and let π denote the quotient map $R[x_1, \dots, x_n] \longrightarrow (R/(I \cap R))[x_1, \dots, x_n]$. Let $G \subseteq I$ be given.

a. Show that if G is a Gröbner basis for I then $\pi(G)$ is a Gröbner basis for $\pi(I)$. [Hint: For all $f \in I$, $f \notin I \cap R$, write $f = f_0 + f_1$ where $\pi(f_0) = 0$ and $f_1 = 0$ or $\pi(\mathrm{lc}(f_1)) \neq 0$.]

b. Show that G is a minimal Gröbner basis for I (see Exercise 4.1.9) if and only if $G \cap R$ is a minimal generating set for $I \cap R$ (see Exercise 4.1.10), $\pi(G - G \cap R)$ is a minimal Gröbner basis for $\pi(I)$, and $\mathrm{lc}(g) \notin I \cap R$ for all $g \in G - G \cap R$.

4.1.12. Let G be a Gröbner basis for a non-zero ideal I of $R[x_1, \dots, x_n]$, where R is an integral domain. Let K be the quotient field of R. Let J be the ideal of $K[x_1, \dots, x_n]$ generated by I. Prove that G is also a Gröbner basis for J with respect to the same order.

4.1.13. Prove Lemma 4.1.3.

4.1.14. Use the ideas in this section and those of Section 3.5 to obtain a theory of Gröbner bases for $R[x_1, \dots, x_n]$-submodules of $R[x_1, \dots, x_n]^m$. In particular state and prove the analog of Theorem 4.1.12 for $R[x_1, \dots, x_n]$-modules.

4.2. Computing Gröbner Bases over Rings. In this section we will give another characterization of Gröbner bases (Theorem 4.2.3) which is similar to the S-polynomial criterion in Theorem 1.7.4, and is the direct analogue of Theorem 3.2.5. Of course, now our syzygy modules are submodules of $(R[x_1, \dots, x_n])^s$. We will then give the analogue of Buchberger's Algorithm (Algorithm 1.7.1) for the case of rings (Algorithm 4.2.1). We will conclude this section by giving an iterative algorithm for computing the syzygy module needed for this generalized Buchberger Algorithm (see Algorithm 4.2.2).

We note, by Theorem 1.1.3, that R Noetherian implies $R[x_1, \dots, x_n]$ is Noetherian. Moreover, by Theorem 3.1.1, $(R[x_1, \dots, x_n])^s$ is Noetherian, and hence any submodule of it is Noetherian and finitely generated.

DEFINITION 4.2.1. *Given power products $X_1, \dots, X_s$ and non-zero elements $c_1, \dots, c_s$ in R set $L = \begin{bmatrix} c_1X_1 & \cdots & c_sX_s \end{bmatrix}$. Then, for a power product X, we call a syzygy $\mathbf{h} = (h_1, \dots, h_s) \in \mathrm{Syz}(L) \subseteq (R[x_1, \dots, x_n])^s$ homogeneous of degree X provided that each h_i is a term (i.e. $\mathrm{lt}(h_i) = h_i$ for all i) and $X_i \,\mathrm{lp}(h_i) = X$ for all i such that $h_i \neq 0$.*

We have the following easy lemma.

LEMMA 4.2.2. *With the notation above, $\mathrm{Syz}(L)$ has a finite generating set of homogeneous syzygies.*

PROOF. As noted above $\mathrm{Syz}(L)$ is finitely generated. Thus it suffices to show that given any syzygy $\boldsymbol{h} = (h_1, \ldots, h_s) \in \mathrm{Syz}(L)$ we may write $\boldsymbol{h}$ as a sum of homogeneous syzygies. Now we know that $h_1c_1X_1 + \cdots + h_sc_sX_s = 0$. Then expanding the polynomials h_i into their individual terms, we see that for any power product X we may collect together all the terms in this last sum whose power product is X and these must also add to zero, since all the c_iX_i are terms. This immediately gives the desired representation. □

We can now give the following theorem whose proof parallels exactly the proof of the similar Theorem 3.2.5 (Exercise 4.2.1).

THEOREM 4.2.3. *Let $G = \{g_1, \ldots, g_t\}$ be a set of non-zero polynomials in A. Let $\mathcal{B}$ be a homogeneous generating set for* $\mathrm{Syz}(\mathrm{lt}(g_1), \ldots, \mathrm{lt}(g_t))$. *Then G is a Gröbner basis for the ideal $\langle g_1, \ldots, g_t \rangle$ if and only if for all $(h_1, \ldots, h_t) \in \mathcal{B}$, we have*

$$h_1g_1 + \cdots + h_tg_t \xrightarrow{G}_+ 0.$$

One can view the expression $h_1g_1 + \cdots + h_tg_t$ above as a generalized S-polynomial, since in that expression the leading terms cancel (this is the basic idea that was used in Section 3.3). Thus we see how we will go about generalizing Buchberger's Algorithm. We first must find a homogeneous generating set for the syzygy module of the leading terms. We then form the generalized S-polynomials and reduce each one of them using the reduction presented in the previous section. If one of these does not reduce to zero, we add the reduction to our set and repeat the procedure.

The next question we must answer is how do we go about constructing a homogeneous generating set for $\mathrm{Syz}(\mathrm{lt}(g_1), \ldots, \mathrm{lt}(g_t))$. Or in general, given power products $X_1, \ldots, X_s$ and non-zero elements $c_1, \ldots, c_s \in R$ how do we construct a homogeneous generating set for $\mathrm{Syz}(c_1X_1, \ldots, c_sX_s)$. In view of Equation (4.1.1) and the surrounding discussion, we make the following

DEFINITION 4.2.4. *For any subset $J \subseteq \{1, \ldots, s\}$, set $X_J = \mathrm{lcm}(X_j \mid j \in J)$. We say that J is* saturated *with respect to $X_1, \ldots, X_s$ provided that for all $j \in \{1, \ldots, s\}$, if X_j divides X_J, then $j \in J$. For any subset $J \subseteq \{1, \ldots, s\}$ we call the* saturation *of J the set J' consisting of all $j \in \{1, \ldots, s\}$ such that X_j divides X_J. (Note that $X_J = X_{J'}$.)*

EXAMPLE 4.2.5. Let $X_1 = xy, X_2 = x^2, X_3 = y$, and $X_4 = x^4$. Then if $J = \{1, 2\}$ we see that $X_J = x^2y$ and J is not saturated since $3 \notin J$, while $X_3 = y$ divides $X_J = x^2y$. On the other hand, if $J = \{1, 2, 3\}$ we see that $X_J = x^2y$ and J is saturated since 4 is the only element of $\{1, 2, 3, 4\}$ not in J and $X_4 = x^4$ does not divide $X_J = x^2y$. Clearly $\{1, 2, 3\}$ is the saturation of $\{1, 2\}$.

We recall the notation for the standard basis vectors

$$\boldsymbol{e}_1 = (1, 0, \dots, 0), \boldsymbol{e}_2 = (0, 1, 0, \dots, 0), \dots, \boldsymbol{e}_s = (0, 0, \dots, 0, 1)$$

for A^s. Given the above notation we are now prepared to state

THEOREM 4.2.6. *For each set $J \subseteq \{1, \dots, s\}$, which is saturated with respect to $X_1, \dots, X_s$, let $\mathcal{B}_J = \{\boldsymbol{b}_{1J}, \dots, \boldsymbol{b}_{\nu_J J}\}$ be a generating set for the R-module of syzygies $\mathrm{Syz}_R(c_j \mid j \in J)$. (Note that each of the vectors $\boldsymbol{b}_{\nu J}$ is in the R-module $R^{|J|}$, where $|J|$ denotes the cardinality of J). For each such $\boldsymbol{b}_{\nu J}$, denote its jth coordinate, for $j \in J$, by $b_{j\nu J}$. Set*

$$\boldsymbol{s}_{\nu J} = \sum_{j \in J} b_{j\nu J} \frac{X_J}{X_j} \boldsymbol{e}_j.$$

(Note that each of the vectors $s_{\nu J}$ is in A^s.) Then the set of vectors $\boldsymbol{s}_{\nu J}$, for J ranging over all such saturated subsets of $\{1, \dots, s\}$ and $1 \le \nu \le \nu_J$, forms a homogeneous generating set for the syzygy module $\mathrm{Syz}(c_1 X_1, \dots, c_s X_s)$.

PROOF. It is first of all clear that each of the vectors $\boldsymbol{s}_{\nu J}$ is homogeneous of degree X_J. Moreover, $\boldsymbol{s}_{\nu J}$ is a syzygy of $\left[\begin{array}{ccc} c_1X_1 & \cdots & c_sX_s \end{array}\right]$, since

$$\begin{aligned} \left[\begin{array}{ccc} c_1X_1 & \cdots & c_sX_s \end{array}\right] \boldsymbol{s}_{\nu J} &= \left[\begin{array}{ccc} c_1X_1 & \cdots & c_sX_s \end{array}\right] \sum_{j \in J} b_{j\nu J} \frac{X_J}{X_j} \boldsymbol{e}_j \\ &= \sum_{j \in J} b_{j\nu J} \frac{X_J}{X_j} c_j X_j = X_J \sum_{j \in J} b_{j\nu J} c_j = 0, \end{aligned}$$

by the definition of $\boldsymbol{b}_{\nu J}$. Now, let $\boldsymbol{h} = (h_1, \dots, h_s) \in \mathrm{Syz}(c_1X_1, \dots, c_sX_s)$. Since, by Lemma 4.2.2, $\mathrm{Syz}(c_1X_1, \dots, c_sX_s)$ is generated by homogeneous syzygies, it suffices to write $\boldsymbol{h}$ as a linear combination of the $\boldsymbol{s}_{\nu J}$'s in the case that $\boldsymbol{h}$ is a homogeneous syzygy, say of degree Y. We write $\boldsymbol{h} = (d_1Y_1, \dots, d_sY_s)$ for $d_1, \dots, d_s \in R$ and power products $Y_1, \dots, Y_s$. Set $J = \{j \mid d_j \neq 0\}$ and let J' be the saturation of J. We have that $Y_jX_j = Y$ for all $j \in J$, since $\boldsymbol{h}$ is homogeneous of degree Y. Then, since $\boldsymbol{h}$ is a syzygy of $\left[\begin{array}{ccc} c_1X_1 & \cdots & c_sX_s \end{array}\right]$, we have $\sum_{j \in J} d_jY_jc_jX_j = Y \sum_{j\in J} d_jc_j = 0$. Thus $(d_j \mid j \in J')$ is a syzygy of $\left[\begin{array}{c} c_j \mid j \in J' \end{array}\right]$ and so by hypothesis

$$(d_j \mid j \in J') = \sum_{\nu=1}^{\nu_{J'}} r_\nu \boldsymbol{b}_{\nu J'},$$

that is, for each $j \in J'$,

$$d_j = \sum_{\nu=1}^{\nu_{J'}} r_\nu b_{j\nu J'},$$

for some $r_\nu \in R$. Now $Y_j X_j = Y$ for all $j \in J$ implies that $X_J = X_{J'}$ divides Y. It now follows that

$$\sum_{\nu=1}^{\nu_{J'}} r_\nu \frac{Y}{X_{J'}} \boldsymbol{s}_{\nu J'} = \sum_{\nu=1}^{\nu_{J'}} \sum_{j \in J'} r_\nu \frac{Y}{X_{J'}} \frac{X_{J'}}{X_j} b_{j\nu J'} \boldsymbol{e}_j$$

$$= \sum_{j \in J'} (\sum_{\nu=1}^{\nu_{J'}} r_\nu b_{j\nu J'}) Y_j \boldsymbol{e}_j = \sum_{j \in J'} d_j Y_j \boldsymbol{e}_j = \sum_{j=1}^{s} d_j Y_j \boldsymbol{e}_j = \boldsymbol{h},$$

as desired. □

We note that we have reduced the problem of computing a generating set for $\mathrm{Syz}(c_1X_1, \ldots, c_sX_s)$ to computing the subsets of $\{1, \ldots, s\}$ which are saturated with respect to $X_1, \ldots, X_s$ (an easy task) and computing syzygies in R (see Definition 4.1.5).

EXAMPLE 4.2.7. We consider $R = \mathbb{Z}$ and let $c_1X_1 = 2xyz, c_2X_2 = 5xy^2, c_3X_3 = 85y^2$, and $c_4X_4 = 6x^2z$. We will assume that the reader can solve the elementary linear diophantine equations that occur in this example (i.e. that the reader can "solve linear equations in $R = \mathbb{Z}$"). One readily checks that the saturated subsets of $\{1, 2, 3, 4\}$ are $\{1\}, \{3\}, \{4\}, \{1, 4\}, \{2, 3\}, \{1, 2, 3\}$, and $\{1, 2, 3, 4\}$. Since $R = \mathbb{Z}$ is an integral domain, the singletons $\{1\}, \{3\}, \{4\}$ do not give rise to any non-zero syzygy. For $J = \{1, 4\}$ we need to solve in $R = \mathbb{Z}$ the equation $2b_1 + 6b_4 = 0$ and one finds a generating set for the solutions of this equation to be $\{(-3, 1)\}$. Since $X_J = x^2yz$, the corresponding syzygy is $-3\frac{x^2yz}{xyz}\boldsymbol{e}_1 + \frac{x^2yz}{x^2z}\boldsymbol{e}_4 = (-3x, 0, 0, y)$. Now for $J = \{2, 3\}$, we need to solve $5b_2 + 85b_3 = 0$ which gives us $\{(-17, 1)\}$ and the syzygy $-17\frac{xy^2}{xy^2}\boldsymbol{e}_2 + \frac{xy^2}{y^2}\boldsymbol{e}_3 = (0, -17, x, 0)$. The set $J = \{1, 2, 3\}$ gives the diophantine equation $2b_1 + 5b_2 + 85b_3 = 0$. This may be solved to yield two generators for the solutions, namely, $(-40, -1, 1)$ and $(-5, 2, 0)$. Then, with $X_J = xy^2z$, we obtain the syzygies $-40\frac{xy^2z}{xyz}\boldsymbol{e}_1 - \frac{xy^2z}{xy^2}\boldsymbol{e}_2 + \frac{xy^2z}{y^2}\boldsymbol{e}_3 = (-40y, -z, xz, 0)$ and $-5\frac{xy^2z}{xyz}\boldsymbol{e}_1 + 2\frac{xy^2z}{xy^2}\boldsymbol{e}_2 = (-5y, 2z, 0, 0)$. Finally for $J = \{1, 2, 3, 4\}$, we get the generators $(-40, -1, 1, 0), (-5, 2, 0, 0)$ and $(-3, 0, 0, 1)$. These will give syzygies that have already been obtained, as is readily checked, and so are not needed in our generating set. So we obtain that

$$\mathrm{Syz}(2xyz, 5xy^2, 85y^2, 6x^2z) = \langle(-3x, 0, 0, y), (0, -17, x, 0),$$

$$(-40y, -z, xz, 0), (-5y, 2z, 0, 0)\rangle.$$

Now that we can compute, by Theorem 4.2.6, a homogeneous generating set for $\mathrm{Syz}(\mathrm{lt}(f_1), \ldots, \mathrm{lt}(f_s))$, for any set of polynomials $\{f_1, \ldots, f_s\}$, we can give Algorithm 4.2.1, the algorithm for computing Gröbner bases for ideals in $A = R[x_1, \ldots, x_n]$.

THEOREM 4.2.8. *If R is a Noetherian ring and linear equations are solvable in R then Algorithm 4.2.1 produces a Gröbner basis for the ideal $\langle f_1, \ldots, f_s\rangle$.*

INPUT: $F = \{f_1, \ldots, f_s\} \subseteq A = R[x_1, \ldots, x_n]$

with $f_i \neq 0$ $(1 \leq i \leq s)$

OUTPUT: $G = \{g_1, \ldots, g_t\}$, a Gröbner basis for $\langle f_1, \ldots, f_s \rangle$

INITIALIZATION: $G := \emptyset, G' := F$

WHILE $G' \neq G$ **DO**

$G := G'$

Let the elements of G be $g_1, \ldots, g_t$

Compute $\mathcal{B}$, a homogeneous generating set

for $\mathrm{Syz}(\mathrm{lt}(g_1), \ldots, \mathrm{lt}(g_t))$

FOR each $\boldsymbol{h} := (h_1, \ldots, h_t) \in \mathcal{B}$ **DO**

Reduce $h_1 g_1 + \cdots + h_t g_t \xrightarrow{G'}_+ r$,

with r minimal with respect to G'

IF $r \neq 0$ **THEN**

$G' := G' \cup \{r\}$

ALGORITHM 4.2.1. *Gröbner Basis Algorithm over Rings.*

PROOF. If the algorithm stops, it clearly, by Theorem 4.2.3, gives a Gröbner basis for the ideal $\langle f_1, \ldots, f_s \rangle$. As the algorithm progresses we add to a set of polynomials G a polynomial r, minimal with respect to G, to obtain a new set G'. By Lemma 4.1.9, $\mathrm{lt}(r) \notin \mathrm{Lt}(G)$ and so $\mathrm{Lt}(G) \subsetneq \mathrm{Lt}(G')$. Thus, since the ring A is Noetherian, the algorithm stops. □

We will give examples of computing Gröbner bases shortly, after we have given a more efficient method for computing the relevant syzygies. We will incorporate this method for computing syzygies into our Gröbner basis Algorithm and so we will actually use Algorithm 4.2.2 to compute these examples (see Examples 4.2.11, 4.2.12, and 4.2.13)

Möller [**Mö88**], has given a method for computing the syzygies that arise in Algorithm 4.2.1. In particular, this method gives an efficient way to avoid duplication in the computation of syzygies (see Algorithm 4.2.2). At each stage of Algorithm 4.2.1 we add one or more polynomials to the generating set for the ideal $\langle f_1, \ldots, f_s \rangle$. So we have to compute a generating set for the syzygy module of this increased set of leading terms. The idea of Möller is to use the computations of syzygies already done in order to compute the new syzygies.

Note that any syzygy of a set of leading terms is automatically a syzygy of a larger set of leading terms (using zeros for the extra leading terms).

Again consider power products $X_1, \dots, X_s$ and non-zero elements $c_1, \dots, c_s \in R$. Let $S_\sigma = \operatorname{Syz}(c_1X_1, \dots, c_\sigma X_\sigma)$ for $1 \leq \sigma \leq s$. We will compute a homogeneous generating set of $S_s = \operatorname{Syz}(c_1X_1, \dots, c_sX_s)$ by inductively constructing generating sets for the S_σ. We note that a homogeneous generating set of S_1 consists of a generating set of the ideal $\{r \in R \mid rc_1 = 0\}$ of R (called the *annihilator* of c_1 and denoted by $\operatorname{ann}(c_1)$) viewed as a subset of A. Also, if we take a homogeneous syzygy $(b_1Y_1, \dots, b_\sigma Y_\sigma)$ in S_σ, there are two possibilities. Either $b_\sigma = 0$ and $(b_1Y_1, \dots, b_{\sigma-1}Y_{\sigma-1})$ is a homogeneous syzygy in $S_{\sigma-1}$; or $b_\sigma \neq 0$ and[2] $b_\sigma \in \langle c_1, \dots, c_{\sigma-1}\rangle_R \colon \langle c_\sigma\rangle_R$.

So we proceed as follows. We consider all subsets J of $\{1, \dots, \sigma\}$, saturated with respect to $X_1, \dots, X_\sigma$, such that $\sigma \in J$. For each such J let $\{b_{1J}, \dots, b_{\mu_J J}\}$ denote a generating set for the ideal in R, $\langle c_j \mid j \in J, j \neq \sigma\rangle_R \colon \langle c_\sigma\rangle_R$. Now for each $b_{\mu J}$ there are $b_j \in R$ such that

$$\sum_{\substack{j\in J\\ j\neq\sigma}} b_jc_j + b_{\mu J}c_\sigma = 0,$$

and we define the homogeneous syzygy in S_σ by

$$\boldsymbol{s}_{\mu J} = \sum_{\substack{j\in J\\ j\neq\sigma}} b_j\frac{X_J}{X_j}\boldsymbol{e}_j + b_{\mu J}\frac{X_J}{X_\sigma}\boldsymbol{e}_\sigma.$$

(The b_j's may not be unique, but any choice will do.) We also consider a homogeneous generating set $\boldsymbol{a}_1, \dots, \boldsymbol{a}_m$ for $S_{\sigma-1}$, which we assume, by induction, has been computed. Define for $1 \leq i \leq m$, $(\boldsymbol{a}_i, 0)$ to be the homogeneous syzygy in S_σ with the coordinates of $\boldsymbol{a}_i$ in the first $\sigma - 1$ coordinates and 0 in the last coordinate. We now can state

THEOREM 4.2.9. *The syzygies $(\boldsymbol{a}_1, 0), \dots, (\boldsymbol{a}_m, 0)$ together with the syzygies $\boldsymbol{s}_{\mu J}$ for $J \subseteq \{1, \dots, \sigma\}$ saturated with respect to $X_1, \dots, X_\sigma$ with $\sigma \in J$, form a homogeneous generating set for the syzygy module S_σ.*

PROOF. Let $\boldsymbol{d} = (d_1Y_1, \dots, d_\sigma Y_\sigma)$, for $d_1, \dots, d_\sigma \in R$ and power products $Y_1, \dots, Y_\sigma$, be a homogeneous syzygy in S_σ of degree Y. If $d_\sigma = 0$ it is clear that $\boldsymbol{d}$ is a linear combination of $(\boldsymbol{a}_1, 0), \dots, (\boldsymbol{a}_m, 0)$. So assume that $d_\sigma \neq 0$. Let J' be the set of all j such that $d_j \neq 0$ and let J be the saturation of J' inside $\{1, \dots, \sigma\}$. Note that $X_J = X_{J'}$ divides Y. Then $d_\sigma \in \langle c_j \mid j \in J, j \neq \sigma\rangle_R \colon \langle c_\sigma\rangle_R$ and so $d_\sigma = \sum_{i=1}^{\mu_J} r_i b_{iJ}$. Then it is easily checked that

$$\boldsymbol{d} - \sum_{\mu=1}^{\mu_J} r_\mu \boldsymbol{s}_{\mu J}\frac{Y}{X_J}$$

[2]Recall that for two ideals $I, J \subseteq R$ the ideal quotient, $I \colon J$, is defined to be $I \colon J = \{r \in R \mid rJ \subseteq I\}$.

is a homogeneous syzygy with a zero in its σth coordinate and so is a linear combination of $(\boldsymbol{a}_1, 0), \ldots, (\boldsymbol{a}_m, 0)$, giving the desired result. $\square$

EXAMPLE 4.2.10. We will redo Example 4.2.7 using this method. We start with $S_1 = \text{Syz}(2xyz) = \langle 0 \rangle$. We now consider $S_2 = \text{Syz}(2xyz, 5xy^2)$. The only saturated subset of $\{1,2\}$ that need be considered is $\{1,2\}$ itself (actually $\{2\}$ is saturated and contains 2 but will only yield the $\mathbf{0}$ syzygy). We note that $\langle 2 \rangle_{\mathbb{Z}} \colon \langle 5 \rangle_{\mathbb{Z}} = \langle 2 \rangle_{\mathbb{Z}}$ giving the syzygy $(-5, 2)$ of $\begin{bmatrix} 2 & 5 \end{bmatrix}$. Thus $S_2 = \langle (-5y, 2z) \rangle$. We now turn to S_3 and note that the saturated subsets of $\{1,2,3\}$ containing 3 are $\{2,3\}$ and $\{1,2,3\}$. Working with the first one, we note that $\langle 5 \rangle_{\mathbb{Z}} \colon \langle 85 \rangle_{\mathbb{Z}} = \langle 1 \rangle_{\mathbb{Z}}$ giving the syzygy $(-17, 1)$ of $\begin{bmatrix} 5 & 85 \end{bmatrix}$ and so the syzygy $(0, -17, x)$ in S_3. For the second set, $\{1,2,3\}$, we note that $\langle 2, 5 \rangle_{\mathbb{Z}} \colon \langle 85 \rangle_{\mathbb{Z}} = \langle 1 \rangle_{\mathbb{Z}}$ and we may use the same syzygy as before. Therefore $S_3 = \langle (-5y, 2z, 0), (0, -17, x) \rangle$. We finally turn our attention to S_4 and note that the saturated subsets of $\{1,2,3,4\}$ containing 4 are $\{1,4\}$ and $\{1,2,3,4\}$. For $\{1,4\}$ we compute $\langle 2 \rangle_{\mathbb{Z}} \colon \langle 6 \rangle_{\mathbb{Z}} = \langle 1 \rangle_{\mathbb{Z}}$ giving the syzygy $(-3, 1)$ of $\begin{bmatrix} 2 & 6 \end{bmatrix}$ and so the syzygy $(-3x, 0, 0, y)$ in S_4. Finally for $\{1,2,3,4\}$ we need to compute $\langle 2, 5, 85 \rangle_{\mathbb{Z}} \colon \langle 6 \rangle_{\mathbb{Z}} = \langle 1 \rangle_{\mathbb{Z}}$ and we may use the same syzygy as we did for $\{1,4\}$. So we obtain

$$S_4 = \text{Syz}(2xyz, 5xy^2, 85y^2, 6x^2z) = \langle (-5y, 2z, 0, 0), (0, -17, x, 0), (-3x, 0, 0, y) \rangle.$$

We note that the syzygy $(-40y, -z, xz, 0)$ is not in this list, but was included in Example 4.2.7. It is not needed, since $(-40y, -z, xz, 0) = z(0, -17, x, 0) + 8(-5y, 2z, 0, 0)$.

We give the improvement of Algorithm 4.2.1 which makes use of Theorem 4.2.9 as Algorithm 4.2.2. We leave the easy proof that it is correct to the exercises (Exercise 4.2.9).

We close this section by giving some examples of Algorithm 4.2.2. Since the polynomials generated by the algorithm are scattered throughout the text of the example, we have put boxes around them for easier reference.

EXAMPLE 4.2.11. We continue with $R = \mathbb{Z}$ and use lex with $x > y$ in $\mathbb{Z}[x, y]$. Let $I = \langle f_1, f_2 \rangle$, where $\boxed{f_1 = 3x^2y + 7y}$ and $\boxed{f_2 = 4xy^2 - 5x.}$ The case $\sigma = 1$ in the algorithm does not generate new polynomials, since $\langle 0 \rangle_{\mathbb{Z}} \colon \langle 3 \rangle_{\mathbb{Z}} = \{0\}$ (we note that, since $R = \mathbb{Z}$ is an integral domain, we never need to consider singleton saturated sets). Now consider the case $\sigma = 2$. The only non-singleton saturated subset containing 2 is $\{1,2\}$. We compute $\langle 3 \rangle_{\mathbb{Z}} \colon \langle 4 \rangle_{\mathbb{Z}} = \langle 3 \rangle_{\mathbb{Z}}$ which gives the syzygy $(4y, -3x)$ in $\text{Syz}(3x^2y, 4xy^2)$. The corresponding S-polynomial, $4yf_1 - 3xf_2 = 15x^2 + 28y^2$, is minimal with respect to G and so we add it to G as $\boxed{f_3 = 15x^2 + 28y^2.}$ To compute new syzygies, we need only consider the saturated subsets $J \subseteq \{1,2,3\}$ containing 3 and they are $\{1,3\}$ and $\{1,2,3\}$. For $\{1,3\}$, we compute $\langle 3 \rangle_{\mathbb{Z}} \colon \langle 15 \rangle_{\mathbb{Z}} = \langle 1 \rangle_{\mathbb{Z}}$ giving the syzygy $(5, 0, -y)$ in $\text{Syz}(3x^2y, 4xy^2, 15x^2)$. The corresponding S-polynomial $5f_1 - yf_3 = -28y^3 + 35y$ is minimal with respect to G and so we add it to G as $\boxed{f_4 = -28y^3 + 35y.}$ We also compute the syzygies corresponding to $\{1,2,3\}$ by computing $\langle 3, 4 \rangle_{\mathbb{Z}} \colon \langle 15 \rangle_{\mathbb{Z}} =$

INPUT: $F = \{f_1, \dots, f_s\} \subseteq R[x_1, \dots, x_n]$ with $f_i \neq 0$ $(1 \leq i \leq s)$

OUTPUT: G a Gröbner basis for $\langle f_1, \dots, f_s \rangle$

INITIALIZATION: $G := F$, $\sigma := 1$, $m := s$

WHILE $\sigma \leq m$ **DO**

Compute $\mathcal{S} = \{$subsets of $\{1, \dots, \sigma\}$, saturated with respect to $\mathrm{lp}(f_1), \dots, \mathrm{lp}(f_\sigma)$, which contain $\sigma\}$

FOR each $J \in \mathcal{S}$ **DO**

$X_J := \mathrm{lcm}(\mathrm{lp}(f_j) | j \in J)$

Compute generators b_{iJ}, $i = 1, \dots, \mu_J$

for $\langle \mathrm{lc}(f_j) | j \in J, j \neq \sigma \rangle_R \colon \langle \mathrm{lc}(f_\sigma) \rangle_R$

FOR $i := 1, \dots, \mu_J$ **DO**

Compute $b_j \in R$, $j \in J, j \neq \sigma$

such that $\displaystyle\sum_{j \in J, j \neq \sigma} b_j \,\mathrm{lc}(f_j) + b_{iJ} \,\mathrm{lc}(f_\sigma) = 0$

Reduce $\displaystyle\sum_{j \in J, j \neq \sigma} b_j \frac{X_J}{\mathrm{lp}(f_j)} f_j + b_{iJ} \frac{X_J}{\mathrm{lp}(f_\sigma)} f_\sigma \xrightarrow{G}_+ r,$

where r is minimal with respect to G

IF $r \neq 0$ **THEN**

$f_{m+1} := r$

$G := G \cup \{f_{m+1}\}$

$m := m + 1$

$\sigma := \sigma + 1$

ALGORITHM 4.2.2. *Gröbner Basis Algorithm in $R[x_1, \dots, x_n]$ using Möller's Technique*

$\langle 1 \rangle_{\mathbb{Z}}$ and we may use the same syzygy $(5, 0, -y)$ as before. The saturated subsets of $\{1, 2, 3, 4\}$ containing 4 are $\{2, 4\}$ and $\{1, 2, 3, 4\}$. For the first we compute $\langle 4 \rangle_{\mathbb{Z}} \colon \langle -28 \rangle_{\mathbb{Z}} = \langle 1 \rangle_{\mathbb{Z}}$ which gives the syzygy $(0, 7y, 0, x)$. The corresponding S-polynomial is $7yf_2 + xf_4 = -35xy + 35xy = 0$. As before the set $\{1, 2, 3, 4\}$ gives no new polynomial. We see now that $\{f_1, f_2, f_3, f_4\}$ is a Gröbner basis for $\langle f_1, f_2 \rangle$. This is a minimal Gröbner basis (see Exercise 4.1.9 for the definition of minimal Gröbner basis).

EXAMPLE 4.2.12. We consider the ideal I in $\mathbb{Z}_{20}[x, y]$ (where $\mathbb{Z}_{20}$ is the ring $\mathbb{Z}/20\mathbb{Z}$) generated by the polynomials $\boxed{f_1 = 4xy + x,}$ and $\boxed{f_2 = 3x^2 + y.}$ We use the lex term order with $x > y$. We again follow Algorithm 4.2.2 to compute a Gröbner basis for I. We first consider $\sigma = 1$. The only saturated subset of $\{1\}$ is $\{1\}$ itself. We compute $\langle 0 \rangle : \langle 4 \rangle = \langle 5 \rangle$ (throughout this example we will use $\langle \cdots \rangle$ for $\langle \cdots \rangle_{\mathbb{Z}_{20}}$). This gives rise to the polynomial $5f_1 = 5x$. This polynomial cannot be reduced so we let $\boxed{f_3 = 5x,}$ and we add it to G. We next compute the saturated subsets of $\{1, 2\}$ which contain 2. These are $\{2\}$, and $\{1, 2\}$. Since 3 is a unit in $\mathbb{Z}_{20}$, $\{2\}$ does not give rise to a new polynomial, i.e. $\langle 0 \rangle : \langle 3 \rangle = \langle 0 \rangle$. For the set $\{1, 2\}$ we compute $\langle 4 \rangle : \langle 3 \rangle = \langle 4 \rangle$ which gives rise to the polynomial $-3xf_1 + 4yf_2 = -3x^2 + 4y^2 \xrightarrow{f_2} 4y^2 + y$. This polynomial cannot be reduced so we let $\boxed{f_4 = 4y^2 + y,}$ and we add it to G. We now compute the saturated subsets of $\{1, 2, 3\}$ which contain 3. These are $\{3\}, \{1, 3\}, \{2, 3\}, \{1, 2, 3\}$. For the set $\{3\}$ we compute $\langle 0 \rangle : \langle 5 \rangle = \langle 4 \rangle$ which gives rise to the polynomial $4f_3 = 0$. For the set $\{1, 3\}$ we compute $\langle 4 \rangle : \langle 5 \rangle = \langle 4 \rangle$ which gives rise to the polynomial $5f_1 - 4yf_3 = 5x \xrightarrow{f_3} 0$. For the set $\{2, 3\}$ we compute $\langle 3 \rangle : \langle 5 \rangle = \langle 1 \rangle$ which gives rise to the polynomial $15f_2 - xf_3 = 15y$ which cannot be reduced. Therefore we let $\boxed{f_5 = 15y,}$ and add it to G. For the set $\{1, 2, 3\}$ we see that $\langle 4, 3 \rangle : \langle 5 \rangle = \langle 3 \rangle : \langle 5 \rangle = \langle 1 \rangle$ and this gives rise to the same polynomial as the set $\{2, 3\}$. We next compute the saturated subsets of $\{1, 2, 3, 4\}$ which contain 4. These are $\{4\}, \{1, 3, 4\}$, and $\{1, 2, 3, 4\}$. For the set $\{4\}$ we compute $\langle 0 \rangle : \langle 4 \rangle = \langle 5 \rangle$ which gives rise to the polynomial $5f_4 = 5y \xrightarrow{f_5} 0$. For the set $\{1, 3, 4\}$ we compute $\langle 4, 5 \rangle : \langle 4 \rangle = \langle 1 \rangle$ which gives rise to the polynomial $yf_1 - xf_4 = 0$. The set $\{1, 2, 3, 4\}$ does not generate a new polynomial, since $\langle 4, 3, 5 \rangle : \langle 4 \rangle = \langle 4, 5 \rangle : \langle 4 \rangle$. We compute the saturated subsets of $\{1, 2, 3, 4, 5\}$ which contain 5. These are $\{5\}, \{1, 3, 5\}$, $\{1, 2, 3, 5\}$, $\{4, 5\}$, $\{1, 3, 4, 5\}$ and $\{1, 2, 3, 4, 5\}$. For the set $\{5\}$ we compute $\langle 0 \rangle : \langle 15 \rangle = \langle 4 \rangle$ which gives rise to the polynomial $4f_5 = 0$. For the set $\{1, 3, 5\}$ we compute $\langle 4, 5 \rangle : \langle 15 \rangle = \langle 1 \rangle$ which gives rise to the polynomial $3yf_3 - xf_5 = 0$. Note then that $\{1, 2, 3, 5\}$, $\{1, 3, 4, 5\}$ and $\{1, 2, 3, 4, 5\}$ do not give rise to any new polynomials. Finally, for the set $\{4, 5\}$ we compute $\langle 4 \rangle : \langle 15 \rangle = \langle 0 \rangle : \langle 15 \rangle = \langle 4 \rangle$ which does not give rise to a new polynomial. Thus the algorithm stops and a Gröbner basis for I is $\{f_1, f_2, f_3, f_4, f_5\}$. Note that we have also shown that

$$\mathrm{Syz}(\mathrm{lt}(f_1), \mathrm{lt}(f_2), \mathrm{lt}(f_3), \mathrm{lt}(f_4), \mathrm{lt}(f_5)) =$$

$$\langle (5, 0, 0, 0, 0), (-3x, 4y, 0, 0, 0), (0, 0, 4, 0, 0), (5, 0, -4y, 0, 0), (0, 15, -x, 0, 0),$$

$$(0, 0, 0, 5, 0), (y, 0, 0, -x, 0), (0, 0, 0, 0, 4), (0, 0, 3y, 0, -x) \rangle.$$

This will be used later.

EXAMPLE 4.2.13. We will now give one more example where the coefficient ring does not have the property that every ideal is principal. We let $R =$

$\mathbb{Z}[\sqrt{-5}] = \{u + v\sqrt{-5} | u, v \in \mathbb{Z}\}$. It is well-known (see, for example, [**AdGo**]) that $\mathbb{Z}[\sqrt{-5}]$ is not a UFD and so not a PID. Indeed it is not hard to show that $2, 3, 1 + \sqrt{-5}$, and $1 - \sqrt{-5}$ are not units and cannot be factored in $\mathbb{Z}[\sqrt{-5}]$, but that $2 \cdot 3 = (1 + \sqrt{-5})(1 - \sqrt{-5})$. According to Theorem 4.2.3, in order to compute a Gröbner basis in such a setting we will need to be able to compute a homogeneous generating set for the syzygies of the leading terms of polynomials. We will follow Algorithm 4.2.2. To do this we need to compute ideals of the form $\langle c_1, \ldots, c_\ell \rangle : \langle c \rangle$ for $c_1, \ldots, c_\ell, c \in \mathbb{Z}[\sqrt{-5}]$ (in this example, of course, $\langle c_1, \ldots, c_\ell \rangle$ means $\langle c_1, \ldots, c_\ell \rangle_{\mathbb{Z}[\sqrt{-5}]}$). One can do this either by using some elementary theory of quadratic fields (see [**AdGo**]) to find generators for these ideals, or one may proceed as follows. We need to find all $\alpha \in \mathbb{Z}[\sqrt{-5}]$ such that $c\alpha = c_1\alpha_1 + \cdots + c_\ell\alpha_\ell$ for some $\alpha_1, \ldots, \alpha_\ell \in \mathbb{Z}[\sqrt{-5}]$. Each of the c's and α's are of the form $u + v\sqrt{-5}$ for integers u, v. So in the desired equation, one can simply equate the real and imaginary parts and obtain a pair of linear Diophantine Equations. These are discussed in elementary number theory courses (see Niven, Zuckerman, and Montgomery [**NZM**]). Many computer algebra systems have the facility to solve such equations. Of course, this method will give a generating set for $\langle c_1, \ldots, c_\ell \rangle : \langle c \rangle$ as a $\mathbb{Z}$-module. This generating set is then also a generating set for $\langle c_1, \ldots, c_\ell \rangle : \langle c \rangle$ as a $\mathbb{Z}[\sqrt{-5}]$-module, but some of the generators might be redundant. In the computations below we will write down only the non-redundant generators but it is possible to verify that these generators are correct by the above procedure. We illustrate this more specifically for the case $\ell = 1$. So in this case we need to solve $c_1\alpha_1 = c\alpha$ for α_1 and α. We write $c_1 = u_1 + v_1\sqrt{-5}$ and $c = u + v\sqrt{-5}$ and $\alpha_1 = \beta_1 + \gamma_1\sqrt{-5}$ and $\alpha = \beta + \gamma\sqrt{-5}$, where u_1, v_1, u, v, β_1, γ_1, β, γ are all integers. Then taking the real and imaginary parts of the equation $c_1\alpha_1 - c\alpha = 0$, we need to solve the pair of equations $u_1\beta_1 - 5v_1\gamma_1 - u\beta + 5v\gamma = 0$ and $v_1\beta_1 + u_1\gamma_1 - v\beta - u\gamma = 0$ for the integers $\beta_1, \gamma_1, \beta, \gamma$. For an alternative method to do these computations, which is in the spirit of this book, see Exercise 4.3.1.

Let us consider the ideal I in $(\mathbb{Z}[\sqrt{-5}])[x, y]$ generated by the two polynomials $\boxed{f_1 = 2xy + \sqrt{-5}y}$ and $\boxed{f_2 = (1 + \sqrt{-5})x^2 - xy.}$ We use the lex ordering with $x > y$. We follow Algorithm 4.2.2 to compute a Gröbner basis for I.

Since $\mathbb{Z}[\sqrt{-5}]$ is an integral domain we do not need to consider saturated sets which are singletons. So the first saturated set we must consider is $\{1, 2\}$. We compute $\langle 2 \rangle : \langle 1 + \sqrt{-5} \rangle = \langle 1 - \sqrt{-5}, 2 \rangle$. Since $3 \cdot 2 - (1 - \sqrt{-5})(1 + \sqrt{-5}) = 0$, the first generator, $1 - \sqrt{-5}$, gives rise to the polynomial $3xf_1 - (1 - \sqrt{-5})yf_2 = (1 - \sqrt{-5})xy^2 + 3\sqrt{-5}xy$, and this cannot be reduced any further using f_1, f_2. We let $\boxed{f_3 = (1 - \sqrt{-5})xy^2 + 3\sqrt{-5}xy.}$ Also, the second generator,2, gives rise to the polynomial

$$(1 + \sqrt{-5})xf_1 - 2yf_2 = 2xy^2 + \sqrt{-5}(1 + \sqrt{-5})xy \xrightarrow{f_1} \sqrt{-5}(1 + \sqrt{-5})xy - \sqrt{-5}y^2.$$

We let $\boxed{f_4 = \sqrt{-5}(1 + \sqrt{-5})xy - \sqrt{-5}y^2.}$

We now consider the saturated subsets of $\{1,2,3\}$ which contain 3. They are $\{1,3\}$ and $\{1,2,3\}$. For the set $\{1,3\}$ we compute $\langle 2\rangle : \langle 1-\sqrt{-5}\rangle = \langle 1+\sqrt{-5}, 2\rangle$. The first generator gives rise to the syzygy $(-3y, 0, (1+\sqrt{-5}))$ and the second to the syzygy $((1-\sqrt{-5})y, 0, -2)$. These give rise to the following polynomials

$$\begin{aligned}
-3yf_1 + (1+\sqrt{-5})f_3 &= 3\sqrt{-5}(1+\sqrt{-5})xy - 3\sqrt{-5}y^2 \xrightarrow{f_4} 0,\\
(1-\sqrt{-5})yf_1 - 2f_3 &= -6\sqrt{-5}xy + \sqrt{-5}(1-\sqrt{-5})y^2\\
&\xrightarrow{f_1} (5+\sqrt{-5})y^2 - 15y.
\end{aligned}$$

We let $\boxed{f_5 = (5+\sqrt{-5})y^2 - 15y.}$ For the saturated set $\{1,2,3\}$ we compute $\langle 2, 1+\sqrt{-5}\rangle : \langle 1-\sqrt{-5}\rangle = \langle 1\rangle$. This gives rise to the polynomial

$$xyf_1 - y^2f_2 - xf_3 = -3\sqrt{-5}x^2y + xy^3 + \sqrt{-5}xy^2,$$

and we let $\boxed{f_6 = -3\sqrt{-5}x^2y + xy^3 + \sqrt{-5}xy^2.}$

We now consider the saturated subsets of $\{1,2,3,4\}$ which contain 4. They are $\{1,4\}$,$\{1,2,4\}$, $\{1,3,4\}$ and $\{1,2,3,4\}$. For $\{1,4\}$ we compute $\langle 2\rangle : \langle -5+\sqrt{-5}\rangle = \langle -1+\sqrt{-5}, 2\rangle$. These give rise to the polynomials

$$\begin{aligned}
3\sqrt{-5}f_1 + (-1+\sqrt{-5})f_4 &= (5+\sqrt{-5})y^2 - 15y \xrightarrow{f_5} 0,\\
(5-\sqrt{-5})f_1 + 2f_4 &= -2\sqrt{-5}y^2 + 5(1+\sqrt{-5})y.
\end{aligned}$$

We let $\boxed{f_7 = -2\sqrt{-5}y^2 + 5(1+\sqrt{-5})y.}$ For the saturated set $\{1,2,4\}$ we compute $\langle 2, 1+\sqrt{-5}\rangle : \langle -5+\sqrt{-5}\rangle = \langle 1\rangle$. This gives rise to the polynomial

$$3xf_1 - yf_2 + xf_4 = (1-\sqrt{-5})xy^2 + 3\sqrt{-5}xy \xrightarrow{f_3} 0.$$

For the saturated set $\{1,3,4\}$ we compute $\langle 2, 1-\sqrt{-5}\rangle : \langle -5+\sqrt{-5}\rangle = \langle 1\rangle$. This gives rise to the polynomial

$$2yf_1 + f_3 + yf_4 = 3\sqrt{-5}xy - \sqrt{-5}y^3 + 2\sqrt{-5}y^2,$$

and we let $\boxed{f_8 = 3\sqrt{-5}xy - \sqrt{-5}y^3 + 2\sqrt{-5}y^2.}$ We note that f_8 cannot be reduced because $3\sqrt{-5} \notin \langle 2, -5+\sqrt{-5}\rangle$. It is easy to see that for the saturated set $\{1,2,3,4\}$ we can use the same syzygy as we did for $\{1,3,4\}$.

We now consider the saturated subsets of $\{1,2,3,4,5\}$ which contain 5. They are $\{1,3,4,5\}$ and $\{1,2,3,4,5\}$. For $\{1,3,4,5\}$ we compute $\langle 2, 1-\sqrt{-5}, -5+\sqrt{-5}\rangle : \langle 5+\sqrt{-5}\rangle = \langle 1\rangle$. This gives rise to the polynomial

$$3yf_1 - f_3 - xf_5 = 3(5-\sqrt{-5})xy + 3\sqrt{-5}y^2 \xrightarrow{f_4} 0.$$

It is easy to see that for the saturated set $\{1,2,3,4,5\}$ we can use the same syzygy as we did for $\{1,3,4,5\}$.

We now consider the saturated subsets of $\{1,2,3,4,5,6\}$ which contain 6. They are $\{1,2,4,6\}$ and $\{1,2,3,4,5,6\}$. For $\{1,2,4,6\}$ we compute $\langle 2, 1+\sqrt{-5}, -5+$

$\sqrt{-5}\rangle : \langle 3\sqrt{-5}\rangle = \langle 2, 1+\sqrt{-5}\rangle : \langle 3\sqrt{-5}\rangle = \langle 2, 1+\sqrt{-5}\rangle$. These generators gives rise to the polynomials

$$\begin{array}{lcl} 3\sqrt{-5}xf_1 + 2f_6 & = & 2xy^3 + 2\sqrt{-5}xy^2 - 15xy \\ & \xrightarrow{f_1} & 2\sqrt{-5}xy^2 - 15xy - \sqrt{-5}y^3 \\ & \xrightarrow{f_1} & -15xy - \sqrt{-5}y^3 + 5y^2 \\ & \xrightarrow{f_8} & -(5+\sqrt{-5})y^3 + 15y^2 \\ & \xrightarrow{f_5} & 0, \\ 3\sqrt{-5}yf_2 + (1+\sqrt{-5})f_6 & = & (1+\sqrt{-5})xy^3 - (5+2\sqrt{-5})xy^2 \\ & \xrightarrow{y^2f_1 - yf_3} & (-5+\sqrt{-5})xy^2 - \sqrt{-5}y^3 \\ & \xrightarrow{f_4} & 0. \end{array}$$

For $\{1,2,3,4,5,6\}$ we compute $\langle 2, 1+\sqrt{-5}, 1-\sqrt{-5}, -5+\sqrt{-5}, 5+\sqrt{-5}\rangle : \langle 3\sqrt{-5}\rangle$ $= \langle 2, 1+\sqrt{-5}\rangle : \langle 3\sqrt{-5}\rangle = \langle 2, 1+\sqrt{-5}\rangle$ and so we can use the same syzygies as in the previous case.

We now consider the saturated subsets of $\{1,2,3,4,5,6,7\}$ which contain 7. They are $\{5,7\}$, $\{1,3,4,5,7\}$ and $\{1,2,3,4,5,6,7\}$. For $\{5,7\}$ we compute $\langle 5+\sqrt{-5}\rangle : \langle -2\sqrt{-5}\rangle = \sqrt{-5}\langle -1+\sqrt{-5}\rangle : \sqrt{-5}\langle -2\rangle = \langle -1+\sqrt{-5}\rangle : \langle 2\rangle = \langle 3, 1-\sqrt{-5}\rangle$. These generators give rise to the polynomials $(1+\sqrt{-5})f_5 + 3f_7 = 0$ and $2f_5 + (1-\sqrt{-5})f_7 = 0$. For $\{1,3,4,5,7\}$ we compute $\langle 2, 1-\sqrt{-5}, -5+\sqrt{-5}, 5+\sqrt{-5}\rangle : \langle -2\sqrt{-5}\rangle = \langle 1\rangle$. We obtain the polynomial

$$\begin{array}{lcl} \sqrt{-5}yf_1 + xf_7 & = & 5(1+\sqrt{-5})xy - 5y^2 \\ & \xrightarrow{15f_1 + 5f_4} & -5(1-\sqrt{-5})y^2 - 15\sqrt{-5}y \\ & \xrightarrow{f_5} & 0. \end{array}$$

It is easy to see that for the saturated set $\{1,2,3,4,5,6,7\}$ we can use the same syzygy as we did for $\{1,3,4,5,7\}$.

We now consider the saturated subsets of $\{1,2,3,4,5,6,7,8\}$ which contain 8. They are $\{1,4,8\}$, $\{1,3,4,5,7,8\}$, $\{1,2,4,6,8\}$ and $\{1,2,3,4,5,6,7,8\}$. For $\{1,4,8\}$ we compute $\langle 2, -5+\sqrt{-5}\rangle : \langle 3\sqrt{-5}\rangle = \langle 2, -5+\sqrt{-5}\rangle$. We obtain the polynomials

$$\begin{array}{lcl} 3\sqrt{-5}f_1 - 2f_8 & = & 2\sqrt{-5}y^3 - 4\sqrt{-5}y^2 - 15y \\ & \xrightarrow{f_7} & (5+\sqrt{-5})y^2 - 15y \\ & \xrightarrow{f_5} & 0, \\ 3\sqrt{-5}f_4 - (-5+\sqrt{-5})f_8 & = & -5(1+\sqrt{-5})y^3 + 5(5+2\sqrt{-5})y^2 \\ & \xrightarrow{-yf_5 + 2yf_7} & 0. \end{array}$$

For $\{1,3,4,5,7,8\}$ we compute $\langle 2, 1-\sqrt{-5}, -5+\sqrt{-5}, 5+\sqrt{-5}, -2\sqrt{-5}\rangle : \langle 3\sqrt{-5}\rangle$ $= \langle 2, -5+\sqrt{-5}\rangle$, and so we may use the same syzygy as in the previous case. For $\{1,2,4,6,8\}$ we compute $\langle 2, 1+\sqrt{-5}, -5+\sqrt{-5}, -3\sqrt{-5}\rangle : \langle 3\sqrt{-5}\rangle = \langle 1\rangle$

We obtain the polynomial

$$f_6 + xf_8 = (1-\sqrt{-5})xy^3 + 3\sqrt{-5}xy^2 \xrightarrow{f_3} 0.$$

It is easy to see that for the saturated set $\{1, 2, 3, 4, 5, 6, 7, 8\}$ we can use the same syzygy as we did for $\{1, 3, 4, 5, 7, 8\}$.

We now have that the polynomials $f_1, f_2, f_3, f_4, f_5, f_6, f_7, f_8$ form a Gröbner basis for I.

We note that the leading coefficients of f_1, f_4 and f_8 form an ideal which is equal to $\langle 1\rangle = \mathbb{Z}[\sqrt{-5}]$, so we can add to the Gröbner basis a polynomial whose leading power product is xy and whose leading coefficient is 1. Indeed, $8\cdot 2 + 3(-5+\sqrt{-5}) + (-1)3\sqrt{-5} = 1$, and so this polynomial is $8f_1 + 3f_4 - f_8 = \boxed{xy + \sqrt{-5}y^3 - 5\sqrt{-5}y^2 + 8\sqrt{-5}y = f_9.}$ It can then be shown that $\{f_2, f_5, f_7, f_9\}$ is also a Gröbner basis for I (Exercise 4.2.10).

Exercises

4.2.1. Prove Theorem 4.2.3.

4.2.2. Prove the following: Let $G = \{g_1, \dots, g_t\}$ be a set of non-zero polynomials in A. Let $\mathcal{B}$ be a homogeneous generating set of $\mathrm{Syz}(\mathrm{lt}(g_1), \dots, \mathrm{lt}(g_t))$. Then G is a Gröbner basis for the ideal $\langle g_1, \dots, g_t\rangle$ if and only if for all $(h_1, \dots, h_t) \in \mathcal{B}$, we have $h_1g_1 + \cdots + h_tg_t = v_1g_1 + \cdots + v_tg_t$ where $\mathrm{lp}(h_1g_1 + \cdots + h_tg_t) = \max(\mathrm{lp}(v_1)\,\mathrm{lp}(g_1), \dots, \mathrm{lp}(v_t)\,\mathrm{lp}(g_t))$. [Hint: See the proof of Theorem 3.2.5.]

4.2.3. Show that in Möller's method of computing syzygies of terms (i.e. in Theorem 4.2.9 and Algorithm 4.2.2), if $J \subseteq J' \subseteq \{1, \dots, \sigma\}$ with $\sigma \in J$, such that

$$\langle c_j \mid j \in J, j \neq \sigma\rangle_R \colon \langle c_\sigma\rangle_R = \langle c_j \mid j \in J', j \neq \sigma\rangle_R \colon \langle c_\sigma\rangle_R$$

and the set J has been used, then the set J' may be ignored.

4.2.4. Show that in Algorithm 4.2.2, if we consider the case where $R = k$ is a field then we may improve the algorithm as follows: For each σ in the main WHILE loop find the minimal number of distinct $j_1, \dots, j_r$, with $1 \leq j_\nu < \sigma$ such that for each $J \in \mathcal{S}$ there is a $j_\nu \in J$. Then in the remainder of the WHILE loop we need only compute the reductions of $S(f_{j_\nu}, f_\sigma)$ for $1 \leq \nu < r$. Use this method to redo Example 3.3.5. Compare this result to the use of crit2 in Example 3.3.5.

4.2.5. Compute generators for the following syzygy modules.

a. For $R = \mathbb{Z}$ compute $\mathrm{Syz}(3x^2y, 5x^2z, 9yz^2, 7xy^2z)$. [An answer: $\{(-5z, 3y, 0, 0), (-3z^2, 0, x^2, 0), (-4yz, y^2, 0, x), (0, 0, -7xy, 9z)\}$.]

b. For $R = \mathbb{Z}_{15}$ compute $\mathrm{Syz}(2x^2y, 5x^2z, 9yz^2, 7xy^2z)$. [An answer: $\{(0, 3, 0, 0), (5z, y, 0, 0), (0, 0, 5, 0), (3z^2, 0, x^2, 0), (0,\ 0,\ -7xy,\ 9z), (-yz,\ -y^2,\ 0,\ x)\}$.]

c. For $R = \mathbb{Z}[i]$ (where $i^2 = -1$) compute $\operatorname{Syz}(3ix^2y, (2+i)x^2z, 5yz^2, 7xy^2z)$. [An answer: $\{(-(2+i)z, 3iy, 0, 0), (0, -(2-i)yz, x^2, 0), (4iyz, (2-i)y^2, 0, x), (0, 0, -7xy, 5z)\}$.]

4.2.6. For the ring $R = \mathbb{Z}$ compute a Gröbner basis for the ideals generated by the given polynomials with respect to the given term order.

a. $f_1 = 2xy - x$, $f_2 = 3y - x^2$ and lex with $x < y$.

b. $f_1 = 3x^2y - 3yz + y$, $f_2 = 5x^2z - 8z^2$ and deglex with $x > y > z$.

c. $f_1 = 6x^2 + y^2$, $f_2 = 10x^2y + 2xy$ and lex with $x > y$.

4.2.7. For the ring $R = \mathbb{Z}_{15}$ compute a Gröbner basis for the ideal generated by the polynomials $f_1 = 2x^2y + 3z$ and $f_2 = 5x^2z + y$ with respect to deglex with $x > y > z$.

4.2.8. For the ring $R = \mathbb{Z}[i]$ (where $i^2 = -1$) compute a Gröbner basis for the ideal generated by the polynomials $f_1 = 3ix^2y + (1+i)z$ and $f_2 = (2+i)x^2z + y$ with respect to deglex with $x > y > z$.

4.2.9. Prove that Algorithm 4.2.2 terminates and has the desired output.

4.2.10. Show that $\{f_2, f_5, f_7, f_9\}$ forms a Gröbner basis as asserted at the end of Example 4.2.13.

4.2.11. Generalize the results in this section to the computation of Gröbner bases for $R[x_1, \ldots, x_n]$-submodules of $(R[x_1, \ldots, x_n])^m$ (see Exercise 4.1.14).

4.3. Applications of Gröbner Bases over Rings. We are interested in applications similar to the ones in the previous two chapters. We have basically seen in Section 4.1 how to solve the ideal membership problem and we will give an example of this. We will then show how to compute a complete set of coset representatives modulo an ideal. This requires more effort than it did in the case of fields because now one has to take into account the ideals in the ring R. We will then explore the use of elimination in this context to compute ideal intersection, ideal quotients and ideals of relations. These applications are very much the same as before and do not require any serious modifications in their statements or proofs. We will close this section by showing how to compute a generating set for the syzygy module of an arbitrary set of polynomials.

So let R be a Noetherian ring in which linear equations are solvable (see Definition 4.1.5) and let $A = R[x_1, \ldots, x_n]$. Let I be a non-zero ideal of A. Suppose that $I = \langle f_1, \ldots, f_s \rangle$, set $F = \{f_1, \ldots, f_s\}$, and let $G = \{g_1, \ldots, g_t\}$ be a Gröbner basis for I. Then by Theorem 4.1.12 we know that $f \in I$ if and only if $f \xrightarrow{G}_+ 0$. Thus we can determine whether or not $f \in I$ (ideal membership problem). Moreover as we saw in Algorithm 4.1.1, if $f \in I$, we can obtain $f = h_1g_1 + \cdots + h_tg_t$ explicitly. Also, using Algorithm 4.2.2, we can find a matrix T such that $(g_1, \ldots, g_t) = T(f_1, \ldots, f_s)$. We can then substitute in for the g_i's to obtain f as a linear combination of the f_j's, provided $f \in I$.

EXAMPLE 4.3.1. We go back to Example 4.2.11. Recall that a Gröbner basis for the ideal generated by $f_1 = 3x^2y + 7y$ and $f_2 = 4xy^2 - 5x$ in $\mathbb{Z}[x, y]$ with respect to the lex ordering with $x > y$ is $\{f_1, f_2, f_3, f_4\}$ where $f_3 = 15x^2 + 28y^2$

and $f_4 = -28y^3 + 35y$. We let $f = 2x^2y^2 - 3x^2y + 5x^2 - 4xy^3 - 12xy^2 + 5xy + 15x + 14y^2 - 7y$. First we verify that $f \in \langle f_1, f_2 \rangle$. We have

$$\begin{array}{rcl} f & \xrightarrow{f_2, f_3} & f - (-7xf_2 + 2y^2 f_3) \\ & = & -3x^2y - 30x^2 - 4xy^3 - 12xy^2 + 5xy + 15x - 56y^4 + 14y^2 - 7y \\ & \xrightarrow{-1, f_1} & -30x^2 - 4xy^3 - 12xy^2 + 5xy + 15x - 56y^4 + 14y^2 \\ & \xrightarrow{-2, f_3} & -4xy^3 - 12xy^2 + 5xy + 15x - 56y^4 + 70y^2 \\ & \xrightarrow{-y, f_2} & -12xy^2 + 15x - 56y^4 + 70y^2 \\ & \xrightarrow{-3, f_2} & -56y^4 + 70y^2 \\ & \xrightarrow{2y, f_4} & 0. \end{array}$$

Thus we have

$$\begin{array}{rcl} f & = & (-7xf_2 + 2y^2 f_3) - f_1 - 2f_3 - yf_2 - 3f_2 + 2yf_4 \\ & = & -f_1 - (7x + y + 3)f_2 + (2y^2 - 2)f_3 + 2yf_4. \end{array}$$

In Example 4.2.11 we saw that

$$\begin{bmatrix} f_1 \\ f_2 \\ f_3 \\ f_4 \end{bmatrix} = \underbrace{\begin{bmatrix} 1 & 0 \\ 0 & 1 \\ 4y & -3x \\ -4y^2 + 5 & 3xy \end{bmatrix}}_{T} \begin{bmatrix} f_1 \\ f_2 \end{bmatrix}.$$

Thus $f = (2y - 1)f_1 + (-x - y - 3)f_2$.

We now consider the problem of determining a complete set of coset representatives for A/I, where I is a non-zero ideal of A. We adapt the method given in Zacharias [**Za**]. In order to do this we must assume that given any ideal $\Im$ of R, we can determine a complete set C of coset representatives of $R/\Im$ and that we have a procedure to find, for all $a \in R$, an element $c \in C$ such that $a \equiv c \pmod{\Im}$. If this is the case we say that R has *effective coset representatives.* Such rings include the integers $\mathbb{Z}$, the finite rings $\mathbb{Z}/n\mathbb{Z}$, the polynomial rings over fields, $\mathbb{Z}[\sqrt{-5}]$, and $\mathbb{Z}[\sqrt{-1}]$. We consider a Gröbner basis $G = \{g_1, \dots, g_t\}$ for the ideal I. With respect to the set $\{\mathrm{lt}(g_1), \dots, \mathrm{lt}(g_t)\}$ of leading terms of G, we consider the saturated subsets $J \subseteq \{1, \dots, t\}$ as defined in Definition 4.2.4. However here we also consider $J = \emptyset$ to be saturated. Then for each saturated subset $J \subseteq \{1, \dots, t\}$, we let I_J denote the ideal of R generated by $\{\mathrm{lc}(g_i) \mid i \in J\}$ (if $J = \emptyset$, then $I_J = \{0\}$). Let C_J denote a complete set of coset representatives for R/I_J. We assume that $0 \in C_J$. Also, for each power product X, let $J_X = \{i \mid \mathrm{lp}(g_i) \text{ divides } X\}$. It is clear that J_X is saturated for all power products X (in the proof of Theorem 4.3.3 we will need C_{J_X} for *all* power products X, and of course J_X could be equal to $\emptyset$ for some X and that is why we had to add $\emptyset$ in our list of saturated subsets).

DEFINITION 4.3.2. *We call a polynomial $r \in A$* totally reduced *provided that for every power product X, if cX is the corresponding term of r (here, of course, $c \in R$ may very well be 0), then $c \in C_{J_X}$. For a given polynomial $f \in A$, we call a polynomial $r \in A$ a* normal form *for f provided that $f \equiv r \pmod{I}$ and r is totally reduced.*

It should be emphasized that the definition of normal form depends not only on the set of polynomials G, but also on the choices of the sets of coset representatives C_J for the set of saturated subsets J.

The complete set of coset representatives is given by the following

THEOREM 4.3.3. *Let G be a Gröbner basis for the non-zero ideal I of A. Assume that for each saturated subset $J \subseteq \{1, \ldots, t\}$, we have chosen, as above, a complete sets of coset representatives C_J for the ideal I_J. Then every $f \in A$ has a unique normal form. The normal form can be computed effectively provided linear equations are solvable in R and R has effective coset representatives.*

PROOF. We will first show the existence of a normal form for f. The proof will be constructive. (We note that this part of the proof does not depend on the fact that G is a Gröbner basis for I.) The proof will be similar in nature to the reduction algorithm. We will use induction on $\operatorname{lp}(f)$ to obtain a totally reduced polynomial r with $f \equiv r \pmod{I}$ satisfying $\operatorname{lp}(r) \leq \operatorname{lp}(f)$. If $f = 0$ then the result is clear. Thus we assume the result for all polynomials whose leading power product is less than $\operatorname{lp}(f)$, where we assume that $f \neq 0$. To simplify the notation we set $J = J_{\operatorname{lp}(f)}$. We may choose $c \in C_J$ such that $\operatorname{lc}(f) \equiv c \pmod{I_J}$. (Note that $c = 0$ if and only if f is reducible.) Then we may write $\operatorname{lc}(f) - c = \sum_{i \in J} c_i \operatorname{lc}(g_i)$ (this can be done effectively by our assumption that linear equations are solvable in R). We consider

$$f_1 = f - \sum_{i \in J} c_i \frac{\operatorname{lp}(f)}{\operatorname{lp}(g_i)} g_i.$$

Write $f_1 = c\operatorname{lp}(f) + f_1'$. Then we have that $\operatorname{lp}(f_1') < \operatorname{lp}(f)$. Thus, by induction, there is a polynomial r_1 which is totally reduced, satisfies $\operatorname{lp}(r_1) \leq \operatorname{lp}(f_1')$ and has the property that $f_1' \equiv r_1 \pmod{I}$. We now have that

$$\begin{aligned} f &= f_1' - r_1 + \sum_{i \in J} c_i \frac{\operatorname{lp}(f)}{\operatorname{lp}(g_i)} g_i + (c\operatorname{lp}(f) + r_1) \\ &\equiv c\operatorname{lp}(f) + r_1 \pmod{I}. \end{aligned}$$

Let $r = c\operatorname{lp}(f) + r_1$. Noting that $\operatorname{lp}(r_1) \leq \operatorname{lp}(f_1') < \operatorname{lp}(f)$, we see that $\operatorname{lp}(r) \leq \operatorname{lp}(f)$. Finally, $\operatorname{lp}(r_1) < \operatorname{lp}(f)$ implies that $r = c\operatorname{lp}(f) + r_1$ is totally reduced and so r is a normal form for f, and the induction is complete.

We next show the uniqueness of normal forms. So suppose that r_1, r_2 are totally reduced, $f \equiv r_1 \pmod{I}$ and $f \equiv r_2 \pmod{I}$. Then we have $r_1 - r_2 \in I$ and, since G is a Gröbner basis for I, we have $r_1 - r_2 \xrightarrow{G}_+ 0$; in particular, if $r_1 \neq r_2$, we have that $r_1 - r_2$ is reducible. Thus if $X = \text{lp}(r_1 - r_2)$ and $c = \text{lc}(r_1 - r_2)$ we have that $c \in I_{J_X}$. Let d_1, d_2 be, respectively, the coefficients of X in r_1, r_2. Then since r_1 and r_2 are totally reduced, $d_1, d_2 \in C_{J_X}$. Thus since $c = d_1 - d_2 \neq 0$ we have that d_1 and d_2 represent distinct cosets mod I_{J_X} and so their difference cannot represent the 0 coset; i.e. $c \notin I_{J_X}$. This is a contradiction. Thus $c = 0$ and $r_1 = r_2$, as desired. □

We note that in the case where $R = k$ is a field, Theorem 4.3.3 asserts that a k-basis for $k[x_1, \ldots, x_n]/I$ consists of all power products X such that $J_X = \emptyset$. This is precisely the statement of Proposition 2.1.6.

EXAMPLE 4.3.4. We go back to Example 4.2.12. Recall that a Gröbner basis for the ideal $I = \langle 4xy + x, 3x^2 + y\rangle$ in $\mathbb{Z}_{20}[x, y]$ with respect to the lex ordering with $x > y$ was computed to be $\{f_1, f_2, f_3, f_4, f_5\}$, where $f_1 = 4xy + x$, $f_2 = 3x^2 + y$, $f_3 = 5x$, $f_4 = 4y^2 + y$, $f_5 = 15y$. We wish to describe a complete set of coset representatives for $\mathbb{Z}_{20}[x, y]/I$. We follow Theorem 4.3.3 and we use the notation given there. First note that if X is any power product not $1, x$, or y, then $I_{J_X} = \mathbb{Z}_{20}$. We also have $I_{J_1} = \{0\}$, and $I_{J_x} = I_{J_y} = \langle 5\rangle$. We now choose a complete set C_{J_X} of coset representatives for $\mathbb{Z}_{20}/I_{J_X}$ for each power product X. Clearly for those ideals I_{J_X} equal to $\mathbb{Z}_{20}$, $C_{J_X} = \{0\}$. We choose $C_{J_x} = C_{J_y} = \{0, 1, 2, 3, 4\}$, and $C_{J_1} = \mathbb{Z}_{20}$. Therefore a complete set of coset representatives for $\mathbb{Z}_{20}[x, y]/I$ is the set $\{a + bx + cy \mid a \in \mathbb{Z}_{20}, b, c \in \{0, 1, 2, 3, 4\}\}$.

Thus, for example if $f = 3x^2y + 2xy + 13y - 5$, we have

$$\begin{aligned} f &\xrightarrow{f_1, f_3} f - (3xyf_3 - 3xf_1) = 3x^2 + 2xy + 13y - 5 \\ &\xrightarrow{f_2} 2xy + 12y - 5 \\ &\xrightarrow{f_1, f_3} 12y - 2x - 5 \\ &\equiv 2y + 3x + 15 \pmod{I}. \end{aligned}$$

As we did in Example 2.1.8 we can also construct a multiplication table for $\mathbb{Z}_{20}[x, y]/I$. Namely, we have $xy \xrightarrow{f_1, f_3} xy - yf_3 + f_1 = x$, $x^2 \xrightarrow{f_2} x^2 - 7f_2 = -7y \equiv 3y \pmod{I}$, and $y^2 \xrightarrow{f_4, f_5} y^2 + f_4 + yf_5 = y$. Thus we obtain

$\times$	1	x	y
1	1	x	y
x	x	$3y$	x
y	y	x	y

So, for example, $(2 + 4x + 2y)(3 + 4x + 4y) = 6 + 14y + 16x^2 + 8y^2 + 4xy \equiv 6 + 14y + 8y + 8y + 4x \equiv 6 + 4x + 10y \equiv 6 + 4x \pmod{I}$.

EXAMPLE 4.3.5. We go back to Example 4.2.13. Recall that a Gröbner basis for the ideal $I = \langle 2xy + \sqrt{-5}y, (1 + \sqrt{-5})x^2 - xy\rangle$ in $(\mathbb{Z}[\sqrt{-5}])[x, y]$ with respect to the lex ordering with $x > y$ was computed to be $\{f_2, f_5, f_7, f_9\}$, where $f_2 = (1 + \sqrt{-5})x^2 - xy$, $f_5 = (5 + \sqrt{-5})y^2 - 15y$, $f_7 = -2\sqrt{-5}y^2 + (5 + 5\sqrt{-5})y$, and $f_9 = xy + \sqrt{-5}y^3 - 5\sqrt{-5}y^2 + 8\sqrt{-5}y$. We wish to describe a complete set of coset representatives for $(\mathbb{Z}[\sqrt{-5}])[x,y]/I$. We follow Theorem 4.3.3 and we use the notation given there. Note first that for every power product X not equal to $1, x^\nu$, or y^μ, $\nu, \mu \geq 1$, we have $I_{J_X} = \mathbb{Z}[\sqrt{-5}]$. We also have $I_{J_1} = I_{J_x} = I_{J_y} = \{0\}$, $I_{J_{x^\nu}} = \langle 1+\sqrt{-5}\rangle$ for $\nu > 1$, and $I_{J_{y^\mu}} = \langle 5+\sqrt{-5}, -2\sqrt{-5}\rangle = \langle 10, 5+\sqrt{-5}\rangle$ for $\mu > 1$. We now choose a complete set C_{J_X} of coset representatives for $\mathbb{Z}[\sqrt{-5}]/I_{J_X}$, for all power products X. Clearly, for those I_{J_X}'s equal to $\mathbb{Z}[\sqrt{-5}]$ we have $C_{J_X} = \{0\}$. We choose $C_{J_{x^\nu}} = \{0,1,2,3,4,5\}$ for $\nu \geq 2$ (note that $6 \in \langle 1+\sqrt{-5}\rangle$ and that $a_1 + a_2\sqrt{-5} \equiv a_1 - a_2 \pmod{1+\sqrt{-5}}$). We choose $C_{J_{y^\mu}} = \{0,1,2,3,4,5,6,7,8,9\}$ for $\mu \geq 2$ (as above we have $a_1 + a_2\sqrt{-5} \equiv a_1 - 5a_2 \pmod{5+\sqrt{-5}}$, and $10 \in I_{J_{y^\mu}}$). Therefore a complete set of coset representatives for $(\mathbb{Z}[\sqrt{-5}])[x,y]/I$ is

$$\Big\{a + bx + cy + \sum_{\nu=2}^{n} d_\nu x^\nu + \sum_{\mu=2}^{m} e_\mu y^\mu \mid a,b,c \in \mathbb{Z}[\sqrt{-5}],$$

$$d_\nu \in \{0,1,2,3,4,5\}, \text{ and } e_\mu \in \{0,1,2,3,4,5,6,7,8,9\}\Big\}.$$

We now turn to applications of the method of elimination (see Sections 2.3 and 3.6). The proofs are very similar to the ones in those two sections, so most of them will be omitted and left to the exercises.

Let $y_1, \ldots, y_m$ be new variables, and consider a non-zero ideal

$$I \subseteq A[y_1, \ldots, y_m] = R[x_1, \ldots, x_n, y_1, \ldots, y_m].$$

We wish to "eliminate" the variables $y_1, \ldots, y_m$, i.e. we wish to compute generators (and a Gröbner basis) for the ideal $I \cap A$. First, we choose an elimination order on the power products of $A[y_1, \ldots, y_m]$ with the y variables larger than the x variables (see Section 2.3). The next result is the analog of Theorems 2.3.4 and 3.6.6 and a generalization of Exercise 4.1.10 **a**.

THEOREM 4.3.6. *With the notation set above, let G be a Gröbner basis for I with respect to an elimination order on $A[y_1, \ldots, y_m]$ with the y variables larger than the x variables. Then $G \cap A$ is a Gröbner basis for $I \cap A$.*

PROOF. We will use the notation Lt_A and $\mathrm{Lt}_{A[y]}$ for the leading term ideals in A and $A[y_1, \ldots, y_m]$ respectively. Then we need to show that $\mathrm{Lt}_A(G \cap A) = \mathrm{Lt}_A(I \cap A)$. So if $0 \neq f \in I \cap A$ then $\mathrm{lt}(f) \in \mathrm{Lt}_{A[y]}(I) = \mathrm{Lt}_{A[y]}(G)$ and so $\mathrm{lt}(f)$ is a linear combination of elements $\mathrm{lt}(g)$ such that $g \in G$ with coefficients in $A[y_1, \ldots, y_m]$. Since $\mathrm{lt}(f)$ is a term involving only the x variables, we may assume each summand is a term and involves only the x variables. Hence for each $g \in G$ appearing in the sum for $\mathrm{lt}(f)$, $\mathrm{lt}(g)$ involves only the x variables and thus, since we are using an elimination order with the y variables larger than the x variables, each term in such a g involves only the x variables, i.e. $g \in G \cap A$. Thus $\mathrm{lt}(f) \in \mathrm{Lt}_A(G \cap A)$, as desired. □

EXAMPLE 4.3.7. We go back to Example 4.2.12. Recall that a Gröbner basis for the ideal $I = \langle 4xy + x, 3x^2 + y\rangle$ in $\mathbb{Z}_{20}[x, y]$ with respect to the lex ordering with $x > y$ was computed to be $\{4xy+x, 3x^2+y, 5x, 4y^2+y, 15y\}$. Using Theorem 4.3.6 and Exercise 4.1.10 **a**, we see that $I \cap \mathbb{Z}_{20}[y] = \langle 4y^2 + y, 15y\rangle$, and that $I \cap \mathbb{Z}_{20} = \{0\}$.

EXAMPLE 4.3.8. We go back to Example 4.2.13. Recall that a Gröbner basis for the ideal $I = \langle 2xy + \sqrt{-5}y, (1 + \sqrt{-5})x^2 - xy\rangle$ in $(\mathbb{Z}[\sqrt{-5}])[x, y]$ with respect to the lex ordering with $x > y$ was computed to be

$$\Big\{2xy + \sqrt{-5}y, (1 + \sqrt{-5})x^2 - xy, (1 - \sqrt{-5})xy^2 + 3\sqrt{-5}xy,$$

$$\sqrt{-5}(1 + \sqrt{-5})xy - \sqrt{-5}y^2, (5 + \sqrt{-5})y^2 - 15y, -3\sqrt{-5}x^2y + xy^3 + \sqrt{-5}xy^2,$$

$$-2\sqrt{-5}y^2 + 5(1 + \sqrt{-5})y, 3\sqrt{-5}xy - \sqrt{-5}y^3 + 2\sqrt{-5}y^2\Big\}.$$

Using Theorem 4.3.6 and Exercise 4.1.10 **a**, we see that

$$I \cap (\mathbb{Z}[\sqrt{-5}])[y] = \langle (5 + \sqrt{-5})y^2 - 15y, -2\sqrt{-5}y^2 + 5(1 + \sqrt{-5})y\rangle$$

and $I \cap \mathbb{Z}[\sqrt{-5}] = \{0\}$.

Our first application will be to compute the intersection of two ideals of A and the ideal quotient of two ideals of A. First, as in the ideal case over fields, Proposition 2.3.5, and as in the module case, Proposition 3.6.8, we have

PROPOSITION 4.3.9. *Let $I = \langle f_1, \ldots, f_s\rangle$ and $J = \langle g_1, \ldots, g_t\rangle$ be ideals of A and let w be a new variable. Consider the ideal*

$$L = \langle wf_1, \ldots, wf_s, (1 - w)g_1, \ldots, (1 - w)g_t\rangle \subseteq A[w] = R[x_1, \ldots, x_n, w].$$

Then $I \cap J = L \cap A$.

As a consequence of this result we obtain a method for computing generators for the ideal $I \cap J \subseteq A$: we first compute a Gröbner basis G for the ideal L in Proposition 4.3.9 with respect to an elimination order on the power products in $A[w]$ with w larger than the x variables; a Gröbner basis for $I \cap J$ is then given by $G \cap A$.

EXAMPLE 4.3.10. In $\mathbb{Z}[x, y]$, we wish to compute $\langle 3x - 2, 5y - 3\rangle \cap \langle xy - 6\rangle$. So following Proposition 4.3.9 and Theorem 4.3.6 we consider the polynomials $f_1 = 3xw - 2w$, $f_2 = 5yw - 3w$ and $f_3 = xyw - 6w - xy + 6$ and compute $\langle f_1, f_2, f_3\rangle \cap \mathbb{Z}[x, y]$. We will outline the computation. We consider the deglex ordering with $x > y$ on the variables x and y and an elimination order between w and x, y with w larger. We follow Algorithm 4.2.2 (observe that much use of Exercise 4.2.3 is made in this computation). The first saturated set to consider is $\{1, 2\}$ which gives rise to $5yf_1 - 3xf_2 \longrightarrow_+ 0$. The only saturated set containing 3 is $\{1, 2, 3\}$ and this gives rise to $f_3 - 2yf_1 + xf_2 \longrightarrow_+ 4wy - 8w - xy + 6 = f_4$. For the saturated sets containing 4 we need to consider $\{2, 4\}$ which gives rise to $-5f_4 + 4f_2 = 28w + 5xy - 30 = f_5$ and $\{1, 2, 3, 4\}$ which gives rise to $-xf_4 + 4f_3 = 8wx - 24w + x^2y - 4xy - 6x + 24 = f_6$. For the saturated sets containing 5 we need to consider $\{1,5\}$ which gives rise to $3xf_5 - 28f_1 \longrightarrow_+ 15x^2y - 10xy - 90x + 60 = f_7$, $\{2,4,5\}$ which gives rise to $yf_5 - 7f_4 \longrightarrow_+ 5xy^2 - 3xy - 30y + 18 = f_8$, and $\{1,2,3,4,5\}$ which gives the same result as the last set (Exercise 4.2.3). For the saturated sets containing 6 we need to consider $\{1,5,6\}$ which gives rise to $f_6 - 12f_1 + xf_5 = 6x^2y - 4xy - 36x + 24 = f_9$ and $\{1,2,3,4,5,6\}$ which gives the same result as the previous set. For the saturated sets containing 7 we need only consider $\{1,2,3,4,5,6,7\}$ which gives rise to $wf_7 - 5xyf_1 \longrightarrow_+ 0$. The remaining cases (only 4 need to be computed) all go to zero and we see that $\{f_1, f_2, f_3, f_4, f_5, f_6, f_7, f_8, f_9\}$ is a Gröbner basis for $\langle f_1, f_2, f_3\rangle$. Thus

$$\langle 3x-2, 5y-3\rangle \cap \langle xy-6\rangle = \langle 5xy^2 - 3xy - 30y + 18, 15x^2y - 10xy - 90x + 60,$$
$$6x^2y - 4xy - 36x + 24\rangle.$$

We also note that $f_{10} = f_7 - 2f_9 = 3x^2y - 2xy - 18x + 12$ is in $\langle f_1, f_2, f_3\rangle$ and we can replace f_7, f_9 by f_{10} so that $\{f_1, f_2, f_3, f_5, f_6, f_8, f_{10}\}$ is also a Gröbner basis for $\langle f_1, f_2, f_3\rangle$. We thus obtain the simpler generating set

$$\langle 3x-2, 5y-3\rangle \cap \langle xy-6\rangle = \langle 5xy^2 - 3xy - 30y + 18, 3x^2y - 2xy - 18x + 12\rangle.$$

The computation of ideal quotients is almost the same as in the previous cases except that one must be careful of possible zero divisors in R and hence in A.

PROPOSITION 4.3.11. *Let $I = \langle f_1, \dots, f_s\rangle$ and $J = \langle g_1, \dots, g_t\rangle$ be ideals of A. Then*

$$I \colon J = \{f \in A \mid fJ \subseteq I\} = \bigcap_{j=1}^{t} I \colon \langle g_j\rangle.$$

Moreover, if

$$I \cap \langle g\rangle = \langle h_1 g, \dots, h_\ell g\rangle$$

and g is not a zero divisor in A, then we have

$$I \colon \langle g\rangle = \langle h_1, \dots, h_\ell\rangle.$$

PROOF. The proof is exactly the same as in the ideal case over a field, see Lemmas 2.3.10 and 2.3.11, until we must verify that $I \cap \langle g\rangle = \langle h_1 g, \dots, h_\ell g\rangle$ implies that $I \colon \langle g\rangle = \langle h_1, \dots, h_\ell\rangle$ which requires that g must be canceled and this requires that g not be a zero divisor. $\square$

As a consequence of this result we obtain a method for computing generators for the ideal $I \colon J \subseteq A$, provided that none of $g_1, \dots, g_t$ are zero divisors.

EXAMPLE 4.3.12. In $\mathbb{Z}[x,y]$, we wish to compute $\langle 3x-2, 5y-3\rangle \colon \langle xy-6\rangle$. From Proposition 4.3.11 we need to compute $\langle 3x-2, 5y-3\rangle \cap \langle xy-6\rangle$ and divide the generators of this ideal by $xy-6$. This intersection was computed in Example 4.3.10 and so dividing the polynomials obtained there by $xy-6$ we obtain immediately that

$$\langle 3x-2, 5y-3\rangle \colon \langle xy-6\rangle = \langle 5y-3, 3x-2\rangle.$$

In Section 2.4 we used elimination to compute the kernel of an algebra homomorphism. Sometimes $R[x_1, \ldots, x_n]$ is called an *R-algebra*, in order to emphasize the special role played by the ring R. In a similar vein, a ring homomorphism between two polynomial rings over R

$$\phi\colon R[y_1, \ldots, y_m] \longrightarrow R[x_1, \ldots, x_n]$$

is called an *R-algebra homomorphism* provided that $\phi(r) = r$ for all $r \in R$. It is then clear that ϕ is uniquely determined by

$$\phi\colon y_i \longmapsto f_i,$$

for $f_1, \ldots, f_m \in R[x_1, \ldots, x_n]$. That is, if we let $h \in R[y_1, \ldots, y_m]$, say $h = \sum_\nu c_\nu y_1^{\nu_1} \cdots y_m^{\nu_m}$, where $c_\nu \in R$, $\nu = (\nu_1, \ldots, \nu_m) \in \mathbb{N}^m$, and only finitely many c_ν's are non-zero, then we have

$$\phi(h) = \sum_\nu c_\nu f_1^{\nu_1} \cdots f_m^{\nu_m} = h(f_1, \ldots, f_m) \in R[x_1, \ldots, x_n].$$

Given this setup we have the analogue of Theorem 2.4.2.

THEOREM 4.3.13. *With the notation above, let $J = \langle y_1 - f_1, \ldots, y_m - f_m\rangle \subseteq R[y_1, \ldots, y_m, x_1, \ldots, x_n]$. Then $ker(\phi) = J \cap R[y_1, \ldots, y_m]$.*

Another way to view Theorem 4.3.13 is that $J \cap R[y_1, \ldots, y_m]$ gives the ideal of relations among the polynomials $f_1, \ldots, f_s$.

EXAMPLE 4.3.14. Consider the $\mathbb{Z}$-algebra homomorphism $\phi\colon \mathbb{Z}[u, v, w] \longrightarrow \mathbb{Z}[x, y]$ defined by $\phi\colon u \longmapsto 3x - 2$, $\phi\colon v \longmapsto 5y - 3$ and $\phi\colon w \longmapsto xy - 6$. We wish to find the kernel of ϕ. Following Theorem 4.3.13 we let $f_1 = 3x - 2 - u$, $f_2 = 5y - 3 - v$, and $f_3 = xy - 6 - w$ and compute $\langle f_1, f_2, f_3\rangle \cap \mathbb{Z}[u, v, w]$. We consider the term orders deglex with $u > v > w$ on u, v, w and deglex with $x > y$ on x, y and an elimination order between x, y and u, v, w with x, y larger. We follow Algorithm 4.2.2. The first saturated set to consider is $\{1, 2\}$ which gives rise to $5yf_1 - 3xf_2 \longrightarrow_+ 0$. The only saturated set containing 3 is $\{1, 2, 3\}$ and this gives rise to $-f_3 + 2yf_1 - xf_2 = vx + 3x - 2uy - 4y + w + 6 = f_4$. For the saturated sets containing 4 we need to consider $\{1, 4\}$ which gives rise to $3f_4 - vf_1 \longrightarrow_+ -6uy - 12y + uv + 3u + 2v + 3w + 24 = f_5$ and $\{1, 2, 3, 4\}$ which gives rise to $yf_4 - vf_3 \longrightarrow_+ -2y^2u - 4y^2 + yu + yw + 8y + vw + 6v = f_6$. There are three saturated sets containing 5 but we need only consider (Exercise 4.2.3) $\{2, 5\}$ which gives rise to $5f_5 + 6uf_2 = -uv - 3u - 2v + 15w + 84 = f_7$ and $\{1, 2, 3, 5\}$ which gives rise to $xf_5 + 6uf_3 \longrightarrow_+ 0$. All remaining polynomials arising from saturated sets containing 6 or 7 reduce to zero. Thus we see that $\ker(\phi) = \langle uv + 3u + 2v - 15w - 84\rangle$.

We conclude this section by giving a method for computing a set of generators for the syzygy module, $\mathrm{Syz}(f_1, \ldots, f_s)$, for a set of non-zero polynomials $\{f_1, \ldots, f_s\}$ in A. We do this by first computing $\mathrm{Syz}(g_1, \ldots, g_t)$ for a Gröbner basis $\{g_1, \ldots, g_t\}$ for the ideal $\langle f_1, \ldots, f_s\rangle$. The theorem describing $\mathrm{Syz}(g_1, \ldots, g_t)$

is the analogue of Theorem 3.4.1. In our case, we will have to replace Theorem 1.7.4 with Theorem 4.2.3.

Let $\{g_1, \dots, g_t\}$ be a Gröbner basis in A. We let $\operatorname{lt}(g_i) = c_i X_i$ for $c_i \in R$ and power products X_i. Let $\mathcal{B} = \{\boldsymbol{h}_1, \dots, \boldsymbol{h}_\ell\}$ be a homogeneous generating set of the syzygy module $\operatorname{Syz}(\operatorname{lt}(g_1), \dots, \operatorname{lt}(g_t)) = \operatorname{Syz}(c_1X_1, \dots, c_tX_t)$. Assume that for $1 \leq j \leq \ell$ we have that $\boldsymbol{h}_j = (d_{j1}Y_{j1}, \dots, c_{jt}Y_{jt})$ for $d_{ji} \in R$ and power products Y_{ji}, where for each j we assume that $\boldsymbol{h}_j$ is homogeneous of degree Z_j; i.e. for all i, j such that $d_{ji} \neq 0$, we have $X_iY_{ji} = Z_j$ (we assume that $Y_{ji} = 0$ if $d_{ji} = 0$). Then by Theorem 4.2.3, for each j, $1 \leq j \leq \ell$ the generalized S-polynomial $\sum_{i=1}^t d_{ji}Y_{ji}g_i$ has the representation

$$\sum_{i=1}^t d_{ji}Y_{ji}g_i = \sum_{\nu=1}^t a_{j\nu}g_\nu,$$

where

$$\max_{1\leq\nu\leq t} \operatorname{lp}(a_{j\nu})\operatorname{lp}(g_\nu) = \operatorname{lp}(\sum_{i=1}^t d_{ji}Y_{ji}g_i) < \max_{1\leq i\leq t} Y_{ji}\operatorname{lp}(g_i).$$

We now define for $1 \leq j \leq \ell$,

$$\boldsymbol{s}_j = \boldsymbol{h}_j - (a_{j1}, \dots, a_{jt}) \in A^t.$$

We note that $\boldsymbol{s}_j \in \operatorname{Syz}(g_1, \dots, g_t)$.

THEOREM 4.3.15. *With the notation above, the collection* $\{\boldsymbol{s}_j \mid 1 \leq j \leq \ell\}$ *is a generating set for* $\operatorname{Syz}(g_1, \dots, g_t)$.

PROOF. Suppose to the contrary that there exists $(u_1, \dots, u_t)$ such that

$$(u_1, \dots, u_t) \in \operatorname{Syz}(g_1, \dots, g_t) - \langle \boldsymbol{s}_j \mid 1 \leq j \leq \ell\rangle.$$

Then we can choose such a $(u_1, \dots, u_t)$ with $X = \max_{1\leq i\leq t}(\operatorname{lp}(u_i)\operatorname{lp}(g_i))$ least. Let

$$S = \{i \in \{1, \dots, t\} \mid \operatorname{lp}(u_i)\operatorname{lp}(g_i) = X\}.$$

Now for each $i \in \{1, \dots, t\}$ we define u_i' as follows:

$$u_i' = \begin{cases} u_i \text{ if } i \notin S \\ u_i - \operatorname{lt}(u_i) \text{ if } i \in S. \end{cases}$$

Of course we have that for all i, $\operatorname{lp}(u_i')\operatorname{lp}(g_i) < X$. Now, for $i \in S$, let $\operatorname{lt}(u_i) = c_i'X_i'$. So for all $i \in S$ we have $X_iX_i' = X$. Since $(u_1, \dots, u_t) \in \operatorname{Syz}(g_1, \dots, g_t)$, we see that

$$\sum_{i\in S} c_i'c_iX_i'X_i = 0,$$

and so

$$\sum_{i\in S} c_i'X_i'\boldsymbol{e}_i \in \operatorname{Syz}(c_1X_1, \dots, c_tX_t).$$

Thus, we may write

$$\sum_{i\in S} c'_i X'_i \boldsymbol{e}_i = \sum_{j=1}^{\ell} v_j \boldsymbol{h}_j,$$

for some $v_j \in A$. We note that, we may assume that either $v_j = 0$ or that $v_j = b_j \frac{X}{Z_j}$, $b_j \in R$. We can see this since, for each $i \in S$, we have $c'_i X'_i = \sum_{j=1}^{\ell} v_j d_{ji} Y_{ji}$ from which we obtain, after multiplying through by X_i, $c'_i X = \sum_{j=1}^{\ell} v_j d_{ji} Z_j$; and thus the power products involved are independent of i. Then we have

$$\begin{aligned}(u_1, \ldots, u_t) &= \sum_{i\in S} c'_i X'_i \boldsymbol{e}_i + (u'_1, \ldots, u'_t) \\ &= \sum_{j=1}^{\ell} b_j \frac{X}{Z_j} \boldsymbol{h}_j + (u'_1, \ldots, u'_t) \\ &= \sum_{j=1}^{\ell} b_j \frac{X}{Z_j} \boldsymbol{s}_j + (u'_1, \ldots, u'_t) + \sum_{j=1}^{\ell} b_j \frac{X}{Z_j} (a_{j1}, \ldots, a_{jt}).\end{aligned}$$

We define

$$(w_1, \ldots, w_t) = (u'_1, \ldots, u'_t) + \sum_{j=1}^{\ell} b_j \frac{X}{Z_j} (a_{j1}, \ldots, a_{jt}).$$

We note that $(w_1, \ldots, w_t)$ is in $\mathrm{Syz}(g_1, \ldots, g_t) - \langle \boldsymbol{s}_j \mid 1 \le j \le \ell \rangle$, since $(u_1, \ldots, u_t)$ and each $\boldsymbol{s}_j$ are in $\mathrm{Syz}(g_1, \ldots, g_t)$ and $(u_1, \ldots, u_t) \notin \langle \boldsymbol{s}_j \mid 1 \le j \le \ell \rangle$. We will obtain the desired contradiction by proving that

$$\max_{1\le\nu\le t} (\mathrm{lp}(w_\nu)\,\mathrm{lp}(g_\nu)) < X.$$

For each $\nu \in \{1, \ldots, t\}$ we have

$$\begin{aligned}\mathrm{lp}(w_\nu)\,\mathrm{lp}(g_\nu) &= \mathrm{lp}(u'_\nu + \sum_{j=1}^{\ell} b_j \frac{X}{Z_j} a_{j\nu}) X_\nu \\ &\le \max(\mathrm{lp}(u'_\nu), \max_{1\le j\le\ell} (\frac{X}{Z_j} \mathrm{lp}(a_{j\nu}))) X_\nu.\end{aligned}$$

But, by definition of u'_ν, we have $\mathrm{lp}(u'_\nu) X_\nu < X$. Also,

$$\frac{X}{Z_j} \mathrm{lp}(a_{j\nu}) X_\nu < \max_{1\le i\le t} \frac{X}{Z_j} Y_{ji} X_i = X.$$

Therefore $\mathrm{lp}(w_\nu)\,\mathrm{lp}(g_\nu) < X$ for each $\nu \in \{1, \ldots, t\}$ violating the condition that $X = \max_{1\le i\le t}(\mathrm{lp}(u_i)\,\mathrm{lp}(g_i))$ is least. □

We now turn our attention to computing $\mathrm{Syz}(f_1, \ldots, f_s)$, for a collection $\{f_1, \ldots, f_s\}$ of non-zero polynomials in A that do not necessarily form a Gröbner basis. We first compute a Gröbner basis $\{g_1, \ldots, g_t\}$ for $\langle f_1, \ldots, f_s \rangle$. Set $F = \begin{bmatrix} f_1 & \cdots & f_s \end{bmatrix}$ and $G = \begin{bmatrix} g_1 & \cdots & g_t \end{bmatrix}$. We can compute a $t \times s$ matrix S and an $s \times t$ matrix T with entries in A such that $F = GS$ and $G = FT$. (Recall that S is obtained using Algorithm 4.1.1 and T is obtained by keeping track of the reductions in Algorithm 4.2.1 or Algorithm 4.2.2.) Now using Theorem 4.3.15, we can compute a generating set $\{\boldsymbol{s}_1, \dots, \boldsymbol{s}_\ell\}$ for $\mathrm{Syz}(G)$. Then exactly as we did in Theorem 3.4.3 we have (Exercise 4.3.12)

THEOREM 4.3.16. *With the notation above we have*

$$\mathrm{Syz}(f_1, \dots, f_s) = \langle T\boldsymbol{s}_1, \dots, T\boldsymbol{s}_\ell, \boldsymbol{r}_1, \dots, \boldsymbol{r}_s \rangle \subseteq A^s,$$

where $\boldsymbol{r}_1, \dots, \boldsymbol{r}_s$ *are the columns of* $I_s - TS$.

EXAMPLE 4.3.17. We go back to Example 4.2.12. Recall that a Gröbner basis for the ideal $I = \langle f_1, f_2 \rangle$ in $\mathbb{Z}_{20}[x, y]$, where $f_1 = 4xy + x$ and $f_2 = 3x^2 + y$, with respect to the lex ordering with $x > y$, was computed to be f_1, f_2, $f_3 = 5x$, $f_4 = 4y^2 + y$, $f_5 = 15y$. We wish to compute a generating set for $\mathrm{Syz}(f_1, f_2, f_3, f_4, f_5)$ and $\mathrm{Syz}(f_1, f_2)$. As in Theorem 4.3.15, we first compute a homogeneous generating set for

$$\mathrm{Syz}(\mathrm{lt}(f_1), \mathrm{lt}(f_2), \mathrm{lt}(f_3), \mathrm{lt}(f_4), \mathrm{lt}(f_5)) = \mathrm{Syz}(4xy, 3x^2, 5x, 4y^2, 15y).$$

In Example 4.2.12 we found that the syzygy module is generated by $(5, 0, 0, 0, 0)$, $(-3x, 4y, 0, 0, 0)$, $(0, 0, 4, 0, 0)$, $(5, 0, -4y, 0, 0)$, $(0, 15, -x, 0, 0)$, $(0, 0, 0, 5, 0)$, $(y, 0, 0, -x, 0)$, $(0, 0, 0, 0, 4)$, $(0, 0, 3y, 0, -x)$. These syzygies give rise to the following polynomials

$$\begin{array}{lcl} 5f_1 & = & 5x = f_3 \\ -3xf_1 + 4yf_2 & = & -3x^2 + 4y^2 \xrightarrow{f_2} 4y^2 + y = f_4 \\ 4f_3 & = & 0 \\ 5f_1 - 4yf_3 & = & 5x = f_3 \\ 15f_2 - xf_3 & = & 15y = f_5 \\ 5f_4 & = & 5y = 3f_5 \\ yf_1 - xf_4 & = & 0 \\ 4f_5 & = & 0 \\ 3yf_3 - xf_5 & = & 0. \end{array}$$

Therefore

$\mathrm{Syz}(f_1, f_2, f_3, f_4, f_5) =$

$$\langle (5, 0, -1, 0, 0), (-3x, 4y + 1, 0, -1, 0), (0, 0, 4, 0, 0), (5, 0, -4y - 1, 0, 0),$$

$$(0, 15, -x, 0, -1), (0, 0, 0, 5, -3), (y, 0, 0, -x, 0), (0, 0, 0, 0, 4), (0, 0, 3y, 0, -x) \rangle.$$

To compute $\mathrm{Syz}(f_1, f_2)$, we first compute the matrices S and T such that

$$\begin{bmatrix} f_1 & f_2 & f_3 & f_4 & f_5 \end{bmatrix} = \begin{bmatrix} f_1 & f_2 \end{bmatrix} T, \text{ and}$$

$$\begin{bmatrix} f_1 & f_2 \end{bmatrix} = \begin{bmatrix} f_1 & f_2 & f_3 & f_4 & f_5 \end{bmatrix} S.$$

It is easy to see that

$$S = \begin{bmatrix} 1 & 0 \\ 0 & 1 \\ 0 & 0 \\ 0 & 0 \\ 0 & 0 \end{bmatrix} \text{ and } T = \begin{bmatrix} 1 & 0 & 5 & -3x & -5x \\ 0 & 1 & 0 & 4y+1 & -5 \end{bmatrix}.$$

We have

$$\begin{array}{lcl} T(5,0,-1,0,0) & = & (0,0) \\ T(-3x,4y+1,0,-1,0) & = & (0,0) \\ T(0,0,4,0,0) & = & (0,0) \\ T(5,0,-4y-1,0,0) & = & (0,0) \\ T(0,15,-x,0,-1) & = & (0,0) \\ T(0,0,0,5,-3) & = & (0,0) \\ T(y,0,0,-x,0) & = & (y+3x^2,-4xy-x) \\ T(0,0,0,0,4) & = & (0,0) \\ T(0,0,3y,0,-x) & = & (15y+5x^2,5x). \end{array}$$

Finally we have

$$I_2 - TS = \begin{bmatrix} 0 & 0 \\ 0 & 0 \end{bmatrix}.$$

Therefore

$$\mathrm{Syz}(f_1, f_2) = \langle (y+3x^2, -4xy-x), (15y+5x^2, 5x) \rangle.$$

Exercises

4.3.1. In this exercise we give another method for computing the ideals in Example 4.2.13. We first observe that $\mathbb{Z}[\sqrt{-5}] \cong \mathbb{Z}[x]/\langle x^2+5\rangle$ under the map $\sqrt{-5} \longmapsto x + \langle x^2+5\rangle$. Thus to find, for example, $\langle 2\rangle : \langle 1+\sqrt{-5}\rangle$, we need to find the syzygies of the matrix $\left[\ 1+x \quad 2 \quad x^2+5\ \right]$ and read off the coefficients of $1+x$. We note that $\{1+x, 2, x^2+5\}$ is a Gröbner basis and we can compute a generating set $\mathrm{Syz}(1+x, 2, x^2+5) = \{(-2, 1+x, 0), (1-x, -3, 1), (0, x^2+5, -2)\}$ which yields $\langle 2\rangle : \langle 1+\sqrt{-5}\rangle = \langle -2, 1-\sqrt{-5}\rangle$. Verify the statements made so far and then go on to use this method to compute all of the ideal quotients in Example 4.2.13.

4.3.2. Generalize Exercise 4.3.1 to the case where $\mathbb{Z}[x]/\langle x^2+5\rangle$ is replaced by $R[x_1, \ldots, x_n]/I$, where R is a commutative ring and I is an ideal in $R[x_1, \ldots, x_n]$, that is, give a method for computing generators for ideal quotients in $R[x_1, \ldots, x_n]/I$.

4.3.3. As in Example 4.2.12, consider the ideal $I \subseteq \mathbb{Z}_{20}[x,y]$ generated by $f_1 = 4xy + x$ and $f_2 = 3x^2 + y$. Show that $f = 12x^3y^2 - x^3y - 10x^3 + 4x^2y^2 - 4x^2y - 4xy^3 - 4xy^2 + 11xy - 6x \in I$ and write f as a linear combination of f_1 and f_2.

4.3.4. In Example 4.2.11 compute a complete set of coset representatives for $\mathbb{Z}[x,y]/I$. Find the totally reduced form of $5x^3y^3$, $3x^3y^3$ and $4x^3y^3$.

4.3.5. Prove Proposition 4.3.9.

4.3.6. In $\mathbb{Z}[x,y]$ compute $\langle 3x-2, xy-6\rangle \cap \langle 5y-3\rangle$. [Answer: $\langle 15xy-9x-10y+6, 5xy^2-48xy+27x, 10y^2-96y+54\rangle$.]

4.3.7. Prove Proposition 4.3.11.

4.3.8. In $\mathbb{Z}[x,y]$ compute $\langle 3x-2, xy-6\rangle : \langle 5y-3\rangle$. [Hint: Look at Exercise 4.3.6.]

4.3.9. Prove Theorem 4.3.13.

4.3.10. Consider the $\mathbb{Z}$-algebra homomorphism $\phi\colon \mathbb{Z}[u,v] \longrightarrow \mathbb{Z}[x,y]$ defined by $\phi\colon u \longmapsto 3x-2$ and $\phi\colon v \longmapsto xy-5$. Compute generators for $\ker(\phi)$.

4.3.11. Consider the $\mathbb{Z}$-algebra homomorphism $\phi\colon \mathbb{Z}[u,v,w] \longrightarrow \mathbb{Z}[x,y]$ defined by $\phi\colon u \longmapsto 3x-2$, $\phi\colon v \longmapsto 2y-5$ and $\phi\colon w \longmapsto x-5y$. Compute generators for $\ker(\phi)$.

4.3.12. Prove Theorem 4.3.16.

4.3.13. Compute generators for the syzygy module of the Gröbner bases constructed in Exercise 4.2.6. Use this to construct the syzygy module for the original polynomials (of course, by unique factorization, this latter problem is trivial). Repeat this exercise for Exercises 4.2.7 and 4.2.8.

4.3.14. Compute $\mathrm{Syz}(f_2, f_5, f_7, f_9)$ for the Gröbner basis in Example 4.2.13 (although this problem is quite doable, it is a long, messy computation).

4.3.15. The following exercise depends on Exercises 4.1.14 and 4.2.11. Generalize the following to the case of $R[x_1, \ldots, x_n]$-submodules of $R[x_1, \ldots, x_n]^m$.

a. Tasks (i) and (ii) at the beginning of Section 3.6.

b. Task (iii) at the beginning of Section 3.6.

c. Theorem 3.6.6.

d. Theorems 3.7.3 and 3.7.6 (of course use the ideas in Theorems 4.3.15 and 4.3.16).

e. As much of Section 3.8 as energy permits.

4.4. A Primality Test. In this section we will give an algorithm that will determine whether a given ideal I in $A = R[x_1, \ldots, x_n]$ is prime or not. Recall that an ideal P is called *prime* if and only if given any polynomials $f, g \in R[x_1, \ldots, x_n]$ with $fg \in P$, we have $f \in P$ or $g \in P$. This is easily seen to be equivalent to the statement that the ring $R[x_1, \ldots, x_n]/P$ is an integral domain. Prime ideals are basic building blocks like prime integers are building blocks in $\mathbb{Z}$. Geometrically, prime ideals in $k[x_1, \ldots, x_n]$, where k is a field, correspond to irreducible varieties (varieties which are not the union of two proper subvarieties) (see [**CLOS**]). Algebraically, any radical ideal is the intersection of prime ideals. The test for primality that we present in this section is taken from the paper of P. Gianni, B. Trager and G. Zacharias [**GTZ**].

Before we give the primality test, we must consider rings of fractions of the ring A. Although this is not strictly necessary, we will assume when discussing rings of fractions that R is an integral domain, since that is all that is needed in order to obtain the primality test (even in the case where R is not an integral domain). Then A is also an integral domain. We have the field of fractions

$$K = \left\{ \frac{f}{g} \mid f, g \in A, g \neq 0 \right\}$$

of A. We consider certain subrings of K. Let $S \subset A$ be a *multiplicative set*; that is, $1 \in S$, $0 \notin S$, and if $f, g \in S$ then $fg \in S$. Examples of multiplicative sets are $S = \{g \in A \mid g \neq 0\}$, $S = \{g \in A \mid g \notin P\}$, where P is a prime ideal in A, and $S = \{g^\nu \mid \nu \in \mathbb{N}\}$ for a fixed $g \in A$.

Given a multiplicative set $S \subset A$ we define the subring $S^{-1}A$ of K by

$$S^{-1}A = \left\{ \frac{f}{g} \in K \mid f \in A \text{ and } g \in S \right\}.$$

It is easy to see that $S^{-1}A$ is a subring of K, which we call the *ring of fractions* of A with respect to S. In the three examples above, if $S = \{g \in A \mid g \neq 0\}$, then $S^{-1}A = K$; if $S = \{g \in A \mid g \notin P\}$, then $S^{-1}A$ is called the *local ring at* P and is denoted by A_P; and finally if $S = \{g^\nu \mid \nu \in \mathbb{N}\}$ then $S^{-1}A = \{\frac{f}{g^\nu} \mid f \in A \text{ and } \nu \in \mathbb{N}\}$ is denoted by A_g. We note that A is a subring of $S^{-1}A$, since for all $f \in A$, we have $f = \frac{f}{1}$ and $1 \in S$, by assumption.

Our main concern is the *saturation* of a non-zero ideal I of A with respect to S, which is defined to be

$$S^{-1}I \cap A,$$

where $S^{-1}I = \{\frac{f}{g} \mid f \in I \text{ and } g \in S\}$. It is readily seen that $S^{-1}I$ is an ideal in $S^{-1}A$ and is, in fact, the ideal in $S^{-1}A$ generated by the set I. It is often also written as $I(S^{-1}A)$, since every element of $S^{-1}I$ is the product of an element of I and an element of $S^{-1}A$. We show that for two multiplicative sets S we can easily compute generators for the saturation of an ideal I with respect to S, namely for the cases where $S = \{g^\nu | \nu = 0, 1, \ldots\}$ (Proposition 4.4.1) and $S = \{r \in R | r \neq 0\}$ (Proposition 4.4.4).

PROPOSITION 4.4.1. *Let R be an integral domain. Let $g \in A$, $g \neq 0$. Let w be a new variable. Consider the ideal $\langle I, wg - 1\rangle$ of $A[w]$. Then*

$$IA_g \cap A = \langle I, wg - 1\rangle \cap A.$$

PROOF. Let $f \in IA_g \cap A$. Then there is a non-negative integer ν such that $g^\nu f \in I$. Then $w^\nu g^\nu f \in IA[w] \subseteq \langle I, wg - 1\rangle$ and so

$$f = w^\nu g^\nu f + (1 - w^\nu g^\nu)f \in \langle I, wg - 1\rangle,$$

as desired. Conversely, let $f \in \langle I, wg - 1\rangle \cap A$. Then,

$$f = f(x_1, \ldots, x_n) = \sum_{\mu=1}^{m} g_\mu(x_1, \ldots, x_n)h_\mu(x_1, \ldots, x_n, w)$$

$$+(wg(x_1, \ldots, x_n) - 1)h_0(x_1, \ldots, x_n, w) \in A = R[x_1, \ldots, x_n],$$

where $g_\mu \in I$ and $h_\mu \in A[w]$ for all μ. Since the variable w does not appear in $f = f(x_1, \ldots, x_n)$ we may substitute $w = \frac{1}{g}$, which shows that $f \in IA_g$. □

Proposition 4.4.1 should be compared with Theorem 2.2.13.

Thus, in order to compute generators for the saturation of an ideal $I \subseteq A = R[x_1, \ldots, x_n]$ with respect to $S = \{g^\nu \mid \nu \in \mathbb{N}\}$ we consider an elimination order on the variables $x_1, \ldots, x_n, w$ with w bigger than the x variables. We compute a Gröbner basis, G, for the ideal $\langle I, wg - 1\rangle$ in $A[w] = R[x_1, \ldots, x_n, w]$. Then $G \cap A$ is a Gröbner basis for this saturation (Theorem 4.3.6).

We will give an example which illustrates Proposition 4.4.1 in Example 4.4.6.

Now let $S = R - \{0\}$, and let k denote the quotient field of R (i.e. $k = S^{-1}R$). In order to compute generators for the saturation of I with respect to S, we need two preliminary results.

For an integral domain R, let S be any multiplicative subset of R. We note that $(S^{-1}R)[x_1, \ldots, x_n] = S^{-1}(R[x_1, \ldots, x_n])$ (this is an easy exercise). In this situation we see that Gröbner bases are well-behaved.

PROPOSITION 4.4.2. *Let R be an integral domain. Let $S \subset R$ be a multiplicative set and let $I \subseteq A$ be a non-zero ideal of A. Let G be a Gröbner basis for I with respect to some term order. Then G is a Gröbner basis for the ideal $S^{-1}I$ in $S^{-1}A$.*

PROOF. This follows easily from the third characterization of Gröbner basis in Theorem 4.1.12. Namely, let $f \in S^{-1}I$. Then there is a $s \in S \subset R$ such that $sf \in I$. Suppose that $G = \{g_1, \ldots, g_t\}$. Then $sf = h_1g_1 + \cdots + h_tg_t$ such that $\mathrm{lp}(sf) = \max_{1\le i\le t}(\mathrm{lp}(h_i)\,\mathrm{lp}(g_i))$. Thus $f = (\frac{1}{s}h_1)g_1 + \cdots + (\frac{1}{s}h_t)g_t$. Since R is an integral domain, we have $\mathrm{lp}(sf) = \mathrm{lp}(f)$ and $\mathrm{lp}(\frac{1}{s}h_i) = \mathrm{lp}(h_i)$ $(1 \le i \le t)$. Thus we have a representation of f in terms of $g_1, \ldots, g_t$ of the desired type showing that $G = \{g_1, \ldots, g_t\}$ is a Gröbner basis for $S^{-1}I$. □

We note that in the following lemma we do not need that R be an integral domain.

LEMMA 4.4.3. *Let $J \subseteq I$ be ideals in A and assume that* $\mathrm{Lt}(I) \subseteq \mathrm{Lt}(J)$. *Then $J = I$.*

PROOF. Of course, since $J \subseteq I$, we have that $\mathrm{Lt}(J) = \mathrm{Lt}(I)$. We observe that the proof of Theorem 4.1.12 did not require that the set G be a finite set. Then we see that J is a (infinite) Gröbner basis for I, and hence by Corollary 4.1.15, J generates I. But J is an ideal and so generates itself, that is, $J = I$. □

Recall that for $s \in R$ we denote by $R_s = \{\frac{r}{s^\nu} \mid r \in R, \nu \in \mathbb{N}\}$. We now give the result which, when combined with Proposition 4.4.1, allows us to compute generators for the saturation of an ideal I with respect to $S = R - \{0\}$.

PROPOSITION 4.4.4. *Let R be an integral domain with k its quotient field. Let $I \subset A = R[x_1, \dots, x_n]$ be a non-zero ideal and let $G = \{g_1, \dots, g_t\}$ be a Gröbner basis for I with respect to some term ordering. Set $s = \mathrm{lc}(g_1)\,\mathrm{lc}(g_2)\cdots\mathrm{lc}(g_t)$. Then*

$$(4.4.1)\ Ik[x_1, \dots, x_n] \cap R[x_1, \dots, x_n] = IR_s[x_1, \dots, x_n] \cap R[x_1, \dots, x_n].$$

PROOF. We will need leading term ideals in $R[x_1, \dots, x_n]$, which we denote by Lt, and in $R_s[x_1, \dots, x_n]$, which we denote by Lt_s. We note that it suffices to show that

$$(4.4.2)\qquad \mathrm{Lt}_s(Ik[x_1, \dots, x_n] \cap R_s[x_1, \dots, x_n]) \subseteq \mathrm{Lt}_s(IR_s[x_1, \dots, x_n]).$$

Indeed, if Equation (4.4.2) holds, we have from Lemma 4.4.3, applied to the ring $R_s[x_1, \dots, x_n]$, that $Ik[x_1, \dots, x_n] \cap R_s[x_1, \dots, x_n] = IR_s[x_1, \dots, x_n]$. The desired result (4.4.1) follows by intersecting this last equation with $R[x_1, \dots, x_n]$.
CLAIM: $\mathrm{Lt}(I)k[x_1, \dots, x_n] \cap R[x_1, \dots, x_n] = \mathrm{Lt}(I)R_s[x_1, \dots, x_n] \cap R[x_1, \dots, x_n]$.

Assuming the Claim, we prove (4.4.2) as follows. We let $f \in Ik[x_1, \dots, x_n] \cap R_s[x_1, \dots, x_n]$. We need to show that $\mathrm{lt}(f)$ is in $\mathrm{Lt}_s(IR_s[x_1, \dots, x_n])$. Since, by Proposition 4.4.2, G is a Gröbner basis for $Ik[x_1, \dots, x_n]$, we may write, by Theorem 4.1.12, $f = h_1g_1 + \cdots + h_tg_t$ where $h_i \in k[x_1, \dots, x_n]$ and $\mathrm{lp}(f) = \max_{1 \le i \le t}(\mathrm{lp}(h_i)\,\mathrm{lp}(g_i))$. Let $V = \{i \mid \mathrm{lp}(f) = \mathrm{lp}(h_i)\,\mathrm{lp}(g_i)\}$. Then $\mathrm{lt}(f) = \sum_{i \in V} \mathrm{lt}(h_i)\,\mathrm{lt}(g_i)$. Since $f \in R_s[x_1, \dots, x_n]$ there is an non-negative integer μ such that $s^\mu f \in R[x_1, \dots, x_n]$. We see that

$$\mathrm{lt}(s^\mu f) = \sum_{i \in V} \mathrm{lt}(s^\mu h_i)\,\mathrm{lt}(g_i) \in \mathrm{Lt}(I)k[x_1, \dots, x_n] \cap R[x_1, \dots, x_n],$$

and so from the Claim

$$\mathrm{lt}(s^\mu f) \in \mathrm{Lt}(I)R_s[x_1, \dots, x_n] \cap R[x_1, \dots, x_n].$$

Thus, recalling that $\mathrm{Lt}(I) = \mathrm{Lt}(G)$, we may write

$$\mathrm{lt}(s^\mu f) = \sum_{i=1}^{t} \frac{a_i}{s^{\nu_i}} X_i\,\mathrm{lt}(g_i),$$

for $a_i \in R$, non-negative integers ν_i, and power products X_i, such that $\mathrm{lp}(f) = X_i\,\mathrm{lp}(g_i)$ for all i such that $a_i \neq 0$. Hence

$$\mathrm{lt}(f) = \sum_{i=1}^{t} \frac{a_i}{s^{\nu_i + \mu}} X_i\,\mathrm{lt}(g_i) \in \mathrm{Lt}_s(IR_s[x_1, \dots, x_n]),$$

since $\mathrm{lt}(g_i) \in \mathrm{Lt}_s(IR_s[x_1, \dots, x_n])$, as desired.

It remains to prove the Claim. We will show that both sides are equal to the ideal, $\langle \mathrm{lp}(g_i) \mid 1 \le i \le t \rangle$ in $R[x_1, \dots, x_n]$. We first note that $R_s \subseteq k$ and so we have immediately that $\mathrm{Lt}(I)R_s[x_1, \dots, x_n] \cap R[x_1, \dots, x_n] \subseteq \mathrm{Lt}(I)k[x_1, \dots, x_n] \cap R[x_1, \dots, x_n]$. We write $\mathrm{lt}(g_i) = c_iX_i$ for $c_i \in R$ and a power product X_i.

Then, since $s = a_i c_i$ for some $a_i \in R$, by the definition of s, we have that $X_i = \frac{a_i}{s}(c_i X_i) \in \mathrm{Lt}(I)R_s[x_1, \ldots, x_n]$ and so we see that

$$\langle X_i \mid 1 \leq i \leq t \rangle \subseteq \mathrm{Lt}(I)R_s[x_1, \ldots, x_n] \cap R[x_1, \ldots, x_n].$$

It remains to show that $\mathrm{Lt}(I)k[x_1, \ldots, x_n] \cap R[x_1, \ldots, x_n] \subseteq \langle X_i \mid 1 \leq i \leq t \rangle$. Since $\mathrm{Lt}(I) = \mathrm{Lt}(G)$ we can express each element f of $\mathrm{Lt}(I)k[x_1, \ldots, x_n] \cap R[x_1, \ldots, x_n]$ as a linear combination of the X_i with coefficients in $k[x_1, \ldots, x_n]$. Let cX be a term in f, where $c \in R$ and X is a power product. Then in order for c to be non-zero we must have an X_i dividing X and then we see that

$$\mathrm{Lt}(I)k[x_1, \ldots, x_n] \cap R[x_1, \ldots, x_n] \subseteq \langle X_i \mid 1 \leq i \leq t \rangle,$$

as desired. □

We note that if R is a UFD, then the element s in Proposition 4.4.4 can be replaced by $s = \mathrm{lcm}(\mathrm{lc}(g_i)|1 \leq i \leq t)$ (Exercise 4.4.1).

COROLLARY 4.4.5. *Let R be an integral domain in which linear equations are solvable and let k be its quotient field. Let $I \subset A = R[x_1, \ldots, x_n]$ be an ideal. Then we can compute generators for the ideal*

$$Ik[x_1, \ldots, x_n] \cap R[x_1, \ldots, x_n].$$

PROOF. This follows from Proposition 4.4.4 and Proposition 4.4.1. □

EXAMPLE 4.4.6. We consider Example 4.2.11 and compute the saturation of $I = \langle 3x^2y + 7y, 4xy^2 - 5x \rangle$ with respect to $\mathbb{Q}[x, y]$ (i.e. with respect to $S = \mathbb{N} - \{0\}$). That is, we compute $I\mathbb{Q}[x, y] \cap \mathbb{Z}[x, y]$. In that example we computed a Gröbner basis for I, with respect to lex with $x > y$, to be $f_1 = 3x^2y + 7y$, $f_2 = 4xy^2 - 5x$, $f_3 = 15x^2 + 28y^2$, and $f_4 = 28y^3 - 35y$. Then, following Proposition 4.4.4 and Exercise 4.4.1, we need to compute $I\mathbb{Z}_{420}[x, y] \cap \mathbb{Z}[x, y]$, since $\mathrm{lcm}(3, 4, 15, 28) = 420$ (here, of course, we are adopting the notation of this section and $\mathbb{Z}_{420}$ does not mean the integers mod 420). Then, from Proposition 4.4.1, we need to compute $\langle f_1, f_2, 420w - 1 \rangle \cap \mathbb{Z}[x, y]$, which we do using Theorem 4.3.6. We, of course, begin with the polynomials f_1, f_2, f_3, f_4 together with $f_5 = 420w - 1$ and observe that, since f_1, f_2, f_3, f_4 is a Gröbner basis, we may start with $\sigma = 5$ in Algorithm 4.2.2, provided we use the lex term ordering with $w > x > y$. We will not go through the computations, but will note that $\{f_1, f_2, f_3, f_4, f_5, 525wx - xy^2, 784wy^2 + x^2, 525wy - y^3, 980wx + x^3, 980wy + x^2y, 4y^3 - 5y, 3x^3 + 28xy^2 - 28x\}$ is a Gröbner basis for $\langle f_1, f_2, 420w - 1 \rangle$, and hence $\{f_1, f_2, f_3, f_4, 4y^3 - 5y, 3x^3 + 28xy^2 - 28x\}$ is a Gröbner basis for $I\mathbb{Q}[x, y] \cap \mathbb{Z}[x, y]$.

Finally we are ready to discuss the issue of primality in rings. We are no longer assuming that R is an integral domain. We need two lemmas from commutative algebra which give criteria for when an ideal in $R[x_1, \ldots, x_n]$ is prime.

LEMMA 4.4.7. *Let $I \subseteq R[x_1, \ldots, x_n]$ be an ideal. Then I is a prime ideal if and only if the image of I in $(R/I \cap R)[x_1, \ldots, x_n]$ is prime. Moreover, in this case $I \cap R$ is prime.*

PROOF. For $r \in R$ we let $\overline{r} = r + I \cap R$ be the image of r in $R/I \cap R$. For $f = \sum_{i=1}^{\ell} a_i X_i \in R[x_1, \ldots, x_n]$, with $a_i \in R$ and power products X_i, we write $\overline{f} = \sum_{i=1}^{\ell} \overline{a}_i X_i$, its image in $(R/I \cap R)[x_1, \ldots, x_n]$. We define a ring homomorphism

$$\phi\colon (R/I \cap R)[x_1, \ldots, x_n] \longrightarrow R[x_1, \ldots, x_n]/I$$

by

$$\phi\colon \sum_{i=1}^{\ell} \overline{a}_i X_i \longmapsto \sum_{i=1}^{\ell} a_i X_i + I.$$

That the map is well-defined follows since if $\sum_{i=1}^{\ell} \overline{a}_i X_i = 0$, then $a_i \in I \cap R$ for all i and so clearly $\sum_{i=1}^{\ell} a_i X_i \in I$. It is then easy to see that ϕ is onto and $\ker(\phi)$ is the image of I in $(R/I \cap R)[x_1, \ldots, x_n]$. It is now immediate that I is a prime ideal if and only if the image of I in $(R/I \cap R)[x_1, \ldots, x_n]$ is prime. Finally, it is trivially seen that if I is prime then so is $I \cap R$. □

LEMMA 4.4.8. *Let R be an integral domain with quotient field k and let $I \subseteq R[x_1, \ldots, x_n]$ be an ideal such that $I \cap R = \{0\}$. Then I is prime if and only if $Ik[x_1, \ldots, x_n]$ is prime in $k[x_1, \ldots, x_n]$ and $I = Ik[x_1, \ldots, x_n] \cap R[x_1, \ldots, x_n]$.*

PROOF. We first assume that I is a prime ideal in $R[x_1, \ldots, x_n]$. Let $f, g \in k[x_1, \ldots, x_n]$ and assume that $fg \in Ik[x_1, \ldots, x_n]$. Then $fg = \frac{h}{r}$ such that $h \in I$ and $r \in R$. Let $d, e \in R$ be such that $df, eg \in R[x_1, \ldots, x_n]$. Then we have $r(df)(eg) = deh \in I$ and so $r(df) \in I$ or $eg \in I$, since I is prime in $R[x_1, \ldots, x_n]$. We then have immediately that $f \in Ik[x_1, \ldots, x_n]$ or $g \in Ik[x_1, \ldots, x_n]$, and so $Ik[x_1, \ldots, x_n]$ is prime, as desired. Moreover, let $\frac{h}{r} \in Ik[x_1, \ldots, x_n] \cap R[x_1, \ldots, x_n]$ with $h \in I$ and $r \in R$, $r \neq 0$. Then $r(\frac{h}{r}) \in I$ and so $r \in I$ or $\frac{h}{r} \in I$ (recall that $\frac{h}{r} \in R[x_1, \ldots, x_n]$). By the hypothesis, $I \cap R = \{0\}$, $r \notin I$ and so $\frac{h}{r} \in I$. The reverse inclusion is trivial.

We now assume that the ideal $Ik[x_1, \ldots, x_n]$ is prime in $k[x_1, \ldots, x_n]$ and $I = Ik[x_1, \ldots, x_n] \cap R[x_1, \ldots, x_n]$. Let $f, g \in R[x_1, \ldots, x_n]$ and assume that $fg \in I$. Then $fg \in Ik[x_1, \ldots, x_n]$ and so by hypothesis, $f \in Ik[x_1, \ldots, x_n]$ or $g \in Ik[x_1, \ldots, x_n]$. We assume that $f \in Ik[x_1, \ldots, x_n]$. Then $f \in Ik[x_1, \ldots, x_n] \cap R[x_1, \ldots, x_n] = I$ and so we see that I is a prime ideal. □

COROLLARY 4.4.9. *Let $I \subseteq R[x]$ be an ideal. Then I is prime if and only if*
  (i) *$I \cap R$ is a prime ideal in R and*
  (ii) *if we let $R' = R/I \cap R$, k' be the quotient field of R', and I' be the image of I in $R'[x]$, then $I'k'[x]$ is a prime ideal of $k'[x]$, and $I' = I'k'[x] \cap R'[x]$.*

PROOF. Noting that $I' \cap R' = \{0\}$, this follows immediately from Lemmas 4.4.7 and 4.4.8. □

Since $R[x_1, \ldots, x_n] = (R[x_1, \ldots, x_{n-1}])[x_n]$, Corollary 4.4.9 gives us an inductive technique of determining whether a given ideal in $R[x_1, \ldots, x_n]$ is prime. The idea is summarized in the next paragraph.

For a single variable x and an ideal $I \subseteq R[x]$ we want to decide if I is a prime ideal. By Corollary 4.4.9, we first need to determine whether $I \cap R$ is prime. We may compute a generating set for $I \cap R$ from Theorem 4.3.6 and then, assuming we can determine whether ideals in R are prime, we can determine whether $I \cap R$ is prime. In this case we will consider the integral domain $R' = R/I \cap R$, its quotient field k', and the ideal I', the image of I in $R'[x]$. Again by Corollary 4.4.9, we then need to determine whether $I'k'[x]$ is a prime ideal for *one* variable x. Since x is a single variable, $k'[x]$ is a PID and, using the Euclidean Algorithm (Algorithm 1.3.2) on the known generators of I', we can determine a single generator f of $I'k'[x]$. Now, assuming that we can determine whether f is irreducible or not in $k'[x]$, we can determine whether $I'k'[x]$ is a prime ideal or not. Finally, by Corollary 4.4.5 we can determine whether or not $I' = I'k'[x] \cap R'[x]$.

Thus we see that in order to have an algorithm determining the primality of ideals in $R[x_1, \ldots, x_n]$, we must assume that we can do similar things in R. Specifically, we must assume that we can determine when an ideal in R is prime. Moreover, we must assume that we can determine, given a prime ideal P of $R[x_1, \ldots, x_n]$, whether polynomials in one variable with coefficients in the quotient field of $R[x_1, \ldots, x_n]/P$ are irreducible or not. For example $\mathbb{Z}$ and $\mathbb{Q}$ have these properties.

The algorithm for determining the primality of an ideal is given as Algorithm 4.4.1.

EXAMPLE 4.4.10. We now give an example of Algorithm 4.4.1. Let $I \subset \mathbb{Q}[x, y, z]$ be the ideal generated by $f_1 = xz - y^2, f_2 = x^3 - yz$ and $f_3 = x^2y - z^2$. As we saw in Exercise 2.5.5, this is the ideal of relations among the three polynomials t^3, t^4, t^5 and hence must be a prime ideal (since then $\mathbb{Q}[x, y, z]/I$ is isomorphic to a subring of the integral domain $\mathbb{Q}[t]$). In this example we will show that I is prime using Algorithm 4.4.1. We use the notation established in that algorithm. We first compute a Gröbner basis G for I using the lex ordering with $x > y > z$ and get $G = \{f_1, f_2, f_3, f_4, f_5\}$, where $f_4 = xy^3 - z^3$ and $f_5 = y^5 - z^4$. We have $n = 3$, $J_4 = J_3 = \langle 0 \rangle$, $J_2 = \langle y^5 - z^4 \rangle$, and $J_1 = I$. We first note that the case $i = 4$ in the algorithm is trivial.

The next case to consider is $i = 3$. We have $J_2 = I \cap \mathbb{Q}[y, z] = \langle y^5 - z^4 \rangle$. Thus, $R' = R_3/J_3 = \mathbb{Q}[z]$, $J' = J_2R'[y] = (y^5 - z^4)\mathbb{Q}[y, z]$, and[3] $k' = \mathbb{Q}(z)$. Then $J'k'[y] = (y^5 - z^4)\mathbb{Q}(z)[y]$ and so $f = y^5 - z^4$. Here f is viewed as a polynomial in y with coefficients in $\mathbb{Q}(z)$. We use "brute force" to see that f is irreducible over $\mathbb{Q}(z)$: It cannot have a root in $\mathbb{Q}(z)$, since there cannot be a rational function in $\mathbb{Q}(z)$ whose fifth power is the fourth power of the prime $z \in \mathbb{Q}[z]$, because of

[3]Here $\mathbb{Q}(z)$ denotes the field of rational functions in z, that is, the quotient field of $\mathbb{Q}[z]$.

INPUT: I, an ideal of $R[x_1, \ldots, x_n]$

OUTPUT: TRUE if I is a prime ideal, FALSE otherwise

Set $R_i := R[x_i, \ldots, x_n]$ for $i = 1, \ldots, n$, and $R_{n+1} := R$

Compute $J_i := I \cap R_i$ for $i = 1, \ldots, n+1$

IF J_{n+1} is not a prime ideal of R **THEN**

  result:=FALSE

ELSE

  result:=TRUE

  $i := n + 1$

  WHILE $i > 1$ **AND** result=TRUE **DO**

    $R' := R_i/J_i$

    $J' :=$ image of J_{i-1} in $R'[x_{i-1}]$

    $k' :=$ quotient field of R'

    Compute the polynomial f such that $J'k'[x_{i-1}] = \langle f \rangle$

    IF f is not zero and reducible over k' **THEN**

      result:=FALSE

    ELSE

      Compute $J'k'[x_{i-1}] \cap R'[x_{i-1}]$

      IF $J'k'[x_{i-1}] \cap R'[x_{i-1}] \neq J'$ **THEN**

        result:=FALSE

      ELSE

        $i := i - 1$

RETURN result

ALGORITHM 4.4.1. *Primality Test in* $R[x_1, \ldots, x_n]$

unique factorization in $\mathbb{Q}[z]$. Moreover, it cannot be the product of a cubic and a quadratic in $\mathbb{Q}(z)[y]$. One way to see this is as follows. If $f = y^5 - z^4$ were the product of a cubic and a quadratic in $\mathbb{Q}(z)[y]$, then we would have a system of 5 polynomial equations in 5 unknowns which would have to have a solution in $\mathbb{Q}(z)$.

If one computes a Gröbner basis for these polynomials one finds that there would have to be a rational function $e(z) \in \mathbb{Q}(z)$ satisfying $e^5 = z^8$ (use lex with $e > z$ being the smallest variables where e is the constant term of the quadratic) and again this is impossible. It is then trivial to check that $(y^5 - z^4)\mathbb{Q}(z)[y] \cap \mathbb{Q}[y, z] = (y^5 - z^4)\mathbb{Q}[y, z]$ (if $(y^5 - z^4)\sum_{\nu=0}^{N} \alpha_\nu(z)y^\nu \in (y^5 - z^4)\mathbb{Q}[y, z]$ with $\alpha_\nu(z) \in \mathbb{Q}(z)$, then a simple induction shows that $\alpha_\nu(z) \in \mathbb{Q}[z]$). Thus we have completed the WHILE loop for $i = 3$ with "result=TRUE".

We now consider the case $i = 2$. In this case $J_1 = I$. We have $R' = \mathbb{Q}[y, z]/\langle y^5 - z^4\rangle$, $J' = I((\mathbb{Q}[y, z]/\langle y^5 - z^4\rangle)[x])$ and $k' =$ quotient field of R'. For polynomials $f \in \mathbb{Q}[x, y, z]$ we denote by $\overline{f}$ the polynomial in $R'[x] = (\mathbb{Q}[y, z]/\langle y^5 - z^4\rangle)[x]$ obtained by reducing the coefficients of the powers of x (these being polynomials in $\mathbb{Q}[y, z]$) modulo $\langle y^5 - z^4\rangle$; i.e. we have a homomorphism

$$\begin{array}{ccc} \mathbb{Q}[x, y, z] & \longrightarrow & (\mathbb{Q}[y, z]/\langle y^5 - z^4\rangle)[x] \\ f & \longmapsto & \overline{f}. \end{array}$$

We also denote the image of an ideal $K \subset \mathbb{Q}[x, y, z]$ in $R'[x]$ by $\overline{K}$. We first must find a generator for $J'k'[x]$, where J' is the ideal generated by $\overline{f}_1, \overline{f}_2, \overline{f}_3, \overline{f}_4 \in R'[x]$. It is, in fact, easy to see that $\overline{f}_2, \overline{f}_3, \overline{f}_4$ are multiples of $\overline{f}_1$ in $k'[x]$. For example, ignoring the "bar" notation for the moment, viewing the following equation as being in $k'[x]$, and noting that z, z^2, and z^3 are non-zero in R', we see that $x^3 - yz = (zx - y^2)(\frac{1}{z}x^2 + \frac{1}{z^2}y^2x + \frac{1}{z^3}y^4)$; that is, $\overline{f}_2$ is a multiple of $\overline{f}_1$. Thus $f = \overline{f}_1$ and it is irreducible over k', since it is of degree 1 in x. It remains to show that $J'k'[x] \cap R'[x] = J'$. We will apply Proposition 4.4.4 and Proposition 4.4.1.

So we first need to compute a Gröbner basis for the ideal $J' = \langle \overline{f}_1, \overline{f}_2, \overline{f}_3, \overline{f}_4\rangle \subset R'[x]$. We will use Algorithm 4.2.2. We make the following general observation which is easily proved: if K is an ideal of $\mathbb{Q}[x, y, z]$ containing $y^5 - z^4$ and $g \in \mathbb{Q}[x, y, z]$, then $\overline{K} \colon \overline{\langle g\rangle} = \overline{K \colon \langle g\rangle}$. We consider the saturated subset $\{1, 2\}$ for which we need to compute $\overline{\langle z\rangle} \colon \overline{\langle 1\rangle} = \overline{\langle z\rangle}$ and the corresponding syzygy is $(x^2, -\overline{z})$ which gives the S-polynomial $x^2\overline{f}_1 - \overline{z}\overline{f}_2 = -\overline{y}^2x^2 + \overline{yz^2} \xrightarrow{\overline{f}_3} 0$. Now the saturated subsets of $\{1, 2, 3\}$ containing 3 are $\{1, 3\}$ and $\{1, 2, 3\}$. For $\{1, 3\}$ we compute $\overline{\langle z\rangle} \colon \overline{\langle y\rangle} = \overline{\langle z, y^5 - z^4\rangle} \colon \overline{\langle y\rangle} = \overline{\langle z, y^5 - z^4\rangle \colon \langle y\rangle} = \overline{\langle z, y^4\rangle}$ (note that this latter is just a computation in $\mathbb{Q}[y, z]$ and so can be done using Lemma 2.3.11). The two syzygies are $(\overline{y}x, -\overline{z})$ and $(\overline{z}^3x, -\overline{y}^4)$. The first gives the S-polynomial $\overline{y}x\overline{f}_1 - \overline{z}\overline{f}_3 = -\overline{y}^3x + \overline{z}^3 \xrightarrow{\overline{f}_4} 0$. The second gives the S-polynomial $\overline{z}^3x\overline{f}_1 - \overline{y}^4\overline{f}_3 = \overline{(z^4 - y^5)}x^2 - \overline{y^2z^3}x + \overline{y^4z^2} \xrightarrow{\overline{f}_1} 0$. The saturated set $\{1, 2, 3\}$ and the saturated subsets of $\{1, 2, 3, 4\}$ containing 4 can be handled in the same way. After this computation, we see that a Gröbner basis for the ideal J' is just the original generating set $\{\overline{f}_1, \overline{f}_2, \overline{f}_3, \overline{f}_4\}$.

Then, as in Proposition 4.4.4, we set $s = \mathrm{lc}(\overline{f}_1)\,\mathrm{lc}(\overline{f}_2)\,\mathrm{lc}(\overline{f}_3)\,\mathrm{lc}(\overline{f}_4) = \overline{z}\overline{1}\overline{y}\overline{y^3} = \overline{y^4z}$, and we need to show that $J'R'_s[x] \cap R'[x] = J'$. We do this using Proposition

4.4.1. Thus we need to compute

$$\overline{\langle J', wzy^4 - 1\rangle \cap R'[x]} = \overline{\langle f_1, f_2, f_3, f_4, y^5 - z^4, wzy^4 - 1\rangle \cap \mathbb{Q}[x, y, z]}.$$

This latter computation can be done in $\mathbb{Q}[x, y, z, w]$. Using lex with $w > x > y > z$, we compute that a Gröbner basis for the ideal $\langle f_1, f_2, f_3, f_4, y^5 - z^4, wzy^4 - 1\rangle$ consists of the 6 polynomials listed above together with $wy^3z^3 - x^2$, $wyz^4 - x$, and $wz^5 - y$. Thus $\langle f_1, f_2, f_3, f_4, y^5 - z^4, wzy^4 - 1\rangle \cap \mathbb{Q}[x, y, z] = \langle f_1, f_2, f_3, f_4, y^5 - z^4\rangle$ and so $\overline{\langle J', wzy^4 - 1\rangle \cap R'[x]} = \overline{\langle f_1, f_2, f_3, f_4, y^5 - z^4, wzy^4 - 1\rangle \cap \mathbb{Q}[x, y, z]} = \langle f_1, f_2, f_3, f_4\rangle = J'$ and the algorithm terminates with "result=TRUE."

Exercises

4.4.1. Prove that if R is a UFD then in Proposition 4.4.4 we can let $s = \operatorname{lcm}(\operatorname{lc}(g_1), \operatorname{lc}(g_2), \ldots, \operatorname{lc}(g_t))$.

4.4.2. Compute the saturation of the following ideals in $\mathbb{Z}[x, y]$ with respect to $\mathbb{Q}[x, y]$ (i.e. with respect to $S = \mathbb{N} - \{0\}$) using Proposition 4.4.4.
 a. $\langle 6x^2 + y^2, 10x^2y + 2xy\rangle$.
 b. $\langle 3x^2y - 3yz + y, 5x^2z - 8z^2\rangle$.

4.4.3. Using Algorithm 4.4.1 and lex with $z > y > x$, show that the ideal $\langle xz - y^3, yz - x^2\rangle \subseteq \mathbb{Q}[x, y, z]$ is not prime.

4.4.4. Using Algorithm 4.4.1, show that the ideal $\langle y^4 - z^3, y^2 - xz, xy^2 - z^2, x^2 - z\rangle \subseteq \mathbb{Q}[x, y, z]$ is prime.

4.5. Gröbner Bases over Principal Ideal Domains. In this section we specialize the results of the previous sections to the case where the coefficient ring R is a *Principal Ideal Domain* (PID). Recall that an integral domain is a PID if every ideal of R is *principal*, that is, if every ideal of R can be generated by a single element. We note that such rings are also Unique Factorization Domains (UFD) (see [**Go, He, Hun**]). We will make extensive use of this fact in this section and in Section 4.6. Examples of such rings include $\mathbb{Z}$, $\mathbb{Z}[\sqrt{2}]$, $\mathbb{Z}[i]$, where $i^2 = -1$, and $k[y]$, where k is a field and y is a single variable. Of course the theory that we have developed so far in this chapter applies to these rings. But, because of the special properties of PID's, we will show that we may construct Gröbner bases using S-polynomials as we did in the case of fields (Algorithm 4.5.1). We will then define strong Gröbner bases which are similar to Gröbner bases when the ring R is a field, and we will show how to compute them in Theorem 4.5.9. We will also describe the structure of strong Gröbner bases in the case of a polynomial ring in one variable over a PID. This will give us a lot of information about the given ideal as we will see in Section 4.6. We will also specialize the ring R to $k[y]$, where k is a field and y is a single variable, and we will describe the relationship among the different notions of Gröbner bases in this case.

Recall that in order to compute a Gröbner bases for $\langle f_1, \ldots, f_s\rangle$ in $A = R[x_1, \ldots, x_n]$, we need to compute a homogeneous generating set for the syzygy module, $\mathrm{Syz}(\mathrm{lt}(f_1), \ldots, \mathrm{lt}(f_s))$ (Theorem 4.2.3). When R is a field, we saw that a generating set $\mathcal{B}$ for $\mathrm{Syz}(\mathrm{lt}(f_1), \ldots, \mathrm{lt}(f_s))$ exists such that every element of $\mathcal{B}$ has exactly two non-zero coordinates (these syzygies correspond to S-polynomials; see Proposition 3.2.3).

DEFINITION 4.5.1. *A generating set* $\mathcal{B}$ *of* $\mathrm{Syz}((\mathrm{lt}(f_1), \ldots, \mathrm{lt}(f_s))$ *is called an* S-basis *if every element of* $\mathcal{B}$ *is homogeneous and has exactly two non-zero coordinates.*

In general, when the coefficient ring is not a field, there is no S-basis for $\mathrm{Syz}((\mathrm{lt}(f_1), \ldots, \mathrm{lt}(f_s))$ (see Example 4.5.4). However, when R is a PID such a generating set exists as the next proposition shows. We assume that $s > 1$ in this entire section, because the case $s = 1$ is trivial, since any single polynomial is automatically a Gröbner basis (note that this is not the case if R has zero divisors; e.g. $\{2x + 1\}$ is not a Gröbner basis in $\mathbb{Z}_6[x]$). We first prove the following identity[4].

LEMMA 4.5.2. *Let* R *be a PID and let* $a, a_1, \ldots, a_\ell$ *be in* $R - \{0\}$. *Then*

$$\langle a_1, \ldots, a_\ell \rangle_R \colon \langle a \rangle_R = \sum_{i=1}^{\ell} \left(\langle a_i \rangle_R \colon \langle a \rangle_R \right).$$

PROOF. Since R is a PID, we have (see Proposition 1.3.8 for the case where $R = k[x]$)

$$\langle a_1, \ldots, a_\ell \rangle_R = \langle \gcd(a_1, \ldots, a_\ell) \rangle_R$$

and (see Lemma 2.3.7 for the case of polynomial rings)

$$\langle a_i \rangle_R \cap \langle a \rangle_R = \langle \mathrm{lcm}(a_i, a) \rangle_R \text{ for all } i = 1, \ldots, \ell.$$

Moreover, it is easy to show that in any UFD we have

$$\mathrm{lcm}(\gcd(a_1, \ldots, a_\ell), a) = \gcd(\mathrm{lcm}(a_1, a), \ldots, \mathrm{lcm}(a_\ell, a)).$$

Thus

$$\begin{aligned} \langle a_1, \ldots, a_\ell \rangle_R \cap \langle a \rangle_R &= \langle \gcd(a_1, \ldots, a_\ell) \rangle_R \cap \langle a \rangle_R \\ = \langle \mathrm{lcm}(\gcd(a_1, \ldots, a_\ell), a) \rangle_R &= \langle \gcd(\mathrm{lcm}(a_1, a), \ldots, \mathrm{lcm}(a_\ell, a)) \rangle_R \\ = \langle \mathrm{lcm}(a_1, a), \ldots, \mathrm{lcm}(a_\ell, a) \rangle_R &= \sum_{i=1}^{\ell} \langle \mathrm{lcm}(a_i, a) \rangle_R = \sum_{i=1}^{\ell} \langle a_i \rangle_R \cap \langle a \rangle_R. \end{aligned}$$

The result now follows easily as in Lemma 2.3.11 and Proposition 4.3.11. □

PROPOSITION 4.5.3. *Let* R *be a PID and let* $f_1, \ldots, f_s$ *be non-zero polynomials in* $R[x_1, \ldots, x_n]$, *with* $s > 1$. *Then* $\mathrm{Syz}((\mathrm{lt}(f_1), \ldots, \mathrm{lt}(f_s))$ *has an S-basis.*

[4]For ideals $I_1, \ldots, I_\ell$ in a ring R we define the ideal $\sum_{i=1}^{\ell} I_i$ to be the ideal in R generated by the ideals $I_1, \ldots, I_\ell$. That is, $\sum_{i=1}^{\ell} I_i = \langle I_1, \ldots, I_\ell \rangle$. It is easy to see that every element $f \in \sum_{i=1}^{\ell} I_i$ can be written as $f = f_1 + \cdots + f_\ell$ where $f_i \in I_i$ for each $i = 1, \ldots, \ell$.

PROOF. For $i = 1, \ldots, s$ let $\operatorname{lt}(f_i) = c_i X_i$, where $c_i \in R$ and $\operatorname{lp}(f_i) = X_i$. For any subset J of $\{1, \ldots, s\}$ we define, as before, $X_J = \operatorname{lcm}(X_j \mid j \in J)$. We use the notation and technique presented in Theorem 4.2.9 to construct the desired generating set. Let $S_\sigma = \operatorname{Syz}(c_1X_1, \ldots, c_\sigma X_\sigma)$ for $1 \leq \sigma \leq s$. We will show by induction on σ that every S_σ has an S-basis. Our induction starts at $\sigma = 2$. Then it is easy to see that S_2 is generated by the syzygy $(\frac{c}{c_1}\frac{X}{X_1}, -\frac{c}{c_2}\frac{X}{X_2})$, where $c = \operatorname{lcm}(c_1, c_2)$, and $X = \operatorname{lcm}(X_1, X_2)$. Now let $\sigma > 2$ and assume by induction that we have computed an S-basis $\mathcal{B}_{\sigma-1}$ for $S_{\sigma-1}$. We now construct an S-basis $\mathcal{B}_\sigma$ for S_σ. Recall from Theorem 4.2.9 that $\mathcal{B}_\sigma$ consists of two groups of elements. First, to each element $\boldsymbol{a}$ of $\mathcal{B}_{\sigma-1}$ corresponds the element $(\boldsymbol{a}, 0)$ in $\mathcal{B}_\sigma$. Clearly $(\boldsymbol{a}, 0)$ has only two non-zero coordinates by the choice of $\mathcal{B}_{\sigma-1}$. The other elements in $\mathcal{B}_\sigma$ are obtained from the ideal $\langle c_j \mid j \in J, j \neq \sigma\rangle_R \colon \langle c_\sigma\rangle_R$, where J is a saturated subset of $\{1, \ldots, \sigma\}$ containing σ. From Lemma 4.5.2 we have

$$\langle c_j \mid j \in J, j \neq \sigma\rangle_R \colon \langle c_\sigma\rangle_R = \sum_{\substack{j \in J \\ j \neq \sigma}} \langle c_j\rangle_R \colon \langle c_\sigma\rangle_R.$$

Now for $j \in J$, $j \neq \sigma$, let d_j be a generator of the ideal $\langle c_j\rangle_R \colon \langle c_\sigma\rangle_R$. Therefore

$$\langle c_j \mid j \in J, j \neq \sigma\rangle_R \colon \langle c_\sigma\rangle_R = \langle d_j \mid j \in J, j \neq \sigma\rangle_R.$$

Moreover, as in Theorem 4.2.9, associated to each d_j we have the element

$$\boldsymbol{s}_{jJ} = \frac{d_j c_\sigma}{c_j}\frac{X_{j\sigma}}{X_j}\boldsymbol{e}_j - d_j\frac{X_{j\sigma}}{X_\sigma}\boldsymbol{e}_\sigma,$$

since $d_j c_\sigma \in \langle c_j\rangle$. Clearly the element $\boldsymbol{s}_{jJ}$ has only two non-zero coordinates.

By Theorem 4.2.9 the vectors $\boldsymbol{s}_{jJ}$, where $j \in J, j \neq \sigma$, and J ranges over all saturated subsets of $\{1, \ldots, \sigma\}$ which contain σ, together with the vectors $(\boldsymbol{a}, 0)$, where $\boldsymbol{a}$ ranges over all elements of a generating set for $S_{\sigma-1}$, forms a generating set for S_σ. Therefore S_σ has an S-basis. □

We now give an example which shows that S-bases do not exist for rings which are not PID's.

EXAMPLE 4.5.4. Consider the ring $R = \mathbb{Z}[z]$. R is not a PID, since, for example, $\langle 2, z\rangle$ is not principal. We note that R is a UFD. Now consider the following polynomials in $(\mathbb{Z}[z])[x, y]$

$$f_1 = 2xy^2 + y, \quad f_2 = zx^2y + x, \quad f_3 = (2 + z)xy + 1.$$

We use the lex order with $x > y$. Then we have

$$\operatorname{lt}(f_1) = 2xy^2, \quad \operatorname{lt}(f_2) = zx^2y, \quad \operatorname{lt}(f_3) = (2 + z)xy.$$

The homogeneous elements of $\operatorname{Syz}(\operatorname{lt}(f_1), \operatorname{lt}(f_2), \operatorname{lt}(f_3))$ with exactly two non-zero coordinates are multiples of $\boldsymbol{s}_1 = (-zx, 2y, 0)$, $\boldsymbol{s}_2 = (-(2 + z), 0, 2y)$, or $\boldsymbol{s}_3 =$

$(0, -(2+z), zx)$. However, the vector $(-x, -y, xy)$ is in $\mathrm{Syz}(\mathrm{lt}(f_1), \mathrm{lt}(f_2), \mathrm{lt}(f_3))$, and so, if $\boldsymbol{s}_1$, $\boldsymbol{s}_2$, $\boldsymbol{s}_3$ generated $\mathrm{Syz}(\mathrm{lt}(f_1), \mathrm{lt}(f_2), \mathrm{lt}(f_3))$, we would have

$$(-x, -y, xy) = h_1(-zx, 2y, 0) + h_2(-(2+z), 0, 2y) + h_3(0, -(2+z), zx),$$

for some $h_1, h_2, h_3 \in (\mathbb{Z}[z])[x, y]$. Then $xy = 2yh_2 + zxh_3$. Note that $h_2 = a_2x$ and $h_3 = a_3y$, for some $a_2, a_3 \in \mathbb{Z}[x, y, z]$. But then $1 = 2a_2 + za_3$, and this is impossible. So $\mathrm{Syz}(\mathrm{lt}(f_1), \mathrm{lt}(f_2), \mathrm{lt}(f_3))$ does not have an S-basis.

In fact, it can be shown that if R is a UFD and if $\mathrm{Syz}(\mathrm{lt}(f_1), \ldots, \mathrm{lt}(f_s))$ has an S-basis for every ideal $I = \langle f_1, \ldots, f_s \rangle$ in $R[x_1, \ldots, x_n]$, then R is a PID (see Exercise 4.5.14). However, there are rings which are not UFD's for which $\mathrm{Syz}(\mathrm{lt}(f_1), \ldots, \mathrm{lt}(f_s))$ has an S-basis for every ideal $I = \langle f_1, \ldots, f_s \rangle$ in $R[x_1, \ldots, x_n]$ (see Exercise 4.5.15).

We note that the syzygies obtained in Proposition 4.5.3 are the analog of the syzygies used to define the S-polynomials in the case where the coefficients are in a field. Indeed, if $\mathrm{lt}(f_i) = c_iX_i, \mathrm{lt}(f_j) = c_jX_j$, where $c_i, c_j \in R$, then the syzygy with exactly two non-zero coordinates corresponding to these two polynomials is

$$\boldsymbol{s}_{ij} = \frac{c}{c_i}\frac{X}{X_i}\boldsymbol{e}_i - \frac{c}{c_j}\frac{X}{X_j}\boldsymbol{e}_j,$$

where $c = \mathrm{lcm}(c_i, c_j)$, and $X = \mathrm{lcm}(X_i, X_j)$. We define the *S-polynomial* of f_i and f_j as

$$S(f_i, f_j) = \frac{c}{c_i}\frac{X}{X_i}f_i - \frac{c}{c_j}\frac{X}{X_j}f_j. \tag{4.5.1}$$

Proposition 4.5.3 can then be used to modify Algorithm 4.2.1 to obtain an algorithm for computing Gröbner bases in $R[x_1, \ldots, x_n]$, where R is a PID. This algorithm is presented as Algorithm 4.5.1. We note that Algorithm 4.5.1 is similar to the algorithm given in the case of $R = k$, a field (see Algorithm 1.7.1).

We now give an example of how Algorithm 4.5.1 is applied.

EXAMPLE 4.5.5. We go back to Example 4.2.11 and use Algorithm 4.5.1 to recompute a Gröbner basis with respect to the lex term ordering with $x > y$ for $I = \langle f_1, f_2 \rangle$, where $\boxed{f_1 = 3x^2y + 7y,}$ $\boxed{f_2 = 4xy^2 - 5x}$ and $R = \mathbb{Z}$. We initialize $G = \{f_1, f_2\}$ and $\mathcal{G} = \{\{f_1, f_2\}\}$. Since $\mathcal{G} \neq \emptyset$, we choose $\{f_1, f_2\} \in \mathcal{G}$, so that now $\mathcal{G} = \emptyset$. We compute $c = \mathrm{lcm}(3, 4) = 12$, $X = \mathrm{lcm}(x^2y, xy^2) = x^2y^2$, so that the corresponding S-polynomial is $S(f_1, f_2) = 4yf_1 - 3xf_2 = 15x^2 + 28y^2$. This polynomial cannot be reduced and we add $\boxed{f_3 = 15x^2 + 28y^2}$ to the basis, so that now $G = \{f_1, f_2, f_3\}$ and $\mathcal{G} = \{\{f_1, f_3\}, \{f_2, f_3\}\}$. Since $\mathcal{G} \neq \emptyset$ we choose $\{f_1, f_3\}$, so that $\mathcal{G} = \{\{f_2, f_3\}\}$. In this case we have $c = 15$ and $X = x^2y$, so that the corresponding S-polynomial is $S(f_1, f_3) = 5f_1 - yf_3 = -28y^3 + 35y$. This polynomial cannot be reduced and we add $\boxed{f_4 = -28y^3 + 35y}$ to the basis, so that $G = \{f_1, f_2, f_3, f_4\}$ and $\mathcal{G} = \{\{f_2, f_3\}, \{f_1, f_4\}, \{f_2, f_4\}, \{f_3, f_4\}\}$. Since

INPUT: $F = \{f_1, \dots, f_s\} \subseteq R[x_1, \dots, x_n]$ with $f_i \neq 0$ $(1 \leq i \leq s)$

OUTPUT: $G = \{g_1, \dots, g_t\}$, a Gröbner basis for $\langle f_1, \dots, f_s \rangle$

INITIALIZATION: $G := F$, $\mathcal{G} := \{\{f, g\} \mid f \neq g \in G\}$

WHILE $\mathcal{G} \neq \emptyset$ **DO**

Choose any $\{f, g\} \in \mathcal{G}$. Let $\text{lt}(f) = c_f X_f, \text{lt}(g) = c_g X_g$

$\mathcal{G} := \mathcal{G} - \{\{f, g\}\}$

Compute $c = \text{lcm}(c_f, c_g)$ and $X = \text{lcm}(X_f, X_g)$

$\frac{c}{c_f}\frac{X}{X_f} f - \frac{c}{c_g}\frac{X}{X_g} g \xrightarrow{G}_+ h$, where h is minimal with respect to G

IF $h \neq 0$ **THEN**

$\mathcal{G} := \mathcal{G} \cup \{\{u, h\} \mid$ for all $u \in G\}$

$G := G \cup \{h\}$

ALGORITHM 4.5.1. *Gröbner Basis Algorithm over a PID*

$\mathcal{G} \neq \emptyset$ we choose $\{f_2, f_3\}$, so that $\mathcal{G} = \{\{f_1, f_4\}, \{f_2, f_4\}, \{f_3, f_4\}\}$. In this case we have $c = 60$ and $X = x^2y^2$, so that the corresponding S-polynomial is

$$S(f_2, f_3) = 15xf_2 - 4y^2f_3 = -75x^2 - 112y^4 \xrightarrow{f_3} -112y^4 + 140y^2 \xrightarrow{f_4} 0.$$

Since $\mathcal{G} \neq \emptyset$, we choose $\{f_1, f_4\}$, so that $\mathcal{G} = \{\{f_2, f_4\}, \{f_3, f_4\}\}$. In this case, $c = 84$, and $X = x^2y^3$, so that the corresponding S-polynomial is

$$S(f_1, f_4) = 28y^2f_1 + 3x^2f_4 = 105x^2y + 196y^3 \xrightarrow{f_1} 196y^3 - 245y \xrightarrow{f_4} 0.$$

It is easy to see that the other two elements of $\mathcal{G}$ also give rise to polynomials which reduce to zero and so do not contribute new polynomials to the basis. Therefore we get (as we did in Example 4.2.11) that $\{f_1, f_2, f_3, f_4\}$ is a Gröbner basis for I.

We note that this computation required more steps than the computation of this Gröbner basis did in Example 4.2.11 using Algorithm 4.2.2. In order to improve the efficiency of Algorithm 4.5.1, Möller gave an analogue of crit2 which eliminates many S-polynomial computations. The interested reader should consult [**Mö88**].

Recall that we defined Gröbner bases in $R[x_1, \dots, x_n]$ the way we did in Definition 4.1.13 because of the problem of dividing by elements of the coefficient ring R. In the case when the coefficient ring is a PID, there is a stronger version of Gröbner bases which is similar to the one we gave when the coefficient ring was a field.

DEFINITION 4.5.6. *Let* $G = \{g_1, \dots, g_t\}$ *be a set of non-zero polynomials in* $R[x_1, \dots, x_n]$. *Then we say that* G *is a* strong Gröbner basis *for* $I = \langle g_1, \dots, g_t \rangle$ *if for each* $f \in I$, *there exists an* $i \in \{1, \dots, t\}$ *such that* $\mathrm{lt}(g_i)$ *divides* $\mathrm{lt}(f)$. *We say that* G *is a* minimal strong Gröbner basis *if no* $\mathrm{lt}(g_i)$ *divides* $\mathrm{lt}(g_j)$ *for* $i \neq j$.

Note that this definition does not in itself require that R be a PID. However strong Gröbner bases exist only in the case when R is a PID (Exercise 4.5.16).

EXAMPLE 4.5.7. Let $R = k[y, z]$, where k is a field. Consider the ideal $I = \langle x, y, z \rangle$ in $R[x]$. This ideal does not have a (finite) strong Gröbner basis. This is because there are an infinite number of non-associate irreducible polynomials in the two variables y, z in I. If a strong Gröbner basis G existed, then each of these irreducible polynomials would have leading term (which is the irreducible polynomial itself) divisible by the leading term of an element of G. This would force an infinite number of elements in G, which violates the definition of a strong Gröbner basis.

The following result is immediate and its proof is left to the reader (Exercise 4.5.6).

LEMMA 4.5.8. *If* $G = \{g_1, \dots, g_t\} \subseteq R[x_1, \dots, x_n]$ *is a strong Gröbner basis, then it is a Gröbner basis.*

We now show how to construct a strong Gröbner basis from a given Gröbner basis. So let $\{f_1, \dots, f_s\}$ be a Gröbner basis for an ideal I in $R[x_1, \dots, x_n]$. Let $\mathrm{lt}(f_i) = c_i X_i$, where $c_i \in R$ and $\mathrm{lp}(f_i) = X_i$. For each saturated subset J of $\{1, \dots, s\}$, let $c_J = \gcd(c_j \mid j \in J)$ and write $c_J = \sum_{j \in J} a_j c_j$ (any such representation will do). Also, let $X_J = \mathrm{lcm}(X_j \mid j \in J)$. Consider the polynomial

$$f_J = \sum_{j \in J} a_j \frac{X_J}{X_j} f_j.$$

THEOREM 4.5.9. *Let* R *be a PID, and* I *be an ideal of* $R[x_1, \dots, x_n]$. *Assume that* $\{f_1, \dots, f_s\}$ *is a Gröbner basis for* I. *Then the set*

$$\{f_J \mid J \text{ is a saturated subset of } \{1, \dots, s\}\}$$

is a strong Gröbner basis for I. *In particular, every non-zero ideal in* $R[x_1, \dots, x_n]$ *has a strong Gröbner basis.*

PROOF. Let $0 \neq f \in I$. Then $\mathrm{lt}(f) \in \mathrm{Lt}(f_1, \dots, f_s) = \langle \mathrm{lt}(f_1), \dots, \mathrm{lt}(f_s) \rangle$. Let $J = \{i \in \{1, \dots, s\} \mid X_i \text{ divides } \mathrm{lp}(f)\}$. It is clear that J is saturated and that X_J divides $\mathrm{lp}(f)$. We also have $\mathrm{lt}(f) \in \langle \mathrm{lt}(f_j) \mid j \in J \rangle$ and so $\mathrm{lc}(f) = \sum_{j \in J} d_j c_j$ for some $d_j \in R$, from which we conclude that c_J divides $\mathrm{lc}(f)$. Therefore $\mathrm{lt}(f_J)$ divides $\mathrm{lt}(f)$. □

We leave it to the exercises to show that every ideal has a minimal strong Gröbner basis (Exercise 4.5.9).

Together with the concept of strong Gröbner bases, we could also define the concept of *strong reduction*. Reduction of f modulo a set $F = \{f_1, \dots, f_s\}$ is performed if $\text{lt}(f) \in \langle \text{lt}(f_1), \dots, \text{lt}(f_s) \rangle$ (see Definition 4.1.1). Strong reduction would require instead that an $\text{lt}(f_i)$ divides $\text{lt}(f)$. One can show that for a set G of non-zero polynomials in $R[x_1, \dots, x_n]$, we have that every element in $\langle G \rangle$ strongly reduces to zero if and only if G is a strong Gröbner basis (Exercise 4.5.1).

Theorem 4.5.9 gives us a method for computing strong Gröbner bases. We give an illustration in the next example.

EXAMPLE 4.5.10. We go back to Example 4.5.5. We saw that a Gröbner basis for the ideal $I = \langle 3x^2y + 7y, 4xy^2 - 5x \rangle \subseteq \mathbb{Z}[x, y]$ with respect to the lex term ordering with $x > y$ is $\{f_1, f_2, f_3, f_4\}$, where $f_1 = 3x^2y + 7y$, $f_2 = 4xy^2 - 5x$, $f_3 = 15x^2 + 28y^2$, and $f_4 = -28y^3 + 35y$. Following the proof of Theorem 4.5.9 we first compute all the saturated subsets of $\{1, 2, 3, 4\}$: $J_1 = \{2\}$, $J_2 = \{3\}$, $J_3 = \{4\}$, $J_4 = \{1, 3\}$, $J_5 = \{2, 4\}$, $J_6 = \{1, 2, 3\}$, $J_7 = \{1, 2, 3, 4\}$. The sets J_1, J_2, and J_3 give rise to the original polynomials $f_{J_1} = f_2$, $f_{J_2} = f_3$, $f_{J_3} = f_4$. Associated with J_4 we have $c_{J_4} = \gcd(3, 15) = 3$, and $X_{J_4} = x^2y$, so $f_{J_4} = f_1$. Associated with J_5 we have $c_{J_5} = \gcd(4, -28) = 4$, and $X_{J_5} = xy^3$. Therefore, $f_{J_5} = yf_2$. Associated with the set J_6 we have $c_{J_6} = \gcd(3, 4, 15) = 1$, and $X_{J_6} = x^2y^2$. Therefore $f_{J_6} = xf_2 - yf_1 = x^2y^2 - 5x^2 - 7y^2$. Associated with the set J_7 we have $c_{J_7} = \gcd(3, 4, 15, -28) = 1$ and $X_{J_7} = x^2y^3$. Therefore $f_{J_7} = xyf_2 - y^2f_1 = yf_{J_6}$. We see we do not need f_{J_5} and f_{J_7} and therefore, a strong Gröbner basis for I is

$$\{4xy^2 - 5x, 15x^2 + 28y^2, -28y^3 + 35y, 3x^2y + 7y, x^2y^2 - 5x^2 - 7y^2\}.$$

To conclude this example we give an example of determining ideal membership using strong reduction. Consider the polynomial $f = 7x^2y^2 - 15x^2y - 35x^2 - 28xy^3 + 35xy - 56y^4 - 28y^3 + 21y^2$. Then we have

$$\begin{aligned}
f \quad &\xrightarrow{7, f_{J_6}} \quad -15x^2y - 28xy^3 + 35xy - 56y^4 - 28y^3 + 70y^2 \\
&\xrightarrow{-5, f_1} \quad -28xy^3 + 35xy - 56y^4 - 28y^3 + 70y^2 + 35y \\
&\xrightarrow{-7y, f_2} \quad -56y^4 - 28y^3 + 70y^2 + 35y \\
&\xrightarrow{2y, f_4} \quad -28y^3 + 35y \xrightarrow{-1, f_4} 0.
\end{aligned}$$

Thus $f = 7f_{J_6} - 5f_1 - 7yf_2 + (2y - 1)f_4$ and since $f_{J_6} = xf_2 - yf_1$ we have $f = (-7y - 5)f_1 + (7x - 7y)f_2 + (2y - 1)f_4$. We note that the first reduction above could not have been done with a single polynomial from f_1, f_2, f_3, f_4.

The next result gives a criterion to determine when we have a strong Gröbner basis.

PROPOSITION 4.5.11. *Let R be a PID. Then $G = \{g_1, \dots, g_t\} \subseteq R[x_1, \dots, x_n]$ is a strong Gröbner basis if and only if*

(i) *G is a Gröbner basis in $R[x_1, \dots, x_n]$ and*

(ii) *for every $J \subseteq \{1, \dots, t\}$, saturated with respect to $\{\mathrm{lp}(g_1), \dots, \mathrm{lp}(g_t)\}$, there exists $i \in J$ such that $\mathrm{lc}(g_i)$ divides $\mathrm{lc}(g_j)$ for all $j \in J$.*

PROOF. Let $I = \langle g_1, \dots, g_t \rangle$. We first prove that (i) and (ii) imply that G is a strong Gröbner basis in $R[x_1, \dots, x_n]$ (this proof is very similar to the proof of Theorem 4.5.9). Let $f \in I$. Since G is a Gröbner basis in $R[x_1, \dots, x_n]$, we have $\mathrm{lt}(f) \in \langle \mathrm{lt}(g_1), \dots, \mathrm{lt}(g_t) \rangle$. Let $J = \{j \in \{1, \dots, t\} \mid \mathrm{lp}(g_j) \text{ divides } \mathrm{lp}(f)\}$. Then J is saturated. Moreover we have

$$\mathrm{lt}(f) = \sum_{j \in J} a_j \, \mathrm{lt}(g_j),$$

for some $a_j \in R[x_1, \dots, x_n]$, which we may assume are all terms with $\mathrm{lp}(f) = \mathrm{lp}(a_j)\,\mathrm{lp}(g_j)$ for all $j \in J$ such that $a_j \neq 0$. Then

$$\mathrm{lc}(f) = \sum_{j \in J} \mathrm{lc}(a_j)\,\mathrm{lc}(g_j).$$

Now choosing $i \in J$ as in (ii) we get $\mathrm{lc}(g_i)$ divides $\mathrm{lc}(f)$ and so, by definition of J, we get $\mathrm{lt}(g_i)$ divides $\mathrm{lt}(f)$ as desired.

Conversely, let G be a strong Gröbner basis in $R[x_1, \dots, x_n]$. Then, by Lemma 4.5.8, G is a Gröbner basis in $R[x_1, \dots, x_n]$. Let J be a saturated subset of $\{1, \dots, t\}$. Let $c = \gcd(\mathrm{lc}(g_j) \mid j \in J)$ and write $c = \sum_{j \in J} d_j \, \mathrm{lc}(g_j)$, for some $d_j \in R$. Also, let $X = \mathrm{lcm}(\mathrm{lp}(g_j) \mid j \in J)$. Now consider the polynomial

$$f = \sum_{j \in J} d_j \frac{X}{\mathrm{lp}(g_j)} g_j.$$

Note that $\mathrm{lt}(f) = cX$ and $f \in I$. Since G is a strong Gröbner basis there is a $g_i \in G$ such that $\mathrm{lt}(g_i)$ divides $\mathrm{lt}(f)$. Therefore $i \in J$ since $\mathrm{lp}(g_i)$ divides X. Moreover, $\mathrm{lc}(g_i)$ divides c and we are done. □

We now consider the case where $R = k[y]$, with k a field and y a single variable. In this setting we have three concepts of Gröbner bases: Gröbner bases in $k[y, x_1, \dots, x_n]$ (as defined in Section 1.6), Gröbner bases in $(k[y])[x_1, \dots, x_n]$ (as defined in Section 4.1), and strong Gröbner bases in $(k[y])[x_1, \dots, x_n]$ (as defined in Definition 4.5.6). Clearly the ring $k[y, x_1, \dots, x_n]$ is the same as the ring $(k[y])[x_1, \dots, x_n]$, but, in the latter ring, polynomials are viewed as polynomials in the variables $x_1, \dots, x_n$ with coefficients in $k[y]$. We have seen in Lemma 4.5.8 that if $\{g_1, \dots, g_t\}$ is a strong Gröbner basis in $(k[y])[x_1, \dots, x_n]$ then it is a Gröbner basis in $(k[y])[x_1, \dots, x_n]$. Moreover, we saw in Theorem 4.1.18 that if $\{g_1, \dots, g_t\}$ is a Gröbner basis with respect to an elimination order with the x variables larger than the y variable in $k[y, x_1, \dots, x_n]$, then it is a

Gröbner basis in $(k[y])[x_1, \ldots, x_n]$. The next theorem strengthens this result considerably. We will first need some notation.

We assume we have an elimination order with the x variables larger than y. We need a notation to distinguish between leading terms in $k[y, x_1, \ldots, x_n]$ and $(k[y])[x_1, \ldots, x_n]$. We use lt, lp, lc for their usual meanings in $k[y, x_1, \ldots, x_n]$, and we will use $\mathrm{lt}_x, \mathrm{lp}_x, \mathrm{lc}_x$ for corresponding objects in $(k[y])[x_1, \ldots, x_n]$. So, for example, if $f \in k[y, x_1, \ldots, x_n]$, then $\mathrm{lc}_x(f)$ is a polynomial in y. Finally for $a \in k[y]$ we use $\mathrm{lt}_y(a)$ for the leading term of a. Note that because of the chosen order, for $f \in k[y, x_1, \ldots, x_n]$ we have $\mathrm{lt}(f) = \mathrm{lt}_y(\mathrm{lc}_x(f))\,\mathrm{lp}_x(f)$.

THEOREM 4.5.12. *Let $R = k[y]$. Then $G = \{g_1, \ldots, g_t\}$ is a Gröbner basis in $k[y, x_1, \ldots, x_n]$ with respect to an elimination order with the x variables larger than y if and only if G is a strong Gröbner basis in $(k[y])[x_1, \ldots, x_n]$.*

PROOF. Let us first assume that $G = \{g_1, \ldots, g_t\}$ is a Gröbner basis in $k[y, x_1, \ldots, x_n]$ with respect to an elimination order with the x variables larger than y. By Theorem 4.1.18 we see that G is a Gröbner basis in $(k[y])[x_1, \ldots, x_n]$. Thus by Proposition 4.5.11 it suffices to show that given a subset J of $\{1, \ldots, t\}$, saturated with respect to $\{\mathrm{lp}_x(g_1), \ldots, \mathrm{lp}_x(g_t)\}$, there exists $i \in J$ such that $\mathrm{lc}_x(g_i)$ divides $\mathrm{lc}_x(g_j)$ for all $j \in J$. So let J be such a saturated set. Let $c = \gcd(\mathrm{lc}_x(g_j) \mid j \in J)$ and $X = \mathrm{lcm}(\mathrm{lp}_x(g_j) \mid j \in J)$. We can write $c = \sum_{j \in J} b_j \,\mathrm{lc}_x(g_j)$, for some $b_j \in k[y]$. Now consider the polynomial

$$f = \sum_{j \in J} b_j \frac{X}{\mathrm{lp}_x(g_j)} g_j \in \langle g_1, \ldots, g_t \rangle.$$

Note that $\mathrm{lt}_x(f) = cX$. Moreover, $\mathrm{lt}(f) = \mathrm{lt}_y(c)X$, since we have an elimination order with the x variables larger than the y variable. Since G is a Gröbner basis in $k[y, x_1, \ldots, x_n]$, there exists $g_i \in G$ such that $\mathrm{lt}(g_i)$ divides $\mathrm{lt}(f) = \mathrm{lt}_y(c)X$. But then $i \in J$ since $\mathrm{lp}_x(g_i)$ divides $\mathrm{lp}_x(f) = X$. Also, $\mathrm{lt}_y(\mathrm{lc}_x(g_i))$ divides $\mathrm{lt}_y(\mathrm{lc}_x(f)) = \mathrm{lt}_y(c)$. Therefore, $\deg(\mathrm{lc}_x(g_i)) \leq \deg(c)$ and so the fact that c divides $\mathrm{lc}_x(g_i)$ implies that c is a non-zero constant in k times $\mathrm{lc}_x(g_i)$, which immediately gives the desired result.

Conversely, let us assume that G is a strong Gröbner basis in $(k[y])[x_1, \ldots, x_n]$. Let $f \in I$. Then there exists $g_i \in G$ such that $\mathrm{lt}_x(g_i) = \mathrm{lc}_x(g_i)\,\mathrm{lp}_x(g_i)$ divides $\mathrm{lt}_x(f) = \mathrm{lc}_x(f)\,\mathrm{lp}_x(f)$. So $\mathrm{lp}_x(g_i)$ divides $\mathrm{lp}_x(f)$ and $\mathrm{lc}_x(g_i)$ divides $\mathrm{lc}_x(f)$. But then $\mathrm{lt}(g_i) = \mathrm{lt}_y(\mathrm{lc}_x(g_i))\,\mathrm{lp}_x(g_i)$ divides $\mathrm{lt}(f) = \mathrm{lt}_y(\mathrm{lc}_x(f))\,\mathrm{lp}_x(f)$. Therefore G is a Gröbner basis in $k[y, x_1, \ldots, x_n]$. □

We now go back to the case of an arbitrary PID R and conclude this section by giving a characterization of strong Gröbner bases in $R[x]$, where x is a single variable. This will be used in Section 4.6 to give the primary decomposition of ideals in $R[x]$. From Lazard [**Laz85**] (see also Szekeres [**Sz**]), we have

THEOREM 4.5.13. *Let R be a PID and let x be a single variable. Assume that $G = \{g_1, \ldots, g_t\}$ is a minimal strong Gröbner basis for an ideal I of $R[x]$ and* *let* $\mathrm{lp}(g_i) = x^{\nu_i}$. *We further assume that we have ordered the* g_i*'s in such a way that* $\nu_1 \leq \nu_2 \leq \cdots \leq \nu_t$. *Let* $g = \gcd(g_1, \ldots, g_t)$. *Then*

$$\begin{aligned} g_1 &= a_2a_3\cdots a_t g \\ g_2 &= a_3\cdots a_t h_2 g \\ &\vdots \\ g_i &= a_{i+1}\cdots a_t h_i g \\ &\vdots \\ g_{t-1} &= a_t h_{t-1} g \\ g_t &= h_t g, \end{aligned}$$

where each $a_i \in R$, *each* h_i *is monic,* $\nu_1 < \nu_2 < \cdots < \nu_t$, *and*

$$h_{i+1} \in \langle h_i, a_i h_{i-1}, \ldots, a_3\cdots a_i h_2, a_2a_3\cdots a_i\rangle,$$

for $i = 2, \ldots, t-1$.

PROOF. From Exercise 4.5.11, we have that $\{g_1, \ldots, g_t\}$ is a strong Gröbner basis if and only if $\left\{\frac{g_1}{g}, \ldots, \frac{g_t}{g}\right\}$ is a strong Gröbner basis. Thus we may assume that $\gcd(g_1, \ldots, g_t) = 1$. Let $\mathrm{lt}(g_i) = c_i x^{\nu_i}$, where $c_i \in R$ $(1 \leq i \leq t)$.

CLAIM 1. $\nu_1 < \nu_2 < \cdots < \nu_t$.

PROOF. Assume to the contrary that we have an i, $1 \leq i < t$, such that $\nu_i = \nu_{i+1}$. Let $c = \gcd(c_i, c_{i+1})$ and write $c = b_i c_i + b_{i+1}c_{i+1}$, for some $b_i, b_{i+1} \in R$. Consider the polynomial $h = b_i g_i + b_{i+1}g_{i+1}$. Then $h \in I$ and $\mathrm{lt}(h) = cx^{\nu_i}$. Since G is a strong Gröbner basis, there exists a $j \in \{1, \ldots, t\}$ such that $\mathrm{lt}(g_j) = c_j x^{\nu_j}$ divides $\mathrm{lt}(h) = cx^{\nu_i}$. Then $\nu_j \leq \nu_i$. But then $\mathrm{lt}(g_j)$ divides $\mathrm{lt}(g_i)$, and so, by our assumption that G is minimal, we have that $i = j$. Hence $c = c_i$, and $\mathrm{lt}(g_i)$ divides $\mathrm{lt}(g_{i+1})$. This contradicts our assumption that G is minimal.

CLAIM 2. c_{i+1} divides c_i, for $i = 1, \ldots, t-1$.

PROOF. Let $i \in \{1, \ldots, t-1\}$. As in Claim 1, let $c = \gcd(c_i, c_{i+1})$ and write $c = b_i c_i + b_{i+1}c_{i+1}$, for some $b_i, b_{i+1} \in R$. Now consider the polynomial $h = b_i x^{\nu_{i+1}-\nu_i} g_i + b_{i+1}g_{i+1} \in I$. Note that $\mathrm{lt}(h) = cx^{\nu_{i+1}}$. Since G is a strong Gröbner basis, there exists a $j \in \{1, \ldots, t\}$, such that $\mathrm{lt}(g_j) = c_j x^{\nu_j}$ divides $\mathrm{lt}(h)$. Thus $\nu_j \leq \nu_{i+1}$, and so $j \leq i+1$, and c_j divides c, and so $\mathrm{lt}(g_j)$ divides $\mathrm{lt}(g_{i+1})$. Since G is minimal, we have that $j = i+1$, and so $c_{i+1} = c = \gcd(c_i, c_{i+1})$. Therefore c_{i+1} divides c_i.

CLAIM 3. $\dfrac{c_i}{c_{i+1}} g_{i+1} \in \langle g_1, \ldots, g_i\rangle$, for $i = 1, \ldots, t-1$.

PROOF. Let $h = \dfrac{c_i}{c_{i+1}} g_{i+1} - x^{\nu_{i+1}-\nu_i} g_i$. Note that $h \in I$ and that $\mathrm{lp}(h) < \mathrm{lp}(g_{i+1})$. Since $\mathrm{lp}(g_{i+1}), \ldots, \mathrm{lp}(g_t)$ are larger than $\mathrm{lp}(h)$, and since $h \xrightarrow{G}_+ 0$, the only polynomials that can be used to reduce h to zero are $g_1, \ldots, g_i$. Therefore $h \in \langle g_1, \ldots, g_i\rangle$, and hence so is $\dfrac{c_i}{c_{i+1}} g_{i+1}$.

CLAIM 4. $g_1 \in R$.

PROOF. Let $c(g_1) \in R$ be the greatest common divisor of the coefficients of the powers of x that appear in g_1. Then $g_1 = c(g_1)p(g_1)$, where $p(g_1) \in R[x]$ and the greatest common divisor of the coefficients of $p(g_1)$ is 1. We show by induction on i that $p(g_1)$ divides g_i. The case $i = 1$ is clear. Now assume that $p(g_1)$ divides $g_1, g_2, \ldots, g_i$. Since $\frac{c_i}{c_{i+1}} g_{i+1} \in \langle g_1, \ldots, g_i \rangle$ by Claim 3, we see that $p(g_1)$ divides $\frac{c_i}{c_{i+1}} g_{i+1}$. But $\gcd\left(\frac{c_i}{c_{i+1}}, p(g_1)\right) = 1$, since any common factor of $\frac{c_i}{c_{i+1}}$ and $p(g_1)$ would have to be in R and would have to be a factor of the coefficients of powers of x appearing in $p(g_1)$. Therefore $p(g_1)$ divides g_{i+1}. Now, since $p(g_1)$ divides every g_i, $p(g_1)$ must be 1 because of the assumption that the g_i's have no factor in common. Therefore $g_1 = c(g_1) \in R$.

CLAIM 5. c_i divides g_i for all $i = 1, \ldots, t$ and $c_t = 1$.

PROOF. We use induction on i. The case $i = 1$ is clear, since $g_1 = c_1$. Assume that c_1 divides g_1, c_2 divides $g_2, \ldots, c_i$ divides g_i. Then, by Claim 2, we have that c_i divides $g_1, g_2, \ldots, g_i$. Since $\frac{c_i}{c_{i+1}} g_{i+1} \in \langle g_1, \ldots, g_i \rangle$ by Claim 3, we see that c_i divides $\frac{c_i}{c_{i+1}} g_{i+1}$. Therefore c_{i+1} divides g_{i+1}. Now, since c_i divides g_i for each i, and since c_t divides c_i for each i, we see that c_t divides g_i for each i. Therefore $c_t = 1$ because of the assumption that the g_i's are relatively prime.

Now set $a_{i+1} = \frac{c_i}{c_{i+1}}$ and $h_i = \frac{g_i}{c_i}$. Then we have

$$
\begin{aligned}
g_t &= c_t h_t = h_t \\
g_{t-1} &= c_{t-1} h_{t-1} = c_t \frac{c_{t-1}}{c_t} h_{t-1} = a_t h_{t-1} \\
g_{t-2} &= c_{t-2} h_{t-2} = c_t \frac{c_{t-2}}{c_{t-1}} \frac{c_{t-1}}{c_t} h_{t-2} = a_{t-1} a_t h_{t-2} \\
&\ \vdots \\
g_i &= a_{i+1} a_{i+2} \cdots a_{t-1} a_t h_i \\
&\ \vdots \\
g_2 &= a_3 \cdots a_{t-1} a_t h_2 \\
g_1 &= a_2 a_3 \cdots a_t.
\end{aligned}
$$

CLAIM 6. $h_{i+1} \in \langle h_i, a_i h_{i-1}, \ldots, a_3 \cdots a_i h_2, a_2 \cdots a_i \rangle$.

PROOF. By Claim 3 we have $\frac{c_i}{c_{i+1}} g_{i+1} \in \langle g_1, \ldots, g_i \rangle$, that is,

$$a_{i+1} a_{i+2} \cdots a_t h_{i+1} \in \langle a_{i+1} \cdots a_t h_i, a_i \cdots a_t h_{i-1}, \ldots, a_3 \cdots a_t h_2, a_2 \cdots a_t \rangle.$$

The result follows after dividing by $a_{i+1} \cdots a_t$. The Theorem is now completely proved. □

We note that the converse is also true. That is, any set of polynomials $g_1, \ldots, g_t$ in $R[x]$ which satisfies the conditions in Theorem 4.5.13 is a strong Gröbner basis and is minimal if no a_i is a unit (Exercise 4.5.12).

COROLLARY 4.5.14. *Let k be a field and let y, x be variables and set $A = k[y,x]$. We assume that we have the lex term ordering with $y < x$. Let $G = \{g_1, \ldots, g_t\}$ be a minimal Gröbner basis for an ideal I of $k[y,x]$ and let $\text{lp}_x(g_i) = x^{\nu_i}$. We assume that we have ordered the g_i's in such a way that $\nu_1 \leq \nu_2 \leq \cdots \leq \nu_t$. Let $g = \gcd(g_1, \ldots, g_t)$. Then*

$$\begin{aligned}
g_1 &= a_2 a_3 \cdots a_t g \\
g_2 &= a_3 \cdots a_t h_2 g \\
&\vdots \\
g_i &= a_{i+1} \cdots a_t h_i g \\
&\vdots \\
g_{t-1} &= a_t h_{t-1} g \\
g_t &= h_t g,
\end{aligned}$$

where $a_i \in k[y]$, $\text{lt}(h_i)$ is a power of x alone, $\nu_1 < \nu_2 < \cdots < \nu_t$, and

$$h_{i+1} \in \langle h_i, a_i h_{i-1}, \ldots, a_3 \cdots a_i h_2, a_2 a_3 \cdots a_i \rangle,$$

for $i = 2, \ldots, t-1$.

PROOF. We have from Theorem 4.5.12, that G is a strong Gröbner basis in $(k[y])[x]$ and so the result follows immediately from Theorem 4.5.13. □

EXAMPLE 4.5.15. We consider the ring $R = \mathbb{Q}[y]$ and the ideal $I = \langle (x+y)(y^2+1), x^2 - x + y + 1 \rangle$ in $R[x]$. We compute a strong Gröbner basis for I by computing a Gröbner basis for I viewed in $\mathbb{Q}[x,y]$ with respect to the lexicographic order with $x > y$ (Theorem 4.5.12). We find

$$\begin{aligned}
g_1 &= y^4 + 2y^3 + 2y^2 + 2y + 1 = (y+1)^2(y^2+1) = a_2 a_3 \\
g_2 &= xy^2 + x + y^3 + y = (y^2+1)(x+y) = a_3 h_2 \\
g_3 &= x^2 - x + y + 1 = h_3.
\end{aligned}$$

EXAMPLE 4.5.16. Consider the three polynomials in $\mathbb{Z}[x]$

$$\begin{aligned}
f_1 &= 630x - 630 = 9 \cdot 5 \cdot 14(x-1) = a_2 a_3 g \\
f_2 &= 70x^2 + 70x - 140 = 5(x+2) \cdot 14(x-1) = a_3 h_2 g \\
f_3 &= 14x^4 + 70x^3 + 196x^2 - 70x - 210 \\
&= (x^3 + 6x^2 + 20x + 15) \cdot 14(x-1) = h_3 g.
\end{aligned}$$

It can be easily verified using Proposition 4.5.11 that $\{f_1, f_2, f_3\}$ is a strong Gröbner basis. We also note that the f_i's have the form required in Theorem 4.5.13, since $h_3 = x^3 + 6x^2 + 20x + 15 = (x+1)(x+2)(x+3) + 9(x+1) \in \langle 9, x+2 \rangle$ and so it also follows from the converse of Theorem 4.5.13 (Exercise 4.5.12) that $\{f_1, f_2, f_3\}$ is a strong Gröbner basis.

Exercises

4.5.1. We define *strong reduction* in $R[x_1, \ldots, x_n]$, for R a PID, with respect to a set $F = \{f_1, \ldots, f_s\}$ of non-zero polynomials in $R[x_1, \ldots, x_n]$ as follows. For $f, g \in R[x_1, \ldots, x_n]$ we write $f \xrightarrow{F}_s g$ provided that for some $i, 1 \leq i \leq s$ we have $\operatorname{lt}(f_i)$ divides $\operatorname{lt}(f)$ and $g = f - \frac{\operatorname{lt}(f)}{\operatorname{lt}(f_i)} f_i$. We write $f \xrightarrow{F}_{+,s} g$, as usual, when we iterate the preceding. Show that the following are equivalent for a set $G = \{g_1, \ldots, g_t\}$ of non-zero polynomials in $R[x_1, \ldots, x_n]$ where we set $I = \langle G \rangle$.

a. G is a strong Gröbner basis for I.

b. For all $f \in I$ we have $f \xrightarrow{G}_{+,s} 0$.

4.5.2. Show that in Exercise 4.5.1 the following statement:

$$\text{for all } i, j \text{ we have } S(g_i, g_j) \xrightarrow{G}_{+,s} 0$$

(see Equation (4.5.1)) does not imply that G is a strong Gröbner basis. [Hint: Look at the polynomials $f_1 = 2x + 1$ and $f_2 = 3y + x$ in $\mathbb{Z}[x, y]$.]

4.5.3. Prove the converse of Theorem 4.5.9. That is, prove that if the set of f_J defined there is a strong Gröbner basis for I then $\{f_1, \ldots, f_s\}$ is a Gröbner basis for I.

4.5.4. Let $f, g \in R[x_1, \ldots, x_n]$, with $f, g \neq 0$, and let $d = \gcd(f, g)$. Prove that $\{f, g\}$ is a Gröbner basis if and only if $\gcd(\operatorname{lt}(\frac{f}{d}), \operatorname{lt}(\frac{g}{d})) = 1$. (This is the analog of crit1.) [Hint: Follow the proof of Lemma 3.3.1.]

4.5.5. For the ring $R = \mathbb{Z}$ use Algorithm 4.5.1 to compute a Gröbner basis for the ideals generated by the given polynomials with respect to the given term order.

a. $f_1 = 2xy - x$, $f_2 = 3y - x^2$ and lex with $x < y$.

b. $f_1 = 3x^2y - 3yz + y$, $f_2 = 5x^2z - 8z^2$ and deglex with $x > y > z$.

c. $f_1 = 6x^2 + y^2$, $f_2 = 10x^2y + 2xy$ and lex with $x > y$.

4.5.6. Prove Lemma 4.5.8.

4.5.7. For the ring $R = \mathbb{Z}$ use Theorem 4.5.9 to compute a strong Gröbner basis for the ideals generated by the given polynomials with respect to the given term order in the exercises in Exercise 4.5.5.

4.5.8. Compute a strong Gröbner basis for the ideal in $\mathbb{Z}[i][x, y, z]$ in Exercise 4.2.5.

4.5.9. Prove that every non-zero ideal of $R[x_1, \ldots, x_n]$, where R is a PID, has a minimal strong Gröbner basis.

4.5.10. Show that for the strong Gröbner basis constructed in Exercise 4.5.7 (for Exercise 4.5.5 part **c**), $-30x^3y^2 + 6x^3 - 5x^2y^3 - 6x^2y^2 + 16x^2y + 5xy^3 + xy^2 + 2xy - 5y^5 + y^3$ strongly reduces to zero.

4.5.11. Prove that if $\{g_1, \ldots, g_t\}$ is a set of non-zero polynomials in $R[x_1, \ldots, x_n]$ where R is a PID and $g = \gcd(g_1, \ldots, g_t)$, then $\{g_1, \ldots, g_t\}$ is a strong Gröbner basis if and only if $\{\frac{g_1}{g}, \ldots, \frac{g_t}{g}\}$ is a strong Gröbner basis.

4.5.12. Prove the converse of Theorem 4.5.13 as stated immediately after the proof of Theorem 4.5.13.

4.5.13. Verify that the Gröbner basis for $\langle -x^2 - xy + x^2y^2 + xy^3, -5y + 5xy - 3xy^2 + 3x^2y^2\rangle \subseteq \mathbb{Q}[x, y]$ with respect to the lex ordering with $x > y$ has the form stated in Corollary 4.5.14.

4.5.14. Prove that if R is a UFD and if for every $\{f_1, \ldots, f_s\} \subseteq R[x_1, \ldots, x_n]$ we have that $\mathrm{Syz}(\mathrm{lt}(f_1), \ldots, \mathrm{lt}(f_s))$ has an S-basis, then R is a PID. [Hint: If R is not a PID, then there exist $a, b \in R$ such that $\gcd(a, b) = 1$ and $1 \notin \langle a, b\rangle$. Note that $(1, 1, -1) \in \mathrm{Syz}(a, b, a + b)$.]

4.5.15. Show that if R is a Dedekind Domain, then for every $\{f_1, \ldots, f_s\} \subseteq R[x_1, \ldots, x_n]$ we have that $\mathrm{Syz}(\mathrm{lt}(f_1), \ldots, \mathrm{lt}(f_s))$ has an S-basis. [Hint: Prove the identity in Lemma 4.5.2 and follow the proof of Proposition 4.5.3 (note that more than one d_j may be needed).]

4.5.16. Prove that if R is a UFD and if for every non-zero ideal $I \subseteq R[x_1, \ldots, x_n]$, I has a strong Gröbner basis, then R is a PID. [Hint: Use the idea in Example 4.5.7. Assume the facts that if every prime ideal in a UFD is principal then it is a PID, and that every non-principal prime ideal in a UFD contains an infinite number of non-associate irreducibles.]

4.6. Primary Decomposition in $R[x]$ for R a PID. In this section we follow Lazard [**Laz85**] and use the results of Section 4.5 to "decompose" ideals in $R[x]$, where R is a PID and x is a single variable. The decomposition we have in mind is one similar to the decomposition of natural numbers into products of powers of prime numbers. In our setting, the analog of a product is an ideal intersection, and the analog of a prime integer is a prime ideal. Recall that an ideal P in a commutative ring A is prime if $fg \in P$ implies that either $f \in P$ or $g \in P$, or equivalently, if the set $S = A - P$ is a multiplicative set (see Section 4.4).

We will need the following elementary fact about Noetherian rings.

LEMMA 4.6.1. *Let A be a Noetherian ring and let $\mathcal{S}$ be any non-empty set of ideals of A. Then $\mathcal{S}$ contains a maximal element, i.e. there is an ideal $I \in \mathcal{S}$ such that there is no ideal $J \in \mathcal{S}$ such that $I \subsetneq J$.*

We first consider the decomposition of the radical $\sqrt{I}$ for an ideal I in A. Recall from Definition 2.2.4 that

$$\sqrt{I} = \{ f \in A \mid f^\nu \in I \text{ for some } \nu \in \mathbb{N} \}.$$

LEMMA 4.6.2. *Let I be an ideal in a Noetherian ring A. We have*

$$\sqrt{I} = \bigcap_{\substack{I \subseteq P \\ P \text{ prime ideal}}} P.$$

PROOF. The inclusion $\sqrt{I} \subseteq \bigcap_{\substack{I \subseteq P \\ P \text{ prime ideal}}} P$ follows from the fact that if a prime ideal P contains I, it also contains $\sqrt{I}$.

We now consider the reverse inclusion. So let us assume that there is an element f in $\bigcap_{\substack{I \subseteq P \\ P \text{ prime ideal}}} P - \sqrt{I}$. Consider the set $S = \{f^\nu \mid \nu = 0, 1, \dots\}$. Note that $S \cap \sqrt{I} = \emptyset$, for otherwise, if $f^\nu \in \sqrt{I}$ for some $\nu \in \mathbb{N}$, then $(f^\nu)^\mu = f^{\nu\mu} \in I$, for some $\mu \in \mathbb{N}$, and this implies that $f \in \sqrt{I}$, which is a contradiction. Now consider the collection $\mathcal{S}$ of all ideals of A which contain $\sqrt{I}$ and have empty intersection with S. Clearly $\mathcal{S}$ is non-empty, since $\sqrt{I} \in \mathcal{S}$. By Lemma 4.6.1, there exists an ideal $M \in \mathcal{S}$ which is maximal in $\mathcal{S}$. In particular, $\sqrt{I} \subseteq M$ and $M \cap S = \emptyset$. We now prove that M is a prime ideal. Let $gh \in M$ and assume that neither g nor h is in M. By maximality of M we have

$$\langle g, M\rangle \cap S \neq \emptyset \text{ and } \langle h, M\rangle \cap S \neq \emptyset.$$

Therefore there exist $\nu, \nu' \in \mathbb{N}$, $m, m' \in M$, and $a, a' \in A$ such that

$$ag + m = f^\nu \text{ and } a'h + m' = f^{\nu'}.$$

But then

$$\underbrace{f^{\nu+\nu'}}_{\in S} = (ag+m)(a'h+m') = (aa')\underbrace{(gh)}_{\in M} + \underbrace{(ag+m)m' + (a'h)m}_{\in M} \in M \cap S,$$

which is a contradiction, and so M is prime. Now we have a prime ideal M which contains $\sqrt{I}$ and hence I, so $f \in \bigcap_{\substack{I \subseteq P \\ P \text{ prime ideal}}} P \subseteq M$ by assumption. But $M \cap S = \emptyset$, and we obtain a contradiction. □

In view of the above lemma, one might think that any ideal in $R[x]$ can be decomposed as the intersection of powers of prime ideals. This is not the case as the following example shows.

EXAMPLE 4.6.3. Consider the ideal $Q = \langle 4, x^2\rangle$ in $\mathbb{Z}[x]$. Any prime ideal which contains Q must contain both 2 and x, and hence must be equal to $M = \langle 2, x\rangle$, since M is a maximal ideal of $\mathbb{Z}[x]$ (since $\mathbb{Z}[x]/\langle 2, x\rangle \cong \mathbb{Z}_2$). Therefore, if Q were the intersection of powers of prime ideals, it would be a power of $\langle 2, x\rangle$. But $M^3 \subsetneq Q \subsetneq M^2$. So Q cannot be a power of M.

The correct analog to powers of primes is the following.

DEFINITION 4.6.4. *An ideal Q of A is called* primary *if $Q \neq A$ and if $fg \in Q$ then either f is in Q or some power of g is in Q.*

It is easy to see that prime ideals are primary. However, powers of prime ideals need not be primary (see [**AtMD**]) although, using Lemma 4.6.13, we see that powers of maximal ideals are primary.

EXAMPLE 4.6.5. The ideal Q given in Example 4.6.3 is primary. To see this, let $fg \in Q = \langle 4, x^2 \rangle$, and let $f \notin Q$. It is easy to see that we can write $f = h_f + r_f$, and $g = h_g + r_g$, where $h_f, h_g \in Q$, and $r_f = a_f x + b_f$, $r_g = a_g x + b_g$, where $a_f, a_g, b_f, b_g = 0, 1, 2$, or 3 (note that r_f and r_g are the totally reduced remainders of f and g as defined in Definition 4.3.2). Moreover, since $f \notin Q$, we have that either a_f or b_f is not equal to 0. Since $fg \in Q$, 4 divides $a_f b_g + a_g b_f$ and $b_f b_g$. Note that if $b_g = 0$, then $g^2 \in Q$, and we would be done. So we may assume that $b_g \neq 0$. If $b_f = 0$ then 4 divides $a_f b_g$, and so since a_f and b_g are non-zero, we have $a_f = b_g = 2$ giving $g^2 \in Q$. Otherwise, $b_f \neq 0$ and so $b_f = b_g = 2$, so again we have $g^2 \in Q$.

LEMMA 4.6.6. *If Q is a primary ideal in a commutative ring A, then $\sqrt{Q}$ is a prime ideal. Moreover $\sqrt{Q}$ is the smallest prime ideal containing Q.*

PROOF. Let $fg \in \sqrt{Q}$. Then $(fg)^\nu = f^\nu g^\nu \in Q$ and so $f^\nu \in Q$ or $(g^\nu)^\mu = g^{\nu\mu} \in Q$. Therefore $f \in \sqrt{Q}$ or $g \in \sqrt{Q}$ and $\sqrt{Q}$ is a prime ideal. The second statement follows from the fact that any prime ideal containing Q also contains $\sqrt{Q}$. □

DEFINITION 4.6.7. *If Q is primary and $\sqrt{Q} = P$, we say that Q is P-*primary.

We can now define what we mean by decomposition.

DEFINITION 4.6.8. *Let $I = \bigcap_{i=1}^r Q_i$, where Q_i is P_i-primary for each i. We call $\bigcap_{i=1}^r Q_i$ a* primary decomposition *of I. If, in addition, the P_i are all distinct and for all i, $1 \leq i \leq r$, we have $\bigcap_{j \neq i} Q_j \not\subseteq Q_i$, we call the primary decomposition* irredundant. *In this latter case, the ideal Q_i is said to be the* primary component *of I which belongs to P_i, and P_i is said to be a* prime component *of I.*

It is easy to prove that if $Q_1, \ldots, Q_r$ are all P-primary then $\bigcap_{i=1}^r Q_i$ is also P-primary. Given this, and the obvious statement that we can remove superfluous Q_i, we see that any ideal that has a primary decomposition also has an irredundant primary decomposition.

We note that in Lemma 4.6.2 we gave a primary decomposition of $\sqrt{I}$ (at least, in the case where that intersection is finite). Also, in Example 4.6.5, the ideal Q was primary, and thus was its own primary decomposition. In general, we have

THEOREM 4.6.9. *Every ideal in a Noetherian ring A has a primary decomposition.*

PROOF. The key to the proof is the concept of *irreducible ideals.* An ideal I is irreducible if $I = I_1 \cap I_2$ implies that $I = I_1$ or $I = I_2$. The proof of the theorem is done in two steps. First we prove that every ideal in A is a finite intersection of irreducible ideals, and then we show that every irreducible ideal is primary.

Let $\mathcal{S}$ be the collection of all ideals in A which are not the intersection of a finite number of irreducible ideals. Assume to the contrary that $\mathcal{S}$ is not empty. By Lemma 4.6.1 there exists a maximal element M in $\mathcal{S}$. Now, since M is in $\mathcal{S}$, M is not irreducible. Therefore there exist ideals M_1 and M_2 such that $M \neq M_1$, $M \neq M_2$ and $M = M_1 \cap M_2$. Thus $M \subsetneq M_1, M_2$. By the maximality of M, we have that M_1 and M_2 are not in $\mathcal{S}$. Therefore they are both a finite intersection of irreducible ideals. But then M is also a finite intersection of irreducible ideals. This is a contradiction, and therefore $\mathcal{S} = \emptyset$.

Now let I be an irreducible ideal. We will show that I is primary. So let $fg \in I$ and let $f \notin I$. Consider the ascending chain

$$I\colon \langle g\rangle \subseteq I\colon \langle g^2\rangle \subseteq \cdots \subseteq I\colon \langle g^\ell\rangle \subseteq \cdots .$$

Since A is Noetherian we have that for some ℓ, $I\colon \langle g^\ell\rangle = I\colon \langle g^{\ell+1}\rangle$.

CLAIM. $(I + \langle g^\ell\rangle) \cap (I + \langle f\rangle) = I$.

PROOF. Clearly $I \subseteq (I + \langle g^\ell\rangle) \cap (I + \langle f\rangle)$. For the reverse inclusion, let $\alpha + rg^\ell = \beta + sf \in (I + \langle g^\ell\rangle) \cap (I + \langle f\rangle)$, where $\alpha, \beta \in I$, $r, s \in A$. Then $rg^{\ell+1} = \underbrace{-g\alpha + g\beta}_{\in I} + \underbrace{sfg}_{\in I} \in I$. Therefore $r \in I\colon \langle g^{\ell+1}\rangle = I\colon \langle g^\ell\rangle$. But then $\alpha + g^\ell r \in I$, and the Claim is proved.

Now by the assumption that I is irreducible, either $I = I + \langle f\rangle$ or $I = I + \langle g^\ell\rangle$. Since $f \notin I$, we have $I = I + \langle g^\ell\rangle$, and hence $g^\ell \in I$. Therefore I is primary. □

We note that the preceding Theorem is purely existential, that is, it gives no indication how to go about computing the primary decomposition of a given ideal in some specific ring. There has been much work done on this problem, see [**GTZ, EHV**]. The main purpose of this section is to use Theorem 4.5.13 to show how to do this explicitly in the ring $R[x]$, where R is a PID in which certain computability assumptions must be made. Namely, we assume that linear equations are solvable in R, that we can factor in R, and that given any prime element $u \in R$ we can factor in $(R/\langle u\rangle)[x]$. Examples of such rings include $\mathbb{Q}[y]$, $\mathbb{Z}$ and $(\mathbb{Z}/p\mathbb{Z})[y]$ for a prime integer p.

Let $I \neq R[x], \{0\}$ be an ideal in $R[x]$ and let $\{g_1, \ldots, g_t\}$ be a minimal strong Gröbner basis for I. For simplicity we will assume that $g = \gcd(g_1, \ldots, g_t) = 1$. In this case we say that the ideal I is *zero-dimensional*[5]. It follows from Theorem 2.2.7 and Corollary 4.5.14 that this definition coincides with the one given before for the special case of $\mathbb{Q}[x, y]$. We will be using Theorem 4.5.13 and the notation set there.

[5]For certain PID's, R, this definition is more restrictive than the one usually found in the literature.

That is, we have

$$\begin{aligned}
g_1 &= a_2a_3\cdots a_t\\
g_2 &= a_3\cdots a_th_2\\
&\vdots\\
g_i &= a_{i+1}\cdots a_th_i\\
&\vdots\\
g_{t-1} &= a_th_{t-1}\\
g_t &= h_t,
\end{aligned}$$

where $a_i \in R$, $\operatorname{lt}(h_i) = x^{\nu_i}$, $\nu_2 < \nu_3 < \cdots < \nu_t$, and

$$h_{i+1} \in \langle h_i, a_ih_{i-1}, \ldots, a_3\cdots a_ih_2, a_2a_3\cdots a_i\rangle,$$

for $i = 2, \ldots, t-1$.

We first show how to compute all of the prime ideals containing I.

THEOREM 4.6.10. *Let I, P be ideals in $R[x]$ with I zero-dimensional and P prime. Then, with the notation above,*

(i) *$I \subseteq P$ if and only if there exists $i \geq 2$ such that $\langle a_i, h_i\rangle \subseteq P$;*

(ii) *Let $i \in \{2, \ldots, t\}$. If $\langle a_i, h_i\rangle \subseteq P$ then $P = \langle u, v\rangle$, where u is an irreducible factor of a_i and v is an irreducible factor of h_i modulo u;*

(iii) *If $I \subseteq P$, then P is maximal.*

PROOF. To prove (i), let $I \subseteq P$. Then $g_1 = a_2a_3\cdots a_t \in P$, so that there exists $i \geq 2$ such that $a_i \in P$, and we choose i largest with that property. Now $g_i = a_{i+1}\cdots a_th_i \in P$, but $a_{i+1}\cdots a_t \notin P$ by the choice of i, and so $h_i \in P$. Therefore $\langle a_i, h_i\rangle \subseteq P$. For the converse, let $i \geq 2$ and assume that $a_i, h_i \in P$. First note that for all $j = 1, \ldots, i$ we have $g_j \in \langle a_i, h_i\rangle$. Now we show that for $j = i+1, \ldots, t$ we have $g_j \in \langle a_i, h_i\rangle$. Since $h_j \in \langle h_{j-1}, a_{j-1}h_{j-2}, \ldots, a_2\cdots a_{j-1}\rangle$, it is an easy induction on j to show that for $j = i+1, \ldots, t$ we have

$$\langle h_{j-1}, a_{j-1}h_{j-2}, \ldots, a_2\cdots a_{j-1}\rangle \subseteq \langle a_i, h_i\rangle.$$

Thus, since g_j is a multiple of h_j we have $g_j \in \langle a_i, h_i\rangle$ for $j = i+1, \ldots, t$. We now see that $I \subseteq \langle a_i, h_i\rangle$. So, since $\langle a_i, h_i\rangle \subseteq P$, we have $I \subseteq P$.

We now prove (ii). Let $\langle a_i, h_i\rangle \subseteq P$. Since $a_i \in P$, an irreducible factor of a_i, say u, is in P. Note that the ideal $\langle u\rangle$ is now a maximal ideal of R, and hence $\widehat{R} = R/\langle u\rangle$ is a field and $\widehat{R}[x]$ is a PID. Let $\widehat{P}$ be the image of P in $\widehat{R}[x]$. Then, since $h_i \in P$, $\widehat{P} \neq \{0\}$ and so $\widehat{P}$ is a maximal ideal of $\widehat{R}[x]$. Since $\widehat{R}[x]$ is a PID, we see that $\widehat{P}$ is generated by an irreducible polynomial $\widehat{v} \in \widehat{R}[x]$, that is, $\widehat{P} = \langle\widehat{v}\rangle$. The image $\widehat{h}_i$ of h_i is in $\widehat{P}$, and so $\widehat{v}$ is an irreducible factor of $\widehat{h}_i$. Let $v \in R[x]$ be a pre-image of $\widehat{v}$. Since $\widehat{v} \in \widehat{P}$, and since $u \in P$, we see that $v \in P$. Therefore $\langle u, v\rangle \subseteq P$. We also have

$$R[x]/\langle u, v\rangle \cong \widehat{R}[x]/\widehat{P} = \widehat{R}[x]/\langle\widehat{v}\rangle.$$

Since $\widehat{R}[x]/\widehat{P}$ is a field, we have that $R[x]/\langle u, v\rangle$ is a field and so $\langle u, v\rangle$ is a maximal ideal. Therefore $P = \langle u, v\rangle$.

Statement (iii) is now immediate. $\square$

LEMMA 4.6.11. *Let I be an ideal in a Noetherian ring A such that every prime ideal containing I is maximal. Let $I = \bigcap_{j=1}^{\ell} Q_j$ be a primary decomposition of I, where Q_j is primary and $\sqrt{Q_j} = M_j$ is maximal. Then for any maximal ideal M such that $I \subseteq M$, there exists $j \in \{1, \ldots, \ell\}$ such that $M = M_j$.*

PROOF. We have

$$\sqrt{I} = \sqrt{\bigcap_{j=1}^{\ell} Q_j} = \bigcap_{j=1}^{\ell} \sqrt{Q_j} = \bigcap_{j=1}^{\ell} M_j.$$

Since $I \subseteq M$, we have

$$\prod_{j=1}^{\ell} M_j \subseteq \bigcap_{j=1}^{\ell} M_j \subseteq M.$$

Therefore there exists $j \in \{1, \ldots, \ell\}$ such that $M_j \subseteq M$. But M_j is maximal, so $M = M_j$. $\square$

Therefore, to compute the primary decomposition of the zero-dimensional ideal $I \subseteq R[x]$, we first have to determine all the maximal ideals containing I. To do this we find all maximal ideals which contain $\langle a_i, h_i\rangle$, for each $i = 2, \ldots, t$. Theorem 4.6.10 gives an explicit method for finding all such maximal ideals: given $\{g_1, \ldots, g_t\}$ a strong Gröbner basis for I as in Theorem 4.5.13, for each $i = 1, \ldots, t$

(i) compute the irreducible factors of a_i;

(ii) for each u computed in (i), compute the irreducible factors of h_i modulo u;

(iii) each u, v computed in (i) and (ii) respectively gives rise to a maximal ideal containing I, namely $M = \langle u, v\rangle$.

Note that the method presented above gives, in fact, a way to compute the primary decomposition of $\sqrt{I}$ for a zero-dimensional ideal I (combining Theorem 4.6.10 and Lemma 4.6.2). We give an example to show how this method is used.

EXAMPLE 4.6.12. We go back to Example 4.5.15 where $R = \mathbb{Q}[y]$ and $I = \langle (x+y)(y^2+1), x^2 - x + y + 1\rangle$. We computed a strong Gröbner basis for I to be

$$\begin{aligned} g_1 &= y^4 + 2y^3 + 2y^2 + 2y + 1 = (y+1)^2(y^2+1) = a_2 a_3 \\ g_2 &= xy^2 + x + y^3 + y = (y^2+1)(x+y) = a_3 h_2 \\ g_3 &= x^2 - x + y + 1 = h_3. \end{aligned}$$

Note that the greatest common divisor of g_1, g_2 and g_3 is 1. We find all maximal ideals containing I. So we find those maximal ideals which contain $\langle a_2, h_2\rangle = \langle (y+1)^2, x+y\rangle$ and those which contain $\langle a_3, h_3\rangle = \langle y^2+1, x^2 - x + y + 1\rangle$.

Maximal ideals which contain $\langle (y+1)^2, x+y\rangle$: The only irreducible factor of $(y+1)^2$ is $u = y+1$. Also, $x+y \equiv x-1 \pmod{y+1}$ is irreducible. We let $v = x-1$. Therefore the only maximal ideal which contains $\langle (y+1)^2, x+y\rangle$ is

$$M_1 = \langle y+1, x-1\rangle.$$

Maximal ideals which contain $\langle y^2+1, x^2-x+y+1\rangle$: Clearly $u = y^2+1$ is irreducible in $\mathbb{Q}[y]$. Now we find the irreducible factors of $x^2-x+y+1$ modulo u, or equivalently, we find the irreducible factors of the image of $x^2-x+y+1$ in the ring $\widehat{R}[x]$, where

$$\widehat{R} = R/\langle u\rangle = \mathbb{Q}[y]/\langle y^2+1\rangle = \mathbb{Q}[i],$$

where $i = \sqrt{-1}$. The image of $x^2-x+y+1$ in $(\mathbb{Q}[i])[x]$ is the polynomial $x^2-x+i+1$. It is easy to see that this last polynomial can be factored as

$$x^2-x+i+1 = (x-i)(x+i-1),$$

and each of the factors in the right-hand side polynomial is irreducible in $(\mathbb{Q}[i])[x]$. We find pre-images of these factors in $R[x]$ and we have

$$x^2-x+y+1 \equiv (x-y)(x+y-1) \pmod{y^2+1}.$$

Therefore we have two maximal ideals in $R[x]$ containing $\langle y^2+1, x^2-x+y+1\rangle$:

$$M_2 = \langle y^2+1, x-y\rangle, \text{ and } M_3 = \langle y^2+1, x+y-1\rangle.$$

We now have the primary decomposition of $\sqrt{I}$:

$$\sqrt{I} = M_1 \cap M_2 \cap M_3 = \langle y+1, x-1\rangle \cap \langle y^2+1, x-y\rangle \cap \langle y^2+1, x+y-1\rangle.$$

Since we now can compute all the maximal ideals containing I, we need to find the primary ideals which correspond to each maximal ideal in order to compute the primary decomposition of I. We do this in two steps. We first give a criterion to determine whether a given ideal Q is M-primary. We then give a criterion to determine which of the M-primary ideals belong to the primary decomposition of I.

LEMMA 4.6.13. *Let A be a Noetherian ring. Let M be a maximal ideal of A and let $Q \subseteq M$ be an ideal of A. Further assume that for each $m \in M$ there exists $\nu \in \mathbb{N}$ such that $m^\nu \in Q$. Then Q is M-primary.*

PROOF. We first prove that $\sqrt{Q} = M$. Since M is prime and $Q \subseteq M$, we have $\sqrt{Q} \subseteq M$. Moreover, if $m \in M$ then $m \in \sqrt{Q}$, since a power of m is in Q, and thus $M \subseteq \sqrt{Q}$ as well.

It remains to show that Q is primary. Let $fg \in Q$ and $f \notin Q$. We show that $g \in M$. Suppose to the contrary that $g \notin M$. Then there exist $h \in A$ and $m \in M$ such that $hg + m = 1$, since M is maximal. Let $\nu \in \mathbb{N}$ such that $m^\nu \in Q$. Then

$$1 = 1^\nu = (hg+m)^\nu = h'\,g + m^\nu,$$

for some $h' \in A$. Then $f = h'gf + m^\nu f \in Q$, a contradiction. Therefore $g \in M$. □

LEMMA 4.6.14. *Let A be a Noetherian ring and let I be an ideal of A. Let $I = \bigcap_{i=1}^{\ell} Q_i$ be an irredundant primary decomposition of I such that Q_i is M_i-primary with M_i maximal. Then for $j = 1, \dots, \ell$*

$$Q_j = \{f \in A \mid I : \langle f \rangle \not\subset M_j\}.$$

PROOF. Let $j \in \{1, \dots, \ell\}$. We denote $\{f \in A \mid I : \langle f \rangle \not\subset M_j\}$ by Q'_j. Let $f \in Q'_j$. Then there exists $g \in A$ such that $g \notin M_j = \sqrt{Q_j}$ and $fg \in I \subseteq Q_j$. Since Q_j is primary, either $f \in Q_j$ or a power of g is in Q_j. But since $g \notin \sqrt{Q_j}$, we must have $f \in Q_j$. Therefore $Q'_j \subseteq Q_j$.

For the reverse inclusion let $f \in Q_j$. For each $i \in \{1, \dots, \ell\}$, $i \neq j$, there exists $s_i \in M_i - M_j$, since M_i and M_j are distinct maximal ideals. Since $M_i = \sqrt{Q_i}$, there exists $\nu_i \in \mathbb{N}$ such that $s_i^{\nu_i} \in Q_i$. We define

$$s = \prod_{\substack{i=1 \\ i \neq j}}^{\ell} s_i^{\nu_i} \in \prod_{\substack{i=1 \\ i \neq j}}^{\ell} Q_i.$$

Note that $s \notin M_j$ by construction. Then $fs \in \prod_{i=1}^{\ell} Q_i \subseteq \bigcap_{i=1}^{\ell} Q_i = I$. Therefore $s \in I : \langle f \rangle$ and $s \notin M_j$, so that $f \in Q'_j$. □

We now return to the case where $A = R[x]$ where R is a PID. We will use the above to give a method for computing the primary decomposition of I. So let $M = \langle u, v \rangle$ be a maximal ideal which contains $\langle a_i, h_i \rangle$, as obtained in Theorem 4.6.10. Then u is an irreducible factor of a_i and so u divides $g_1 = a_2 \cdots a_t$. Let m be the largest integer such that u^m divides g_1 (so $m \geq 1$). Now, we know that the image $\widehat{v}$ of v in the ring $(R/\langle u \rangle)[x]$ is an irreducible factor of the image $\widehat{h}_i$ of h_i. Therefore $\widehat{v}$ divides the image $\widehat{g}_t$ of g_t, since $g_t = h_t \in \langle a_i, h_i \rangle$ and u divides a_i. Let n be the largest integer such that $\widehat{v}^n$ divides $\widehat{g}_t$ (note that $n \geq 1$). Then we have $g_t \equiv v^n w \pmod{u}$, for some $w \in R[x]$, such that the image $\widehat{w}$ of w is not divisible by $\widehat{v}$.

THEOREM 4.6.15. *We use the notation above. Let $V, W \in R[x]$ be such that*

$$g_t \equiv VW \pmod{u^m}, \quad V \equiv v^n \pmod{u}, \quad \textit{and } W \equiv w \pmod{u}.$$

Then $Q = \langle u^m, g_2, \dots, g_{t-1}, V \rangle$ is M-primary and is the M-primary component of I.

PROOF. We first note that $I \subseteq Q$. Indeed, we can write $g_t = VW + hu^m$, for some $h \in R[x]$, and so $g_t \in Q$. Moreover $g_1 \in Q$, since u^m divides g_1. Therefore $I \subseteq Q$.

We now show that Q is M-primary using Lemma 4.6.13. Clearly $Q \subseteq M$, since u^m, V, and $g_2, \ldots, g_{t-1}$ are in M (recall that we noted in the proof of Theorem 4.6.10 that $g_2, \ldots, g_{t-1} \in \langle a_i, h_i \rangle \subseteq M$). To conclude, it is sufficient to show that some power of u and v are in Q. Clearly $u^m \in Q$. Now we can write $V = v^n + h' u$, for some $h' \in R[x]$. Therefore $v^n = V - h' u$, and hence $v^{nm} = (V - h' u)^m \in Q$.

We now show that Q is the M-primary component of I using Lemma 4.6.14. We define $Q' = \{f \in R[x] \mid I : \langle f \rangle \not\subseteq M\}$. Let $f \in Q'$ and let $g \in I : \langle f \rangle - M$. Then $fg \in I \subseteq Q$. Since Q is M-primary, either $f \in Q$ or $g^\nu \in Q$, for some $\nu \in \mathbb{N}$. But if $g^\nu \in Q$, then $g \in \sqrt{Q} = M$, a contradiction. Therefore $f \in Q$, and $Q' \subseteq Q$.

We now prove the reverse inclusion. First note that $\dfrac{g_1}{u^m}$ is relatively prime to u because of the choice of m. Therefore $\dfrac{g_1}{u^m} \notin M$, for otherwise, since $u \in M$, we would have $1 \in M$ (recall that g_1 and u are in R and that R is a PID). Also, $\dfrac{g_1}{u^m} u^m = g_1 \in I$, and so $u^m \in Q'$. Now let j be such that $2 \leq j \leq t-1$. Then $1 \notin M$, and $1 g_j \in I$, so that $g_j \in Q'$. It remains to show that $V \in Q'$. To see this we note that $W \notin M$, for otherwise, $W = hu + h' v$ for some $h, h' \in R[x]$, and so $w \equiv W \equiv h' v \pmod{u}$. But this means that $\widehat{v}$ divides $\widehat{w}$, and this is a contradiction. Then $W \dfrac{g_1}{u^m} \notin M$ and $VW = g_t - h_1 u^m$ for some $h_1 \in R[x]$ imply

$$VW \frac{g_1}{u^m} = (g_t - h_1 u^m) \frac{g_1}{u^m} = g_t \frac{g_1}{u^m} - h_1 g_1 \in I.$$

Therefore $V \in Q'$. $\square$

The last thing that remains to be done, in order to compute the primary decomposition of I, is to compute the polynomials V and W required in the last theorem. This is always possible and it follows from a result known as Hensel's Lemma. We will state the result and show how it is used in examples, but we will not provide a proof of it. However, the examples that follow illustrate all the essential ideas needed for the proof. The interested reader should consult [**Coh**].

THEOREM 4.6.16 (HENSEL'S LEMMA). *Let u be an irreducible element of the PID R and let $f \in R[x]$. Let $g^{(1)}, h^{(1)} \in R[x]$ be two polynomials such that their images in $(R/\langle u \rangle)[x]$ are relatively prime and such that*

$$f \equiv g^{(1)} h^{(1)} \pmod{u}.$$

Then for any m there exist polynomials $g^{(m)}$ and $h^{(m)}$ in $R[x]$ such that

$$f \equiv g^{(m)} h^{(m)} \pmod{u^m}, \quad g^{(m)} \equiv g^{(1)} \pmod{u}, \quad h^{(m)} \equiv h^{(1)} \pmod{u}.$$

If linear equations are solvable in R then these polynomials are computable.

We will now illustrate the technique presented in this section in three examples. These examples will also illustrate how the Hensel lifting technique works.

EXAMPLE 4.6.17. We go back to Example 4.6.12. We found that $I = \langle (x+y)(y^2+1), x^2-x+y+1\rangle \subseteq (\mathbb{Q}[y])[x]$ has the following strong Gröbner basis

$$\begin{aligned} g_1 &= y^4+2y^3+2y^2+2y+1 = (y+1)^2(y^2+1) = a_2a_3 \\ g_2 &= xy^2+x+y^3+y = (y^2+1)(x+y) = a_3h_2 \\ g_3 &= x^2-x+y+1 = h_3, \end{aligned}$$

and that the maximal ideals which contain I are

$$M_1 = \langle y+1, x-1\rangle, M_2 = \langle y^2+1, x-y\rangle, M_3 = \langle y^2+1, x+y-1\rangle.$$

We now find the primary components Q_1, Q_2, and Q_3 corresponding to M_1, M_2, and M_3 which belong to I. We use Theorem 4.6.15 and the notation set there to find Q_1, Q_2, and Q_3.

Primary component which corresponds to M_1: In this case we have $u = y+1$ and $v = x-1$. The largest integer m such that u^m divides $g_1 = (y+1)^2(y^2+1)$ is $m=2$. Now we need to factor g_3 modulo u. We have

$$g_3 = x^2-x+y+1 \equiv x^2-x \equiv (x-1)x \pmod{y+1}.$$

Therefore the largest n such that v^n divides g_3 modulo u is $n=1$. We also have $w = x$. Now we need to find V and W such that

$$VW \equiv x^2-x+y+1 \pmod{(y+1)^2},\ V \equiv x-1 \pmod{y+1},$$
$$\text{and } W \equiv x \pmod{y+1}.$$

To do this we find polynomials h and h' in $\mathbb{Q}[x,y]$ such that

$$V = (x-1)+(y+1)h \text{ and } W = x+(y+1)\,h',$$

and which satisfy the congruence

$$\begin{aligned} g_3 &= x^2-x+y+1 \\ &= x(x-1)+(y+1) \\ &\equiv VW \pmod{(y+1)^2} \\ &\equiv x(x-1)+x(y+1)h+(x-1)(y+1)\,h' \\ &\quad +(y+1)^2h\,h' \pmod{(y+1)^2} \\ &\equiv x(x-1)+x(y+1)h+(x-1)(y+1)\,h' \pmod{(y+1)^2}. \end{aligned}$$

Therefore $y+1 \equiv x(y+1)h+(x-1)(y+1)\,h' \pmod{(y+1)^2}$, or equivalently $1 \equiv xh+(x-1)\,h' \pmod{y+1}$. One obvious solution to this equation is $h=1$ and $h'=-1$. Therefore we have

$$V = (x-1)+(y+1) = x+y, \text{ and } W = x-(y+1) = x-y-1.$$

By Theorem 4.6.15 the primary component of I which corresponds to the maximal ideal M_1 is

$$Q_1 = \langle u^2, g_2, V\rangle = \langle (y+1)^2, (y^2+1)(x+y), x+y\rangle = \langle (y+1)^2, x+y\rangle.$$

Primary component which corresponds to M_2: In this case we have $u = y^2+1$ and $v = x - y$. The largest integer m such that u^m divides $g_1 = (y+1)^2(y^2+1)$ is $m = 1$. Now we need to factor g_3 modulo u. We have

$$g_3 = x^2 - x + y + 1 \equiv (x-y)(x+y-1) \pmod{y^2+1}.$$

Therefore the largest n such that v^n divides g_3 modulo u is $n = 1$. We also have $w = x+y-1$. Since $m = 1$ we may let $V = v$ and $W = w$. Therefore the primary ideal corresponding to M_2 and which belongs to I is

$$Q_2 = \langle y^2+1, (y^2+1)(x+y), x-y\rangle = M_2.$$

Primary component which corresponds to M_3: This computation is the same as the previous one and we get that the primary ideal corresponding to M_3 and which belongs to I is

$$Q_3 = \langle y^2+1, (y^2+1)(x+y), x+y-1\rangle = M_3.$$

Therefore we have

$$\begin{aligned} I &= \langle (x+y)(y^2+1), x^2 - x + y + 1\rangle = Q_1 \cap Q_2 \cap Q_3 \\ &= \langle (y+1)^2, x+y\rangle \cap \langle y^2+1, x-y\rangle \cap \langle y^2+1, x+y-1\rangle. \end{aligned}$$

The next example illustrates how the Hensel lifting technique is applied repeatedly.

EXAMPLE 4.6.18. In this example we again consider the ring $R = \mathbb{Q}[y]$. Let I be the ideal $\langle (x+y)^2(y-1), x^2+x+y\rangle$ of $R[x]$. Again, to compute the strong Gröbner basis for I we compute the Gröbner basis for I viewed as an ideal in $\mathbb{Q}[x,y]$ with respect to the lex term ordering with $x > y$. We get

$$\begin{aligned} g_1 &= y^5 - y^4 = y^4(y-1) = a_2a_3 \\ g_2 &= xy - x + 2y^4 - y^3 - y = (y-1)(x+2y^3+y^2+y) = a_3h_2 \\ g_3 &= x^2 + x + y = h_3. \end{aligned}$$

Note that the greatest common divisor of g_1, g_2, and g_3 is 1. As in Example 4.6.12 we first compute the maximal ideals which contain I. We find those maximal ideals which contain $\langle a_2, h_2\rangle = \langle y^4, x+2y^3+y^2+y\rangle$ and those which contain $\langle a_3, h_3\rangle = \langle y-1, x^2+x+y\rangle$.

Maximal ideals which contain $\langle y^4, x+2y^3+y^2+y\rangle$: The only irreducible factor of $a_2 = y^4$ is $u = y$. The only irreducible factor of $x+2y^3+y^2+y$ modulo y is $v = x$. Therefore the maximal ideal which contains $\langle y^4, x+2y^3+y^2+y\rangle$ is

$$M_1 = \langle y, x\rangle.$$

Maximal ideals which contain $\langle y-1, x^2+x+y\rangle$: It is easy to see that the only maximal ideal that contains $\langle y-1, x^2+x+y\rangle$ is

$$M_2 = \langle y-1, x^2+x+1\rangle.$$

That is, we have $u = y - 1$ and $v = x^2 + x + 1$, since $x^2 + x + 1$ is irreducible in the ring $\widehat{R}[x]$, where $\widehat{R} = R/\langle u \rangle = \mathbb{Q}[y]/\langle y - 1 \rangle \cong \mathbb{Q}$. So the original ideal $\langle y - 1, x^2 + x + y \rangle = M_2$ is itself maximal.

Thus, the primary decomposition of $\sqrt{I}$ is

$$\sqrt{I} = M_1 \cap M_2 = \langle y, x \rangle \cap \langle y - 1, x^2 + x + 1 \rangle.$$

We now find the primary components that belong to I and which correspond to M_1 and M_2. We use Theorem 4.6.15 and the notation set there to find Q_1 and Q_2.

Primary component which corresponds to M_1: In this case we have $u = y$ and $v = x$. The largest integer m such that u^m divides $g_1 = y^4(y - 1)$ is $m = 4$. Now we need to factor g_3 modulo u. We have

$$g_3 = x^2 + x + y \equiv x^2 + x \equiv x(x + 1) \pmod{y}.$$

Therefore the largest n such that v^n divides g_3 modulo u is $n = 1$. We also have $w = x + 1$. Thus we need to find V and W such that

$$VW \equiv x^2 + x + y \pmod{y^4}, \; V \equiv x \pmod{y}, \text{ and } W \equiv x + 1 \pmod{y}.$$

To do this we use the Hensel lifting technique three times to find $V^{(i)}, W^{(i)}$ such that at each stage we have

$$V^{(i)}W^{(i)} \equiv x^2 + x + y \pmod{y^i}, \; V^{(i)} \equiv x \pmod{y}, \; W^{(i)} \equiv x + 1 \pmod{y},$$

for $i = 2, 3, 4$. This is done inductively by constructing $V^{(i+1)}$ and $W^{(i+1)}$ in terms of $V^{(i)}$ and $W^{(i)}$ respectively. We start with lifting modulo y^2. To do this we must find polynomials $h^{(2)}$ and $h'^{(2)}$ in $\mathbb{Q}[x, y]$ such that

$$V^{(2)} = x + yh^{(2)} \text{ and } W^{(2)} = x + 1 + yh'^{(2)},$$

and which satisfy the following congruence

$$\begin{aligned} g_3 &= x^2 + x + y \\ &\equiv V^{(2)}W^{(2)} \pmod{y^2} \\ &\equiv x^2 + x + (x + 1)yh^{(2)} + xyh'^{(2)} \pmod{y^2}. \end{aligned}$$

Therefore we have $y \equiv (x + 1)yh^{(2)} + xy\,h'^{(2)} \pmod{y^2}$, or equivalently $1 \equiv (x + 1)h^{(2)} + x\,h'^{(2)} \pmod{y}$. An obvious solution to this congruence is $h^{(2)} = 1$ and $h'^{(2)} = -1$. Therefore we have

$$V^{(2)} = x + y \text{ and } W^{(2)} = x - y + 1.$$

To lift modulo y^3, we find $h^{(3)}$ and $h'^{(3)}$ such that

$$V^{(3)} = V^{(2)} + y^2h^{(3)} = x + y + y^2h^{(3)},$$

$$W^{(3)} = W^{(2)} + y^2h'^{(3)} = x - y + 1 + y^2h'^{(3)}$$

and which satisfy the congruence

$$\begin{aligned} g_3 &= x^2+x+y \\ &\equiv V^{(3)}W^{(3)} \pmod{y^3} \\ &\equiv x^2+x+y-y^2+(x+y)y^2h'^{(3)}+(x-y+1)y^2h^{(3)} \pmod{y^3}. \end{aligned}$$

Canceling y^2 we have

$$(x+y)h'^{(3)}+(x-y+1)h^{(3)} \equiv 1 \pmod{y}.$$

A solution to this congruence is $h^{(3)}=1$ and $h'^{(3)}=-1$. Therefore we have

$$V^{(3)}=x+y^2+y \text{ and } W^{(3)}=x-y^2-y+1.$$

Finally we lift modulo y^4. We find $h^{(4)}$ and $h'^{(4)}$ such that

$$V=V^{(4)}=V^{(3)}+y^3h^{(4)}=x+y^2+y+y^3h^{(4)},$$

$$W=W^{(4)}=W^{(3)}+y^3h'^{(4)}=x-y^2-y+1+y^3h'^{(4)}$$

and which satisfy the congruence

$$\begin{aligned} g_3 &= x^2+x+y \\ &\equiv V^{(4)}W^{(4)} \pmod{y^4} \\ &\equiv x^2+x+y-2y^3+(x+y^2+y)y^3h'^{(4)} \\ &\quad +(x-y^2-y+1)y^3h^{(4)} \pmod{y^4}. \end{aligned}$$

Canceling y^3 we have

$$(x+y^2+y)h'^{(4)}+(x-y^2-y+1)h^{(4)} \equiv 2 \pmod{y}.$$

A solution to this congruence is $h^{(4)}=2$ and $h'^{(4)}=-2$. Therefore we have

$$V=V^{(4)}=x+2y^3+y^2+y \text{ and } W=W^{(4)}=x-2y^3-y^2-y+1.$$

Now that we have V and W we can find the primary ideal Q_1 corresponding to M_1 and which belongs to I,

$$\begin{aligned} Q_1=\langle u^m, g_2, V\rangle &= \langle y^4,(y-1)(x+2y^3+y^2+y), x+2y^3+y^2+y\rangle \\ &= \langle y^4, x+2y^3+y^2+y\rangle. \end{aligned}$$

Primary component which corresponds to M_2: In this case we have $u=y-1$ and $v=x^2+x+1$. The largest integer m such that u^m divides $g_1=y^4(y-1)$ is $m=1$. We thus have that $Q_2=\langle y-1,x^2+x+1\rangle=M_2$.

Therefore the primary decomposition of I is

$$I=Q_1\cap M_2=\langle y^4,x+2y^3+y^2+y\rangle\cap\langle y-1,x^2+x+1\rangle.$$

We conclude this section with an example in $\mathbb{Z}[x]$.

EXAMPLE 4.6.19. The polynomials

$$\begin{aligned} g_1 &= 45 = 9 \cdot 5 = a_2 a_3 \\ g_2 &= 5x + 10 = 5(x+2) = a_3 h_2 \\ g_3 &= x^3 + 6x^2 + 20x + 15 = h_3. \end{aligned}$$

form a minimal strong Gröbner basis in $\mathbb{Z}[x]$. We will compute the primary decomposition of $I = \langle g_1, g_2, g_3 \rangle$. We first compute the maximal ideals which contain $\langle a_2, h_2 \rangle$ and $\langle a_3, h_3 \rangle$ respectively.

Maximal ideals which contain $\langle 9, x+2 \rangle$: The only irreducible factor of 9 is $u = 3$. Also, $v = x + 2$ is irreducible modulo 3. Therefore there is only one maximal ideal which contains $\langle 9, x+2 \rangle$, and it is

$$M_1 = \langle 3, x+2 \rangle.$$

Maximal ideals which contain $\langle 5, x^3 + 6x^2 + 20x + 15 \rangle$: The only irreducible factor of 5 is $u = 5$. Also,

$$x^3 + 6x^2 + 20x + 15 \equiv x^3 + x^2 \equiv x^2(x+1) \pmod{5}.$$

Therefore there are two irreducible factors modulo 5, namely x and $x + 1$. Thus there are two maximal ideals which contain $\langle 5, x^3 + 6x^2 + 20x + 15 \rangle$,

$$M_2 = \langle 5, x \rangle \text{ and } M_3 = \langle 5, x+1 \rangle.$$

The primary decomposition of $\sqrt{I}$ is

$$\sqrt{I} = M_1 \cap M_2 \cap M_3 = \langle 3, x+2 \rangle \cap \langle 5, x \rangle \cap \langle 5, x+1 \rangle.$$

Now we find the primary component for each of these maximal ideals.

Primary component which corresponds to M_1: In this case we have $u = 3$ and $v = x + 2$. The largest integer m such that u^m divides $g_1 = 45$ is $m = 2$. Now we need to factor g_3 modulo u. We have

$$g_3 = x^3 + 6x^2 + 20x + 15 \equiv x^3 + 2x \equiv x(x^2 + 2) \equiv x(x+1)(x+2) \pmod{3}.$$

Therefore the largest n such that v^n divides g_3 modulo u is $n = 1$. We also have $w = x(x+1)$. Now we need to find V and W such that

$$VW \equiv x^3 + 6x^2 + 20x + 15 \pmod{9}, \; V \equiv x + 2 \pmod{3},$$

$$\text{and } W \equiv x(x+1) \pmod{3}.$$

As before we use the Hensel lifting technique. We find polynomials h and h' in $\mathbb{Z}[x]$ such that

$$V = (x+2) + 3h \text{ and } W = x(x+1) + 3\,h',$$

and which satisfy the following congruence

$$\begin{aligned} g_3 &= x^3 + 6x^2 + 20x + 15 \\ &\equiv VW \pmod 9 \\ &\equiv x^3 + 3x^2 + 2x + 3x(x+1)h + 3(x+2)\,h' \pmod 9. \end{aligned}$$

Therefore we have $3x^2 + 6 \equiv 3x(x+1)h + 3(x+2)\,h' \pmod 9$, or equivalently, $x^2 + 2 \equiv x(x+1)h + (x+2)\,h' \pmod 3$. A solution to this congruence is $h = 0$ and $h' = x + 1$. Thus we have

$$V = x + 2 \text{ and } W = x(x+1) + 3(x+1) = x^2 + 4x + 3.$$

We now can find the primary ideal Q_1 corresponding to M_1 and which belongs to I,

$$Q_1 = \langle u^m, g_2, V\rangle = \langle 9, 5(x+2), x+2\rangle = \langle 9, x+2\rangle.$$

Primary component which corresponds to M_2: In this case we have $u = 5$ and $v = x$. The largest integer m such that u^m divides $g_1 = 45$ is $m = 1$. Now we factor g_3 modulo u. We have

$$g_3 = x^3 + 6x^2 + 20x + 15 \equiv x^3 + x^2 \equiv x^2(x+1) \pmod 5.$$

Therefore the largest n such that v^n divides g_3 modulo u is $n = 2$. We also have $w = x + 1$. Since $m = 1$ we may let $V = v$ and $W = w$. The primary ideal Q_2 corresponding to M_2 and which belongs to I is

$$Q_2 = \langle u^m, g_2, V\rangle = \langle 5, 5(x+2), x^2\rangle = \langle 5, x^2\rangle.$$

Primary component which corresponds to M_3: In this case we have $u = 5$ and $v = x+1$. As above the largest integer m such that u^m divides $g_1 = 45$ is $m = 1$. Also, as above $g_3 \equiv x^2(x+1) \pmod 5$, and so the largest n such that v^n divides g_3 modulo u is $n = 1$. We also have $w = x^2$. As above we can choose V and W equal to v and w respectively, and so the primary ideal Q_3 corresponding to M_3 and which belongs to I is

$$Q_3 = \langle u^m, g_2, V\rangle = \langle 5, 5(x+2), x+1\rangle = \langle 5, x+1\rangle.$$

Therefore we have the following primary decomposition of I

$$I = Q_1 \cap Q_2 \cap Q_3 = \langle 9, x+2\rangle \cap \langle 5, x^2\rangle \cap \langle 5, x+1\rangle.$$

Exercises

4.6.1. Consider the following three polynomials in $\mathbb{Q}[x,y]$

$$\begin{aligned} g_1 &= (y^2 - y + 1)^2(y-1) = a_2a_3 \\ g_2 &= (y-1)(x^2+y) = a_3h_2 \\ g_3 &= x^8 + 2x^6 + 3x^4 + x^2 - y + 1 = h_3. \end{aligned}$$

Let $I = \langle g_1, g_2, g_3 \rangle \subseteq (\mathbb{Q}[x])[y]$. Verify that $\{g_1, g_2, g_3\}$ is a strong Gröbner basis for I and find the primary decomposition of I.

4.6.2. Consider the following three polynomials in $\mathbb{Z}[x]$

$$\begin{aligned} g_1 &= 4 \cdot 7 = a_2 a_3 \\ g_2 &= 7(x^2+1) = a_3 h_2 \\ g_3 &= x^4 + 4x + 3 = h_3. \end{aligned}$$

Let $I = \langle g_1, g_2, g_3 \rangle \subseteq \mathbb{Z}[x]$. Verify that $\{g_1, g_2, g_3\}$ is a strong Gröbner basis for I and find the primary decomposition of I.

Appendix A. Computations and Algorithms

Computations. There are many Computer Algebra Systems which have a Gröbner basis package, for example AXIOM, MAPLE, and MATHEMATICA. There are other packages which are entirely devoted to computing in polynomial rings and which have an extensive list of commands to perform some of the computations presented in this book. In particular, we mention CoCoA and MACAULAY. Most of the computations in the examples and exercises in this book (except in Chapter 4) were performed using CoCoA. Many of these computations could not have been done with the other systems listed above. MAPLE and MATHEMATICA do not allow computations in modules and have only a limited choice of orders. These systems allow the user to program and the algorithms presented in this book are, in principle, programmable. However, any practical implementation of these algorithms requires a lot of material not included in this book and many hours of work. MACAULAY computes only with homogeneous polynomials and focuses on computations applied to algebraic geometry. Most of our examples are non-homogeneous. Some of the computations could still be done using MACAULAY, but with care (see Exercises 1.4.9 and 1.6.18). None of the above systems have an implementation for the computation of Gröbner bases over rings which are not fields. Of course the algorithms over rings could also, in principle, be implemented using some of these systems.

CoCoA provides commands for the algebraic manipulation of polynomials in $k[x_1, \dots, x_n]$, ideals of $k[x_1, \dots, x_n]$, and submodules of the free modules $(k[x_1, \dots, x_n])^m$, where k is a field. Moreover, the user can choose among predefined term orders or define a custom ordering. The algebraic procedures include the computation of the intersection of ideals and modules, ideal quotients, normal forms, syzygy modules, elimination ideals and modules, free resolutions, and, of course, Gröbner bases for ideals and modules. When starting a computation with CoCoA the user specifies the characteristic of the field. Even when computing over $\mathbb{Q}$, CoCoA performs arithmetic modulo a large prime number, so errors may occur and the user must keep this in mind.

CoCoA was developed at the University of Genova, Italy, by Antonio Capani, Alessandro Giovini, Gianfranco Niesi, and Lorenzo Robbiano. To obtain a copy of CoCoA send a message to the developers at

`cocoa@dima.unige.it`.

Algorithms. The algorithms in this book are presented in pseudo-code, and the format is as follows. We always start our algorithms by specifying the input (which follows the word **INPUT**) and the output (which follows the word **OUTPUT**). We then always specify how we initialize the variables involved in the algorithm (this follows the word **INITIALIZATION**).

When we need to assign an expression to a variable we use the instruction

variable := *expression*

For example, if the current value of the variable ℓ is 4 and the current value of the variable G is $\{1, 2, 3\}$, and if the instructions

$$G := G \cup \{\ell\}$$
$$\ell := \ell + 1$$

are executed, then the operation $\{1, 2, 3\} \cup \{4\} = \{1, 2, 3, 4\}$ is performed, and G takes on its new value $\{1, 2, 3, 4\}$, and then the operation $4 + 1 = 5$ is performed, and ℓ takes on its new value 5.

We use the following conditional structure:

```
IF condition THEN
        action 1
ELSE
        action 2
```

This means that if *condition* is true, then *action 1* is performed, and if *condition* is false then *action 2* is performed. Note that the truth of *condition* depends on the current value of the variables in the algorithm. Sometimes we omit the **ELSE** statement which always means that *action 2* is simply "do nothing." The indentation always indicates what *action 1* and *action 2* are and when the conditional structure terminates.

We also use two loop structures:

```
WHILE condition DO
        action
```

and

```
FOR each item in a set S DO
        action
```

In the **WHILE** loop, *action* is repeated as long as *condition* holds and is not performed if *condition* does not hold. In the **FOR** loop, *action* is performed once for each item in the set S. Note that, in many instances, an order on the set S is prescribed, and the **FOR** loop must be executed in that order. No other instruction is executed until the entire **WHILE** or **FOR** loop is completed. The indentation always indicates what the *action* is and when the loops terminate.

Appendix B. Well-ordering and Induction

In this book we use a form of "proof by induction" which may be unfamiliar to the reader. It is the purpose of this appendix to briefly describe this process.

We consider a non-empty set T which we assume has a *total order* "$\leq$" on it. That is, we assume we have a relation "$\leq$" on T satisfying the following properties:

- the relation $\leq$ is reflexive: for all $t \in T$, $t \leq t$;
- the relation $\leq$ is transitive: for all $s, t, u \in T$, if $s \leq t$ and $t \leq u$ then $s \leq u$;
- the relation $\leq$ is antisymmetric: for all $s, t \in T$, if $s \leq t$ and $t \leq s$ then $s = t$;
- the relation $\leq$ is total: for all $s, t \in T$, either $s \leq t$ or $t \leq s$.

We say that a total order $\leq$ is a *well-ordering* provided that we have the additional property

- Let $S \subseteq T$ such that $S \neq \emptyset$. Then S contains a smallest element. That is, there is an $s \in S$ such that for all $t \in S$ we have $s \leq t$.

We will use two examples of well-ordered sets. There is, of course, the one familiar to most people, the set of natural numbers $\mathbb{N} = \{0, 1, 2, \dots\}$. The one we will use most in this book is the set $\mathbb{T}^n$ of power products in the variables $x_1, \dots, x_n$ on which we will put various orders (see Section 1.4); all of them will be well-orderings.

Let T be a well-ordered set. Since T is a subset of itself, it has a smallest element which we will denote by 1 (of course, in $\mathbb{N}$ this element is 0).

We assume that we are given a set $\{\mathcal{P}_t \mid t \in T\}$ of statements to be proved. We go about this in one of two ways.

- We argue by contradiction. If one of the statements is false, we can let S be the set of all $t \in T$ such that $\mathcal{P}_t$ is false. Since $S \neq \emptyset$ we may choose $s \in S$ least. Another way to say this is that if some of the $\mathcal{P}_t$'s are false then we may choose a smallest $s \in T$ for which $\mathcal{P}_s$ is false. The idea then is to find $t < s$ for which $\mathcal{P}_t$ is also false and so arrive at a contradiction and conclude that all $\mathcal{P}_t$'s are true.
- The second way to proceed is in the more traditional "induction argu-

ment" style (commonly referred to as strong induction). We assume that

(i) $\mathcal{P}_1$ is true;

(ii) For all $t \in T$ the truth of $\mathcal{P}_s$ for all $s < t$ implies the truth of $\mathcal{P}_t$.

We conclude from this that all of the $\mathcal{P}_t$'s are true.

The validity of the first method of reasoning is obvious. Moreover the proof that these two forms of induction are equivalent proceeds exactly as it does in the case $T = \mathbb{N}$.

References

[AdGo] W.W. Adams and L.J. Goldstein, *Introduction to Number Theory,* Prentice Hall, Englewood Cliffs, NJ, 1976.

[AtMD] M.F. Atiyah and I.G. MacDonald, *Introduction to Commutative Algebra,* Addison-Wesley, Reading, MA, 1969.

[Ba] D. Bayer, *The Division Algorithm and the Hilbert Scheme,* Ph.D. Thesis, Harvard University, Cambridge, MA, 1982.

[BaSt88] D.A. Bayer and M. Stillman, *On the complexity of computing syzygies,* J. Symb. Comp. **6** (1988), 135–147.

[BaSt90] D.A. Bayer and M. Stillman, *Macaulay: A system for computation in algebraic geometry and commutative algebra.* User's manual.

[BeWe] T. Becker and V. Weispfenning, *Gröbner Bases: A Computational Approach to Commutative Algebra,* Springer Verlag, Berlin and New York, 1993.

[Bu65] B. Buchberger, *Ein Algorithmus zum Auffinden der Basiselemente des Restklassenringes nach einem nulldimensionalen Polynomideal,* Ph.D. Thesis, Inst. University of Innsbruck, Innsbruck, Austria, 1965.

[Bu79] B. Buchberger, *A criterion for detecting unnecessary reductions in the construction of Gröbner bases,* in EUROSAM'79, An International Symposium on Symbolic and Algebraic Manipulation (E.W. Ng, ed.), Lecture Notes in Comput. Sci., vol. 72, Springer Verlag, Berlin and New York, 1979, 3–21.

[Bu83] B. Buchberger, *A note on the complexity of constructing Gröbner bases,* in EUROCAL'83, European Computer Algebra Conference (J.A. van Hulzen ed.), Lecture Notes in Comput. Sci., vol. 162, Springer Verlag, Berlin and New York, 1983, 137–145.

[Bu85] B. Buchberger, *Gröbner bases: An algorithmic method in polynomial ideal theory,* in Multidimensional Systems Theory (N.K. Bose ed.), Reidel, Dordrecht, 1985, 184–232.

[CGGLMW] B. Char, K. Geddes, G. Gonnet, B. Leong, M. Monogan, and S.M. Watt, *Maple V Library Reference Manual,* Springer Verlag, Berlin and New York, 1991.

[Coh] H. Cohen, *A Course in Computational Algebraic Number Theory,* Springer Verlag, Berlin and New York, 1993.

[CoTr] P. Conti and C. Traverso, *Buchberger algorithm and integer programming,* in Applied Algebra, Algebraic Algorithms, and Error-Correcting Codes AAECC'9 (H.F. Mattson, T. Mora, and T.R.N Rao eds.), Lecture Notes in Comput. Sci., vol. 539, Springer verlag, Berlin and New York, 1991, 130–139.

[CLOS] D. Cox, J. Little, and D. O'Shea, *Ideals, Varieties, and Algorithms: An Introduction to Computational Algebraic Geometry and Commutative Algebra,* Springer Verlag, Berlin and New York, 1992.

[Cz] S.R. Czapor, *Gröbner basis methods for solving algebraic equations,* Ph.D. Thesis, University of Waterloo, Canada, 1988.

[EHV] D. Eisenbud, C. Huneke, and W. Vasconcelos, *Direct methods for primary decomposition,* Invent. Math. **110** (1992), 207-235.

[FGLM] J.C. Faugère, P. Gianni, D. Lazard, and T. Mora, *Efficient computation of zero-dimensional Gröbner bases by change of ordering,* J. Symb. Comp. **16** (1993), 329–344.

[Ga] D. Gale, *The Theory of Linear Economic Models,* McGraw-Hill, New York, 1960.

[GTZ] P. Gianni, B. Trager, and G. Zacharias, *Gröbner bases and primary decomposition of polynomial ideals,* J. Symb. Comp. **6** (1988), 149–167.

[GMNRT] A. Giovini, T. Mora, G. Niesi, L. Robbiano, and C. Traverso, *"One sugar cube, please" or selection strategies in the Buchberger algorithm,* in Proc. International Symposium on Symbolic and Algebraic Computation ISSAC'91 (S.M. Watt ed.), ACM Press, New York, 1991, 49–54.

[GiNi] A. Giovini and G. Niesi, *CoCoA user manual,* 1990.

[Go] L. J. Goldstein, *Abstract Algebra, a First Course,* Prentice Hall, Englewood Cliffs, NJ, 1973.

[He] I.N. Herstein, *Topics in Algebra,* Blaisdell, New York, 1964.

[Hi] H. Hironaka, *Resolution of singularities of an algebraic variety over a field of characteristic zero,* Ann. Math. **79** (1964), 109–326.

[Hun] T.W. Hungerford, *Algebra,* Springer Verlag, Berlin and New York, 1974.

[Huy] D.T. Huynh, *A superexponential lower bound for Gröbner bases and Church-Rosser commutative Thue systems,* Inf. Contr. **68** (1986), 196–206.

[JeSu] R.D. Jenks and R.S. Sutor, *AXIOM: The Scientific Computation System,* Springer Verlag, Berlin and New York, 1992.

[Lak] Y.N. Lakshman, *On the complexity of computing Gröbner bases for zero-dimensional polynomials ideals,* Ph.D. Thesis, Rensselaer Polytechnic Institute, Troy, NY, 1990.

[Lan] S. Lang, *Algebra,* Addison-Wesley, Reading, MA, 1970.

[Laz85] D. Lazard, *Ideal bases and primary decomposition: Case of two variables,* J. Symb. Comp. **1** (1985), 261–270.

[Laz91] D. Lazard, *Systems of algebraic equations (algorithms and complexity),* Computational Algebraic Geometry and Commutative Algebra (D. Eisenbud and L. Robbiano, eds.), Cambridge University Press, London, 1993.

[Mac] F.S. Macaulay, *Some properties of enumeration in the theory of modular systems,* Proc. London Math. Soc. **26** (1927), 531–555.

[MaMe] E.W. Mayr and A.R Meyer, *The complexity of the word problems for commutative semigroups and polynomial ideals,* Adv. Math **46** (1982), 305–329.

[Mi] B. Mishra, *Algorithmic Algebra,* Springer Verlag, Berlin and New York, 1993.

[Mö88] H.M. Möller, *On the construction of Gröbner bases using syzygies,* J. Symb. Comp. **6** (1988), 345–359.

[Mö90] H.M. Möller, *Computing syzygies a la Gauss Jordan,* in Effective Methods in Algebraic Geometry (T. Mora and C. Traverso, eds.), Progress in Mathematics **94**, Birkhäuser Verlag, Basel, 1990, 335–345.

[MöMo] H.M. Möller and T. Mora, *New constructive methods in classical ideal theory,* J. Algebra **100** (1986), 138–178.

[MoRo] T. Mora and L. Robbiano, *The Gröbner fan of an ideal,* J. Symb. Comp. **6** (1988), 183–208.

[NZM] I. Niven, H.S. Zuckerman, and H.L. Montgomery, *An Introduction to the Theory of Numbers,* 5th ed., Wiley, New York, 1991.

[RoSw] L. Robbiano and M. Sweedler, *Subalgebra bases,* in Proc. Commutative Algebra Salvador (W. Burns and A. Simis eds.), Lecture Notes in Math., vol. 1430, Springer Verlag, Berlin and New York, 1988, 61–87.

[Schre] F.O. Schreyer, *Die Berechnung von Syzygien mit dem verallgemeinerten Weierstrasschen Divisionsatz,* Diplomarbeit, Hamburg, 1980.

[Schri] A. Schrijver, *Theory of Linear and Integer Programming,* Wiley, Chichester, NY, 1986.

[Se] A. Seidenberg, *Constructions in algebra,* Trans. Amer. Math. Soc. **197** (1974), 272–313.

[ShSw] D. Shannon and M. Sweedler, *Using Gröbner bases to determine algebra membership, split surjective algebra homomorphisms determine birational equivalence,* J. Symb. Comp. **6** (1988), 267–273.

[Sz] G. Szekeres, *A canonical basis for the ideals of a polynomial domain,* Am. Math. Mon. **59** (1952), 379–386.

[Wo] S. Wolfram, *Mathematica: A System for Doing Mathematics by Computer,* Addison-Wesley, Reading, MA, 1991.

[Za] G. Zacharias, *Generalized Gröbner bases in commutative polynomial rings,* Bachelor's Thesis, MIT, Cambridge, MA, 1978.

List of Symbols

$\mathbb{N}$	natural numbers
$\mathbb{Z}$	ring of integers
$\mathbb{Z}_n$	ring of integers modulo n
$\mathbb{Q}$	field of rational numbers
$\mathbb{R}$	field of real numbers
$\mathbb{C}$	field of complex numbers
$\mathbb{T}^n$	set of power products in the variables $x_1, \dots, x_n$
k	a field
k^n	affine space
$k[x_1, \dots, x_n]$	ring of polynomials in the variables $x_1, \dots, x_n$ with coefficients in the field k
$\langle f_1, \dots, f_s \rangle$	ideal (submodule) generated by $f_1, \dots, f_s$
$f \equiv g \pmod{I}$	f congruent to g modulo I
$k[x_1, \dots, x_n]/I$	quotient ring of $k[x_1, \dots, x_n]$ by the ideal I
$f + I$	coset of f modulo I
$V(S)$	variety in k^n defined by the set of polynomials S
$V(f_1, \dots, f_s)$	variety in k^n defined by the polynomials $f_1, \dots, f_s$
$I(V)$	ideal in $k[x_1, \dots, x_n]$ defined by $V \subseteq k^n$
$\langle S \rangle$	ideal generated by the polynomials in the set S
gcd	greatest common divisor
lcm	least common multiple
$\dim_k(L)$	dimension of the k-vector space L
$\mathrm{lt}(f)$	leading term of f
$\mathrm{lp}(f)$	leading power product of f
$\mathrm{lc}(f)$	leading coefficient of f
$\boldsymbol{x}^{\boldsymbol{\alpha}}$	$x_1^{\alpha_1} \cdots x_n^{\alpha_n}$
$f \xrightarrow{G} h$	f reduces to h modulo G in one step
$f \xrightarrow{G}_+ h$	f reduces to h modulo G

$f \xrightarrow{X,g} h$	f reduces to h and $h = f - Xg$
$\mathrm{Lt}(S)$	leading term ideal (submodule) defined by S
$S(f, g)$	S-polynomial of f and g
$N_G(f)$	normal form of f with respect to G
$V_K(S)$	variety in K^n defined by $S \subseteq k[x_1, \ldots, x_n]$
$\overline{k}$	algebraic closure of the field k
$\sqrt{I}$	radical of the ideal I
$I \cap J$	intersection of I and J
$I : J$	ideal quotient of I by J
$\mathrm{im}(\phi)$	image of the map ϕ
$\ker(\phi)$	kernel of the map ϕ
$k[f_1, \ldots, f_s]$	k-algebra generated by the polynomials $f_1, \ldots, f_s$
$k[F]$	k-algebra generated by the polynomials in the set F
$k(x_1, \ldots, x_n)$	rational function field
$k(\alpha)$	field extension of k obtained by adjoining α
$k(\alpha_1, \ldots, \alpha_n)$	field extension of k obtained by adjoining $\alpha_1, \ldots, \alpha_n$
A^m	set of column vectors with entries in the ring A
M/N	quotient module of M by N
$\cong$	isomorphic to
$\boldsymbol{e}_1, \ldots, \boldsymbol{e}_m$	standard basis for the free module A^m
$\left[\begin{array}{ccc} f_1 & \cdots & f_s \end{array}\right]$	$1 \times s$ matrix whose entries are polynomials $f_1, \ldots, f_s$
$\left[\begin{array}{ccc} \boldsymbol{f}_1 & \cdots & \boldsymbol{f}_s \end{array}\right]$	$m \times s$ matrix whose columns are vectors $\boldsymbol{f}_1, \ldots, \boldsymbol{f}_s$
$\mathrm{Syz}(f_1, \ldots, f_s)$	syzygy module of the matrix $\left[\begin{array}{ccc} f_1 & \cdots & f_s \end{array}\right]$
$\mathrm{Syz}(F)$	syzygy module of the matrix F
$\mathrm{lm}(\boldsymbol{f})$	leading monomial of the vector $\boldsymbol{f}$
$F \oplus G$	direct sum of the matrices F and G
$[F \vert G]$	concatenation of the matrices F and G
${}^tS$	transpose of the matrix S
$\mathrm{ann}(M)$	annihilator of the module M
$F \otimes G$	tensor product of the matrices F and G
$\mathrm{Hom}(M, N)$	set of all A-module homomorphisms between M and N
$\langle U \rangle$	module generated by the columns of the matrix U
$S^{-1}A$	localization of the ring A at the multiplicative set S
A_P	localization of the ring A at the prime ideal P
A_g	localization of the ring A at the set $\{1, g, g^2, g^3, \ldots\}$

Index